ADVANCES IN FISH PROCESSING TECHNOLOGIES

Preservation, Waste Utilization, and Safety Assurance

ADVANCES IN FISH PROCESSING TECHNOLOGIES

Preservation, Waste Utilization, and Safety Assurance

Edited by

Ranendra K. Majumder, PhD

Amjad K. Balange, PhD

APPLE ACADEMIC PRESS

First edition published 2023

Apple Academic Press Inc.
1265 Goldenrod Circle, NE,
Palm Bay, FL 32905 USA

760 Laurentian Drive, Unit 19,
Burlington, ON L7N 0A4, CANADA

CRC Press
6000 Broken Sound Parkway NW,
Suite 300, Boca Raton, FL 33487-2742 USA

4 Park Square, Milton Park,
Abingdon, Oxon, OX14 4RN UK

Library and Archives Canada Cataloguing in Publication

Title: Advances in fish processing technologies : preservation, waste utilization, and safety assurance / edited by Ranendra K. Majumder, PhD, Amjad K. Balange, PhD.
Names: Majumder, Ranendra K., editor. | Balange, Amjad K., editor.
Description: First edition. | Includes bibliographical references and index.
Identifiers: Canadiana (print) 20220399697 | Canadiana (ebook) 20220399735 | ISBN 9781774911075 (hardcover) | ISBN 9781774911082 (softcover) | ISBN 9781003300595 (ebook)
Subjects: LCSH: Fishery processing. | LCSH: Fishery technology.
Classification: LCC SH335 .A38 2023 | DDC 664/.949—dc23

Library of Congress Cataloging-in-Publication Data

Names: Majumder, Ranendra K., editor. | Balange, Amjad K., editor.

Title: Advances in fish processing technologies : preservation, waste utilization, and safety assurance / edited by Ranendra K. Majumder, PhD, Amjad K. Balange, PhD.

Description: First edition. | Palm Bay, FL, USA : Apple Academic Press ; Boca Raton, FL, USA : CRC Press, 2023. | Includes bibliographical references and index.

Identifiers: LCCN 2022036439 (print) | LCCN 2022036440 (ebook) | ISBN 9781774911075 (hbk) | ISBN 9781774911082 (pbk) | ISBN 9781003300595 (ebk)

Subjects: LCSH: Fishery processing. | Fishery technology.

Classification: LCC SH335 .A33 2023 (print) | LCC SH335 (ebook) | DDC 664/.94--dc23/eng/20220829

LC record available at https://lccn.loc.gov/2022036439

LC ebook record available at https://lccn.loc.gov/2022036440

ISBN: 978-1-77491-107-5 (hbk)
ISBN: 978-1-77491-108-2 (pbk)
ISBN: 978-1-00330-059-5 (ebk)

About the Editors

Ranendra K. Majumder, PhD
Professor and Head, Department of Fish Processing Technology and Engineering, College of Fisheries of Central Agricultural University (I), Lembucherra, Tripura, India

Ranendra K. Majumder, PhD, is presently working as Professor and Head in the Department of Fish Processing Technology and Engineering, College of Fisheries of Central Agricultural University (I), Lembucherra, Tripura, India. He has in total 23 years of experience in different academic capacities, including eight years as Professor. He has expertise in the field of ethnic and traditional fish products of India and Bangladesh, fish biochemistry, fisheries waste management, and value addition of low-cost fish. He is also working in the field of entrepreneurship development of youths through value addition to fish. He has satisfactorily completed four externally funded research projects and nine university-funded research projects. Prof. Majumder is guiding three PhD scholars and has already guided 19 master's theses. He has presented papers at a number of international and national seminars. During his tenure of academic and research life, he has publication in 35 international and 28 national journals, as well as 12 book chapters. He has developed several technologies like different value-added fish products, low-cost solar tent dryers, and consumer-friendly portable mini fish smoking kiln. Dr. Majumder's keen interest in ethnic fermented fish products has given him an identity in the international arena, and he is presently working for the development of their standards as assigned by the Bureau of Indian Standards (BIS) and Food Safety and Standards Authority of India (FSSAI). Dr. Majumder earned his doctoral degree from the Central Institute of Fisheries Education (Deemed University), Mumbai, India.

Amjad K. Balange, PhD

Principal Scientist, Department of Post-Harvest Technology, ICAR-Central Institute of Fisheries Education (Deemed University), Mumbai, Maharashtra, India

Amjad K. Balange, PhD, is currently Principal Scientist in the Department of Post-Harvest Technology of the ICAR-Central Institute of Fisheries Education (Deemed University), Mumbai, Maharashtra, India, where he has been on the faculty for over 8 years. He also worked in the capacities of Biochemist and Assistant Professor in the Fish Processing Technology Department of the College of Fisheries, Ratnagiri, Maharashtra, India. He has over over 21 years in teaching, research, and extension experience. Dr. Balange's research and teaching interests revolve around fish processing technology and microbiology, and heh as published over 100 refereed journal publications in the areas of fish post-harvest management, fish biochemistry, value addition of fish, and fish waste management. He has supervised the completion of 10 PhD and 15 master's research programs and is also a recipient of a fellowship for the international PhD program from the Faculty of Agro-Industry, Prince of Songkla University, Thailand. He received an excellent doctorate thesis award for the academic year 2008.

Contents

Contributors

Parvathi Ammini
CSIR–National Institute of Oceanography, Regional Center, Kochi – 682018, Ernakulam, Kerala, India

Amjad K. Balange
Department of Post-Harvest Technology; Fish Processing Technology, ICAR-Central Institute of Fisheries Education, Versova, Mumbai – 400061, Maharashtra, India, E-mail: amjadbalange@cife.edu.in

Soottawat Benjakul
Department of Food Technology, Faculty of Agro-Industry, Prince of Songkla University, Hat Yai, Songkhla – 90110, Thailand, E-mail: Soottawat.b@psu.ac.th

J. Bindu
ICAR-Central Institute of Fisheries Technology (CIFT), Cochin, Ernakulam, Kerala, India, E-mail: bindujaganath@gmail.com

Niladri S. Chatterjee
Biochemistry and Nutrition Division, Central Institute of Fisheries Technology, Cochin – 29, Ernakulam, Kerala, India

Kasturi Chattopadhyay
Department of Post-Harvest Technology, Fishery Resource Harvest and Postharvest Management Division, ICAR-Central Institute of Fisheries Education, Versova, Mumbai – 400061, Maharashtra, India

Basanta Kumar Das
ICAR-Central Inland Fisheries Research Institute, Barrackpore, Kolkata – 700120, West Bengal, India

R. P. Deepitha
Department of Post-Harvest Technology, ICAR-Central Institute of Fisheries Education, Versova, Mumbai – 400061, Maharashtra, India; Division of Fisheries Science, ICAR-Central Island Agricultural Research Institute, Garacharma, Port Blair – 744105, Andaman and Nicobar Islands, India

K. Dhanapal
Department of Fish Processing Technology, College of Fishery Science, Andhra Pradesh Fisheries University, Muthukur – 524344, Andhra Pradesh, India, E-mail: kdhanpal@gmail.com

K. Elavarasan
ICAR-Central Institute of Fisheries Technology, Kochi – 682029, Ernakulam, Kerala, India

Satabdi Ganguly
ICAR-Central Inland Fisheries Research Institute, Barrackpore, Kolkata – 700120, West Bengal, India

A. Jeyakumari
Mumbai Research Center of ICAR-Central Institute of Fisheries Technology, CIDCO Admin Building, Sector I, Vashi, Navi Mumbai – 400703, Maharashtra, India

G. Jeyasekaran
Department of Fish Quality Assurance and Management, Tamil Nadu Dr. J. Jayalalithaa Fisheries University, Tuticorin – 628008, Tamil Nadu, India, E-mail: jeyasekarang@gmail.com

S. A. Joshi
Assistant Professor, College of Fishery Science, MAFSU, Nagpur – 440006, Maharashtra, India

Sathish Kumar Kannaiyan
ICAR-Central Institute of Fisheries Technology (CIFT), Cochin, Ernakulam, Kerala, India

G. Praveen Kumar
Department of Fish Processing Technology, College of Fishery Science, Andhra Pradesh Fisheries University, Muthukur – 524344, Andhra Pradesh, India

H. Sanath Kumar
Department of Post-Harvest Technology, ICAR-Central Institute of Fisheries Education, Mumbai – 400061, Maharashtra, India, E-mail: sanathkumar@cife.edu.in

Lekshmi R. G. Kumar
Biochemistry and Nutrition Division, Central Institute of Fisheries Technology, Cochin – 29, Ernakulam, Kerala, India

Sanath H. Kumar
QC Laboratory, Post-Harvest Technology Department, ICAR-Central Institute of Fisheries Education, Mumbai – 400061, Maharashtra, India, E-mail: sanathkumar@cife.edu.in

P. Layana
Department of Post-Harvest Technology, ICAR-Central Institute of Fisheries Education, Versova, Mumbai – 400061, Maharashtra, India, E-mail: layanap@cife.edu.in

Manjusha Lekshmi
QC Laboratory, Post-Harvest Technology Department, ICAR-Central Institute of Fisheries Education, Mumbai – 400061, Maharashtra, India, E-mail: Manjusha@cife.edu.in

Arabinda Mahanty
ICAR-Central Inland Fisheries Research Institute, Barrackpore, Kolkata – 700120, West Bengal, India; Crop Protection Division, ICAR – National Rice Research Institute, Cuttack – 753006, Odisha, India

Ranendra K. Majumder
College of Fisheries, Central Agricultural University (Imphal), Lembucherra – 799210, Tripura, India, E-mail: drrkmcof@gmail.com

L. Manjusha
Department of Post-Harvest Technology, ICAR-Central Institute of Fisheries Education, Mumbai – 400061, Maharashtra, India

Suseela Mathew
Biochemistry and Nutrition Division, Central Institute of Fisheries Technology, Cochin – 29, Ernakulam, Kerala, India, E-mail: suseela1962@gmail.com

Tandrima Mitra
ICAR-Central Inland Fisheries Research Institute, Barrackpore, Kolkata – 700120, West Bengal, India

C. O. Mohan
ICAR-Central Institute of Fisheries Technology, Kochi – 682029, Ernakulam, Kerala, India, E-mail: comohan@gmail.com

Bimal Prasanna Mohanty
ICAR-Central Inland Fisheries Research Institute, Barrackpore, Kolkata – 700120, West Bengal, India; Fisheries Science Division – ICAR, KAB II, Pusa, New Delhi – 110012, India, E-mail: bimal.mohanty@icar.gov.in

Sasmita Mohanty
Faculty of Science and Technology, Department of Biotechnology, Rama Devi Women's University, Bhubaneswar – 751022, Odisha, India

L. Narasimha Murthy
Mumbai Research Center of ICAR-Central Institute of Fisheries Technology, CIDCO Admin Building, Sector I, Vashi, Navi Mumbai – 400703, Maharashtra, India, E-mail: murthycift@gmail.com

Binaya Bhusan Nayak
Department of Post-Harvest Technology; QC Laboratory, Post-Harvest Technology Department, ICAR-Central Institute of Fisheries Education, Mumbai – 400061, Maharashtra, India

George Ninan
ICAR-Central Institute of Fisheries Technology, Cochin – 682029, Ernakulam, Kerala, India, E-mail: georgr66jiji@gmail.com

Girija G. Phadke
Mumbai Research Center of ICAR-Central Institute of Fisheries Technology, CIDCO Admin Building, Sector I, Vashi, Navi Mumbai – 400703, Maharashtra, India

R. Ramesh
Regional Research Center of ICAR-Central Institute of Freshwater Aquaculture, Fish Seed Farm, Penamaluru, Poranki P. O., Vijayawada – 521137, Andhra Pradesh, India

C. N. Ravishankar
ICAR-Central Institute of Fisheries Education, Mumbai 400 061, India, E-mail: cnrs20002000@yahoo.co.in

S. Remya
ICAR-Central Institute of Fisheries Technology, Kochi – 682029, Ernakulam, Kerala, India

Ajit Kumar Roy
Ex. Pr. Scientist and Head, Social Sciences, ICAR-CIFA, Bhubaneswar – 751002, Odisha, India, E-mail: akroy1946@yahoo.co.in

Thanasak Sae-Leaw
Department of Food Technology, Faculty of Agro-Industry, Prince of Songkla University, Hat Yai, Songkhla – 90110, Thailand

T. V. Sankar
ICAR-Central Institute of Fisheries Technology, Cochin – 682029, Ernakulam, Kerala, India, E-mail: tvsankar.cift@gmail.com

R. Jeya Shakila
Department of Fish Quality Assurance and Management, Tamil Nadu Dr. J. Jayalalithaa Fisheries University, Tuticorin – 628008, Tamil Nadu, India

B. A. Shamasundar
Former Professor and Head, Department of Fish Processing Technology, College of Fisheries, Mangalore – 575002, Karnataka, India, E-mail: bashamasundar@rediffmail.com

Avtar Singh
Department of Food Technology, Faculty of Agro-Industry, Prince of Songkla University, Hat Yai, Songkhla – 90110, Thailand

K. R. Sreelakshmi
ICAR-Central Institute of Fisheries Technology, Kochi – 682029, Ernakulam, Kerala, India

Jerusha Stephen
QC Laboratory, Post-Harvest Technology Department, ICAR-Central Institute of Fisheries Education, Mumbai – 400061, Maharashtra, India

G. Venkateshwarlu
Education Division, Indian Council of Agricultural Research, KAB-II, New Delhi – 110012, India, E-mail: venkateshwarlu.icar@gmail.com

S. Vijayakumar
Fisheries Research Station and Information Center, Buthnal, KAFSU, Bidar, Karnataka, India

K. A. Martin Xavier
Department of Post-Harvest Technology, Fishery Resource Harvest and Postharvest Management Division, ICAR-Central Institute of Fisheries Education, Versova, Mumbai – 400061, Maharashtra, India, E-mail: martinxavier@cife.edu.in

B. M. Yadav
Associate Professor, College of Fisheries, Dr. B. S. K. K. V., Dapoli, Ratmagiri – 415629, Maharashtra, India

Abbreviations

ABI	agri-business incubation center
ACE	angiotensin-converting enzyme
AD	Alzheimer's disease
ADA	American Diabetic Association
ADHD	attention-deficit hyperactivity disorder
ADV	adenoviruses
AGVs	automated guided vehicles
AHA	American Heart Association
AI	artificial intelligence
AMD	age-related macular degeneration
AMG	acetylated monoglycerides
AP	active packaging
APW	alkaline peptone water
ASP	amnesic shellfish poisoning
ATSDR	Agency for Toxic Substances and Disease Registry
AZP	azaspiracid shellfish poisoning
BHA	butylated hydroxyanisole
BHI	brain heart infusion
BHT	butylated hydroxytoluene
BIPs	border inspection posts
BSE	bovine spongiform encephalopathy
CA	competent authority
CAC	Codex Alimentarius Commission
CAF	conventional air freezing
CAGR	compound annual growth rate
CAT	cefoperazone, amphotericin B and teicoplanin
CAZ	ceftazidime
CCA	Central Competent Authority
CCFFP	codex committee for fish and fishery products
CCFH	codex committee on food hygiene
CCPs	critical control points
CDC	center for disease control
cDNA	complementary double-stranded DNA
CFC	chlorofluorocarbons

CFR	code of federal regulations
CFTRS	Central Fisheries Technological Research Station
CIFT	Central Institute of Fisheries Technology
CLE	Chamuang leaf extract
CMC	carboxymethylcellulose
COFI	committee on fisheries
COOL	country of origin labeling
COVIS	cholera and other vibrio information service
CP	critical point
CPD	cefpodoxime
CRE	carbapenem-resistant Enterobacteriaceae
CSW	chilled seawater
CTX	cefotaxime
DAG	diacylglycerol
DEAE	diethylaminoethylsepharose
DES	deep eutectic solvents
DG-SANTE	Directorate-General for Health and Food Safety
DH	degree of hydrolysis
DHA	docosahexaenoic acid
DMA	dimethylamine
DSP	diarrhetic shellfish poisoning
DTX1	dinophysistoxin-1
DTX2	dinophysistoxin -2
EC	epicatechin
ECG	epicatechin gallate
ECHE	ethanolic coconut husk extract
ECs	edible coatings
EFs	edible films
EFSA	European Food Safety Authority
EGC	epigallocatechin
EGCG	epigallocatechin gallate
EIAs	Export Inspection Agencies
EIC	Export Inspection Council
EPA	eicosapentaenoic acid
ESBL	extended-spectrum β-lactamase
EU	European Union
EVOH	ethylene-vinyl alcohol
FA	fatty acid
FAO	Food and Agriculture Organization

FBOs	food business operators
FDA	Food and Drug Administration
FFA	free fatty acid
FFDCA	Federal Food, Drug, and Cosmetic Act
FPH	fish protein hydrolysates
FPI	fish protein isolate
FSMA	Food Safety Modernization Act
FSMS	food safety management system
FSSAI	Food Safety Standards Authority of India
FVO	food and veterinary office
GBF	glycol/water bath freezing
GDP	gross domestic product
GHPs	good hygiene practices
GIF	gel degradation inducing factor
GMO	genetically modified organisms
GMPs	good manufacturing practices
GRAS	generally recognized as safe
GTE	green tea extract
HA	hepatitis A
HACCP	hazard analysis and critical control point
HAV	hepatitis A viruses
HBA	hydrogen bond acceptor
HBD	hydrogen bond donor
HDPP	high-density polypropylene
HETE	hydroxyeicosatetraenoic acid
HGT	horizontal gene transfer
HL	headless
HP	high pressure
HPAF	high-pressure-assisted freezing
HPP	high-pressure processing
HPSF	high-pressure-shift freezing
HuNoVs	human noroviruses
IAEA	International Atomic Energy Agency
ICAR-CIFT	ICAR-Central Institute of Fisheries Technology
ICMSF	International Commission for Microbiological Specifications for Foods
IFT	Institute of Food Technologists
IMP	imipenemase
IoT	internet of things

IQF	individual quick freezing
ISD	inspection and safety division
ISO	International Organization for Standardization
IUU	illegal, unreported and unregulated
JECFA	Joint FAO/WHO Expert Committee on Food Additives
KGM	Konjac glucomannan
LBG	locust bean gum
LDL	low-density lipoprotein
LOX	lipoxygenase
MAE	microwave-assisted extraction
MAFF	Ministry of Agriculture, Forestry, and Fisheries
MAG	monoacylglycerol
MAP	modified atmosphere packaging
MBLs	metallo-β-Lactamases
MDR	multiple drug-resistant
MFRD	Marine Fisheries Research Department
MHC	myosin heavy chain
MHLW	Ministry of Health, Labor and Welfare
MIPs	modori inducing proteases
MMPs	matrix metalloproteases
MMT	million metric ton
MOU	memorandum of understanding
MRL	maximum residual limit
MRSA	methicillin-resistant Staphylococcus aureus
MSC	Marine Stewardship Council
MSPs	matrix serine proteases
MTGase	microbial transglutaminase
MUFA	monounsaturated fatty acid
MW	molecular weight
n-3 PUFAs	*n*-3 polyunsaturated fatty acids
NAIF	National Agriculture Innovation Fund
NGOs	non-governmental organizations
NIID	National Institute of Infectious Diseases
NIOSH	National Institute for Occupational Safety and Health
NMFS	National Marine Fisheries Service
NOAEL	no observable adverse effect level
NSP	neurologic shellfish poisoning
NSSP	National Shellfish Sanitation Program
OFA	oxygenated FA

OIE	international office of epizootics
OSHA	Occupational Safety and Health Administration
OXA	oxacillinases
PAHs	polycyclic aromatic hydrocarbons
PAI	pathogenicity islands
PCA	principal component analysis
PCBs	polychlorinated biphenyls
PCDDs	polychlorinated dibenzo-p-dioxines
PCR	polymerase chain reaction
PD	peeled and deveined
PE	polyethylene
PEG	polyethylene glycol
PL	phospholipids
PLE	pressurized liquid extraction
PMQR	plasmid-mediated quinolone resistance
PPE	pomegranate peel extract
PPO	polyphenol oxidase
PRPs	prerequisite programs
PSF	pressure shift freezing
PSP	paralytic shellfish poisoning
PTO	peeled tail on
PUD	peeled and un-deveined
PUFA	polyunsaturated fatty acids
QRDR	quinolone-resistance determining region
RASFF	rapid alert system for food and feed
RDI	recommendation daily intake
RF	radio frequency
ROS	reactive oxidative species
RSW	refrigerated seawater
RTE	ready-to-eat
SCF	supercritical fluid
SDG	sustainable development goal
SEAFDEC	Southeast Asian Fisheries Development Center
SFAs	saturated fatty acids
SFE	supercritical fluid extraction
SFP	scombrotoxin fish poisoning
SHS	switchable hydrophilicity solvent
SMS	sodium metabisulfite
SS	switchable solvents

SSOPs	sanitation standard operating procedures
TAG	triacylglycerols
TBT	technical barriers to trade
TDH	thermostable direct hemolysin
TGase	transglutaminase
TMAO	trimethylamine oxide
TMAOase	trimethylamine-N-oxide demethylase
TRH	TDH-related hemolysin
TRL	technology readiness level
UAE	ultrasound-assisted extraction
UN	United Nations
USDA	U.S. Department of Agriculture
USFDA	U.S. Food and Drug Administration
VIM	Verona Integron-encoded metallo-β-lactamase
WHO	World Health Organization
WTO	World Trade Organization
ZTMC	Zonal Technology Management Center

Foreword 1

The fishery sector has witnessed the development of various innovative approaches, particularly in the field of processing, preservation, and waste utilization. It is the time to promote a 'Fish for All and Forever' movement to augment nutritional security. Fish processing is playing a significant role in providing safe and nutritious products free from contaminants to satisfy consumer demand and in the promotion of export. The technological advancements have increased fish production from both inland and marine sectors. Higher production needs to be coupled with much improved and efficient post-harvest management with innovative approaches. Food loss and waste is a serious issue globally. Fish processing waste, especially from shore-based facilities, can cause harm to the marine environment in the surrounding area as well as to the terrestrial environment as a potent bio-pollutant. Therefore, recycling fish processing waste is necessary to contribute to better health of the environment and the generation of more wealth.

The book *Advances in Fish Processing Technologies: Preservation, Waste Utilization, and Safety Assurance* is a very good timely initiative. I hope that the book will be a handy reference for students, researchers, entrepreneurs, policymakers, and other stakeholders. I congratulate the Editors for their effort in bringing out the publication.

—Trilochan Mohapatra, PhD
Secretary & Director General
Department of Agricultural Research & Education
and
Indian Council of Agricultural Research
Ministry of Agriculture and Farmers Welfare
Krishi Bhavan, New Delhi – 110 001, India
Tel.: 23382629; 23386711; Fax: 91-11-23384773
E-mail: dg.icar@nic.in

Foreword 2

Fish was not an expensive food product until the second half of the 20th century. The development of improved processing, packaging, and marketing techniques has made it possible to send fish to remote areas. Until then, fish was mostly a food eaten by people of coastal areas. Today fish is the most expensive animal food in the world. The cost of fish per kg is much higher than that of chicken and other animal meat.

In the 1950s many low-values fish were thrown back into the sea from shrimp trawlers. They are usually called by-catches. But times have changed. In many countries, these previously low-value fish enjoy a good market and command high prices.

Fish is now considered a healthy food. It contains high amounts of protein-rich essential amino acids. Fish is also a good source of omega-3 fatty acids, vital for the synthesis of several compounds like prostaglandins, thromboxanes, and biologically active compounds needed for human health and medical application. Many compounds extracted from fish waste now enjoy a good commercial market. Chitin, chitosan, and glucosamine from shrimp shells now have an established market. Chondroitin sulfate extracted from shark bones is now widely used for the treatment of osteoarthritis.

This book *Advances in Fish Processing Technologies: Preservation, Waste Utilization, and Safety Assurance* discusses all aspects of fish processing. The book has 21 chapters. All the contributors are very experienced scientists and professors. In fact, some of them have reputed research articles in international journals to their credit.

Prof. Ranendra K. Majumder and Dr. Amjad K. Balange have done outstanding work in organizing and editing this book. I hope this book will be well received by students and research scientists in fish processing.

—K. Gopakumar, PhD

Professor of Eminence, Department of Food Science and Technology, Kerala University of Fisheries and Ocean Studies, Kochi, Ernakulam, Kerala, India

Former Director, Central Institute of Fisheries Technology (CIFT), Kochi and Deputy Director General (Fisheries), Indian Council of Agricultural Science ICAR), New Delhi, India,

E-mail: kgopa@yahoo.com

Preface

The idea of writing this book emerged from discussions with students, researchers, and teachers working in the fisheries colleges and research institutes. Being updated with new information has become the need of the hour, and the fish processing sector globally and in India is also changing and upgrading itself to fulfill the ever-changing demand of consumers. Fisheries in India have become one of the important economic activities, with total fish production of 13.42 million metric tons during 2018–2019 and export of the fish products valued at Rs. 46,589 crores (2018–2019). With the increase in fish production, challenges of supplying the fish in the freshest possible state are also increasing. Consumer demand for safe, nutritious, healthy, and conveniently ready-to-eat fish products and industry demand of zero wastage during processing paved the way for several need-based research in the area of preservation, processing, waste utilization, and safety assurance of fish and fish products. This brought our attention to compiling the recent research recommendations on various aspects of preservation, processing, waste utilization, and safety of fish and fish products.

Fish processing industries in India had developed gradually from quite a few freezing plants in 1950–1951 to more than 700 freezing plants by 2020. To date, freezing and canning are the two most practiced modern methods of fish processing in India. Though there are other advanced innovative and minimal processing methods like high-pressure processing, freeze-drying, pulsed light preservation, microwave preservation, pulsed electric field processing, etc., which have been adopted by some of the developed countries, it may take some time for these to be adopted by the industries in India. Hence continuous awareness and popularization on the recent advances in fish processing have become very important, and efforts should be made in that direction.

The first section of the book covers the advancement in fish processing technology, green technologies for extracting nutraceuticals, the role of endogenous enzymes in the quality of fish/shellfish and their products, disruptive technologies, and restructured product-based technologies. Topics like innovative methods for the utilization of fish waste have been given due attention. Similarly, quality and safety aspects of fish and fish products have been covered aptly with reference to antimicrobial resistance bacteria and developments in safety and quality management systems of fish and fish products.

The chapters received for this book are from the eminent scientists and teachers working in the area of fish processing for quite a good time now. All the chapters are prepared covering the recent updates in the fish processing, preservation, value addition, waste utilization, quality, and safety aspects of fish and fishery products. We expect that the topics of greater significance have been covered adequately. We also believe that this book can be a good source of recent information for students, scientists, teachers, technologists, and entrepreneurs working in the field of fish processing.

—Ranendra K. Majumder, PhD
Amjad K. Balange, PhD

Part I

Advances in Fish Preservation, Processing, and Packaging Technologies

CHAPTER 1

Advancement in Fish Processing Technology: The Indian Scenario

C. N. RAVISHANKAR[1] and C. O. MOHAN[2]

[1] *ICAR-Central Institute of Fisheries Education, Mumbai 400 061, India*

[2] *ICAR-Central Institute of Fisheries Technology, Kochi – 682029, Ernakulam, Kerala, India*

ABSTRACT

Fish and shellfishes are highly sought after due to their proven health benefits. They are a rich source of easily digestible proteins, healthful polyunsaturated fatty acids (FAs), and minerals. Their production, as well as consumption, is on the rise globally. Over the years, Indian fish production has increased many folds, and its share in export is also on the rise. However, a very large quantity of fish produced in India is lost due to improper handling and preservation. Loss of this highly nutritive food commodity is not good for a country with so many undernourished populations. Improved handling and processing, and preservation technologies are to be adopted to overcome this loss. The present chapter provides an insight into the fish processing scenario in India. The various methods developed for fish preservation, starting from chilling, freezing, drying to thermal processing, extrusion, etc., are described apart from various packaging interventions.

1.1 INTRODUCTION

Over the years, the demand for fish has been on the rise all over the world due to its health benefits. However, fish is highly susceptible to spoilage due to autolysis and bacterial action when compared to other agricultural food

Advances in Fish Processing Technologies: Preservation, Waste Utilization, and Safety Assurance.
Ranendra K. Majumder, PhD & Amjad K. Balange, PhD (Eds.)

products. With the advancement in modern fish-catching methods and gear systems and improvements in the culture methods, fish production showed a drastic increase, necessitating the adoption of post-harvest preservation techniques. Major technological advancements in post-harvest aspects of fish over the last six decades are in the method of fish preservation, processing, packaging, improved attention for quality, and improvements in processing machinery. Initially, some basic processing aspects like fish curing, fish byproducts, fundamental microbiology of marine environments, ice-preserved fish, and biochemical composition of fresh fish were carried out in different organizations in India. However, there existed a wide gap in organized research activities on fishery technology in the country even up to a decade after our independence. A Fisheries Research Committee was appointed by Mr. Rafi Ahmed Kidwai, Honorable Union Minister of Food and Agriculture, in 1954 under the chairmanship of Dr. B.N. Chopra, the then Fisheries Development Adviser, to study the gap analysis in fisheries research. The committee studied the field situation in detail and recommended the formation of the Central Institute of Fisheries Technology (then named Central Fisheries Technological Research Station) with the objective to undertake studies on craft and gear development and processing and byproducts technology. The committee also suggested seeking expert advice in planning the work from the Fisheries Division of the Food and Agriculture Organization (FAO). Though the institute was formally initiated to meet the requirements of craft and gear development in the country for the efficient harvest of fish resources, it was felt that Post-harvest technology is also equally and perhaps more importantly, since the success of the former depends on efficient and economical utilization of the harvested fish.

1.2 ROLE OF ICAR-CIFT IN DEVELOPMENT OF FISH PROCESSING TECHNOLOGY IN INDIA

ICAR-Central Institute of Fisheries Technology (ICAR-CIFT) was established as Central Fisheries Technological Research Station (CFTRS) on 29th April 1957, and later named as CIFT in April 1961. The institute holds a unique position as the National Center for research in harvest and post-harvest technologies of fisheries. Since its inception, ICAR-CIFT has remained as the premier Institution in the country which offers standardized technologies for sustainable utilization of aquatic resources through preservation, processing, and value addition. With the establishment of CIFT, particularly the Fish Processing Division, the export of seafood products improved as it translated the research outcomes to the industry. Intimate contact with the fish processing industry to solve their problems and to advise the industry in the right direction helped the industry to reach a record export of fishery products. Initially, only frozen shrimps were exported from India, and to some extent, canned shrimps were exported. It was only in the late 1960s the export of other varieties of fish, squid, cuttlefish, octopus, crabs, clams, and mussels, mainly in frozen form, was initiated with the research backup from the institute.

Pioneering works were carried out at ICAR-CIFT on fresh and cured fish and fishery products to ensure them are free from chemical and microbial hazards for safety and longer shelf life. The institute has been in the forefront for the application of cutting edge technologies in food processing sector; high pressure processing, pulsed light preservation, active, and intelligent packaging are to name a few. Utilization of secondary raw materials from fish and shellfish processing for developing high value products is one of the key areas where ICAR-CIFT has made significant contributions. Chitin, chitosan, chitin derivatives, oyster peptide extract, collagen chitin films, collagen peptide, PUFA concentrate are some of the commercially successful products for which the institute has developed and optimized technologies. Standard protocols and packaging systems for the preservation and utilization of commercially important indigenous and exotic aquatic resources have been developed. The achievements of the institute over the six decades are highlighted in subsections.

1.2.1 LIVE FISHERY PRODUCTS

There is a great demand for live fish and shellfishes, the world over. These products fetches maximum price compared to all the other forms of value added products as it maintains freshness. The candidate species for live transportation include high value species, cultured grouper, red snapper, seabreams, seabass, red tilapia, reef fish, air breathing fishes, shrimp, crabs, lobster, clams, oyster, and mussels. These are normally transported in air cargo maintained at low temperature in order to lessen the metabolic activities of the animals. Preliminary work on transportation of crabs and clams in live condition has been attempted by the institute. For crabs, chelating legs and other legs are tied and placed on moistened sawdust in corrugated box, maintaining low temperature. The method proposed has helped in transporting crab in live condition for many days. In recent days, the live crabs are transported along with a frozen water bottle to reduce the temperature without sawdust (Figure 1.1). The method is still persistent in many parts of the country, especially in Tamil Nadu.

FIGURE 1.1 Transportation of crab in live condition.

1.2.2 HANDLING OF FISH

Proper handling of fish and shellfish is of utmost importance as it affects the quality to a greater extent. After an initial field survey and observing the prevailing bad handling practices, the institute has taken up extensive work on method of handling on the quality of chilled and frozen fish products. The institute has developed better handling practices and trained the preprocessing and processing personnel across India, which helped the industry, to a great extent. A study was undertaken to preserve fishes onboard using refrigerated seawater (RSW) and chilled seawater (CSW) system. Initial studies to preserve fish onboard were carried out to store oil sardine, mackerel, prawns, seer, jewfish, and ribbonfish in refrigerated seawater (RSW). Whole sardine and mackerel could be kept better in RSW up to 3–4 days, compared to ice. However, blackening of meat along the backbone was a drawback. Eviscerated sardine and mackerel developed rancidity quickly. Eviscerated seer fish maintained better quality up to 10 days in RSW storage. Non-fatty fishes such as jewfish and ribbonfish maintained better condition in RSW than in ice. However, prawns developed black spot rapidly. Addition of NaCl to RSW up to 7% did not improve the shelf life of sardine and mackerel. Introduction of Carbon dioxide in gas form in RSW improved the shelf life of both fatty and lean varieties of fishes. CSW was also found to improve the shelf life of fishes. The method proposed by the institute was adopted in the initial days for preserving the bulk quantity of fishes onboard. However, due to the disadvantages of these methods and with the advent of other methods of chilling, the RSW and CSW techniques are preferred less these days. The process of hand filleting of suitable fishes has been worked out and transferred to the industry.

1.2.3 CHILLED FISHERY PRODUCTS

Chilling is an effective method of maintaining the freshness of fish products. This normally involves keeping fishes in melting ice or slurry ice to maintain the fish temperature around 1–4°C, which delays the enzymatic action and microbial activity, thereby extending the shelf life of the products. Traditionally, chilling is carried out using melting ice, either flake ice or crushed block ice. Of late, slurry ice has been introduced for chilling. The institute has studied in detail, the properties of various types of ice used for different fish products and their advantages. A wide range of fish and shellfish products varying from whole, headless, peeled gutted, headless gutted fish, fillets, steaks, loins, cubes can be preserved by chilling. Shelf life of fishes from different environment has been studied by the institute extensively. Shelf life of 12–15 days has been achieved for seer fish and black pomfret. Indian Mackerel and Indian oil sardine had very short shelf life in ice (3–7 days), due to rancidity and belly bursting. Tilapia from freshwater and brackish water showed significant difference in shelf life when stored in ice. The former kept longer (14–15 days) than the latter (8–10 days). In general, shrimps had a shelf life of 8 days, compared to 7–10 days for crabs and 8–9 days for cephalopods. Pre-processed crab meat was found to be acceptable up to 6 days, whereas fresh and cooked mussel meat remained in acceptable condition up to 3–4 days. Whole fish had better quality retention in chilled storage compared to whole cleaned fish. Studies on quality changes of fish preserved by using different types of ice have shown that tube ice and slurry ice performed better in maintaining quality. Work on the sashimi-grade tuna was also initiated in the institute as a considerable quantity of tuna is landed in the country. It was observed that chilled products were superior to frozen products in terms of its texture and taste, albeit limited shelf life. Chilled storage studies of whole cleaned squid (*Loligo duvauceli*) indicated a shelf life of 9 and 11 days in 12 μM polyester laminated with 75 μM low density polyethylene flexible pouches and shrink packs, respectively. Shelf life of cuttlefish (*Sepia pharaonis*) in skinless gutted form packed in 12 μM polyester laminated with 75 μM low density polyethylene flexible pouches stored under different temperature (1–2°C; 7°C and room temperature, 28°C) indicated that the shelf life of cuttlefish at room temperature was less than a day whereas it was 18 days under ice and only 5–6 days for the samples stored at 7°C. For fishes with thick skin, the effect of de-skinning on the shelf life was evaluated. For de-skinned fish steaks of leather jacket (*Aluterus monoceros*) packed in LLDPE and stored in iced condition, a shelf life of 11 days was observed compared to only 7 days for steaks with skin. As consumers prefer fresh fish products without chemical preservatives, studies were undertaken to find alternative natural preservatives to enhance the quality of fish. Chitosan edible coating, treatment with plant extract and animal extracts have given considerable lead in this area. The chitosan treatment further extended the eating quality of skinless leather jacket steaks up to 16–19 days, compared to 11 days for control samples. Chitosan treatment for double fileted Indian oil sardine and ribbonfish enhanced the shelf life considerably. Treatment with plant extracts like rosemary (*Rosmarinus officinalis*) essential oil, curry leaf (*Murraya koenigii*) essential oil, ginger (*Zingiber officinale*) essential oil, oregano (*Origanum vulgare*) essential oil, essential oil from fresh lemongrass leaves (*Cymbopogon citratus*) and lemongrass extract, extracts of sea grape (*Caulerpa lentillifera*), moringa gum (*Moringa oleifera*), gum ghatti or Indian gum obtained from the species *Anogeissus latifolia*, tragacanth gum (*Astragalus* spp.), guar gum from guar beans (*Cyamopsis tetragonoloba*), mint (*Mentha arvensis*) leaves, citrus (*Citrus aurantium*) peels, pomegranate (*Punica granatum*) peel and seed extract and grape seed extract alone or in different combinations improved the quality and shelf life of chilled fish products as they exhibit antimicrobial and antioxidant properties. These can be used by the industry as natural preservatives instead of chemical preservatives which may be harmful to human health in the long term.

1.2.4 ADVANCED PACKAGING METHODS FOR CHILLED FISHES

Survey of packaging materials used in the fish processing industry was carried out to assess the quality of packaging materials used in the industry. Various physical properties such as water vapor transmission rate, tearing strength, gas permeability, bursting strength, puncture resistance, heat seal strength, tensile strength and elongation at break of packaging materials like polythene, paper, laminates, and corrugated box were tested. Preliminary work on packing the fish in ice in thermocol-lined plywood boxes developed under the All India Coordinated Research Project showed that the containers can advantageously be used for long distance transportation of chilled or frozen fish (Figure 1.2). Though chilled fish products are highly appreciated by the consumers, their limited shelf life pose a threat to long-term preservation. To overcome this, various packaging materials including laminated packaging materials have been tried and optimized. The normal polybag used in the industry is LDPE with very high permeability for water and gasses which limits the shelf life of fish products. The use of packaging materials by laminating two or three packaging materials will improve the properties leading to improved shelf life. One such packaging material identified is polyester laminated with low density polyethylene for chilled fishes, which has given considerable improvement in the quality and shelf life of fish products. Apart from interventions in packaging material, advanced technologies like vacuum packaging, modified atmosphere, and active packaging (AP) technologies have been developed by the institute, for a number of fishes. In vacuum packaging, the air inside the package is removed by applying vacuum using machineries. The entrapped air in the ordinary air pack will accelerate the oxidation of fishery products and encourage the growth of aerobic spoilage organisms. This can be minimized by adopting vacuum packaging technique. However, it creates an anaerobic environment which encourages the growth of anaerobic spoilers and the most dangerous *Clostridium botulinum* growth and toxin production. This further can be minimized by using good quality raw material and maintaining the temperature of storage less than 3.3°C throughout, which prevents the growth and toxin production of anaerobic spoiler microorganism. In Modified Atmosphere Packaging (MAP), the air inside the package is replaced with favorable gasses like Carbon dioxide, nitrogen, and oxygen, depending on the fat content of the fishes. Extensive studies on both vacuum and modified atmosphere on the fishes have been carried out in the institute. Shelf life of ghol steaks packed under control air packs were compared with the vacuum packaging and samples were in acceptable condition only up to eight days for air packs, compared to 18 days for vacuum-packed samples. Shelf life of vacuum and control air packed horse mackerel (*Megalapsis cordyla*) steaks packed in high impact polypropylene trays stored at 2–4°C indicated a shelf life of 9–10 days for control air packs compared to 14 days for vacuum-packed samples. Quality and shelf life of mahi-mahi or dolphinfish (*Coryphaena hippurus*) packed in air and vacuum pack stored under chilled condition was undertaken. A shelf life of 10 days was observed for air packed samples compared to 18 days for vacuum packed samples. A combination of 60: 40 (CO_2: N_2) for fatty varieties and 40: 30: 30 (CO_2 : N_2: O_2) for lean fishes has been established. Apart from this, different gas compositions have also been suggested for marine, brackish water and freshwater fishes.

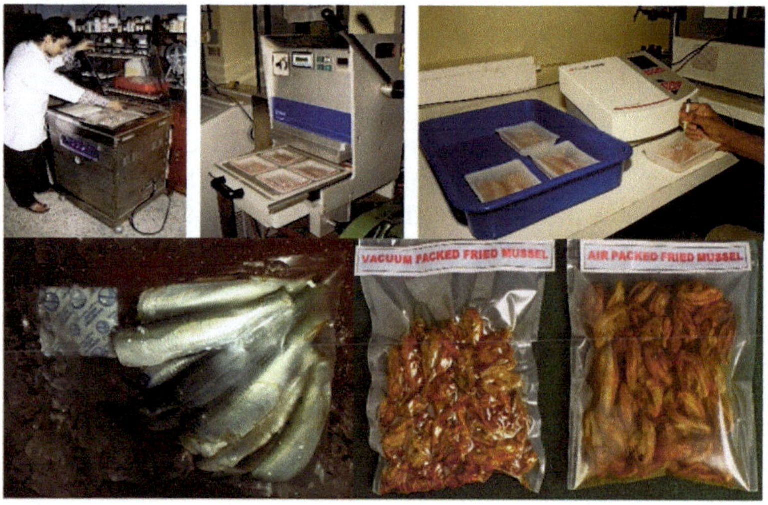

FIGURE 1.2 Advanced packaging (VP, MAP, and active packaging) methods developed for fish products.

However, the vacuum packaging and MAP are cost-intensive as they require costly equipment and pure gasses, which can be overcome by adopting AP technology. AP techniques, mainly O_2 scavenger, CO_2 emitters, antimicrobial, and antioxidant packaging material have been developed by the institute. Quality changes of yellowfin tuna slices packed with a CO_2 emitting sachet indicated a shelf life of 13 days, compared to only seven days for air packed samples. Shelf-life studies of chilled stored ghol, mahi-mahi, and horse mackerel packed under air and with O_2 absorber indicated shelf life of 24, 24, and 20 days for active packed samples, compared to only eight, 10 and 9–10 days for air packed samples, respectively. Shelf life assessment of longtail tuna (*Thunnus tonggol*) chunks packed under air and O_2 absorber during iced storage conditions indicated a shelf life of 5–6 and 9–10 days for air packed and O_2 scavenger packs, respectively. Combined effect of chitosan edible coating and AP like O_2 absorber and CO_2 emitters on the eating quality of leatherjacket (*Aluterus monoceros*) steaks packed in Ethylene Vinyl Alcohol (EVOH) pouches was studied under chilled storage condition. Various quality parameters assessed indicated that control fish was in good acceptable state only up to 11 days, compared to 22 and 29 days for CO_2 emitter packs and O_2 scavenger packs, respectively. Biodegradable AP films like antimicrobial and antioxidant packaging materials were developed using chitosan and plant extracts (Figure 1.3). Active antimicrobial packaging films prepared from chitosan incorporated with ginger essential oil exhibited antibacterial property against *Staphylococcus aureus* and *Escherichia coli* and extended the shelf life of barracuda by 1.6 times, compared to air packed samples. Combination of chitosan-based antimicrobial film incorporated with ginger essential oil and oxygen scavenger further enhanced the shelf life of fish in chilled condition. Chitosan-based films prepared by incorporating rosemary essential oil and oregano extract exhibited antioxidant properties and enhanced the quality of fishes packed in these films. The studies indicate significant quality improvement for fishes stored using these advanced packaging methods either alone or in combination with natural treatments which can be adopted by the industry to improve the storage life of fish considerably.

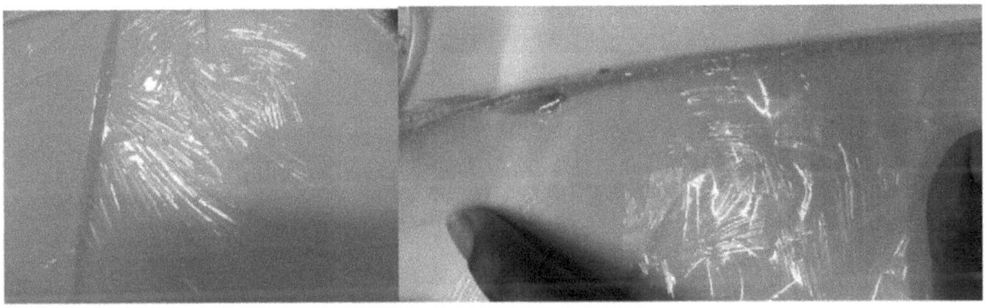

FIGURE 1.3 Biodegradable packaging films developed by the institute using shrimp shell.

1.2.5 *SOUS-VIDE* PRODUCTS

Processing conditions for cook chilled and vacuum cooking, commonly known as *sous-vide*, products has been optimized. Cook chilling at desired temperature improves the shelf life of cobia (*Rachycentron canadum*) considerably, compared to control samples. Shelf-life studies of *sous-vide* packed cobia was 25 days, compared to only 12 days for vacuum packs. Condiment incorporated Indian white shrimp had a shelf life of 28 days, compared to eight and 15 days for air and vacuum packed samples in chilled condition. Introduction of flash-frying process further enhanced the eating quality of Indian white shrimp up to 34 days.

1.2.6 FROZEN FISHERY PRODUCTS

Freezing is an age old practice to retain the quality and freshness of fishery products for a long time. This involves the conversion of water present in fishery products to ice, i.e., a phase change from liquid to solid phase takes place in freezing. This retards the microbial and enzymatic action by reducing the water available for their action. The institute has developed methods for freezing shrimps initially, as shrimps were the major export commodity during the 1950s. Later, the method was extended to almost the entire spectrum of fish species and fish based products. The institute established the advantages of quick freezing over slow freezing, in terms of the quality changes. Slow freezing affects the quality, whereas quick freezing preserves the quality. Quick freezing is normally accomplished by using any of the following four methods: air blast freezing, indirect contact freezing (plate freezing), immersion freezing and cryogenic freezing. Normally products are frozen till it attains a core temperature of –18°C or lower and are stored in cold storage maintained at this temperature. The freezing and frozen storage of fish have been largely used to retain their sensory and nutritional properties. Frozen products form one of the largest portions of fishery products traded all over the world. Freezing time for different freezing methods, subsequent storage conditions, treatments to overcome problems during storage, packaging methods, etc., have been optimized by the institute. Wrapping with polyethylene increased the storage life of fishery products. Frozen storage characteristics of body and claw meat of crab exhibited significant differences, claw meat having longer shelf life. Frozen storage of mussel meat showed that cooked frozen meat kept for longer periods in better condition than the fresh meat. The shelf life of fishes differed depending on the fat content and period of ice storage before freezing. Generally, at a temperature of –18°C, for sardines with 10–25% fat (dry weight basis), a shelf life of 12–16 weeks was obtained while those with 25–40% fat (dry weight basis) had only 10 weeks of shelf life. When frozen in blocks, with heavy glaze, the raw material being very fresh, oil sardine with 20% fat (dry weight basis) remained in good condition for 6–8 months. Tilapia from freshwater had better keeping quality in frozen condition compared to brackish water tilapia. Freezing time for big size fishes using air blast freezing was worked out in commercial units, and quality changes during subsequent storage have been characterized for various fishes. Effect of different freezing methods like plate freezing, air blast freezing and brine freezing methods on quality of tiger shrimp (*Peneaus monodon*) and Pacific white shrimp (*Littopenaeus vannamei*) indicated better textural and biochemical quality for brine frozen samples whereas salt content (NaCl) was slightly higher compared to plate and air blast frozen samples. Effect of different freezing methods on the quality of cultured tiger shrimp (*P. monodon*) and Pacific white shrimp (*L. vannamei*) were examined. Freezing conditions for different grades of shrimps, squids, cuttlefish, octopus were also optimized by the institute. Studies were undertaken to minimize the thaw drip loss from shrimps and other categories of seafood and popularized among the industry. Various chemical agents and natural ingredients have been evaluated for this purpose. Rohu filets treated with 1% chitosan and gelatin and packed in polyester laminated with LDPE and frozen in air blast freezer prevented thaw drip formation considerably. Procedure for blanching and cooking has been standardized for different styles of shrimps and cephalopods. A re-cooking step before final packing of crab meat prior to freezing was introduced to reduce the bacterial load. Studies on quality comparison of Individually Quick Frozen (IQF) squid (*Loligo duvauceli*) rings and cooked shrimps stored at –18°C indicated a minimum shelf life of 18 months. Optimized recipe for shrimp samosa incorporating varying amount of shrimp (30 to 50%) and its quality changes during frozen storage indicated a shelf life of nine months. A wide variety of fish, shrimp, and cephalopods in various forms are preserved in frozen form and exported from India. Among fish products, whole, whole gutted, headed, and gutted, fileted, loins, steaks, breaded, and battered, and minced fish are frozen stored. Shrimp-based products ranges from head-on IQF to headless, peeled, and deveined, peeled, and un-deveined, easy-peel shrimp, butterfly style, stretched shrimp, skewered shrimp, breaded or tempura shrimps and cooked shrimps. Various IQF packs of whole squid, squid tubes, squid filets, rings, stuffed squid, cuttlefish, octopus, lobster whole and tail, sushi, and sashimi, squid filets, pine cut blanched and cooked products, seafood mix and analog products from surimi are frozen and marketed. Consumer retail packs is the recent trend in the marketing of frozen products. Protocols were developed for frozen fish steaks, fillets, and fish fingers, cleaned, and blanched shrimps and cephalopods in consumer packs. Apart from these, various frozen value-added ready-to-fry products like balls, cutlets, specialty products and curry products has also been developed.

1.2.7 DRIED AND SALTED FISHERY PRODUCTS

Drying is probably one of the oldest methods of food preservation. It consists of removal of water to a final desired concentration, which in turn reduces the water activity of the product, thereby assuring microbial stability and extended shelf-life of the product. In some cases, common table salt (Sodium chloride) is also used to prolong the shelf life of fish. Salt absorbs much of the water in the food and makes it difficult for micro-organisms to survive. Though drying has been widely practiced in the country, scientific method of processing for drying has not been practiced. The intervention from the institute in establishing processing methods for salting and drying to ensure optimum quality has helped the industry, to a greater extent. Generally, small-sized fishes such as anchovies, Bombay duck, lizardfish shark, ribbonfish, threadfin bream and lesser sardines; small freshwater fishes; medium and small-sized shrimps are dried whole after proper washing, whereas some large species of fishes such as sharks and catfishes are washed, degutted, and spread open to dry for human consumption. Some fish species are also dried for use as poultry feed. These are dried without any pre-treatment. Quality levels of dried small-sized fishes were surveyed in various markets and the common defects identified have been inadequate drying, contamination with sand and reddish discoloration due to infestation with red halophilic bacteria. Remedies for these problems have been worked out in the institute and popularized among the stakeholders. In the majority of the countries, sun drying has been the method predominantly used for drying. Due to poor quality, these products generally fetch low prices in the local as well as international markets. However, with the introduction of modern automatic hygienic drying techniques and attractive packaging, it has been possible to fetch better prices for dried fish products. A method has been standardized for preparing good quality cured fish. In this process, after proper evisceration and cleaning, the fish is salted in a specified way and kept in clean cement containers for about 24 h. The salted fish is dipped in saturated brine containing 5% calcium propionate for 2–5 min., followed by drying (Figure 1.4). The dried fish is then packed in polyethylene bags for retail marketing and in synthetic woven sacks for bulk marketing. Effect of open sun drying and use of solar dryer on the quality of different shrimp and fish species have been established. Drying time varied from 10–11 h for solar drying and 20–22 h for sun dried shrimps. For fishes like ribbonfish, jewfish, and lizardfish, drying time was 15–18 h in solar dryer compared to 28–32 h for sun drying. Solar dried samples had good sensory attributes and better acceptability.

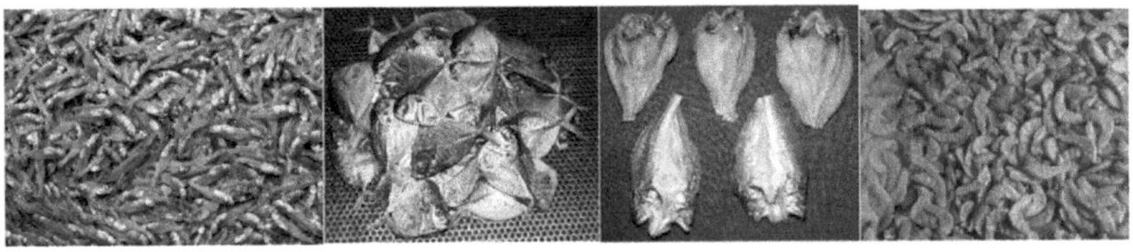

FIGURE 1.4 Dried fish and shellfish products hygienically prepared by the institute.

Red halophilic bacterial and fungal attacks are major issues during the storage of cured fish. A large quantity of such processed products has been lost every year in the country leading to financial loss and wastage of nutritional food. A simple technique of treating with calcium propionate was developed to overcome this defect. The treated dried fish can be kept for 8–10 months, compared to only two months in the case of untreated samples. Commercially dried Bombay duck is very poor in quality and is usually consumed by economically poor people. A process has been formulated to prepare laminated Bombay duck. The fresh fish, after surface drying and removal of head, tail, and fins is split open and dried. The dried material is pressed through a roller press, sides trimmed and further dried to about 10% moisture level, followed by packing in polyethylene bags. Laminated Bombay duck, thus prepared, can be stored for at least one year. The insect infestation is another major problem in dried fish. A detailed survey on the insect and flies of dried fish was carried out by the institute and a method to overcome this has been established.

Jellyfish is a marine species abundant in the east and west coasts of India. It has hitherto been unutilized even though some of the species are edible. The institute has worked out a suitable method for processing jellyfish in dehydrated form. In this, the umbrella portion alone is taken, trimmed, cleaned, washed well and then treated in four solutions of salt and alum of different concentrations and drained till the moisture content is reduced to less than 60%. It is then graded, packed, and stored in chilled condition at around 0°C. The method developed has been popularized in the industry and is extensively used in Jakhau part of Gujarat with slight modifications in which instead of umbrella alone, whole jelly fish is used.

Beche-de-mer is the commercial name for cured Holothurians, commonly known as sea cucumbers. This is a high value dried seafood item almost entirely exported. An improved method for processing *beche-de-mer* was developed which involves evisceration of fresh sea cucumbers, cooking in boiling water, drying, and packing in gunny bags, which has been adopted by the industry.

1.2.8 SMOKED FISHERY PRODUCTS

Smoking is one of the most widely used traditional fish processing methods employed in many countries to preserve fish. The preservation effect of the smoke is a result of drying of the product during the smoking as well as due to smoke particle absorption into the flesh. The smoke particles, mainly phenolic compounds, carbonyl, and organic acids, being absorbed by the product, inhibit bacterial growth on the surface of the product. The smoke particles also have a positive effect on the taste and color of the product and in many instances, smoking is normally practiced to improve these sensory characteristics. Smoking conditions for different fishes have been established by the institute and the quality attributes during storage have been characterized. These include Indian mackerel, tuna, rainbow trout, mahseer, clam, oyster, and freshwater fishes mainly by hot smoking technique. Studies indicated that smoked product had better sensory attributes and improved the shelf life. A detailed study on the quality changes of smoked rainbow trout packed in air, vacuum, and with O_2 absorber in EVOH pouches indicated a shelf life of 53 days compared to 49 days in vacuum and only 34 days in air packs. The improved method for the preparation of masmin, a traditional product of Lakshadweep Island has been established (Figure 1.5). Masmin chunks, mas flakes and mas powder preparation is also optimized. A method was developed to prevent the insect infestation in masmin. Due to the health risks of hot smoked products, studies on liquid smoke have been initiated. A method was developed to produce liquid smoke from different sources of wood. Liquid smoke treatment for different fishes has been also optimized by the institute.

FIGURE 1.5 Hot and liquid smoked fish chunks and masmin chunks.

Design, fabrication, installation, and demonstration of different models of eco-friendly, energy-efficient Community Fish Smoking Kilns christened as "COFISKI" was done in several locations in the country. The installations of COFISKI in all eight Northeastern states *viz*., Assam, Arunachal Pradesh, Manipur, Meghalaya, Mizoram, Nagaland, Tripura, Sikkim, and eastern states of Odisha and West Bengal has resulted in production of smoke-cured fish that is microbiologically safe, hygienic, high quality, longer shelf life, and lower carbon footprint. This has generated substantial sustainable income for thousands of hinterland women fishers belonging to economically under-privileged SC and ST groups. Establishment of "Mini-Fish Processing Facility" in fisheries sector is expected to minimize post-harvest losses and improve seafood safety, by developing entrepreneurial skills among small-scale fishers and empower them to contribute more effectively to the needs of multi-trillion rupee fisheries sector of the country.

1.2.9 THERMAL PROCESSED FISHERY PRODUCTS

Thermal sterilization is one of the most efficient methods of food preservation widely practiced worldwide. The main objective of the thermal processing is to achieve long-term shelf stability. Thermal processing generally involves heating the food products packaged in hermetically sealed containers for a pre-determined time at a pre-selected temperature to eliminate the pathogens of public health significance as well as those microorganisms and enzymes that deteriorate the food during storage. Work carried out in this area include, fish species, both in conventional form like natural pack, oil, brine, and sauce, to ethnic varieties like curries of regional importance (Figure 1.6). Processes have been standardized for the production of sardine, mackerel, tuna, and cobia canned in oil, brine, tomato sauce and in its own juice, with satisfactory products. The product in its own juice had natural flavor and the process was much cheaper than others. Canning of smoked sardines, canning of frog legs in brine and tomato sauce; canning of crab meat; squid masala; sardine curry; mackerel curry; seer fish molee; tuna curry; mussel and clam in masala, curry, and oil medium and region-specific curries, are some of the notable products developed. In canned crab meat in brine, blackening was formed due to the presence of ions of copper and iron. A suitable method was devised to control the blackening in crab canning. Canning conditions for tuna in different mediums like oil, brine, sauce, and with vegetables like broccoli, green peas and baby corn have been optimized. Thermal processing of Ready-to-Eat (RTE) fish products like traditional Kerala style mackerel curry; rohu curry; seer fish curry; seer fish molee; prawn kuruma; shrimp in curry, brine, and natural pack, were also developed. Procedure for canning mussel and clam meat in oil, brine, sauce, and curry medium were developed. Purification methods to reduce the grit and bacterial contamination in shucked meat of mussels and clams have been evolved. Canning of boneless, skinless fillets of mackerel was developed. Apart from these initiatives, canning of smoked fish products in different packaging material has been developed. A formula has been developed for a good lacquer for coating the cans for packing fish. The lacquer developed has good adhesive property, hardness, flexibility, and resistance towards alkalis and acids. The lacquer overcomes many drawbacks of other lacquers. Extensive work on retortable pouches for RTE products have been carried out, over-pressure conditions for processing of such products in retortable pouches have been optimized, and many RTE products have been developed. RTE fish products fortified with calcium, iron, and fish oil have also been developed. Process conditions for developing a seafood mix consisting of fish and shellfishes in different mediums were optimized. The technology has been transferred to industry for the production of RTE products for both export and domestic markets.

FIGURE 1.6 Thermal processed ready-to-eat fish products developed by the institute.

1.2.10 *SPECIALTY FISHERY PRODUCTS*

The institute has been instrumental in the standardization of a variety of value-added products from low-value fishes and specialty products (Figures 1.7(a)–1.7(d)). These include fish mince, *surimi*, balls, cutlets, fingers, patties, burger, coated products and many imitation products, specialty shrimp based products like whole shrimp, peeled shrimp, peeled, and deveined (PD), peeled, and un-deveined (PUD), cooked shrimp, stretched shrimp (Nobashi), cooked butterfly shrimp, skewered shrimp, shrimp head-on and center peeled, shrimp head-on center peeled and cooked, coated shrimps in different forms, coated squid rings, coated squid tube and coated fish filets. The institute also developed a method to extract natural pigments from plant sources for imparting color to the bread crumbs used for coating fishery products. The range of products also includes coated oyster and mussels, wafers, and fish with mixed vegetable products. Initial investigations on the processes for the production of fish soup powder, fish, and prawn flakes, fish hydrolysates have been pursued under the All India Coordinated Research Project on Trash Fish Utilization. Methods to extract shark fin rays have been developed. Methods for the production, packing, and storage of spiced pickles from fish, clams, and mussel have been developed and popularized widely.

FIGURE 1.7(A) Shrimp based value added products developed by the institute.

FIGURE 1.7(B) Natural plant extracts developed by the institute for preparing colored bread crumbs and coating fishery products.

Preparation of fish incorporated bakery products like rusk, bread, and *khari* (a Gujarati bakery product) were optimized by incorporating rohu mince at different levels which enhanced the protein content of the products. Seaweed and fish protein-incorporated biscuits were developed which enhanced the fiber content and protein content, respectively. Fish mince is prepared by concentrating only edible muscle part removing all the other parts like scale, skin, gut, and bones. Fish mince acts as base material for the majority of the value-added products. It is prepared by split opening the fish ventrally and passing them through the meat-bone separator. Larger bones are normally removed manually, and fine pin bones are then separated by passing the mince through a pin bone remover so that the bone content in the final mince would not exceed 2% as it affects the quality of mince. Due to the very fine nature of the mince, the surface area will be more and hence it is susceptible to spoilage. Chilled and frozen storage characteristics of mince from different fishes have been characterized. Incorporation of chitosan improved the quality of mince considerably. Fish mince is normally used for the preparation of *surimi*, fish balls, fish finger and other specialty products. Fish fingers are regular rectangular sized fish portions made from either frozen fish filet or fish mince. Skinless and boneless filet are partially frozen to get the correct shape of the finger. A method to prepare fish finger using minced meat was optimized. Methods of preparation and storage conditions for fish ball, cutlet, nuggets, and burger have been also optimized. Batter mixture ingredients were optimized for imparting better quality to the coated products. Processing conditions for the preparation of stuffed fish-based product 'Momo' was optimized. The effect of cooking methods like steaming and frying on the quality changes of fish based products have been characterized. Filling ingredients were standardized to develop a novel fish-based stuffed product in capsicum and tomato for high value markets. Extruded snack products, such as noodles incorporating fish mince, shrimp, and squid was developed. A process for developing nutrient rich energy bar incorporating fish meat with a variety of nuts has been developed.

FIGURE 1.7(C) Specialty products like fish ball, cutlet, roll, and fish finger developed by the institute.

FIGURE 1.7(D) Fish-based extruded products developed.

FIGURE 1.8 Analog products developed by the institute.

1.2.11 SURIMI

Surimi is a myofibrillar protein concentrate obtained from mechanical deboning of fish flesh which is washed with chilled water and with added cryoprotectants. Methodology of *surimi* preparation from marine, freshwater, and brackish water fishes were developed and popularized in the industry (Figure 1.8). Physical properties of *surimi* from different fishes have been established. The *surimi* can be used as base material for the preparation of textured imitation products like lobster tail, crab legs, shrimp, scallop, and crab stick. A method for the preparation of analog products like shrimp, crab stick, lobster, etc., have been developed by the institute. Studies on the effect of gelatin incorporation on gel strength of freshwater fish (*Cyprinus carpio*) *surimi* were undertaken. In this gelatin prepared from rohu (*Labeo rohita*) skin was incorporated into water washed minced meat at 1 and 2% along with cryoprotectants (sucrose:sorbitol:STPP in 4:4:0.2%) and mixed well and gel strength and other parameters were compared during frozen storage. Gel strength of 1 and 2% gelatin incorporated sample increased to 2.1 and 2.8 times compared to control surimi.

1.2.12 FISH SAUSAGE

Sausages are usually prepared from minced meat mixed with ingredients like fat, binders, fillers, spices, and salt, according to the consumer preference. The mixed and comminuted meat is stuffed into cylindrical casings, either natural or synthetic, of desired size and marketed either as fresh, cooked, smoked or dried. A recipe for fish sausage preparation has been optimized by the institute. Studies on the development of functional fish sausage incorporating various functional ingredients like β-carotene, microencapsulated fish oil, chitosan, glucosamine hydrochloride and Malabar tamarind extract has also been standardized. Process for the development of fiber rich fish sausage has been developed incorporating fiber from seaweed. Process for fish sausage in dried form which facilitates storage under room temperature has been developed (Figure 1.9).

FIGURE 1.9 Different stages in fish sausage preparation.

1.2.13 TRANSPORTATION OF HIGH-VALUE PRODUCTS USING DRY ICE

Transportation of fishery products to distant places is normally done using refrigerated systems maintaining the cold chain throughout the transportation. High-value products like frozen crab sticks stored in dry ice (solid CO_2; –78°C) at 1:1 ratio in thermocol boxes indicated that the dry ice can maintain the internal product temperature of –20°C up to 129 h. The study indicated that the dry ice can be used for transporting high-value products to distant markets, maintaining the desired product temperature.

1.2.14 FISH BY-PRODUCTS

Huge quantities of waste are generated during the processing of fish and shellfishes. In the early days, this waste was simply dumped in barren lands, posing health and environmental problems. The institute has developed many high-value products utilizing this waste material. The major ones are highlighted in subsections.

1.2.14.1 FISH MEAL AND FISH OIL

Methods have been developed to utilize the waste from fish processing units, into fish meal and oil. The processing waste from tuna, mackerel, and sardine has been utilized for fish meal preparation, whereas whole sardines have been used for fish oil extraction. A large quantity of body and liver oil from fish is extracted in the fishing villages by conventional and crude methods resulting in a reduction in the quality of the fish oil. The institute has developed and popularized improved methods for extracting high-quality oil from body and liver of fish.

1.2.14.2 FISH FEED

With the advancement of scientific fish and shrimp farming in India, the need for good quality feed has increased. The indigenous feed used by farmers does not often meet the nutritional requirements of the fish in the early period of growth. Other types of feed available in the market are very costly. Formulations have been worked out at the institute for preparation of fish and prawn feed at reasonable cost by incorporating miscellaneous fish and fishery wastes like prawn shell and squilla, groundnut oil cake, groundnut oil, starch, and salt mixture, in a specified proportion (Figure 1.10). A method to prepare the feed with minimum facilities was popularized among the farmers.

FIGURE 1.10 Fish feed developed by the institute.

1.2.14.3 FISH ENSILAGE

A practical, easy, and economical way of making up the nutritional deficiency in cattle feed is to incorporate a suitable amount of fish protein in an acceptable form into the feed. Two methods have been developed for the preparation of fish ensilage which is a form of fish protein. In the first, any non-oily trash fish is minced, formic acid added to it and mixed well and kept covered for some time till the whole mass becomes a homogeneous slurry (Figure 1.11). The ensilage thus prepared is kept in stoppered containers. In the second method, 10% by weight of molasses is added to the minced fish, the material made into a slurry by adding water and the whole mass is cooked and then cooled to the room temperature. A pure culture of *Lactobacillus plantarum* is added to the mass, stirred well and allowed to ferment. The product obtained after fermentation is fish ensilage.

FIGURE 1.11 Ensilage from fish and cuttlefish waste developed by the institute.

Considerable work has been carried out in the institute to develop fish protein hydrolysate from fish meat and seafood processing waste. The fish hydrolysate has been characterized for its properties and various products have been developed incorporating fish protein hydrolysate.

1.2.14.4 SHRIMP EXTRACT FROM PRAWN WASTE

Initially, large quantities of prawn waste were thrown away or used for converting into manure. The high percentage of protein present in the prawn waste can be isolated for human consumption or for the preparation of animal feed. The institute has developed a process by which the prawn waste can be converted into a protein-rich product. The fresh prawn head and shell waste is minced after thorough washing, boiled with Sodium hydroxide, the boiled mass filtered and the filtrate neutralized with Hydrochloric acid and Acetic acid. The neutralized filtrate is then concentrated to obtain a semi-solid mass. The product is then either packed in metal containers or polyethylene bags or stored under frozen condition. A method of developing prawn powder from small prawns has been developed for use as food flavoring.

1.2.14.5 CHITIN AND CHITOSAN FROM PRAWN SHELL WASTE

The huge quantity of prawn and other crustacean shell waste available from the processing units of India can be utilized by converting them into chitin and chitosan. The institute has developed a process for the preparation of chitin and chitosan from prawn shell waste and transferred the technology to the industry (Figure 1.12). The use of chitin for the production of glucosamine hydrochloride has also been established. The process for preparation of water-soluble chitosan has been worked out. Process conditions have been optimized for producing high viscosity grade chitosan from prawn shell waste.

1.2.14.6 GELATIN AND COLLAGEN EXTRACTION

The skin, bone, and scales from processing discards can be used for the preparation of gelatin and collagen which have wide applications. A method has been optimized for the production of gelatin from freshwater fishes, tuna, and other species of fish. Collagen extraction method was also optimized from processing discards and the product quality has been characterized.

FIGURE 1.12 Chitin from prawn shell waste.

1.2.14.7 SHARK CARTILAGE

The skeleton of a shark is made of cartilaginous bone which is about 10–15% of body weight. This cartilage is rich in chondroitin sulfate which has got application in medicine. A procedure has been developed for the processing of the cartilage into a clean, dry, white attractive material without any characteristic smell.

1.2.14.8 SQUALENE FROM SHARK LIVER OIL

Squalene is a hydrocarbon extracted from the liver oil of shark by fractional distillation under vacuum. An improved method has been developed to extract squalene from shark liver oil (Figure 1.13).

FIGURE 1.13 Commercialized squalene-based pharma product from the institute.

1.2.14.9 EXTRACTION OF SHARK FIN RAYS

Shark fin rays are valuable products of export from India. Earlier, only dried shark fins were being exported (Figure 1.14). The institute has developed a method to extract shark fin rays from fins using dilute acetic acid. The extracted shark fin rays are dried and packed in polythene bags.

1.2.14.10 FISH MAWS

The institute has developed a method to prepare fish maws from air bladder. In this method, the fresh air bladder is taken out, the internal membrane is removed and the bladder is well cleaned and dried under controlled conditions. The dried maws are packed in polyethylene bags and stored.

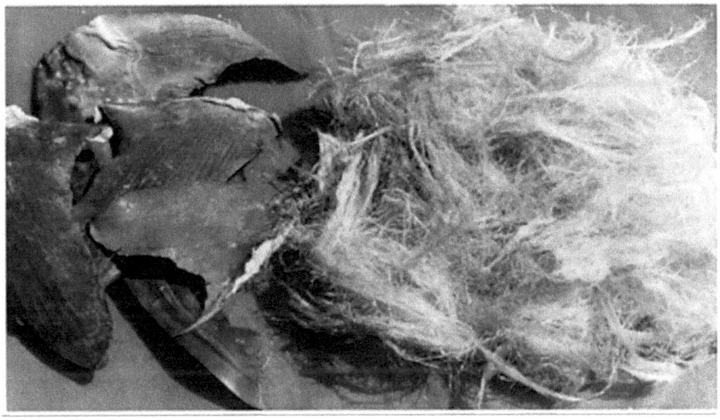

FIGURE 1.14 Shark fin rays.

1.2.14.11 LIME, CALCIUM, AND HYDROXYAPATITE FROM FISHERY RESOURCES

A considerable quantity of mussels is landed along the coast of India. Even though the meat is utilized for human consumption, the shell is discarded as waste. A simple technology has been developed for converting the mussel shell into lime. The same technology can be adopted for clams as well. A method to produce calcium and hydroxyapatite from fishery waste is also optimized by the institute.

1.2.14.12 AGAR FROM SEAWEEDS

Seaweeds are gaining more and more importance due to their wider applications. The institute has developed a process for producing good quality agar with high gel strength by alkali treatment instead of the conventional acid treatment method. In this method, the dried seaweeds are soaked in water, agitated to remove foreign particles, drained, treated with Sodium carbonate solution and heated. It is then washed repeatedly to free it from the alkali and the agar extracted by boiling with dilute hydrochloric acid. The extracted gel is frozen and then thawed and the thaw water drained off. It is then powdered and stored in plastic bags.

1.2.14.13 ANTISEPTIC OINTMENT FOR USE BY PRAWN HANDLERS

Constant contact with prawns, water, and ice, produces certain lesions on the palms of workers engaged in prawn peeling. On contact with soap and chlorine, such eruptions develop burning sensation and workers suffering from such skin lesions hesitate to keep their hands clean by washing and disinfection. Contact with unclean hands will adversely affect the quality of the processed products. The institute has developed an effective antiseptic ointment to heal such skin lesions and thereby avoid contamination of processed products. The main constituents of the ointment are benzoic acid, salicylic acid, potash alum, coconut oil and white soft paraffin.

Apart from these achievements, the institute has also taken a lead role in establishing advanced processing facilities such as high pressure processing, pulsed light preservation and microwave processing, for processing of fishery products. The effect of high pressure processing (HPP) on the shelf life of various fish and shellfish have been established. A method for shucking shells of mussels and clams has been developed. Application of pulsed light for surface decontamination of fish has been established.

The modern fish processing facility and business incubation service, coupled with pilot plant facilities and production line for chilled and frozen fish, coated fish products, RTE fish products, extruded fish products, sausage, spray drier, fluidized bed drying, advanced packaging methods like vacuum packaging, MAP, production of chitin and chitosan and glucosamine hydrochloride, supported by the state of the art research laboratory with sophisticated analytical instruments, is well equipped to take up any emerging challenge in the field of post-harvest technology of fish.

1.3 CONCLUSION

In the present era, consumers demand high-quality processed fish with minimal changes in nutritional and sensory properties. Apart from the above, globally several alternative or novel fish processing technologies are being explored and implemented to provide safe, fresher-tasting, nutritive foods including seafood without the use of heat or chemical preservatives. To consumers, the most important attributes of a food product are its sensory characteristics, and the goal of food manufacturers is to develop and employ processing technologies that retain or create desirable sensory qualities or reduce undesirable changes in food due to processing. Recent developments have improved techniques in handling, product development, packaging, preservation, and storage. Entrepreneurs shall be benefited with a method capable of increasing the safety and shelf life of their foods.

KEYWORDS

- **canning**
- **chilling**
- **ethylene-vinyl alcohol**
- **extrusion**
- **fish**
- **freezing**
- **modified atmosphere packaging**
- **waste utilization**

REFERENCES

Balachandran, K. K., (2001). Canning. In: *Post-harvest Technology of Fish and Fish Products* (pp. 158–220). Daya Publishing House, New Delhi.

Gopal, T. K. S., & Ravishankar, C. N., (2001). Packaging of value added fish products. *Indian Food Industry, 20*(3), 64–67.

Gopal, T. K. S., Antony, K. P., & Govindan, T. K., (1981). Packaging of fish and fishery products, present status and future prospects. *Seafood Export J., XIII*, 15–22.

Gopal, T. K. S., Ravishankar, C. N., Vijayan, P. K., Madhavan, P., & Balachandran, K. K., (2002). Heat processing of seer fish curry in retort pouch. In: Boopendranath, M. R., Meenakumari, B., Joseph, J., Sankar, T. V., Pravin, P., & Edwin, L., (eds.), *Riverine and Reservoir Fisheries of India* (pp. 211–216). Society of Fisheries Technologists (India), Cochin.

CHAPTER 2

Surimi and Surimi-Based Products

B. A. SHAMASUNDAR

Former Professor and Head, Department of Fish Processing Technology, College of Fisheries, Mangalore – 575002, Karnataka, India

ABSTRACT

Fisheries resources have enormous potential to enhance the livelihood of low-income groups and at the same time fulfill the requirements of nutritional security so as to alleviate malnutrition problem. Surimi is a water washed fish mince and partially dehydrated. Processing of low economic value fish to surimi is gaining importance in India mostly for the export market. A dramatic growth in surimi-based product industry began in the mid-1960s. In India, threadfin breams, croaker, ribbonfish, and bigeye snappers fishes contributed significantly as raw material for surimi production. At present world, 500,000 tons of surimi are marketed across the globe, of which India's share is about 45,000 tons. The type of fish normally used for surimi production is lean variety of fishes. However, with modified technology any fish can be used for surimi production. The final quality of surimi is judged by three important parameters: moisture content, gel-forming ability, and color are important. The microbial safety and quality of surimi largely depend on the quality of fish used for production. The major challenges for the surimi industry are the supply of raw material at a constant price throughout the fishing season. The profitability gets eroded if the raw material cost is fluctuating. Also, the vagaries of nature make a very poor fishing season.

Advances in Fish Processing Technologies: Preservation, Waste Utilization, and Safety Assurance.
Ranendra K. Majumder, PhD & Amjad K. Balange, PhD (Eds.)

2.1 INTRODUCTION

Broadly surimi refers to the separated fish mince subjected to repeated water washing, removal of excess water, addition of certain additives refereed as cryoprotectants and block frozen. Surimi originated in Japan, the science and technology mainly derived from traditional products such as Kamaboko and other paste products. In order to produce surimi, the fish is dressed (removal of head, entrails, and other inedible parts), meat is separated by mechanical meat separator. The separated fish meat is subjected to water washing for 3–5 times, and excess water is removed by mechanical means. The mince-meat is mixed with certain additives like sucrose, sodium tri-polyphosphate and sorbitol and then packed in plastic bags of required weight. The meat along with additives is frozen and frozen stored at –20°C. The frozen blocks obtained are termed surimi. The surimi is different from the minced meat, wherein, the separated fish meat is not subjected to water washing. The surimi is used for production of a variety of products such as Kamaboko, fish sausage, analog products like crab sticks and molded products. The utilization of fish for surimi production has distinct advantages. They are (a) identity of fish species for human consumption is lost so that all the underutilized fishes which otherwise would not have been used for human consumption will be used effectively (b) range of novel products can be prepared by application of new technologies (c) consistent quality of surimi with scale-up operation using different species (d) increased consumer demand for surimi based products has enabled the industry to grow at rapid pace. However, a few disadvantages are also encountered in surimi production like (a) loss of nutrients (water-soluble proteins) as a result of washing process (b) heavy dependence on potable water this is becoming scarce commodity (c) environmental pollution if the process waste is not treated.

2.2 RAW MATERIAL RESOURCES FOR SURIMI PRODUCTION

The potential of fisheries resources to livelihood of marginal fishermen have been well recognized. It is one of the sectors that have grown rapidly in terms of production so as to alleviate the malnutrition problem. With increase in world population especially in developing countries, it is important that supply of protein from different sources is of paramount importance for a well-being of state. The utilization of animal protein, especially that from fish may be the answer in offering nutritional security of many developing countries.

The world's total fish production during the year 2018 was 178.5 million metric tons which comprises of 96.4 million metric tons of capture production and 82.1 million metric tons of culture production (FAO, 2020). In India, total fish production during the year 2018–2019 was 12.6 million metric tons. Recently, fish production from aquaculture growing at a very fast pace and contributes nearly 65% of total fish production. Nevertheless, the demand for fish is increasing rapidly with awareness of fish being health food. In order to meet ever-growing demand for fish, the strategies employed by planners is to either to increase the fish production or reduce the postharvest losses, which accounts for 15–20% of total fish landings. Alternately, adaption of novel technology for utilization of low economic value fish and making available different forms of processed fish are being given importance. It is also anticipated that demand for convenience fish products are likely to increase with improvement in the living standards of Indian population.

Processing of low economic value fish to surimi is gaining importance in India mostly for the export market. Surimi is a traditional product of Japan consumed worldwide and production has reached about 830,000 MT during 2017 with the major contribution of 400,000 MT from South Asian and southeast Asian countries. Alaska Pollack is the single major species contributing 21% of the total surimi production whereas the tropical species like threadfin bream, croaker, lizard fish, ribbon fish, and bigeye snapper together contributing about 60% (www.future-seafood.com). Increasing consumption and demand of surimi-based products may be attributed to the wholesomeness, nutritious nature, and affordable price.

2.2.1 *SURIMI PRODUCTION IN INDIA*

There are about 10 surimi processing plants in India and are mostly situated on the western coast of India. About 8–10% of the world's production is produced in India. The production of surimi during the last five years is around 60,000 t. The Karnataka state has four surimi plants, out of which three are active. Two surimi plants are situated in Mangalore, and the other two are in the Udupi district. The fish species used for surimi production in India include threadfin breams, croakers, ribbon fish, bulls eye fish and to some extent lesser sardines. The plant capacities of surimi plant varies from 30–100 t of final product/day. Normally the yield of surimi varies from 30–45% depending on the size of fish. On an average the yield varies from 35–40%.

2.3 HISTORY OF DEVELOPMENT OF SURIMI INDUSTRY

A dramatic growth in surimi-based product industry began in the mid-1960s. This high growth rate of surimi industry was made possible as surimi was used for the traditional Japanese products called Kamaboko. This tremendous growth in the industry was primarily made possible by an increase in a stable supply of raw material in the form of frozen surimi.

Prior to the development of frozen surimi, the production of each kamaboko plant was limited by the amount of raw surimi that could be produced from fish delivered daily. The surimi-based product plants could stockpile their raw material to assure production throughout the year on steady schedules. The compact frozen surimi block was ready to use as an intermediate raw material and saved the cumbersome task of preparing the raw material, along with its accompanying waste-disposal problem.

2.3.1 BACKGROUND

Prior to 1960, the Japanese Alaska Pollock catch of 300,000 tons, mostly from the surrounding waters of Hokkaido Island, was based on its yield of roe, a highly-prized and gourmet food item. The disposal of remaining meat was a major challenge as it accumulated in huge quantities. It was further compounded that the meat of Alaska pollock (*Theragra chalcogramma*) lost its functional properties very rapidly during iced storage as well as during frozen storage. Thus, maintenance of the initial quality after the catch was the key to its efficient utilization.

2.3.2 INVENTION

In 1959, Nishiya's research team discovered a unique technique of stabilizing the muscle protein of surimi during frozen storage. The innovation included water washing the mince, thereby removing the water-soluble proteins and to achieve the frozen storage stability addition of additives such as sucrose, sorbitol, and polyphosphates were standardized. Another important invention was the development of frozen salted surimi which contained 2% salt and 10% sugar. Sugar compounds were the common cryoprotectant for the two types of frozen surimi. However, the salt-added surimi developed by Ikeuchi group did not develop the popularity of the non-salt type developed by Nishiya group, due to its inferior storage life.

To stabilize muscle protein in frozen surimi, the Nishiya group used 5% glucose initially. Glucose however was replaced by sucrose and sorbitol, since glucose causes a brownish discoloration in the final surimi-based product due to the amino-carbonyl (Millard browing) reaction. The level of sugars was subsequently increased to 8–10% to give better stability to the exhibited by other compounds, such as salts of glutamate and citrate. However, these compounds have not found application in the commercial production of frozen surimi.

2.3.3 GROWTH OF THE INDUSTRY

Based on Nishiya's patent, a frozen surimi industry on Alaska Pollock began in 1960 with four small-scale plants in Hokkaido. The fish sausage industry, facing a heavy shortage of raw material, was the first to accept surimi as a raw material as the quality was not yet good enough for most types of kamaboko. As the quality of the surimi was improved, the traditional kamaboko industry also gradually began to accept the same as raw material. By 1966, the deep sea trawlers off the Kamchatka Peninsula in 1966, the production of surimi by shore plants increased markedly and amounted to 205,000 tons in 1973.

Frozen surimi, production at sea began with two mothership-trawler fleets established in 1965 in the Bering Sea. The large fishing companies had already sent factory ship to the Bering Sea to manufacture fish meal, and they had noticed the presence of huge resource of Alaska Pollock in the area. The excellent frozen shelf life and availability of higher frozen storage space in comparison to storage of whole fish or filets permitted the factory ship to manufacture surimi in far distant waters from Japan.

They succeeded in making freshwater on-board and established a technique of washing the minced flesh with minimum quantity of water than was normally used in land-based plants. A better design of the plant with a limited number of workers in the confined space aboard required the development of a continuous and well-controlled system. The productivity of surimi production on board was improved. Later, three more mothership fleets and about 20 factory trawlers engaged in surimi production. The initial production of 8,000 tons of frozen surimi on factory ships in 1965 increased to a record of 218,000 tons in 1973. About half of the Japanese frozen surimi supply was then processed on factory ships, with the remainder produced in the shore plants. However, the quality of the factory ship product was generally superior, since these plants were able to process fish on the same day of catch while the shore plants had access only to fish of 1–5 days age postharvest.

2.3.4 PROCESSING MACHINERY AND EQUIPMENT

In manufacture frozen surimi, the initial steps are identical to those for preparing the traditional kamaboko products: removal of head, entrails, and separation of flesh from bones and skin, water washing of separated mince, dewatering, and straining of the minced, leached flesh. The difference is in the production scale. A typical shore plant has a production capacity of 10 to 20 tons of frozen surimi per day from 50 to 100 tons of fish, whereas the traditional kamaboko plants could process at most only 10 tons of fish daily. The equipment and machinery for surimi production had been developed through improvements in kamaboko processing machinery. This included such pieces of equipment as the fish washer, descaler, various types of meat separators of large capacity, washing tanks, rotary sieves, refiner, screw press, and decanter. The introduction of the rotary sieve in the washing process increased the leaching efficiency. The screw press took the place of basket-type centrifuges which necessitated a batch type operation, and made the dewatering operation continuous and efficient. The refiner replaced the strainer for removing small bones and tendons remaining in the washed flesh. The refiner operates on the leached flesh of high water content and thus generates little heat, whereas the strainer acted on the dewatered leached flesh and incurred heat buildup with consequential denaturation of the proteins. The decanter centrifuge will help in the recovery of suspended particles in the wash water.

2.3.5 GROWTH OF THE KAMABOKO INDUSTRY

An abundant supply of raw material in the form of frozen surimi and rapid growth of the Japanese national economy stimulated the kamaboko industry's activity. In 1973, the surimi-based products industry produced 1,187,000 tons of these products, including 179,000 tons of fish sausage. Introduction of frozen surimi clearly changed the surimi-based product industry. The kamaboko plants were now freed from the cumbersome task of preparing raw surimi and the trouble of waste disposition, resulting in a number of unit operations in the process, Surimi-based product plants grew in size and changed the processing system to accommodate the new starting material.

Traditionally, in kamaboko manufacturing units, the thawing of frozen surimi was achieved by holding the surimi at ambient temperature for long periods. Later, high frequency microwave radiation was introduced for continuous and short-time thawing. However, this equipment was expensive and its maintenance costly. Modern surimi-based product plants generally thaw surimi in air-conditioned tempering rooms, or by contact-plate generally thaw surimi in air-conditioned tempering rooms, or by contact-plate units. In the latter, the frozen surimi is placed between shelves in which warm water is circulated, and it can be thawed in a very short time period.

The grinding of surimi with additives to prepare traditional kamaboko products was achieved by stone mortar. However, the stone mortar was replaced by the silent cutter because of its inferior efficiency. When raw surimi is used, the silent cutter generates a considerable amount of heat, resulting in a marked decrease in the functional properties of the myofibrillar protein. For this reason, traditional kamaboko plants never installed silent cutters. However, the low temperature of frozen surimi overcomes the heat-generation problems of the silent cutter. The efficiency of silent cutter was greatly improved with an additional blending device and this type of improved cutter was rapidly adopted by the surimi-based products plants.

The surimi production was introduced in South Asian countries by the Marine Fisheries Research Department (MFRD) of the Southeast Asian fisheries development center (SEAFDEC). Since then, there has been tremendous development not only in the processing of surimi for export but also in the upgrading of the local fish jelly product industries in countries like Thailand, Malaysia, and Singapore.

At present 12 surimi factories are in Thailand with a total production of about 60,000 metric tons per year with 90% exported to Japan and South Korea the remainder is exported to Singapore and Malaysia. These factories produce surimi from threadfin bream, big eye snapper and croaker. Some factories also produce lizardfish surimi, although it is generally of lower quality.

The surimi industry in Malaysia started later, and by 1990 there were three surimi factories (each with a capacity of 5 metric tons per day). In addition, 5-metric ton surimi factory started operations in the east coast of peninsular Malaysia in 1995. A Japanese surimi crabstick industry was established in Malaysia in the late 1980s.

Recently, the local fish ball-fish cake industry in Malaysia has realized the advantage of using surimi and has acquired equipment for the preparation of leached fish meat for local use. This, coupled with the use of locally produced surimi or Thai surimi, has made it possible to higher production capacities of about 2–3 tons/day capacities.

The growth of fish ball and fish cake industry in Singapore was rapid in the early 1980s. This rapid growth was possible due to improved productivity and mechanization. During this period, the MFRD introduced technology, including the use of related machinery, for the production and use of surimi in fish ball and fish cakes. Coupled with the development of the surimi industry in Thailand, the fishball manufactures were able to change from the use of fresh fish to frozen surimi. Singapore now imports 8,000–10,000 metric tons of surimi per year. This has enabled them to upgrade their processing technology and invest in machinery to increase their productivity as well. There are now about 30 fish ball-fish cake manufacturers producing about 50 metric tons of products per day, mainly for local consumption. The largest factory now uses surimi from Thailand and Malaysia to produce about 15 metric tons of products per day. This factory has also invested in the use of silent cutters, automated forming machines such as fish cake forming machines, chikuwa forming machines, and automated fish ball forming–setting cooking and chilling lines. A local manufacturer produces about 2–3 metric tons of leached meat per day from threadfin bream and big eye snappers for daily distribution to local fish ball manufacturers. Consequently, to remain competitive, because of the high cost of labor in the Singapore, manufactures must automate to increase productivity and product quality.

Manufactures of frozen fish ball and cuttlefish from Singapore were looking at the potential to export to Australia, Japan, and the United States. The manufacturers of fish balls and cake from Singapore started investing overseas and have established in Malaysia and China.

The surimi production in other countries increased with the success stories from Singapore, Malaysia, and Thailand. Vietnam already has several surimi factories producing low grade lizardfish surimi, mainly for export to Korea, India on the other hand, has several surimi factories producing tropical fish surimi for export to Japan and Singapore.

2.4 PRESENT GLOBAL STATUS OF SURIMI INDUSTRY

The production of surimi world over especially in Southeast Asian countries steadily increased from 1980 onwards. During the same period there was a decline in the surimi production Japan as stocks of Alaks Pollack fish drastically reduced. This paved way for diversification fisheries resources for surimi production in Southeast Asian countries. Totally there were more than

80 surimi plants in Thailand, Vietnam, Myanmar, Indonesia, and Vietnam (SAFDC, 2009). According to FAO statistics, 2004 global production of surimi has been between 860,000 and 1,150,000 metric tons, depending on calculation modes. However, total output as evaluated by industry insiders is lower, at 750,000 metric tons. Currently, about 2–3 million metric tons of total Martine catches are being used for surmi production (www.Eurofish–Surimi, 2006). The database from this website provides a general overview on all aspects of surimi production and marketing by companies of each producer country. Further, the future development and potential of surimi world over.

The major producers of surimi are USA, Japan, and Thailand. The contribution of China for surimi production is also showing increasing trend. The other major surimi producing countries are Vietnam, India, Argentina, Chile, and a few counties from Asia (Figure 2.1).

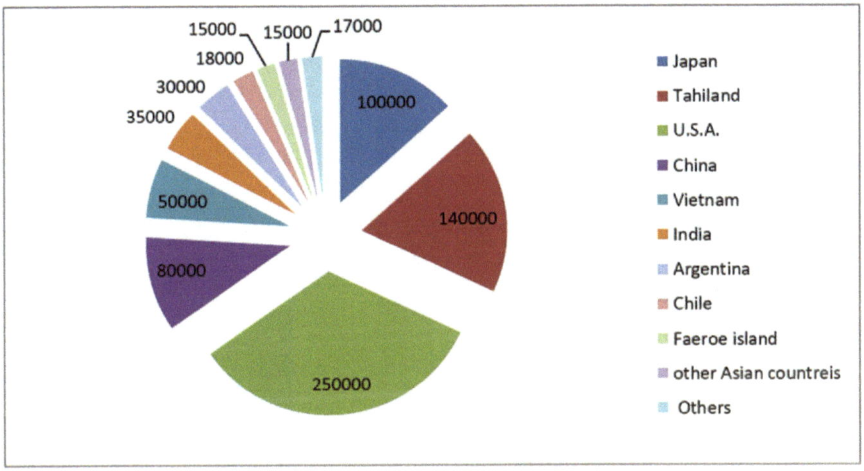

FIGURE 2.1 Estimated World Surimi Production in 2005 in metric tons.
Source: Estimated output based on evaluation by industry insiders (www.Eurofish–Surimi, 2006).

2.5 SURIMI INDUSTRY IN INDIA

With increased trawl catches, the resources were used for surimi production. As per the Ministry of Food Processing Industries Annual report of 2006, there are 11 surimi plants in different parts of India (Annual Report, 2005–2006). The species used for surimi production include threadfin breams, croaker, ribbonfish, and bigeye snappers are used as main raw materials. At present world, 500,000 tons of surimi are marketed across the globe, of which India's share is about 9%, or 45,000 tons. About 65% surimi is exported to only Japan from India, and from the last 34 years, China and EU countries too have started importing surimi from India.

2.6 SURIMI FROM LIGHT MUSCLE FISH

Surimi processing consists of several unit operations and variations in the operations depends on the type of fish species used. Based on fat content fishes are classified into fatty fish (fat content more than 5%) and lean variety (fat content is less than 4%). The color of fish meat can be white/light or dark. The lean variety fishes normally have light or white meat. The various steps involved in the production of surimi from light muscle fish is given in Figure 2.2. The fish used for surimi production should be of prime quality and chilled till processed. Removal of head and entrails is to be done immediately and chilled further. The block frozen surimi is depicted in Figure 2.3.

Fresh Whole fish

↓

Heading and gutting

↓

Meat separation

↓

Water washing (leaching)
(3 cycles)(1:2 - mince: water, 5-10°C)

↓

Dewatering (straining)

↓

Mixing with Cryoprotectants

↓

Forming into blocks and freezing (-40°C)

↓

Packaging into cardboard box

↓

Frozen storage (-20± 2°C)

FIGURE 2.2 Flow chart for the preparation of surimi.

Source: Adapted from Shaviklo and Johannsson, 2006.

FIGURE 2.3 Frozen block of surimi (10 Kg block).

Source: Reprinted from Shaviklo and Johannsson, 2006.

2.6.1 FISH SPECIES USED FOR SURIMI

Any finfish can be used for surimi production. But the preferred characteristics of fish meat to be used for high grade surimi production are: (a) meat color should be white or light; (b) fish meat should have good gel-forming ability; (c) the fat content should be less than 4%; (d) the fish should be available throughout the season; and (e) the cost of fish should be economical for industrial production. With the development of new technologies for surimi production, any type of fish can be used. In the earlier phase of surimi industry development Alaska Pollack was widely used as raw material. A decrease in Alaska Pollack harvests caused by stricter fisheries management has paved the way for the use of new species in the surimi industry. Economically, it is mostly produced from the Alaska Pollack (*Theragra chalcoramma*) which is a lean variety and meat is white in color. There are more than 50 different species are currently being used for surimi production. They include cod (*Gadus morhua*), herring (*Clupea harengus*), whiting (*Merlangius merlangus*), mackerel (*Scomber scombrus*), sardine (*Sardina pilchardus*), anchovy (*Engraulis encrasicholus*) and goby (*Gobius* spp.), etc., have been used in surimi production (Lee, 1984). In a tropical country like India, fishes used for surimi production include pink perch (threadfin bream), croaker, big eye snapper, ribbonfish, and leather jacket are widely used. It is important that the fish used for surimi production should be of prime quality. The production of surimi can be on board the factory ship or land factory. The use of shellfish such as shrimp, lobsters, oysters, and clams are not favored for surimi production because of intrinsic meat character.

2.6.1.1 MECHANICAL SEPARATION OF FISH MINCE

The first step in surimi production is to remove the head, gut, and bones and separate the fish meat for further processing. A schematic presentation of mechanical fish bone separator is given in Figure 2.4, The design of fishbone separators have been improved to maximize the yield of mince. The initial preparation whole for deboning is done either by removal of head, entrails or by filets. Heading and gutting should be done without disturbing internal organs so as to avoid the entry of enzymes into the meat system. The enzymes which are proteolytic in nature can alter the properties of major protein fraction during frozen storage leading to quality impairment of surimi.

During the deboning step, plenty of potable water is used to remove the undesirable parts of fish. The factory trawlers use refrigerated seawater due to the paucity of fresh water on board.

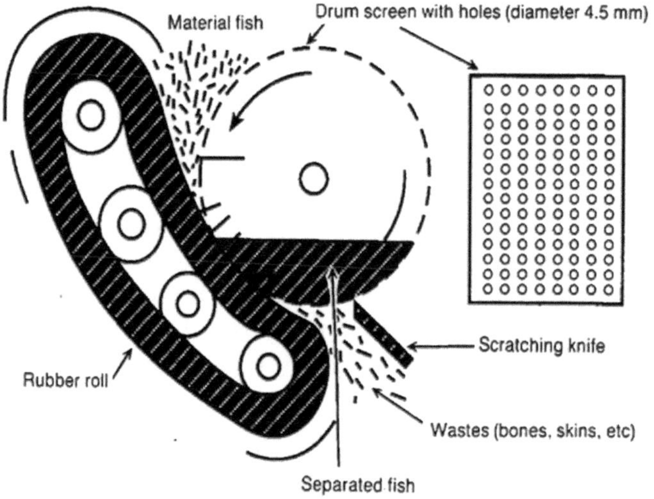

FIGURE 2.4 Steel-net-roll type meat separator.
Source: Redrawn from Miyake, Hirasawa, and Miyanabe (1985); Original drawing courtesy of Ikeuchi Iron Works.

2.6.1.2 MINCING

All deboning machines work on the principle that the soft fish meat is forced through drum perforations leaving behind skin and bones to be scraped out. The size of meat particles will be around 3–5 mm. A Headed and gutted carcass, on the other hand, results in a higher yield of minced flesh, but the quality is relatively low. Another method is skin on filet (butterfly shape), which increases the yield and loses quality to a minimum. The dressed fish (after removal of head and gut) will lead to a higher yield of mince with better meat quality.

2.6.1.3 WASHING AND DEWATERING

Water washing of separated fish mince is critical unit operation in surimi production which will ensure better gelling ability and improved color. The water washing of mince will remove pigments, odor components responsible for fishy odor, reduces lipid content and removes sarcoplasmic protein fractions and concentrate myofibrillar components. Hence surimi is also referred to as concentrated wet myofibrillar protein concentrate. The viscoelastic nature of surimi-based products mainly comes from myofibrillar protein fractions. The removal of sarcoplasmic protein fractions will enable improved frozen storage stability of surimi. The water washing process involves mixing of mince and cold water with slow agitation. The slurry is allowed to settle and decanted to remove the water. The decanted water is made to pass through screen to filter meat particles for recycling. The quantity of water used in the washing process is in the ratio of 1:3 (Fish mince: water) and normally carried out for three cycles. The excess water in the mince is removed by screw press. Prior to dewatering, small fins and bones in the mince is removed by using a refiner. The perforations in the screw press will be around 1–2 mm diameter which allows meat to get squeezed out and retains small fins and finer particles. To facilitate easy removal of water during the dewatering process it is common to add salts like sodium chloride and calcium chloride at 0.1–0.2%. At the end of final wash cycle, the ratio of fish mince to water used depends on the species and freshness. It may vary from 1:3 to 1:5 and the number of wash cycles also varied from 3–5. For a shoreside operation 29.1 kg of wastewater was generated to produce 1 kg of surimi. Consequently, the least amount of water is used in factory trawler operations, where freshwater is limited.

In line washing in to the second rinse stage, eliminating the necessity of a third rinse, by placing static mixing devices at intervals with in the pipe line. This technique minimizes the contact time between wash water and flesh while significantly reducing the amount of wash water required. However, since the first pilot-scale demonstration in 1985 at a surimi plant in Kodiak, AK, no manufacturers have commercially attempted to install this system. Consequently, there is a need for further studies on the efficient water washing process so that a large quantity of freshwater could be saved.

2.6.1.4 STABILIZING SURIMI WITH CRYOPROTECTANTS

The washed and partially dehydrated meat are mixed with certain additives like sucrose, sorbitol, and sodium tri-polyphosphate. These additives are termed cryoprotectants. The main purpose of adding cryoprotectants is to minimize the freeze denaturation reaction during freezing and frozen storage. The freeze denaturation process will be faster in the frozen minced meat affecting overall end quality of surimi. In order to minimize this undesirable reaction cryoprotectants are added so that the final quality of surimi will be good. Commercially there are different cryoprotectants available and the concentration varies from one manufacturer to other depending on the requirement of the end-user. The cryoprotectants are either added in combination or singly based on the end usage. The mixing of cryoprotectants to the mince is carried out using a kneader or bowl chopper. The latter is preferred for uniform mixing and minimum changes in temperature. The mixing of cryoprotectants in the bowl chopper are carried out at 10°C. After mixing the cryoprotectants, the mince is formed into 10 Kg blocks and then taken for freezing.

Surimi prepared from a few fish species are rich in endogenous proteases, which may affect the quality of surimi and surimi-based products. To arrest the activity of these enzymes, specific inhibitors are used. The most commonly used inhibitors in surimi industry are egg white, whey protein concentrate, potato extract and beef plasma. It is common that the enzyme inhibitors is formulated with sucrose, sorbitol, sodium tri polyphosphates, tetrasodium pyrophosphates, calcium compounds (calcium lactate, calcium sulfate, calcium citrate, calcium caseinate), sodium bicarbonate, monoglyceride or diglyceride, and partially hydrogenated canola oil. The composition of the additives (cryoprotectants and other additives) may vary from one manufacturer to other depending on the end-use. It is a common practice to add enzyme inhibitor only during final product formulation from frozen surimi. The addition of calcium compounds should be treated with caution as it may enhance protein denaturation, which will impair the final quality of surimi.

2.6.1.5 *RAPID FREEZING AND PACKAGING*

The most common method of freezing is by contact plate freezer. The total time taken for plate freezing is approximately $2^{1}/_{2}$ hr. or until the core temperature reaches –25°C. Another method of freezing of surimi is drum freezer which is less commonly used. The frozen blocks of surimi are checked for metal contamination by passing through the metal detector and then packed to corrugated card board boxes. Drum freezing of surimi, on the other hand, offers the prospect of rapid freezing, which enhances surimi quality and results in frozen surimi chips, which is a more convenient product form. However, drum freezing may not be preferred by at-sea processors where storage space is limited. A typical yield pattern during processing of fish into surimi is given in Table 2.1.

TABLE 2.1 Yield During Surimi During Different Production Steps

Production Steps	Amount (g)	(%)
Whole fish	2,234	100.00
Headed, gutted, and boned cleaned fish	1,423	63.70
Minced fish	1,382	61.90
Dehydrated mince (raw surimi)	500	22.40
Cryoprotectants added mince (surimi)	534	24.00

Source: Kaba (2006).

2.7 UTILIZATION OF DARK MUSCLE FISH FOR SURIMI PROCESSING

The color of fish muscle can be either dark or light depending on the concentration of pigments. The preferred choice of raw material for surimi production is light colored fish muscle, nevertheless dark colored fish muscle is also used with slight modification in the process. Nearly 50% of total marine catches in the world is constituted by fish species with dark color muscle. There is great interest in using the large quantities of this low value, fatty, pelagic fish, which are currently available, for human food. Cost-effective ways to improve the quality of the material, however, are first required. The advantages of a process for isolating a high-quality human food from these species are obvious. To upgrade products made from the pelagic species would not only add economic value, but it would also be a more responsible use of an important resource. The meat characteristics of pelagic fish are special in nature. Hence, the unit operations involved in the production of surimi from dark muscle is different from that of light muscle. The greatest portion of the problems associated with producing surimi from small pelagic species is the content of red muscle. Further, the pelagic fish have high fat content, greater concentration of heme pigments and sarcoplasmic proteins have been suggested to contribute to the difficulties in making high quality surimi.

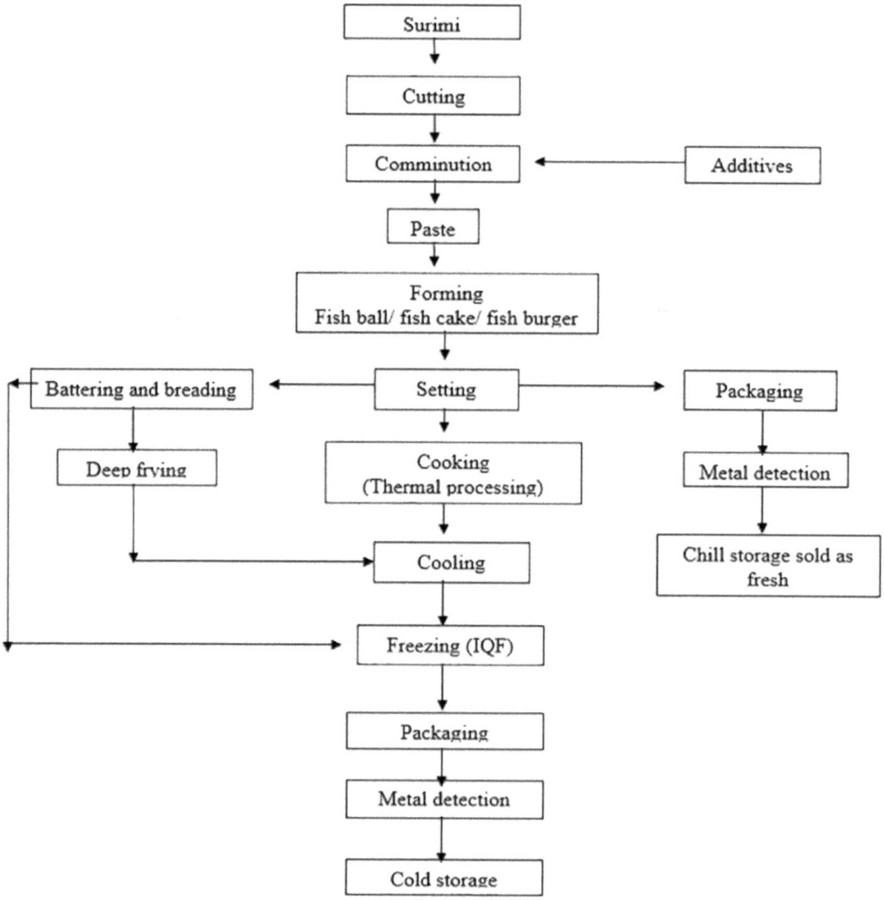

FIGURE 2.5 Flow chart for the preparation of surimi from dark muscle fish.

Source: Adapted from Shaviklo and Johannsson, 2006.

The production of surimi from dark muscle includes washing of mince with 0.5% sodium bicarbonate (w/v) during water washing process. This will facilitate easy removal of fat. The quality of surimi obtained from dark muscle will be of inferior quality than that from light muscle. One more improved process, a mixture of sodium pyrophosphate and sodium bicarbonate was used to wash for 20–30 min, under vacuum fish muscle that had been homogenized into fine fragments the pyrophosphate was added to detach actomyosin. These processes gives better gelation, better color and remove more fat, and have better flavor than the standard process. The schematic representation of the production of surimi from dark muscle is given in Figure 2.5. The volume of water used and number of washing cycles will vary from 1: 5 (mince: water) and 5, respectively.

2.8 QUALITY EVALUATION OF SURIMI

The final quality of surimi is judged by three important parameters. Moisture content, gel-forming ability and color are important. In addition, impurities present in the surimi like fins, scales, bones, filth, and metals are given due consideration before grading. The gels forming ability is determined by preparing a gel using surimi and then determines gel strength by folding test and instrument. The color of surimi is determined by colorimeter based on reflectance. Many different instruments are available for measuring gel strength, the most common being the Instron Universal Testing Machine. This instrument measures force versus displacement (distance).

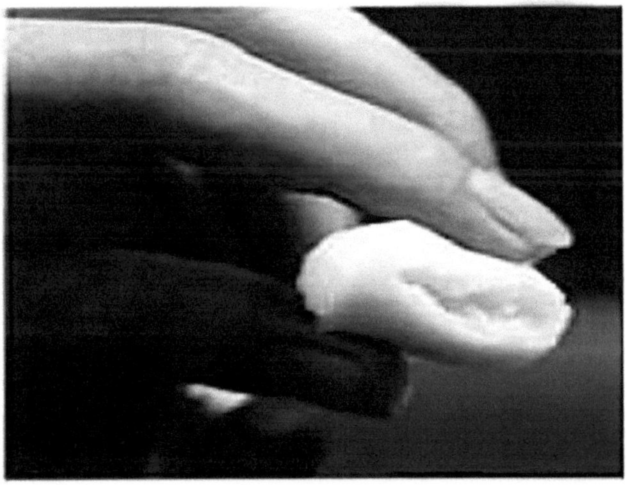

FIGURE 2.6 Folding test used for measuring the gel strength of surimi gel.

Source: Reprinted from Shaviklo and Johannsson, 2006.

The folding test (Figure 2.6) is conducted by folding a 3 mm thick slice of gel slowly in half and in half again while examining it for signs of structural failure (cracks) of surimi gel. The surimi is graded as 'AA,' 'A,' 'B,' 'C,' and 'D' on the basis of the cracks that form as the slice is folded. 'AA' grade surimi is the one that does not crack after folding twice and 'D' grade surimi breaks by finger pressing even before folding.

Other quality parameters for surimi are water content and color. Water content of surimi is measured usually with an infrared moisture meter (Table 2.2).

TABLE 2.2 The Different Classes of Surimi based on Moisture and Gel Forming Ability

Quality of Surimi	Moisture (%)	Gel Forming Ability (Folding Test)
Superclass (SA)	79	AA (without starch)
I (A) class	80%	AA (with 3% starch)
II (B) class	81%	AA (with 5% starch)
Off grade (C) class	82.%	AA (with 8% starch)

The color, whiteness of surimi, is another important factor which affects the quality and its acceptability in especially for East Asian markets. The surimi gel (Figure 2.7) was cut into flat and smooth slices 15 mm in thickness or more. The samples were evaluated immediately with a reflectance colorimeter on a Hunter scale, where whiteness is reported as the L values in the "L a, b" system. The "L" scale describes whiteness or blackness, the "a" scale describes red or green and "b" describes yellow or blue. Whiteness, as an index for the general appearance of surimi gel, can be calculated as: Whiteness = $L^* - 3b^*$.

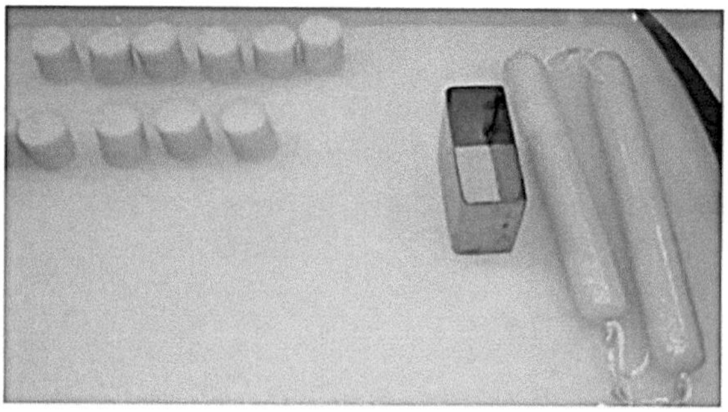

FIGURE 2.7 Sample for measuring the gel strength and whiteness.

Source: Reprinted from Shaviklo and Johannsson, 2006.

2.8.1 MICROBIAL QUALITY OF SURIMI

Commercial quality of surimi during processing is evaluated by various microbial qualities. The quality and safety of surimi seafood products depend largely on the contaminant load in the raw surimi seafood ingredients. Variability in the bacteriological data exists for surimi seafood operations that involve different number of steps, degrees of handling and product specification. Coliforms and fecal coliforms, present in surimi and other ingredients or contributed through handling, range from low to absent in pasteurized surimi seafood. Bacterial biofilms can develop on equipment during extended processing. In addition, inadequate cleaning and sanitation can lead to product quality and potential seafood safety concerns.

The bacteriological standards that are applicable for fish and shellfish are also applicable for surimi and surimi products. The aerobic plate count, *Escherichia coli, Salmonella, Staphylococcus aureus, Listeria monocygenes* are normally enumerated for assessing microbiological quality.

2.9 SURIMI-BASED PRODUCTS

The frozen surimi is most commonly used for products such as crab leg analogs, shrimp, and lobster molded products and scallop analog products. The processing technology and ingredients used are the key factors for successful development of surimi-based products. Surimi-based products can be categorized by the amount and arrangement of internal fibrous structure in the product. The most sophisticated shellfish analog products are composed of parallel-aligned fibers, which is similar to the structure of crab leg meat. This final product is composite-molded products also exhibit an internal fibrous nature. But the fibrous are randomly aligned. Many Japanese kamaboko products except those containing vegetables are smooth gels with little or no fibrous nature. As the manufacturing processes for these three categories of products vary somewhat, each category will be considered separately.

Fish is an integral part of the Asian diet and surimi-based products are equally popular. The surimi-based products are also very popular in the Asian diet and demand for the same is showing increasing trend. The increased consumption of surimi-based products in America and Europe is mainly due to better nutrition, competitive prices and promoting innovation in the development of new products. This led to phenomenal demand for surimi-based products. This is evident from the consumption pattern wherein, the sales have grown from zero in 1980 to 25,000 tons in 2004. Due to their healthy nutritional values, functional versatility, and competitive prices in relation to their natural counterparts, the consumption of these products in the west has reached almost 250,000 metric tons. Phenomenal growth has also been experienced in the USA and Europe, including Russia, where sales have grown from virtually zero in 1980 to 900,000 metric tons and 150,000 metric tons in 2004≥ The other surimi based products such as fish cakes, fish balls, fish burgers, molded shrimp and lobsters and fish sausages are finding its way into different markets of the world (Park et al., 2005a). Processing steps of surimi-based products are described in Figure 2.8.

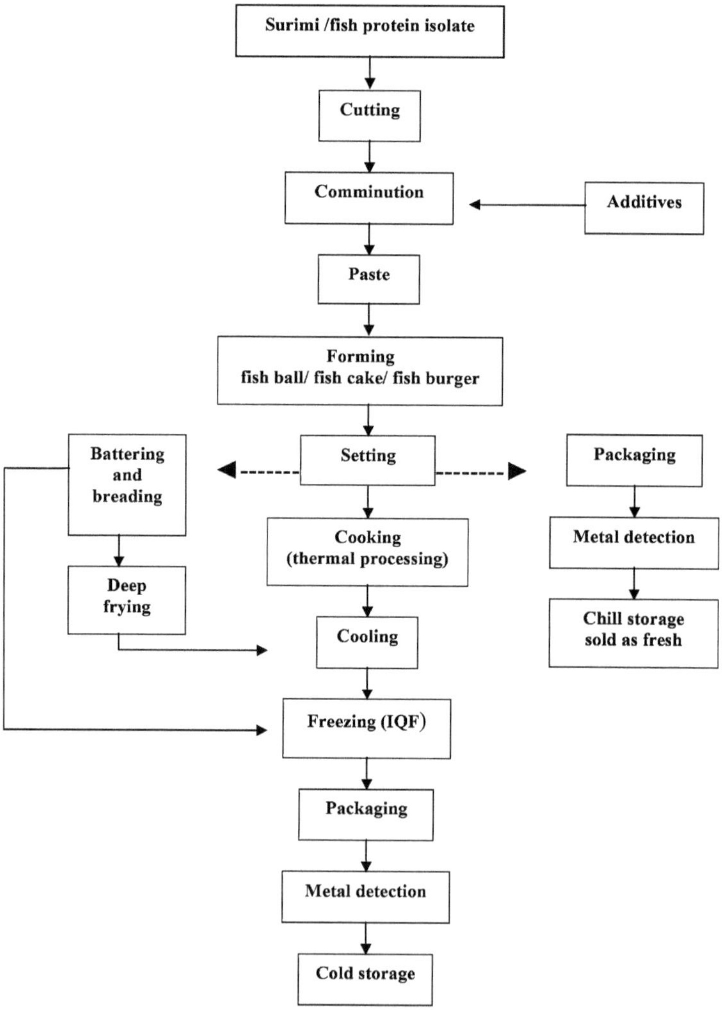

FIGURE 2.8 Processing flowchart for products from surimi.
Source: https://www.grocentre.is/static/gro/publication/7/document/reza06prf.pdf

2.10 CHALLENGES AND OPPORTUNITIES IN REALIZING THE HIGHER PRICE FOR SURIMI

2.10.1 *CHALLENGES*

1. The major challenges for surimi industry are the supply of raw material at a constant price for more than 200 days of operation. The profitability gets eroded if the raw material cost is fluctuating. Also, the vagaries of nature make a very poor fishing season. The quality of raw material arriving at surimi plant will determine the export value. Most often, the multi-day trawl boats are used wherein the fishing extends for 10–15 days, impacting the quality of raw material. Though insulated boxes are used during fishing, a better strategy needs to be evolved for keeping the quality of fishes.
2. The International prices too often fluctuate from USD 2–5/Kg. The stocks of frozen surimi cannot be held for a longer period, as cold storage costs become prohibitive.
3. If the raw material is not available at the factory site, it has to be obtained from other states which make raw material expensive.
4. The non-existence of a domestic market for surimi is another reason for over-dependence on the international market leading to price fluctuations.
5. The surimi industry is highly capital intensive, and many times, the margins are so thin, and it is the volume that matters for trade.
6. The stringent quality standards stipulated by importing countries will add to operational costs.
7. As surimi industry requires huge quantity of fresh water, efficient recycling and reuse for the processes will be critical.

2.10.2 *OPPORTUNITIES*

1. Innovations in improving the yield by 3–5% will increase the profit margin;
2. Use of different raw material for surimi production is a potential source for innovation;
3. Efficient waste utilization for high value product like gelatin, enzymes, and fish protein hydrolysates (FPH);

4. Newer and efficient ETP should be looked into to minimize environmental pollution in and around the plant.

KEYWORDS

- **challenges in surimi industry**
- **export market**
- **surimi**
- **surimi industry**
- **surimi quality**
- ***Theragra chalcogramma***

REFERENCES

Chaijan, M., Klomklao, S., & Benjakul, S., (2013). Characterization of muscles from Frigate mackerel (*Auxisthazard*) and catfish (*Clarias macrocephalus*). *Food Chem., 139*, 414–419.

Chaijan, M., Panpipat, W., & Benjakul, S., (2010). Physicochemical properties and gel-forming ability of surimi from three species of mackerel caught in Southern Thailand. *Food Chem., 121*, 85–92.

Lanier, T. C., & Lee, C. M., (1992). *Surimi Technology* (pp. 167–179). Marcel Dekker, New York.

Miyake, Y., Hirasawa, Y., & Miyanabe, M., (1985). Technology of surimi manufacturing. *Infofish Marketing Digest, 5*, 29–32.

Kaba, N., (2006). *Turkish Journal of Fisheries and Aquatic Sciences, 6*, 29–35.

Pangsorn, S., Laong-Manee, P., & Siriraksophon, S., (2007). *Status of Surimi industry in the Southeast Asia (*pp. 1–28). SEAFDEC Capture Fishery Technology Division.

Park, J. W., & Lin, T. M., (2005). Surimi: Manufacturing and evaluation. In: Park, J. W., (ed.), *Surimi and Surimi Seafood* (pp. 33–106). Taylor & Francis Group, Boca Raton, Florida.

Shaviklo, G.R., Johannsson, R. (2006). Quality Assessment of Fish Protein Isolates Using Surimi Standard Methods. In UNU - Fisheries Training Programme.

Venugopal, V., (2009). Seafood proteins: Functional properties and protein supplements. In: Venugopal, V., (ed.), *Marine Products for Healthcare: Functional and Bioactive Nutraceutical Compounds from the Ocean* (pp. 51–102). CRC Press, Boca Raton.

CHAPTER 3

Freezing of Fish and Fishery Products: Basics and Advancements

K. DHANAPAL and G. PRAVEEN KUMAR

Department of Fish Processing Technology, College of Fishery Science, Andhra Pradesh Fisheries University, Muthukur – 524344, Andhra Pradesh, India

ABSTRACT

The advancement in the food industry is due to the application of different new technologies in machinery, preservation methods, packaging, labeling, and transportation. With the busy lifestyle of people and change in food habits, there is an increased demand for ready-to-eat and ready-to-cook products leading to the increased concern about the safety and nutritional quality of food supplied to consumers. Freezing alone contributes to 62% of processed foods. India is the 4th major exporter of seafood, mainly in frozen form. In the modern food industry, freezing of foods has been well established as a preservation method to increase shelf life. Though the basic freezing methods like contact plate freezers, air blast freezers, and immersion freezers can preserve the food, there are some deteriorative changes in quality during freezing and storage. The advanced methods of freezing utilize the dipole moment of water and freeze the product very rapidly to minimize the deteriorative changes in quality and nutritive value to a maximum extent. Hence, this chapter will guide the academicians and industrialists to understand the concepts of freezing and guide them in adopting the best method of freezing as it helps in enhancing shelf life, quick, convenient, and effective one for aquatic food products.

Advances in Fish Processing Technologies: Preservation, Waste Utilization, and Safety Assurance.
Ranendra K. Majumder, PhD & Amjad K. Balange, PhD (Eds.)

3.1 INTRODUCTION

Fish is an important source of omega 3 fatty acids (FAs), protein, vitamin B complex, and some minerals. Fish is highly perishable commodity due to presence of high moisture content which causes physical, chemical, microbiological, and histological changes along with the development of odors. Spoilage of fish can be reduced by adopting different preservation techniques like traditional drying, smoking, salting, fermentation, marination, etc., and by the recent methods like canning, freezing, icing, irradiation, freeze-drying, pulsed electric fields, high-pressure processing, ohmic heating, microwave heating, ultrasound processing, etc. Though there are different preservation techniques, selecting an appropriate preservation technique to maintain the quality and extending shelf life of the product is an important job of any processor.

In 2018, global fish production has reached about 179 million tons, of which 156 million tons was used for human consumption. Among 156 million tons, 44% was consumed in the form of live, fresh or chilled fish, 35% in the form of frozen foods, 11% as prepared and preserved foods and 10% as cured products. Among these freezing accounts for 62% of all processed fish for human consumption, becoming the main method of preservation (excluding fresh, live or chilled fish) and with regard to trade, the frozen products had the highest share (90% of trade) in quantity consisting of processed foods. Of the total production, about 2/3rd of fish production that is intended for human consumption is used in frozen forms (prepared and preserved) in Europe and North America. The share of frozen fish for human consumption in developed countries was 58% in 2018.

India is the fourth major exporter of seafood, following China, Norway, and Vietnam, due to an increase in farmed shrimp production (FAO, 2020). Seafood worth 6.68 billion US Dollars and quantity 12,89,651 MT was exported by India during 2019–2020. Of the total exports, the contribution of frozen shrimp is 50.58%, frozen fish is 17.32%, and frozen cuttlefish 5.50% in quantity. The contribution of frozen foods during 2019–2020 for frozen shrimp, frozen fish, frozen cuttlefish, and frozen squid were 73.21%, 7.69%, 4.29% and 4.71% respectively in terms of value.

Freezing is a method of preservation where the water present in fish transforms its state to form ice crystals hence preventing the availability of water for the growth of microorganisms and the action of enzymes on food. During freezing the temperature of the product is initially reduced to refrigerated temperature (0 to 2°C) followed by the conversion of water molecules into ice crystals at –1°C called thermal arrest period, is the most critical step in freezing and finally the temperature of product gets reduced till all water molecules get converted to ice. This method of preservation was used from the ancient times in colder areas where they gather food and used to freeze and thereby quality was maintained. Technologically in 1861, the developments in freezing began with Thomas Sutcliffe Mort by starting the first freezing plant in the world at Darling Harbor in Sydney, Australia designed by Eugene Dominic Nicolle. The company trailed shipments of frozen meat to London in 1868 which was not successful due to the product became tasteless and more water leakage after defrost. During 1912 to 1917, Birdseye, popularly called the father of freezing a native of Brooklyn lived in Chilly Labrador, Canada has observed that fish immediately after catching gets frozen in frigid air and when the fish was cooked the taste was similar to fresh fish. In the United States, he worked in US Fisheries Association as assistant to the President. He developed a quick freezing process in 1924 in two methods. In one method, package was placed between two metal belts chilled at –40°F to –45°F using calcium chloride solution and in second method, packed product was placed between two hollow metal plates chilled to –25°F by evaporating ammonia.

3.2 STAGES OF FREEZING

Freezing usually takes place in three main steps, namely pre-cooling (water sensible cooling), phase transition (nucleation and formation of dendritic ice) and tempering (solidification). In pre-cooling stage, the sensible heat is removed from the product, thereby lowering the temperature of the product. Cooling continues even after reaching the freezing point and phase transition is not initiated immediately. Rapid cooling can force the water to cool below the freezing point without freezing or converting into solid. In this stage, water sample enters supercooled region and here only sensible heat can be removed resulting in decrease in temperature. The difference in temperature of water compared to the freezing point of water denotes the degree of supercooling which is usually negative. The degree of supercooling indicates that the phase transition has occurred in the supercooled water. At a certain degree of nucleation, The water molecules will lose sensible heat and become more compacted together due to sudden nucleation of water crystals occurs. Later the dendrites will grow from the nucleus. The time period between nucleation of ice and formation of dendritic ice is less than one second. The phase transition is instantaneous and accompanied by a sudden increase in temperature as latent heat is released during the nucleation and dendrite formation. With the increase in temperature, freezing enters into tempering stage where water undergoes complete solidification or crystallization usually represented by a continuous decrease in temperature till it forms a plateau horizontally. The solidification of ice is a slow process which requires more time than phase transition stage during freezing.

During the process of ice crystal nucleation and growth, several changes take place in fish. When the temperature of fish is sufficiently lowered the ice crystallization will start from surface to center of cell leading to supersaturation of dispersed substances. Formation of very fine ice crystals which are in equilibrium with surrounding air and act as a site for further growth and expansion of ice crystal is called nucleation. Nucleation process acts as a base for the development and growth of ice crystal that occurs with the decrease in temperature. As the ice crystal size increases the concentration of solutes increases leading to a reduction in the chemical potential of water. Hence for ice crystals to form, the temperature of the product should be reduced further so that all the water present in the cell gets frozen. The freezing process is said to be completed only when most of the water at the center gets frozen or converted to ice. More than 80% of water gets frozen at –15°C and the system is divided into crystalline phase of pure water and amorphous phase containing solutes and residual water. Further decrease in temperature increases the viscosity of interstitial fluid rapidly due to increase in concentration and decrease in temperature. Vitrification (solidification) occurs at a viscosity of 1,011–1,012 Pa. where the concentrated phase surrounding ice crystals becomes a glass. The temperature at which this transition to glass occurs is called Glass Transition Temperature (Tg') where all the molecules in cell get frozen. At this temperature freezing of water is stopped. Even at this temperature, some water remains unfrozen, which is called 'unfreezable water.'

Generally two types of nucleation occur during the formation of ice crystals. They are homogenous nucleation and heterogamous nucleation. Homogeneous nucleation occurs in pure water due to random orientation and combination of water molecules in the food. Heterogeneous nucleation usually occurs in solid foods. In solid foods nucleation occurs on the cell surface or on cell wall and then proceeds inwards of cell. The rate of heat transfer decides the nucleation and growth of ice crystals. With the increase in the rate of heat transfer, smaller, and numerous ice crystals are formed. While freezing different foods with the same dimensions and rate of heat transfer, the ice crystals formed may be of different size depending on the availability of water for freezing.

3.2.1 SLOW FREEZING AND FAST FREEZING

Freezing can be divided as slow freezing and quick freezing depending on the rate of crystallization. In slow freezing, large uneven ice crystals will be formed. Due to the larger ice crystals, the cells get damaged leading to higher drip loss. In quick freezing smaller and even-sized ice crystals will be formed. Due to smaller ice crystals formed during freezing, there will be less damage to cell and less thaw drip. As the formation of small ice crystals will result in less cell damage, less thaw drip and thereby the quality of frozen foods is enhanced. Hence a lot of research is being carried out in different areas to reduce the ice crystal size and crystallization to occur very fast within seconds by using technologies.

Due to the disadvantages of slow freezing, most of the products are prepared by using quick freezers. Even among the quick freezers, the fish can be frozen in batches or continuously where the fish is arranged on shelves and is conveyed for freezing. With the advancement in technology and increase in freezing rate individually quick frozen products are in great demand. For efficient freezing of any product we should know the characteristics of the product, conditions required for freezing and type of freezer that can be used. Hence knowledge on the different types of freezers is necessary to understand their mechanism and operation.

3.3 TYPES OF FREEZERS

Based on the medium in contact, there are three types of freezers. They are discussed in subsections.

3.3.1 CONTACT PLATE FREEZERS

In this type of freezers, the fish comes in contact with the cooled plates made of aluminum alloy. The contact plate freezers consist of metal plates, and the refrigerant flows in metal plates so direct contact of refrigerant to food is avoided. In this type of freezers, the products are positioned between the plates either in metallic trays or kept directly. Later these plates are moved using hydraulic ram creating a close contact between plates and product. This type of freezer is suitable for blocks or regular products, flat, and thin products like fillets and fingers. Based on the arrangement of plates, Contact plate freezers are classified as Horizontal plate freezers and Vertical plate freezers.

3.3.1.1 HORIZONTAL PLATE FREEZER

Horizontal freezers are the most commonly used freezers in many fish processing plants to prepare regular packed blocks of fish fillets, fish fingers and shrimp or prawn. The freezer usually consists of 15–20 plates; the product arranged in metal trays with the addition of little glaze water is placed between plates manually. Once all the trays are arranged between the plates, the metal trays are maintained in close contact with top and bottom plate by using a hydraulic system facilitating quick heat transfer. In some cases, the trays may be lined with polythene to facilitate easy removal of the formed blocks. The freezing temperature is maintained between –35 to –40°C. For freezing a block of 3–4 cm thick at –35°C, it takes around 1 to 1.5 hours. For freezing fish it takes 2 to 2.5 hours whereas for consumer pack it takes 30 minutes and for industrial blocks it takes 2 hours. Defrosting of plates is required to separate metal trays from plates which requires about 30 minutes or more even after using hot gas which is the quickest method to defrost. The horizontal plate freezer has the advantage of producing uniform blocks with good freezing requiring less floor space and also minimizes the problems like product dehydration and bulging.

The operation of horizontal plate freezer is labor-intensive and time consuming during filling of product between plates and removal of metal trays after freezing. This can be reduced by using automatic or semi-continuous systems for loading, unloading, and stacking of products. For automatic operation the plates are moved up and down in elevator system and at the level of conveyor the plates are separated so that metal trays can be placed between the plates. During the process, the frozen trays will be removed from the other side when the trays are pushed between plates. Hence both loading and unloading can be easy.

3.3.1.2 VERTICAL PLATE FREEZER

Vertical plate freezer consists of plates arranged in vertical position and the product to be frozen is usually placed between the plates from top either by weight or volume. These types of freezers are mainly used for freezing at sea. The spaces between the plates are called stations. The frozen block can be removed from one side upwards or from down through the bottom. The plates are positioned by using a hydraulic system which helps in closing or opening valves and movement of plates.

Defrosting by using hot gas is done to separate the block from plates and compressed air is also used to push the block sideways or down. The distance between plates is 75–100 mm, and 5–6 Kg or more can be placed between the plates in the closed position as fishes placed in open position may undergo pressing mainly fatty fishes and soft fishes. After initial freezing, filling of water or lid may reduce pressure and rancidity problems during storage but thawing and weighing may become difficult. The maximum size of the block formed by this method is 1,070 × 535 mm.

3.4 AIR BLAST FREEZERS

3.4.1 FREEZING BY CONTACT WITH COLD AIR

Industrially forced convection is used to freeze the product than natural convection as natural convection has lower freezing rates and forced convection results in quick freezing. In this method, uniform air temperature is maintained and air velocity is controlled, but the freezing time depends on the product characteristics. The air temperature of a typical refrigeration system varies between –32 to –40°C. The different types of freezers in this category are still air freezing, batch air blast freezing, continuous air blast freezing, tunnel freezer, belt freezer, and fluidized bed freezer.

3.4.2 FREEZING IN STILL AIR

This type of freezer consists of an insulated cabinet maintained at –28 to –45°C has shelves made of pipes or coils through which refrigerant is circulated. The product to be frozen is packed or kept as such on aluminum trays and placed on shelves. The freezing time is influenced by the temperature of the freezer, size, and shape of product, packed or unpacked, load in the freezer and arrangement in the freezer. This is the least expensive but freezing will is slow. It takes around 12 hours or more for freezing of product depending on characteristics of the product.

3.4.3 BATCH AIR BLAST FREEZER

Air blast freezer is an insulated room or cabinet consisting of fans that blow cold air through finned cooling coils of refrigeration system. As the cold air is blown over the product, the heat is picked up from product and walls of cabinet thereby freezing the product. A temperature of –35 to –40°C is maintained in the cabinet. The foods to be frozen can be stacked in racks or hung. Care must be taken that the cold air must pass over the surface of each product individually for quicker and proper freezing to happen. In the case of packed foods, the freezing time increases as the packaging material acts as a barrier between cold air and product and there won't be proper circulation of air between product and cold air. It's a versatile method applicable to any type and shape of fish producing individually quick frozen products. The major disadvantage of blast freezing is that the unpacked product undergoes excessive dehydration, freezer occupies more space and consume more energy, and sometimes the product will have bulged appearance if not placed well between rigid surfaces during freezing.

3.4.4 CONTINUOUS AIR BLAST FREEZER

This is an improved method in blast freezing where the trays with product to be frozen are stacked on trolleys, or the food is moved into the freezer or tunnel with the help of conveyors. The conveyor belt that is used should be flexible, easily cleanable, non-corroding, must be suitable for direct contact with foods and should not have any effect on freezing time and product quality. The most commonly used conveyor belts are stainless steel mesh link belts or chain-link belts but these are expensive and affect the appearance of product (when fish is placed on belt, it produces crinkled or indented appearance that is not acceptable by consumer). By using open mesh belts, it becomes difficult to remove the product after freezing and there may be a loss in weight due to physical damage. The skin on fillets can be removed easily than skinless fillets with fish portions sticking to belt leading to a reduction in weight.

The continuous air blast freezers can be constructed such that the air flows either crossflow or series flow air circulation. In cross-flow arrangement, the cold air flows in opposite direction of movement of product. In series flow arrangement, the product and cold airflow in the same direction and hence the coldest fish meets the coolest air. If air velocity is maintained at 150–300 m/sec can create intimate contact with product therefore freezing occur faster.

3.4.5 *TUNNEL FREEZER*

Tunnel freezers are an improvement of air-blast freezing. The tunnel freezer consists of an insulated tunnel where chilled air is circulated in opposite direction of product by using fans at a speed of 1 to 20 m/sec and temperature is maintained between –18 to –40°C. The air temperature maintained at –30°C with speed of 10–12 m/sec is the optimum range for operation of tunnel freezer. Typically the tunnel freezer may be either batch operated or continuous system. In the batch operated system, the fish are arranged on trays of size 100 × 200 cm^2, and these trays are stacked one above the other till 2 meters high on a cart. These carts are loaded into the tunnel containing fans, one or more refrigeration coils and baffles that direct the airflow evenly over the product. The capacity of the freezer ranges from 200 to 500 Kg/h.

The racks can be moved in and out either manually or fork lift trucks in case of stationary tunnels or pushed through the tunnel by pushing mechanism as in push-through tunnel or carried by driving equipment, chain drive as in carrier freezer or slide through as in sliding tray freezer. Guide devices present in freezer properly locate the tray of food for uniform freezing of food. These freezers are very flexible by freezing products of any size and shape, either packed or unpacked in batch or continuous operated tunnels. Individual quick freezing (IQF) is preferred to obtain free flow of product, improved handling and to increase freezing rate.

3.4.6 *BELT FREEZERS*

Belt freezers are commonly used for freezing of unpacked products like individual quick frozen products and are suitable for freezing foods that need careful handling. Belt freezers may consist of single or multiple belts. Single belt freezer consists of only one belt. In multi-belt freezers, the belts can be arranged one above other running either in the same direction or opposite direction to increase the rate of production and reduce the floor space. In spiral belt freezer the belt is wound round to a rotating drum stacking up to 30 tiers of belt above each other. As the belt made of wire mesh remains inside the freezer, there is a need of ancillary equipment for feeding the trays or product in and out. In the modern belt freezers the air flows vertically so that the air is forced through the product layer. When single belt freezers are operated with high air velocities, the products may agitate. Major care that must be taken while operating any type of belt freezer is that the product should be spread uniformly across the belt in order to avoid channeling a condition where the air stream bypasses the product.

Spiral belt freezer is a type of belt freezer and a modification of continuous tunnel freezer containing continuous belts for carrying products. If we examine the design it consists of self-stacking and self-enclosing spiral belt that provides compact and controlled airflow over the product. The number of tiers or levels of the belt varies with the capacity to handle and layout of the production line. The product to be frozen is placed on the belt outside the freezer and these products are later conveyed into freezer through the belt. The direction of airflow can be either vertical or horizontal, but the vertical airflow is more efficient. While freezing foods in belt freezers, straight-line belt freezers are used for particulate foods like shrimp, whereas spiral freezers are used to freeze products that require longer retention time like fillets and other prepared foods. But in both the systems, the cold air is circulated through the mesh belt that conveys the product to the freezer. Spiral freezers provide higher freezing capacity in limited floor space. The advanced spiral freezers are provided with improved conveyor belt, auto washing system, drive technology which requires less maintenance and suitable for installation in remote areas also. The problem of sticking of product to conveyor can be eliminated by initially pulling the product across a polyethylene film conveyor at −40°C or colder freezer plates which can freeze the bottom surface of fillet to a thickness of 1 mm in 1 min approximately. By using this technology the fillets appear glass-smooth instead of wire mesh marks of conveyor belt. Then the product can be sent to belt freezer.

3.4.7 FLUIDIZED BED FREEZER

Fluidized bed freezer is an improvement of continuous air blast freezer. These freezers have efficient heat transfer, faster freezing rates and lower dehydration of product. Fluidization is a process where the particles of fairly uniform size and shape when subjected to airstream exhibit fluid-like behavior partially supported by rising column of cold air. This freezer consists of perforated bed through which the cold air is blown vertically upward at high velocities by using powerful fans. When particulate foods of uniform size enters the freezer, the products will float individually in the air stream blown from the bottom so that each particle will be surrounded by air which keeps them separated from each other. In this stage, the product will act like liquid and the product gets frozen before it leaves the freezer. Fluidized bed freezers are more suitable for small foods less than 15 mm with uniform size and shape. The shape and size of the product determines the air velocity required for fluidization.

If the product is conveyed to an inclined trough where product is conveyed on upper end, the fluidized mass will move towards lower end as long as the product is added and thus the product is frozen and conveyed by the air simultaneously without the help of mechanical conveyor. For fluidizing shrimp an air velocity of 2 m/s at –35°C is commonly employed. As the bed depth depends on size, shape, and uniformity of product and ease of operation, a bed depth higher than 3 cm is suitable for small prawns and a bed depth of 20–25 cm is required for fish fillets. It takes around 30–35 min to reduce the temperature from 30 to –18°C while freezing fillets of 3 cm thick.

A two-stage technique can be employed in some fluidized bed freezer where the product is initially frozen by air blast freezer in order to set the surface of product followed by freezing in fluidized bed freezer. In the modern equipment, a pre-cooling section present at the entrance of the freezer can increase the quality by minimizing clumping of soft, sticky products like raw peeled shrimp. Stainless steel contact parts with easily accessible coils can be used in order to increase hygienic quality of product. If the coils are placed beside the conveyor belt instead of placing below the conveyor belt, the product will not fall on coils, and the space below the conveyor belt will be available for cleaning. In sequential defrost system the evaporator coils can be defrosted and cleaned one section at a time thereby allowing for continuous operation of the freezer.

The major advantage of fluidized bed freezer when compared to belt freezers is that the product is always individually quick frozen, is independent of load fluctuations (as the air distribution is same for full loaded or partly loaded) and if the product is overloaded the product will flow to floor and reliable even when using wet products as the fluidized bed can accept products with high water content.

3.4.8 IMPINGEMENT FREEZING

Impingement freezer is an advanced method of freezing that uses jets of air to freeze the product. Impingement is a process of directing jets of fluid onto a surface at very high velocity (up to 50 m/s) to break up the static surface boundary layer of gas surrounding the product effecting a change in the product. The medium surrounding the product is more turbulent and heat exchange through this zone is more effective producing more effective and faster freezing. Impingement freezing is effective to freeze products with high surface area to weight ratio like fish fillets or burgers. The product is effective for products that need a very rapid surface freezing called crust freezing. The advantages of Impingement freezing over conventional freezing are depending on product thickness. Impingement freezing has 62–79% faster freezing times and weight losses are reduced by 36–72%. However, the drawback of this system is that it requires high capital costs than conventional freezers and energy requirements are more due to higher fan power consumption.

3.5 LIQUID IMMERSION FREEZING

In this freezer, the freezing is achieved either by spraying or immersing the product in the refrigerant solution. In this method product comes in direct contact with the refrigerant ensuring efficient heat transfer between the product and refrigerant. The refrigerant moves through the baskets of product in one direction and returns through the refrigeration coils in another direction with the help of propellers or high volume low pressure pumps. If this freezer is used for onboard freezing then tanks should be shorter, deeper or compartmentalized in order to prevent spillage of brine. The refrigerant used should be inexpensive, nontoxic, stable, should have low viscosity, low freezing point, low vapor pressure, high thermal conductivity but it should not produce any undesirable effects on the product. Different refrigerants used in this method are sodium chloride, glycerol, purified chlorofluorocarbons (CFC), propylene glycol, calcium chloride, salt/sugar mixtures, water/glycerol/alcohol mixtures. Though CFC is most efficient in heat transfer, the use of CFC has been reduced due to environmental issues.

Sodium chloride is the common refrigerant that is used onboard fishing vessels either by immersing the product in brine or by spraying brine on the product. The saturated brine generally freezes at –21°C but practically it can be frozen only up to –18°C. In order to achieve still lower temperatures, they have to be transferred to cold store and frozen by maintaining the required temperature. Freezing using sodium chloride is rapid and can be used for continuous operation. But the major problem by using sodium chloride is that the fish absorbs a little salt depending on the type and conditions of fish and surroundings. This problem can be reduced by using a combination of glucose or corn syrup and salt as refrigerant as glucose salt solution provides a protective glaze to product and frozen product will stick to each other. The salt uptake by-product can also be reduced by dipping the product in freshwater before loading the fish in brine freezer.

3.5.1 HYDROFLUIDIZATION FREEZING

It is a form of immersion freezing and is analogous to liquid Impingement freezing. It consists of a circulating system which pumps refrigerated liquid upwards through orifices or nozzles in a refrigerated vessel producing agitating jets. The agitating jets of highly turbulent liquid and moving products produce extremely high surface heat transfer coefficients resulting in freezing the products rapidly. In this method unfreezable liquid refrigerating media like aqueous sodium chloride are used as fluidizing agents as higher refrigeration temperatures can be attained even at slight jet agitation. Pumpable ice slurries also called two-phase slurries like pumpable ice or fluid ice, binary ice, flo-ice, slurry-ice, liquid ice, are environment friendly secondary coolants circulated to heat transfer equipment instead of harmful refrigerants like CFC or HCFC based refrigerants. The minute ice particles absorb the latent heat while thawing on surface of product providing high surface heat transfer coefficient in short time and uniform temperature distribution in the freezing chamber. The major advantage in this method is the use of unfreezable liquids or pumpable ice slurries as fluidizing agents. Addition of ethanol, salts, glycerol, etc., to binary or ternary aqueous solutions containing soluble carbohydrates like sucrose, glucose, fructose, invert sugar, other mono and disaccharides are the possible immersion media. Its advantages over the conventional freezers include, very high heat transfer rate with a small temperature difference between product and medium, critical zone of water crystallization is quickly reached producing fine ice crytals in food preventing cellular damage during freezing. Fluidized state is achieved with low velocity and pressure of fluid jets due to Archimedes forces and buoyancy of the products leading to energy savings and minimum mechanical action on foods. Lastly, its operation is continuous, easy to maintain, convenient for automation and reduced labor costs.

3.6 CRYOGENIC FREEZER

Quick freezing in cryogenic freezing is achieved by exposing the product to extremely cool freezant due to a change in state. Due to lower freezing temperature and high heat transfer rates, small ice crystals are formed with reduced cell damage, flavor, and moisture losses. In this freezing the food directly comes in contact with freezant in three different ways by spraying the cryogenic liquid on the food in tunnel dryer, the cryogenic liquid will be vaporized and blown over the product as in spiral freezer or batch freezers or by immersing in cryogenic liquid as in immersion freezer. Among these methods, spraying of cryogenic liquid over the product while the product is conveyed in tunnels is the most common method used. In the principle, when the product is exposed to temperatures below –60°C, the cryogenic liquid undergoes a change in state to remove the heat from the product. In liquid immersion freezer there won't be any change in state of freezant whereas in cryogenic freezer the freezant will undergo a change in state to effectively remove heat from product.

The colorless, odorless, and chemically stable carbon dioxide with boiling Point –78.5°C (liquid or solid), purified fluorocarbons and liquid nitrogen with boiling point –196°C are the common refrigerants used in cryogenic freezing. When compared to conventional air blast freezer or plate freezer the freezing rates of cryogenic freezer are more rapid, but the freezing rates of cryogenic freezer are moderately greater than liquid immersion and fluidized bed freezer. To freeze shrimp it takes around 9 minutes in liquid nitrogen freezer whereas it takes 12 minutes in fluidized bed freezer. For freezing a 3 cm^3 size shrimp packed in polyethylene it takes around 151.3 minutes using air blast freezer and only 14.6 minutes by using liquid nitrogen tunnel freezer. This high difference in freezing times is due to the enormous temperature-driving force available for heat transfer.

The major advantages of using cryogenic freezers include lower time required for freezing, higher rate of production (low floor space), flexible (adaptable to different products), low dehydration loss (<1%), absence of contact with oxygen and oxidative changes so improved quality of product in terms of flavor, appearance, low drip loss, simplicity of equipment, possibility of freezing unpacked or thin packed products, lower handling losses, reduced manpower requirement, reduced maintenance cost, etc. For setting up of liquid nitrogen-based freezer, the capital investment is 1/3rd of mechanical refrigeration unit as drive motors, compressors, and chilling units are not required. The high cost of liquid nitrogen is the only drawback in this type of freezer.

Based on the refrigerant used, the cryogenic freezers are of three main types: liquid nitrogen freezer, carbon dioxide freezer, and liquid fluorocarbon freezer.

3.6.1 LIQUID NITROGEN FREEZER

In this type of freezer, the liquid nitrogen at –196°C is sprayed over the product present on conveyor belt of tunnel freezer, moving in a countercurrent direction such that the fish gets pre-cooled before reaching the nitrogen spray. As the product comes below the liquid nitrogen sprayer, the sprayed nitrogen will get evaporate immediately after contacting the product and the heat removal from the product is due to evaporative cooling of cryogens. The liquid nitrogen vapors produced after freezing the product are used for pre-cooling of the product.

The liquid nitrogen freezer is an insulated chamber with three chambers. The product is placed on the conveyor belt and conveyed to zone A where the product gets pre-cooled with the help of slightly cool nitrogen vapors (–60°C) moving in a countercurrent direction. In zone B, the liquid nitrogen is sprayed over the product and the product gets frozen. Then the product moves into zone C where the product is allowed to equilibrate to the final desired temperature.

Due to higher freezing rates, a product with improved texture can be produced by using this freezer. If proper precautions are not taken, there may be a chance of cracking of product. Liquid nitrogen freezers are used for surface freezing like immersion freezers. If liquid nitrogen freezers are used for final freezing of product, it requires about 1.0 to 1.5 kg of liquid nitrogen per Kg of product making production expensive. Except this, because of the low investment and simple operation, this method makes this method economical in some production lines like inline processes.

3.6.2 CARBON DIOXIDE FREEZER

In this freezer, carbon dioxide used as freezant is sprayed in liquid form on the surface of product conveyed to tunnel freezer through a conveyor belt. Once the liquid carbon dioxide is sprayed from the nozzles, the pressure of liquid carbon dioxide decreases, leading to conversion of carbon dioxide liquid (approx. 50%) to tiny snow particles that absorb the heat from surrounding and gets converted to gas-producing rapid cooling of the product. Freezing can also be achieved by using solid carbon dioxide and this can be used in transportation of frozen product in air. Though carbon dioxide freezers are having more advantages than liquid nitrogen freezer, the major disadvantage of using carbon dioxide freezer is that the unpacked product absorb or entrap carbon dioxide to cause swelling of the product.

3.6.3 LIQUID FREON FREEZANT OR LIQUID FREEZANT FREEZER (LFF)

LFF freezer uses specially purified dichlorodifluoromethane (fluorocarbon) having boiling point of –30°C at atmospheric pressure. In the design, the equipment consists of a container with openings on the top of the container. The product entering the container is dropped into a flowing system of freezant. Because of very good heat transfer, the surface of the product is frozen instantaneously which is later conveyed to horizontal freezing belt where the surface frozen product is sprayed with freezant till it gets frozen completely. In this freezer, a discharge conveyor is used to bring the product up and put of the freezer. Immediately after contact with the product the freezant evaporates and the vapors can be recovered by condensation on surface of refrigerator with only a slight loss and the remaining freezant remains in container with only small loss to atmosphere. There is no measurable weight loss due to dehydration by using this method. Small traces of fluorocarbon may be present in the product, but experiments in this area are still continuing.

3.7 MECHANICAL METHODS OF INDUCED FREEZING

These are the freezing methods which utilize the dipole moment of water to enhance the freezing by rotation of molecules in the field that is applied. Different methods under this group are discussed in subsections.

3.7.1 ELECTRO FREEZING

Freezing process assisted by charge flow or external electric fields (DC and AC fields) is called electro freezing. The water consists of dipole molecules. The water molecule consists of two hydrogen atoms and one oxygen atom in its structure. The oxygen side of the molecule is partially negative and the hydrogen side of the water molecule is partially positive. When an electric field is applied in water, the electric field tends to weaken or destructs the hydrogen bonding network of water. This enhances the free rotation of hydrogen atoms of water molecules about the direction of the applied field and tends to realign themselves along the direction of the field. Two inferences can be made from the molecular realignment or reorientation. The reorientation breaks the existing hydrogen bonding making liquid or frozen water less structured. The electric field makes water molecules more polarized in the direction of the electric field and the degree of polarization depends on temperature and field strength. This helps in inducing spontaneous nucleation in supercooled water. There are two components associated with enhancement of reorientation of water molecules, namely horizontal and transverse components. Horizontal component is associated with the tumbling molecular motions, and transverse component is associated with both tumbling and spinning motions about field axis. When temperature is below melting temperature, both strong and weak fields enhance spinning but not tumbling whereas when temperature is above melting temperature tumbling motions are dominated than spinning.

Electric charge flow assisted freezing is caused by application of either continuous or pulsed high DC voltage between narrowly spaced electrodes which are in direct contact with the sample. This method is used to assist the nucleation at higher nucleation temperatures compared to spontaneous nucleation temperature in supercooled solutions. When direct current is applied, the nucleation of ice crystals occurs under small supercooling than the spontaneous nucleation. But Alternating current and electrostatic field have less effect on nucleation form melt.

3.7.2 MAGNETIC RESONANCE FREEZING

As water is diamagnetic material, it is susceptible to be magnetized by an electric field. Under a magnetic field, more ordered and stable configuration is formed due to the stronger hydrogen bonds between water molecules. Stability of water molecules under a magnetic field is due to the higher hydrogen bonding. This type of ordering of water molecules with high hydrogen bonding changes the properties of water like increase in melting point and heat conductivity.

During quick freezing method by applying a magnetic field, the surrounding temperature of the object will be reduced to –30 to –100°C. When the unidirectional magnetic field is applied, the magnetic field can direct the magnetic moment caused by electron-spin of molecules present in product and water in one direction. Therefore, the cool temperature can be transmitted into the inner portion of product quickly. The difference in internal and external temperature of product will be lowered so that uneven cooling will be considerably reduced to produce quick freezing. Hence freezing occurs in a uniform and simultaneous manner and does not start from the outer surface. When the unidirectional magnetic field is applied, the imbalance in electron spin creates thermal vibration thereby preventing freezing and water molecules will be in a metastable state. When magnetic field is removed the electron spin will cancel each other and there won't be any imbalance in electron spin leading to freezing of water in product which will lead to removal of latent heat of freezing from water entering into frozen state. When the temperature is further lowered or magnetic field is suddenly removed, freezing will occur rapidly. Magnetic field vibrates the molecules thereby breaking the hydrogen bond between water molecules and prevent freezing. Advantages of magnetic resonance freezing are formation of more uniform ice crystals, less mechanical damage to product, reduced freezing time, etc. Although, there is one major disadvantage such as application of only magnetic field will not produce effective freezing and it has to be combined with other methods.

3.7.3 ULTRASOUND-ASSISTED FREEZING

Ultrasound is usually classified into two fields namely high frequency low energy (2–10 MHz) and low frequency high energy power ultrasound (20–100 kHz). In low energy high frequency power up to 10 W is used. At this low power, ultrasound won't cause any physical or chemical alterations in properties of material; hence it is termed as the non-destructive method used for quality assurance, process control, and to determine characteristics of food like texture, fluid level, structure, thickness, rate flow, etc. Frequencies at 5–10 MHz are used in diagnostic purposes. The power ultrasound alters the material properties like disrupting the physical integrity, acceleration of certain chemical reactions by generating immense pressure, shear, and temperature gradient in order to propagate power sound waves.

Ultrasound travels in a series of series of compressions and rarefactions produced in a medium like any sound wave. When high power is applied rare fractions exceed the attractive forces in the liquid medium leading to the formation of cavitation bubbles. Each bubble will interfere with neighboring bubble and affect the localized field resulting in instability of cavitation bubble and collapses releasing energy required for many mechanical and chemical reactions. It's well-known that during freezing ice crystal formation usually occurs in two steps, namely nucleation and crystal growth. The cavitation bubbles after reaching the critical nucleus size can act as nuclei for ice nucleation and the number of nuclei increases. When the oscillating bubbles produce a circulatory motion, micro streaming associated with cavitations occur setting up strong eddy currents, which are used to enhance mass and heat transfer in the freezing process at the ice/liquid interface thereby increasing the freezing rate. When ultrasound is passed in a dense and incompressible medium, it has resulted in fracture of ice formed on the cold surface (dendritic front) where ice gets fragmented and dispersed into unfrozen bulk liquid. The cavitation bubbles in the product will produce even ice nucleation and fragment ice crystals will be converted to smaller crystals which results in increase in conductive heat transfer in cooling medium thereby speeding up of freezing process.

There are three mechanisms of freezing acceleration by ultrasound. Acoustic cavitations which forms the micro streaming leading to intensive collision between microparticles in the medium and thereby generating thinner solid-liquid interface. The generation of cavitation bubbles in transport media as well as collapse of bubbles will contribute to production of ultra-high pressure (100 MPa). The high pressure creates high supercooling promoting instantaneous nucleation. In the third method, primary nucleation is induced by cavitation bubbles of a certain size and secondary nucleation is produced by small pieces from large dendritic ice crystals. Based on frequencies, the ultrasound is further classified into power ultrasound (2–100 kHz), high-frequency ultrasound (20 kHz–2 MHz) and diagnostic ultrasound (>1 MHz) and based on utilization, ultrasound can also be classified into two kinds: low intensity (<1 W/cm^2) and high intensity (10–1,000 W/cm^2).

3.8 HIGH-PRESSURE FREEZING

The principle of this method is that the solid (ordered) phase can't exist in liquid (disordered) domain, but it is possible in metastable state which is enhanced by pressure and this process is called supercooling or undercooling. At high freezing rates, the particles and large molecules in water serving as cores for heterogeneous nucleation (water becomes solid in glassy or vitreous state but doesn't show crystalline structure). During freezing at atmospheric pressure, the freezing rates can produce an ice crystal of 5–25 μM when the initial freezing point of water is reduced by increasing the ambient pressure or cryoprotectants. The homogenous nucleation temperature of water can be reduced more than the melting point of water by increasing the pressure up to 210 MPa. By applying the pressure the metastable stage can be extended. Hence the temperature at which water can be supercooled may be reduced from –40°C at atmospheric pressure to –92°C at 210 MPa. The freezing point can be reduced to –22°C at a pressure of 200 MPa enabling a vitrification depth of 200 μM and the objects of thickness of 0.4–0.6 mm could be well frozen. The food is cooled under subzero temperatures but a phase change will not happen till pressure is released. When pressure is released, rapid nucleation will take place throughout the food resulting in the formation of small, uniform, homogeneously distributed ice crystals. During this change, the duration of phase transition is reduced, reduced mechanical stress during formation of smaller ice crystals with uniform distribution are formed throughout the food.

Based on the phase transition, there are three types of high pressure freezing viz., High-pressure-assisted freezing (HPAF), High-pressure-shift freezing (HPSF) and High-pressure-induced freezing (HPIF). In pressure assisted freezing phase transition occurs under constant pressure. In pressure shift freezing, phase transition occurs due to a pressure release. In pressure-induced freezing, the phase transition is initiated by a pressure increase that is continued at constant pressure.

In high-pressure-assisted freezing process (HPAF), under constant pressure above atmospheric pressure, the phase transition occurs where the temperature is reduced below the freezing point of water. This method is similar to traditional freezing under atmospheric pressure but the only difference is that this process takes place under pressure. In high pressure shift freezing (HPSF), the product is initially kept in a liquid state at subzero temperature using high pressure. As pressure is released phase transition occurs resulting in a large degree of supercooling and rapid ice nucleation. In high pressure induced freezing process (HPIF), the phase transition is initiated by an increase in pressure and completed under pressure. In order to freeze a product it is initially compressed to an intermediate pressure level and cooled under pressure where the sample remains in unfrozen state. Later the pressure is increased to a desired level and phase transition occurs. Though this is interesting, but it has the disadvantages of ice polymorphs into higher forms than ice from HPAF. So after high pressure induced freezing only higher ice polymorphs will be produced. Hence the frozen products should be stored and thawed under pressure to maintain the benefit of higher ice polymorphs.

3.9 MICROWAVE-ASSISTED FREEZING

In principle, it was observed that microwave-assisted freezing utilizes the dipole moment of water. When microwaves are sent through food material they increase the dipole rotation of water to disturb the nucleation and formation of ice during freezing.

3.10 RADIOFREQUENCY ASSISTED FREEZING

The principle is similar to microwave freezing that is utilization of dipole rotation of water which is induced by application of radiofrequency waves to disturb the nucleation and formation of ice during freezing. It will also reduce freezing point; thereby, more nucleation sites are produced.

3.11 CONCLUSION

Quality plays an important role during the consumption of any type of food product. With industrialization and urbanization, there is a change in the food habits of people from consuming fresh produce to ready-to-eat/serve products. With this change, there is a great demand for frozen products. Though there are different freezing methods available, choosing appropriate freezing equipment for different products is very important. With the advancement in industry, there are new freezing techniques which can freeze the products very faster and thereby increasing the shelf life of the product significantly. Hence, there is a need to utilize these new freezing methods appropriately to increase the shelf life of the product and reduce the cost of processing.

KEYWORDS

- **chlorofluorocarbons**
- **convenient products**
- **freezing**
- **high-pressure-shift freezing**
- **nutritive value**
- **quality**
- **shelf life**

REFERENCES

Alexandre, E. M., Brandão, T. R., & Silva, C. L. M., (2013). Frozen food and technology. In: Visakh, P. M., Thomas, S., Iturriaga, L. B., & Ribotta, P. D., (eds.), *Advances in Food Science and Technology* (pp. 123–150). John Wiley & Sons, Inc. Hoboken, New Jersey, and Scrivener Publishing LLC, Salem, Massachusetts.

Balachandran, K. K., (2001). Freezing. In: *Post-Harvest Technology of Fish and Fish Products*. Daya Publishing House, Delhi.

Blond, G., & Le Meste, M., (2004). Principles of frozen storage. In: Hui, Y. E., Cornillon, P., Legaretta, I. G., Lim, M. H., Murrell, K. D., & Nip, W., (eds.), *Handbook of Frozen Foods* (pp. 25–53). Marcel Dekker, New York.

Cheng, X., Zhang, M., Xu, B., Adhikari, B., & Sun, J., (2015). The principles of ultrasound and its application in freezing related processes of food materials: A review. *Ultrasonics Sonochemistry, 27*, 576–585.

Evans, J., (2016). Emerging refrigeration and freezing technologies for food preservation. *Innovation and Future Trends in Food Manufacturing and Supply Chain Technologies* (pp. 175–201). Woodhead Publishing, UK.

James, C., Purnell, G., & James, S. J., (2015). A review of novel and innovative food freezing technologies. *Food Bioproc. Technol., 8*(8), 1616–1634.

Jha, P. K., Xanthakis, E., Jury, V., Havet, M., & Le-Bail, A., (2018). Advances of electro-freezing in food processing. *Curr. Opi. Food Sci., 23*, 85–89.

Li, B., & Sun, D. W., (2002b). Novel methods for rapid freezing and thawing of foods: A review. *J. Food Engg., 54*(3), 175–182.

Magnussen, O. M., Hemmingsen, A. K., Hardarsson, V., Nordtvedt, T. S., & Eikevik, T. M., (2008). Freezing of fish. In: Evans, J. A., (ed.), *Frozen Food Science and Technology* (pp. 151–164). Blackwell Publishing Ltd, Garingston Road, Oxford OX4 2DQ, UK.

Ninawe, A. S., & Rathnakumar, K., (2008). Freezing technology. In: *Fish Processing Technology and Product Development* (pp. 148–212). New Delhi, India, NBH.

Sanz, P. D., & Otero, L., (2005). High pressure freezing. In: Sun, D., (ed.), *Emerging Technologies for Food Processing* (2nd edn., pp. 505–538). Academic Press, Massachusetts, USA.

Venugopal, V., (2006). Quick freezing and individually quick frozen products. *Seafood Processing Adding Value Through Quick Freezing Retortable Pouch Packaging and Cook-Chilling* (pp. 95–140). CRC press, Boca Raton Florida.

Woo, M. W., & Mujumdar, A. S., (2010). Effects of electric and magnetic field on freezing and possible relevance in freeze-drying. *Dry. Technol., 28*(4), 433–443.

Zaritzky, N., (2012). Physical-chemical principles in freezing. In: Sun, D., (ed.), *Handbook of Frozen Food Processing and Packaging* (2nd edn.). CRC Press, Taylor and Francis group London, New York.

CHAPTER 4

Emerging Fish Processing and Packaging Technologies: An Overview

C. O. MOHAN,[1] S. REMYA,[1] K. R. SREELAKSHMI,[1] K. ELAVARASAN,[1] and C. N. RAVISHANKAR[2]

[1] *ICAR-Central Institute of Fisheries Technology, Kochi – 682029, Ernakulam, Kerala, India*

[2] *ICAR-Central Institute of Fisheries Education, Mumbai 400 061, India*

ABSTRACT

The principle used in food processing is altering their state and its atmosphere suitably to overcome the microbial as well as biochemical reactions so that the product shelf life can be extended. Food processing also helps in making the food fit for consumption by modifying taste, aroma, and texture and improves the nutritional value. Fresh food is highly perishable and can be preserved by use of different conventional processing technologies. However, in spite of many conventional processing technologies, food industry struggles to meet the consumers demand for convenient, nutritious, wholesome safe and quality food. Emerging food processing and packaging technologies will help in achieving this. This chapter provides a detailed note on the available emerging technologies for food processing and preservation.

4.1 INTRODUCTION

Research on food processing have attracted more due to huge demand in supply of healthy and safe food products. Health, nutrition, and convenience are the major factors driving the global food industry in this era. Fish products have attracted considerable attention as a source of high amounts of

Advances in Fish Processing Technologies: Preservation, Waste Utilization, and Safety Assurance.
Ranendra K. Majumder, PhD & Amjad K. Balange, PhD (Eds.)

important nutritional Components like high-quality protein, essential vitamins, minerals, and healthful polyunsaturated fatty acids (FAs) to the human diet. As a result of this the fresh fish and seafood's rank third among the food categories with the fastest overall growth worldwide, next to drinkable yogurt (18%), and fresh soup (18%). Consumption of both freshwater and seawater fish is expected to increase in the future. As fish is highly nutritious, it is also highly susceptible to spoilage, due to intrinsic and extrinsic factors. Proper processing and packaging help in maintaining the eating quality of fish for extended period. Worldwide, an array of processing and packaging methods are followed. This ranges from simple chilled or ice storage, salted, and drying to most recent and advanced high pressure and electromagnetic field applications, which attracts opportunities from both small-scale and industrial-level entrepreneurs. Fish products in live, fresh chilled, whole cleaned, filets steaks, battered, and breaded products, variety of dried products, smoked fish, fish sausage, and traditional products are the range of low-cost processing methods which can be readily adopted by small-scale fishers. The processing methods like canning or heat processing, freezing, vacuum, and modified atmosphere packaging, analog products, high-pressure processing, pulsed light processing, irradiation, electromagnetic field, etc., are the processing methods which requires higher investments can be adopted by large scale entrepreneurs, apart from the above-mentioned processing methods.

4.1.1 BENEFITS OF PROCESSING

- Converts raw food into edible, usable, and palatable form;
- Helps in preservation and storage of perishable and semi-perishable agricultural commodities;
- Helps in avoiding glut in the market and reduces post-harvest losses and make the produce available during off-season;
- Generates employment;
- Development of ready-to-consume convenient products which saves time for cooking;
- Helps in improving palatability and organoleptic quality of the produce by value addition and helps in inhibiting anti-nutritional factors;
- Helps in easing marketing and distribution tasks;
- Enables transportation of delicate perishable foods across long distances;

- Makes foods safe for consumption by controlling pathogenic microorganisms;
- Modern food processing also improves the quality of living by way of healthy foods developed for special people who are allergic to certain ingredients, diabetic, etc., who cannot consume some common food elements;
- Food processing can also bring nutritional and food security;
- Provides potential for export to fetch foreign exchange.

4.1.2 AIMS OF PRESERVATION/PROCESSING

Based on the perishability and the extent of preservation required, foods may be classified as:

1. **Perishable Foods:** Those that deteriorate readily (Seafood, meat, fruits, and vegetables) unless special methods of preservation are employed.
2. **Semi-Perishable Foods:** Those that contain natural inhibitors of spoilage (root vegetables) or those that have received some type of mild treatment which creates greater tolerance to the environmental conditions and abuses during distribution and handling (such as pickled meat and vegetables).
3. **Non-Perishable Foods (Shelf-Stable):** Those that are non-perishable at room temperature (cereal grains, sugar, nuts). Some have been made shelf stable by suitable means (canning) or processed to reduce their moisture content (dried fish and shellfishes, raisins). Food preservation in the broad sense, refers to all the measures taken against any kind of spoilage in food.

Fish is regarded as one of the highly nutritious foods. It is rich in easily digestible protein, healthful FAs, minerals, low in cholesterol and sodium content. Wide variety of fish and shellfish is another advantage for its consumers. However, fish is also highly perishable in nature due to its near neutral muscle pH, high levels of non-protein nitrogen fraction, easily digestible protein and high water activity. Various processing and preservation techniques have been adopted to overcome its spoilage. These include conventional icing, chilling or refrigeration, freezing, curing, and drying, heat processing, extrusion, etc. Apart from these emerging technologies like microwave processing, ultrasound processing, ohmic heating, pulsed electric field, pulse light technology, high-pressure processing, infra-red, and radio-frequency processing, biopreservation, etc., are being used for preservation. In addition, supercritical fluid extraction and sub-critical fluid extraction technologies are being applied for extracting valuable compounds from fishery resources. Further, advanced packaging technologies like high barrier packaging materials, vacuum packaging, modified atmosphere packaging, sous-vide processing, active, and intelligent packaging will complement the processing technologies adopted.

4.2 CONVENTIONAL FISH PRESERVATION METHODS

Chilling is an effective method of maintaining the freshness of fish products. This normally involves keeping fishes in melting ice or slurry ice to maintain the fish temperature around 1–2°C, which delays the enzymatic action and microbial activity, thereby extending the shelf life of the products. Traditionally, chilling is carried out using melting ice, either flake ice, or crushed block ice. Of late, slurry ice has been introduced for chilling. A wide range of fish and shellfish products varying from whole, headless, peeled gutted, headless gutted fish, filets, steaks, loins, cubes can be preserved by chilling. Shelf life of fishes from different environments has been studied extensively. Shelf life of 12–15 days has been achieved for seer fish and black pomfret. Indian Mackerel and Indian oil sardine had very short shelf life in ice (3–7 days), due to rancidity and belly bursting. Tilapia from freshwater and brackish water showed significant differences in shelf life when stored in ice. The former kept longer (14–15 days) than latter (8–10 days). In refrigeration or air circulated chilling methods, a temperature of 2–4°C is monitored. In some cases super chilling, i.e., reducing the temperature of fish to sub-zero temperatures (–1 to –7°C), is also practiced to enhance the shelf life. Freezing is an age-old practice to retain the quality and freshness of fishery products for a long time. This involves the conversion of water present in fishery products to ice, i.e., a phase change from liquid to solid phase takes place in freezing. This retards the microbial and enzymatic action by reducing the water available for their action. This involves exposing fish products to very low temperatures (<–35°C) to enable freezing of free water and maintained at –18°C till it is consumed. Plate freezing, air blast freezing, cryogenic freezing, and individual quick freezing are the methods adopted by the industry to preserve food products. Drying is probably one of the oldest methods of food preservation. It consists of removal of water to a final desired concentration, which in turn reduces the water activity of the product, thereby assuring microbial stability and extended shelf-life of the product. In some cases, common table salt (Sodium chloride) is also used to prolong the shelf life of fish. Salt absorbs much of the water in the food and makes it difficult for microorganisms to survive. Smoking is one of the most widely used traditional fish processing methods employed in many countries to preserve fish. The preservation effect of the smoke is a result of drying of the product during the smoking as well as due to smoke particle absorption into the flesh. The smoke particles, mainly phenolic compounds, carbonyl, and organic acids, being absorbed by the product, inhibit bacterial growth on the surface of the product. The smoke particles also have a positive effect on the taste and color of the product and in many instances, smoking is normally practiced to improve these sensory characteristics. Heat processing, which is also known as thermal processing is another age-old processing method to achieve very long shelf life at normal room temperature storage. In this, product is packed and hermetically sealed and processed at high temperature and pressure and cooled and stored. Packaging materials used for this purpose include, glass, metal can, rigid containers, flexible retortable pouches. Besides cost reduction, retort pouch packages have unique advantages like reduced process time, improved nutrient retention, boil in bag facility, ease of opening, reduced weight, and do not require refrigeration for storage. Processed food products can be kept for long periods at ambient temperature. The energy-saving is more in processing in flexible pouches compared to cans. On a comparison of total costs, including energy, warehousing, and shipping, the pouch looks even more favorable. There is 30 to 40% reduction in processing time compared to cans, solids fill is greater per unit, empty warehousing is 85% smaller and weight of the empty package is substantially smaller. In order to improve the utilization of underutilized fisheries resources, there is a need to minimize the post-harvest losses, develop innovative processing technologies and utilize processing waste for industrial and human use. One such technology, which will be suitable for utilization of low value fish or by catch, is extrusion technology.

Use of fish mince with cereals for extrusion process will enable production of shelf-stable products at ambient temperature. Extrusion cooking is used in the manufacture of food products such as ready-to-eat breakfast cereals, expanded snacks, pasta, fat-bread, soup, and drink bases. The raw material in the form of powder at ambient temperature is fed into extruder at a known feeding rate. The material first gets compacted and then softens and gelatinizes and/or melts to form a plasticized material, which flows downstream into extruder channel. Basically, an extruder is a pump, heat exchanger, and bio-reactor that simultaneously transfer, mixes, heats, shears, stretches, shapes, and transforms chemically and physically at elevated pressure and temperature in a short time. At times, the extrusion cooking process is also referred to as the High-Temperature Short Time process. In the extrusion process gelatinization of starch and denaturation of protein ingredient is achieved by combined effect of temperature and mechanical shear. The conversion of raw starch to cook and digestible materials by the application of heat and moisture is called gelatinization. During extrusion the conditions that prevail are high temperature, high shear rate and low moisture available for starch may lead to breakdown of starch molecules to dextrins. Irradiation is a physical treatment that consists of exposing foods to the direct action of electronic, electromagnetic rays to assure the innocuity of foods and to prolong the shelf life. Irradiation of food can control insect infestation, reduce the numbers of pathogenic or spoilage microorganisms, and delay or eliminate natural biological processes such as ripening, germination, or sprouting in fresh food. Like all preservation methods, irradiation should supplement rather than replace good food hygiene, handling, and preparation practices.

Three types of ionizing radiation are used in commercial radiation to process products such as foods and medical and pharmaceutical devices (International Atomic Energy Agency (IAEA), radiation from high-energy gamma rays, X-rays, and accelerated electrons:

- Gamma rays, which are produced by radioactive substances (called radioisotopes). The approved sources of gamma rays for food irradiation are the radionuclides cobalt-60 (^{60}Co; the most common) and cesium-137 (^{137}Cs). They contain energy levels of 1.17 and 1.33 MeV (^{60}Co) and 0.662 MeV (^{137}Cs).
- Electron beams, which are produced in accelerators, such as in a linear accelerator (linac) or a Van de Graaff generator at nearly the speed of light. Maximum quantum energy is not to exceed 10 MeV.
- X-rays or decelerating rays, which can be likewise produced in accelerators. Maximum quantum energy of the electrons is not to exceed 5 MeV.

Different forms of irradiation treatment are radurization (for shelf life extension), radicidation (for elimination of target pathogens), and radappertization (for sterilization). Radiation processing is widely used for medical product sterilization and food irradiation. Moreover, the use of irradiation has become a standard treatment to sterilize packages in aseptic processing of foods and pharmaceuticals. Irradiation produces some chemical changes, which, although lethal to foodborne bacteria, do not affect the nutritional quality of the food but lead to the production of small amounts of radiolytic products. Gamma irradiation has been considered as an interesting method of preservation to extend the shelf life of fish and also to reduce qualitatively and quantitatively the microbial population in fish and fish products. Irradiation doses of 2–7 kGy can reduce important food pathogens such as Salmonella, *Listeria*, and *Vibrio* spp., as well as many fish-specific spoilers such as *Pseudomonaceae* and *Enterobacteriaceae* that can be significantly decreased in number.

4.3 EMERGING PRESERVATION TECHNOLOGIES

In conventional methods, fish is processed and preserved by applying low temperature, high temperature or by reducing moisture level. Generally in the low temperature preservation, food is exposed to chilled or sub-zero temperature to reduce the activity of microorganisms and biochemical and enzyme activity. In thermal processing, vegetative microorganisms will be destroyed during mild heat treatment and spore-forming gram-positive anaerobic bacteria *Clostridium botulinum* in the ready-to-eat food processing methods. However, these thermal treatments and freezing triggers many unwanted biochemical reactions leading to undesirable organoleptic characteristics of the product. These can be overcome by employing emerging advanced processing and preservation technologies. These apart from preserving fish products they ensure high safety and better nutrient retention compared to conventional methods. These technologies are discussed in detail in subsections.

4.3.1 MICROWAVE PROCESSING

Microwave heating of foods is attractive due to its instantaneous and rapid increase in temperature, controllable heat transmission, and easy clean-up opportunities. It is currently being used for a variety of domestic and industrial food processing applications. The permitted microwave frequencies are 915, 2,450, 5,800 MHz. The dielectric properties of these materials are important in the design of electrical and electronics equipment and various techniques for measuring the dielectric properties of different applications have been developed. The largest use of industrial microwave processing of food has been for tempering of meat for further processing. Microwave tempering is defined as taking a product from freezer temperature to a condition (between –4 to –2°C) in which the product is not frozen but is still firm. Tempering at low temperatures results in minimal quality deterioration and can help in saving energy. Frozen blocks of meat at –18°C is to cut or separated into individual pieces. Conventional tempering techniques take a lot of time with considerable drip loss resulting in loss of protein and quality and economic loss. Microwaves can easily penetrate the whole frozen product, thus effectively reaching the inner regions within a short time. The microwave tempering can be performed in few minutes for a large amount of frozen products (5–10 min for 20–40 kg). Microwaves are easily absorbed by the meat; its intensity is reduced by the penetration depth. Surface layers retain more energy and heat up faster compared to the inner regions of the product. The lower frequency (915 MHz band) has an advantage for tempering of thick products because of its deeper penetration and longer wavelength compared to the higher frequency (2,450 MHz) microwave. Currently, most food industries use microwave at 915 MHz for tempering purposes. Precooking of bacon is the second-largest application of microwave heating in the food industry. Microwave heating is found to be an ideal system for cooking bacon compared to conventional grilling. The third-largest application of microwave processing is in sausage cooking. The quality of the sausage patty is improved along with better yield by using the microwave process. In sausage cooking also, microwave processing is used to reduce drip loss, fat, nutrients, and flavor. Microwaveable foods in suitable packaging materials are being developed by food processors to meet the growing demand. These convenience foods are microwaveable for use at home and away. High-density polypropylene (HDPP) is suitable for microwave processes over other materials since it can withstand high temperature.

4.3.2 RADIO FREQUENCY HEATING

Radio frequency heating refers to the use of electromagnetic waves of certain frequencies to generate heat in the material. Radiofrequency heating can be performed in any of the three frequencies (13.56 MHz ± 6.68 kHz; 27.12 MHz ± 160 kHz and 40.68 MHz ± 20 kHz). The principle of RF heating relies on the fact that, the heat is generated in most dielectric materials when a high frequency electric field is applied across them. The amount of heat generated in the product is determined by the frequency, the square of the applied voltage, dimensions of the product and the dielectric loss factor of the material, which is essentially a measure of the ease with which the material can be heated by radio frequency waves. It is reported that the RF applications have shown a number of benefits over conventional heating methods. The advantages of RF heating are increased heating and processing speed, improved product quality and yield because of uniform heating, usage of only 1/3rd of the floor space of conventional heating units, instant on/off and temperature change and higher energy savings up to 60–70%. Another important advantage claimed for RF heating is its 'self-limiting' property, which controls the consumption of RF energy according to the workload.

4.3.2.1 RADIO FREQUENCY THAWING

Radio frequency thawing is similar to microwave ovens where fish products passing through the oven (heater), are subjected to a direct or volumetric heating process in the form of a radio frequency (RF) energy source. The RF heating process depends upon the ionic conductivity of the material being heated. Radio-frequency thawing systems are also available, where the frozen product is placed between two parallel electrodes and alternating radiofrequency energy is applied to the electrodes. Temperature rise within the product is relatively uniform, the degree of uniformity being dependent on the size and composition of the product. It is suggested that 5 cm blocks of fish can be thawed rapidly. However, radio frequency treatments have more promising attributes for processing seafood. At the lower frequencies of RF, penetration of the RF energy into foods is much greater and enables the temperature of blocks to increase from –20°C to –2 or 0°C. Radio-frequency systems are available in both batch and continuous methods. Batch RF systems operate from 40 to 350 kg/hour while continuous RF systems can operate from 900 to 3,000 kg/hour.

4.3.3 ULTRASOUND PROCESSING

Ultrasound refers to sound that is just above the range of human hearing, i.e., above frequency of 20 MHz. Ultrasound when propagated through a biological structure induces compressions and depressions of the medium particles imparting a high amount of energy to the material. The sound ranges for food applications employed can be divided into two, namely, low energy, high frequency diagnostic ultrasound and high energy low frequency power ultrasound. Low energy applications involve the use of ultrasound in the frequency range of 5–10 MHz at intensities below 1 W/cm^2. Ultrasonic waves at this range are capable of causing physical, mechanical, or chemical changes in the material leading to disrupting the physical integrity, acceleration of certain chemical reactions through generation of immense pressure, shear, and temperature gradient in the medium. Ultrasonics has been successfully used to inactivate *Salmonella* spp., *Escherichia coli, Listeria monocytogenes, Staphylococcus aureus*, and other pathogens.

Ultrasound is probably the simplest and most versatile method for the disruption of cells and for the production of extracts. It is efficient safe and reliable. Ultrasound cavitation creates shear forces that break cell walls mechanically and improve material transfer. This effect is being used in the extraction of liquid compounds from solid cells. The compound to be dissolved into a solvent is enclosed in an insoluble structure. In order to extract it, the cell membrane must be destructed. For the purpose, ultrasound is faster and more completed than maceration or stirring. The particle size reduction by the ultrasonic cavitation increases the surface area in contact between the solid and liquid phase, significantly. The mechanical activity of this technique enhances the diffusion of the solvent into the tissue; Ultrasound breaks the cell wall mechanically by the cavitation shear forces at it facilitate the transfer from the cell into the solvent. This technique has potential advantages over other techniques including freedom from radiation hazards, which may appear in some of the existing non-destructive methods. The presence of the small gas bubbles in a sample can greatly attenuate ultrasound making signal detection impossible. This can be solved by using reflection measurements rather than transmission measurement. The advantages of ultrasonics in food processing are lower energy consumption, reduced processing time, higher throughput, reduced flavor loss, greater homogeneity, and significant energy saving.

4.3.3.1 ULTRASOUND THAWING OF FROZEN FOODS

Effective acoustic or ultrasound thawing relies on the selection of an appropriate frequency and intensity to defrost the food efficiently without excessive heating near the surface. When ultrasound was applied directly to meat (beef, pork, or fish), excessive heating near the surface was particularly a problem at high intensities (1–3 W/cm^2) and over a range of frequencies (220 kHz to 3.3 MHz), with cavitation causing problems at lower frequencies, while high attenuation caused excessive heating at higher frequencies. A combination of 500 kHz and 0.5 W/cm^2 was found to offer effective thawing while minimizing surface heating. Acoustic defrosting (1,500 Hz) of fish blocks in an agitated water bath (18°C, 3.8 m/s water velocity at block surface) reduced the time to go from –29°C to –1°C by approximately 25%–70% (depending on ultrasound power input level), while larger reductions were seen going from –5°C to –1°C (approximately 32%–82%). Some rapid thawing techniques can cause excessive heating at the product surface leading to loss of product quality. However, combined acoustic and water bath thawing gave the same surface temperature as water bath thawing alone. US patent No: US 4464401 A on acoustic thawing of frozen food in 1982 explains the use of ultrasound waves in thawing frozen foods.

4.3.4 OHMIC HEATING

Ohmic heating is an emerging technology with large number of actual and future applications. Ohmic heating technology is considered a major advance in the continuous processing of particulate food products. Ohmic heating is direct resistance heating by the flow of an electrical current through foods, so that heating is by internal heat generation. Ohmic heating is defined as a process wherein electric current is passed through materials with the primary purpose of heating the object. During ohmic heating, heating occurs in the form of internal energy transformation (from electric to thermal) within the material. Therefore, it can be explained as an internal thermal energy generation technology and it enables the material to heat at extremely rapid rates from a few seconds to a few minutes. Ohmic heating have a large number of actual and potential future applications, including its use in blanching, evaporation, dehydration, fermentation, extraction, sterilization, pasteurization, and heating of foods. The microbial inactivation due to ohmic heating can be explained by the presence of electric field. The additional effect of ohmic treatment may be its low frequency (usually 50 to 60 Hz), which allows cell walls to build up charges and form pores. As a main consequence of this effect, the D value observed for the microbial inactivation under ohmic heating is reduced when compared to traditional heating methods. More research is needed to completely understand all effects produced by ohmic heating to food products, effects of applied electric field, the applied electric frequency during ohmic heating over different microorganisms and foods, cold spot determination, etc.

4.3.5 INFRARED AND RADIOFREQUENCY PROCESSING TECHNOLOGIES

Electromagnetic radiation is a form of energy that is transmitted through the space at an enormous velocity (speed of light). The heat generation in material exposed to EMR could be due to vibrational movement (as in case IR) or rotational movement (as in case of RF and MW) of molecules. Application of EMR heating is gaining popularity in food processing because of its definite advantages over the conventional processes. Faster and efficient heat transfer, low processing cost, uniform product heating and better organoleptic and nutritional value in the processed material are some of the important features of EMR processing. In conventional heating system like hot air heating, the heat is applied at the surface which is carried inwards through conduction mode of heating. In case of EMR/dielectric heating, the waves can penetrate the material to be absorbed by inner layers. The quick energy absorption causes rapid heat and mass transfer leading to reduced processing time and better product quality.

The main advantage of electromagnetic heating over conventional electric and gas oven-based heating is its high thermal efficiency in converting the electrical energy to heat in the food. In ordinary ovens, a major portion of the energy is lost in heating the air that surrounds the food, a fairly good amount escapes through the vent, besides being lost through the conduction to the outside air. In contrast, almost all the heat generated by electromagnetic radiations, which reaches the interior of the oven, is produced inside the food material itself. According to the reports, the energy efficiency of EMR based systems is 40–70%, as compared to approximately 7–14% in the case of conventional electric and gas ovens.

4.3.6 HIGH-PRESSURE PROCESSING

High pressure (HP) processing is an emerging and innovative technology that has a great potential for extending the shelf-life with minimal or no heat treatment. High-pressure processing is also called high hydrostatic pressure processing, pascalization, or high-pressure pasteurization. It is also effective in preserving the organoleptic attributes of many foods. The early efforts for

HP technology have been initiated in Japan, but now have been commercialized in North America, Europe, and in China. The first high-pressure processing line was introduced in Japan for jam manufacture in 1990's and has since been upgraded to several food products. High-pressure processed foods were first commercialized in Japan in 1992. Machines are now available with operating pressures in the range 400–700 MPa. HP processing affects mainly the non-covalent bonds of the food and quality characteristics of foods such as color, flavor, and nutrients generally remain unaffected. There are two general scientific principles of direct relevance to the use of high pressure in food processing. The first is Le Chatelier's Principle, which applies to all physical processes and states that, when a system at equilibrium is disturbed the system responds in a way that tends to minimize the disturbance. This means that HP stimulates reactions that result in a decrease in volume but opposes reactions that involve an increase in volume. Any phenomenon (e.g., phase transition, change in molecular configuration, chemical reaction) that is accompanied by a decrease in volume will be enhanced by pressure. Secondly, the Isostatic Rule states that pressure is instantaneously and uniformly transmitted throughout a sample under pressure, whether the sample is in direct contact with the pressure medium or hermetically sealed in a flexible package that transmits pressure. Pressure is transmitted in a uniform (isostatic) manner throughout the sample; the time necessary for pressure processing is therefore independent of sample size, in contrast to thermal processing.

High-pressure processing (HPP) holds the potential for preserving foods by combining elevated pressures (up to 900 MPa or approximately 9,000 atmospheres) and moderate temperatures (up to 120°C) over a short period. Other advantages of the technology include uniform pressure application, minimal heat damage to food and potential for altering functional properties of foods. The possibility of extending shelf-life without heating the food for prolonged periods greatly helps to satisfy consumer demand for fresher and higher quality heat-sensitive foods that are otherwise difficult to process using conventional food preservation methods. Effects of HPP on microorganisms in surimi paste showed that all of the microbes were destroyed at 300–400 MPa; fungi showed highest sensitivity to HPP, followed by Gram-negative and Gram-positive bacteria. Pressure resistant bacteria, *Moraxella* spp. was viable at 200 MPa and *Acinetobacter calcoaceticus* was viable at 300 MPa. In Japan, hydrostatic pressure has been used to induce the gelation of different kinds of surimi from pollack, sardine, skipjack tuna and squid. High pressure at levels of 300–3,740 atm for 30 min to formulate gels from bluefish meat paste, and the properties of the resulting gels were compared with those of heat-induced gels formulated at 90°C for 20 min or 60°C for 60 min. When shrimp was pressure-treated at levels of 100, 270, and 435 MPa for 5 min at room temperature (25±2°C), a shelf life was extension of 15 days was possible in high pressure treated shrimp at 435 MPa compared with 5 days in untreated sample. Pressure level of 220 MPa and a 30 min holding time were optimal and most effective in prolonging the storage period of tuna muscle (up to 9 days), as well as in reducing the proteolysis activity, texture degradation, TVBN, and histamine formation. High-pressure processing increased pH, protected from lipid oxidation, maintained low microorganisms levels, changed the color of the muscle, and induced the formation of high molecular weight polypeptides most likely through disulfide bonding promoting texture improvement and thereby improve the shelf life of minced albacore muscle for more than 22 days at 4°C and more than 93 days at −20°C when high pressure of 275 and 310 MPa applied for 2, 4, and 6 sec to minced albacore muscle.

4.3.6.1 PRESSURE SHIFT FREEZING

Pressure shift freezing (PSF) has the potential benefits for producing fine and uniform ice crystals within the food systems, particularly meat. PSF helps to reduce tissue deformation and shrinkage since these two novel processes create smaller and uniformly distributed ice crystals and well isotropic spread of ice crystals in meat tissues. Therefore, this method can be used to freeze foods with uniform ice crystal distribution and small size of ice crystal, resulting in a significant improvement of product quality. Studies conducted on Atlantic salmon (*Salmo salar*) indicated that pressure shift freezing is advantageous over glycol/water bath freezing (GBF) and conventional air freezing (CAF). In the study, Atlantic salmon samples were frozen by pressure shift freezing (PSF) at 100 MPa (–8.4°C), 150 MPa (–14°C) and 200 MPa (–20°C), by CAF at –30°C and GBF at –20°C. CAF created larger and irregular ice crystals, and resulted in irreversible damage to muscle tissues. Due to its higher freezing rate, GBF produced smaller ice crystals than CAF, but the cross-section area and roundness values had larger deviations. The PSF process produced large amounts of fine and regular intracellular ice crystals that were homogeneously distributed throughout the sample. Microscopic images clearly showed that the muscle fibers were well maintained in the PSF treated samples as compared with unfrozen muscle structures.

4.3.6.2 PRESSURE ASSISTED THAWING

It is well recognized that the success of freezing preservation depends on careful pre-freezing preparation, control of the freezing rate, storage conditions and thawing conditions. Only limited investigations have aimed at improving the thawing rate. Due to microbial growth during thawing, minimal ambient temperature must be ensured. The use of high pressure (HP) for thawing offers a new alternative to this process. The use of HP for thawing biological substances offers a unique alternative to conventional processes in that it decreases the phase change temperature of pure water (down to –22°C at 220 MPa). The lowering of the melting point of ice under HP allows the temperature gap between the heat source and the phase change front to increase, and thus enhances the heat flux rate in the case of pressure assisted thawing. The phase change temperature increases with pressure above 220 MPa, and thus 220 MPa appears as the maximum pressure of interest in thawing with respect to the heat transfer criterion.

4.3.7 PULSED LIGHT TECHNOLOGY

Pulsed light (PL) is an emerging non-thermal technology for decontamination of food surfaces and food packages, consisting of short time high-peak pulses of broad-spectrum white light. The term light is generally used to mean radiations having wavelength ranging from 180 to 1,100 nm, which includes ultraviolet rays (UV 180–400 nm, roughly subdivided into UV-A, 315–400 nm, UV-B, 280–315 nm, UV-C, 180–280 nm), visible light (400–700 nm) and infrared rays (IR 700–1,100 nm). Pulsed light is produced using technologies that multiply power manifold. It is used for the rapid inactivation of microorganisms on food surfaces, equipment, and food packaging materials. The effect on microorganisms is mostly due to the photochemical action of the Ultra Violet part of the light spectrum that causes thymine dimerization in the DNA chain preventing replication and ultimately leading to cell death. PL treatment of foods has been approved by the FDA in 1996 under the code 21CFR179.41.

Many research findings have shown that this can be used for the reduction of microbial load of *Listeria monocytogenes, Salmonella enteritidis, Pseudomonas aeruginosa, Bacillus cereus*, and *Staphylococcus aureus, Escherichia coli*, etc. Many works have shown extended shelf life in shrimp treated with pulsed light. Shrimp treated with pulsed light and stored at refrigerated condition for 7 days remained edible whereas untreated shrimp showed extensive microbial degradation and were discolored, foul smelling and not edible. Pulsed light was able to reduce the psychotroph and coliform population on the surface of summer flounder filets. The sensory attributes indicates the filets remained acceptable after 15 days of refrigerated storage.

There have been findings that indicate 1 log reduction of *Escherichia coli* O157:H7 or *Listeria monocytogenes* on salmon filets, when pulsed light treatment ($5.6\ Jcm^{-1}$) per 60 second treatment was done at 8 cm distance without affecting the quality. Studies conducted at McGill University, Canada show promise for pulsed light treatment for cold smoked vacuum packed salmon to control *Listeria monocytogenes* and *Clostridium botulinum* A and E.

Although many studies are related to microbial reduction in dairy products, vegetables, and fruits, there are limited studies about effectiveness of pulsed light on fish micro flora and shelf life. Studies need to be conducted to assess the effects of PL treatments on food properties beyond safety and spoilage. Packaging plays in important role in the preservation, transporting, and marketing of the products. Suitable packaging materials are to be selected for packaging of animal products for PL treatment. Pulsed light can be used to inactivate microorganisms on the surface of food packaging materials, and potentially on the surface of products packaged in UV transparent materials. The use of pulsed light could lead to a reduction in the need for preservatives or chemical sterilizing agents. Chemical surface sterilizers such as hydrogen peroxide, propylene oxide or peracetic acid may leave a residue or require time to reduce to an acceptable level but pulsed light has the advantage of not leaving undesirable residue after treatment. The inactivation of microbial load by pulsed light is a fast process and takes place in a very short period. It is a green technology because the energy consumption is very less during its application. Due to the absence of any harmful residues, chemical, and toxic by-products in pulsed treated foods, this technology is proven to be safe. It does not affect the nutritional and sensory quality of the products. The concerns of ionized radicals and radioactive by-products in foods by consumers are removed in PL due to its non-ionizing spectrum. Low penetration power and lipid oxidation are a major constraint in meat industry. Packing material having high penetration of pulsed light should be used while treatment of packed food. The limited control over food heating still remains the main concern in pulsed light technology. Sample heating is perhaps the most important limiting factor of PL for practical applications.

Pulsed light treatment done on pearlspot (*Etroplus suratensis*) filets for 12 sec with a total energy of 25 J/cm^2 indicated a shelf life extension of 6 days when compared to control. Yellowfin tuna (*Thunnus albacares*) steaks packed in cast polypropylene was found to give maximum bacterial reduction. Pulse treatment for 6 sec with an energy output of 11.5 Jcm^{-2} was found to be acceptable microbiologically and with regard to sensory attributes. Dip treatment in 2% sodium acetate, 2% potassium sorbate and a combination of 2% sodium acetate-potassium sorbate solution with pulsed light treatment resulted better sensory and L*a*b* color values of pulse treated tuna steaks than dip and pulsed treated combinations. Shelf life extension of 13 days for pulsed treated and pulsed-dip combinations was obtained, however organoleptically; the pulse treated samples were rated superior. Pulsed light treatment is effective in reducing the microorganisms load considerably. Masmin treated with pulse light resulted in a reduction of 0.7 log fungal counts. Challenge studies were conducted to assess the efficiency of the pulse light treatment on the pathogenic bacteria. A reduction of 2.66 log was observed for *Escherichia coli* O157 in tuna homogenate for 30 sec pulse light treatment. The D value was found to be 12.38 sec. A 0.90 log reduction on 30 sec of treatment was found when *Staphylococcus aureus* ATCC 6538 was inoculated in Tuna Homogenate. The D value was found to be 35.90 sec.

4.3.8 PULSED ELECTRIC FIELD

PEF processing consists of the application of very short electric pulses (1–100 μs) at electric field intensities in the range of 0.1–1 kV cm−1 (reversible permeabilization for stress induction in plant cells), 0.5–3 $kVcm^{-1}$ (irreversible permeabilization of plant and animal tissue), and 15–40 kV cm−1 for the irreversible permeabilization of microbial cells. The PEF technology is considered a non-thermal cell disintegration or preservation process. It is possible that a rapid rupture occurs in limited areas without affecting the entire cell membrane but that depends on the external electric field applied. It provides an alternative to mechanical, thermal, or enzymatic cell disintegration of plant and animal raw materials, providing a short-time (milliseconds), low energy treatment, as well as to the traditional thermal pasteurization of liquid food products Electrical stimulation of meat, an electric field of 5–10 V/cm is applied as alternating current (AC) pulses to the sample through electrodes fixed at opposite ends of the long axis of the muscle. It is generally accepted that the primary effect of PEF on biological cells is related to local structural changes and the breakdown of the cell membrane, which is a highly important component of the biological cell as it acts as a semipermeable barrier responsible for mass transfer and plays an important role in the synthesis of RNA and DNA, protein, and cell wall components as well as many other complex metabolic activities. Disruption of intracellular organelles and other structural changes have also been described. The electrical breakdown of cellular membranes has been explored based on model systems such as phospholipid vesicles and planar bilayers as well as microorganisms.

Microbial inactivation by PEF has been extensively investigated within the last few decades initially in batch treatment systems and model foods which are free of fat or proteins. The low energy consumption (1–2 kJ/kg for stress induction and 5–10 kJ/kg for plant cell permeabilization) and the continuous operability of this short-time, waste free membrane permeabilization technique are key advantages and allow the development of innovative, cost effective and sustainable processing concepts in the food and drink industry as well as in the biotechnology and pharmaceutical industry. PEF processing offers high quality fresh-like liquid foods with excellent flavor, nutritional value, and shelf-life. Since it preserves foods without using heat, foods treated this way retain their fresh aroma, taste, and appearance. Application of PEF technology has been successfully demonstrated for the pasteurization of foods fish soups, tomato juice and liquid eggs. Application of PEF processing is restricted to food products with no air bubbles and with low electrical conductivity. PEF is a continuous processing method, which is not suitable for solid food products that cannot be pumped.

4.3.9 COLD ATMOSPHERIC PLASMA

Cold atmospheric plasma which is also commonly known as cold plasma or non-equilibrium or non-thermal plasma is another emerging technique which is finding various application. In this, highly energetic and reactive gases are used mainly to inactivate contaminating microbes on the surface of food. These plasmas are good sources of highly reactive oxidative and reductive species and plasma electrons. Using these oxidative and reactive species, electrical energy can be directed into favorable gas chemistry for decontaminating surface of food or fragmenting larger molecules into smaller ones. Cold plasma can be effectively used for decontaminating the foods which are not suitable for chemical treatment or food with fragile surfaces. Treatment time ranges from 3 to 120 seconds depending on the food to be treated and the process conditions.

4.3.10 ELECTROLYZED WATER

Electrolyzed water which is also known as electro-chemically activated water, electro-activated water or electrolyzed oxidizing water is very useful for cleaning food contact surfaces and to for surface decontamination of food products. Electrolyzed water is produced by electrolyzing a weak salt solution like sodium chloride solution which produces sodium hypochlorite, a disinfectant.

4.3.11 BIO-PRESERVATION

Bio-preservation is gaining popularity due to its health benefits and its safety features. It is the use of natural microflora or controlled homogeneous or heterogeneous group of microorganisms for preserving the food and for extending the shelf life. The use of desirable microflora or the products of fermentation by these organisms can act as biopreservation to control the spoilage as well as affects food borne pathogens activity. Bacteriocins are a heterogeneous group of antibacterial proteins that vary in spectrum of activity, mode of action, molecular weight, genetic origin and biochemical properties. Various spices and essential oils have preservative properties and have been used to extend the storage life of fish and fishery products. Natural compounds such as essential oils, chitosan, nisin, and lysozyme, bacteriocins have been investigated to replace chemical preservatives and to obtain green label products. Enzymes have been used for the production of various cured and fermented fish products from centuries. Because of their appreciable activity at moderate temperature, products, and process have emerged that utilizes enzymes in a deliberate and controlled fashion in the field of food processing. Cold active enzymes including elastase, collagenase, chymotrypsin extracted from Atlantic cod were used in various food processing applications. The other applications of cold active enzymes include caviar production, extraction of carotenoprotein, etc. Treatment with protease under mild treatment conditions extending for a few hours can result in the recovery of the proteins from fish frame or shrimp shell waste. The role of transglutaminase in surimi production is well established. The gel strength of surimi can be improved by the application of extracellular microbial transglutaminase. Lipase extracted from *Pseudomonas* spp. can be used to produce PUFA enriched cod liver oil. Enzymatic de-skinning of fish filets was done by partial denaturation of skin collagen using a gentle heat treatment followed by immersion in enzyme solution for several hours at low temperature (0–10°C). De-skinning of tuna, Herrin, Squid were also carried out by using different enzyme technology.

4.3.12 SUPER CRITICAL EXTRACTION

Supercritical carbon dioxide is a fluid state of carbon dioxide where it is held at or above its critical temperature and critical pressure. Carbon dioxide usually behaves as a gas in air at STP or as a solid called dry ice when frozen. If the temperature and pressure are both increased from STP to be at or above the critical point (CP) for carbon dioxide, it can adopt properties midway between a gas and a liquid. It behaves as a supercritical fluid above its critical temperature (31.1°C) and critical pressure (72.9 atm/7.39 MPa), expanding to fill its container like a gas but with a density like that of a liquid. Supercritical CO_2 is becoming an important commercial and industrial solvent due to its role in chemical extraction in addition to its low toxicity and environmental impact. The relatively low temperature of the process and the stability of CO_2 also allow most compounds to be extracted with little damage or denaturing. The use of supercritical fluids as an extraction media provides a powerful alternative to traditional chemical extraction methods. SFE is relatively cheap, nontoxic, and nonflammable with zero ozone-depletion potential. Well known processes such as the decaffeination of coffee, or the fractionation of hops for the flavoring of beer, have been abetted by more recent processes producing an array of food-related products. These products have included spice and flavor extracts, defatted or reduced-cholesterol products, natural antioxidants, and specialized oil-derived products for the nutritional market. However, the research works done on the application of SFE in processing fish and fish products are relatively scanty. The future appears bright for further extension of the critical fluid technology platform towards the processing of fish and fishery products for the health food, and nutraceutical markets. Some additional benefits can occur when using high pressure carbon dioxide or water for processing seafood products, such as the deactivation of harmful microbes or enzymes, allowing sterilization or stabilization of the resultant end products, without resorting to the extreme temperatures and pressures required in ultra-high pressure seafood processing.

4.4 ADVANCED PACKAGING TECHNOLOGIES

4.4.1 *APPROPRIATE BARRIER FILM FOR FOOD PRESERVATION*

Air packaging has limited shelf life of fishery products. Shelf life of fish can be enhanced by using appropriate packaging materials. Petroleum-based packaging materials barrier properties depend on their composition. Single layered packaging material will have very low barrier properties which can be improved by suitable combination of many films. Among the films used in food industry, low density polyethylene has the lowest barrier properties for oxygen, i.e., 1,250 $mL/m^2/24$ h whereas water vapor transmission rate is 2 $gm/m^2/24$ h. Aluminum foil has very high barrier properties, i.e., OTR, and WVTR of <0.1 $mL/m^2/24$ h and $gm/m^2/24$ h, respectively. Ethylene vinyl alcohol (EVOH) also is a good barrier for OTR (~0.15 $mL/m^2/24$ h) but poor barrier for water vapor (22 $gm/m^2/24$ h). Combination of nylon, polyester, polyethylene, and polypropylene will improve the barrier properties significantly, which can be used for vacuum packaging and modified atmosphere packaging applications.

4.4.2 *VACUUM PACKAGING*

Vacuum packaging involves the removal of air from the package and the application of a hermetic seal. The air removal creates a vacuum inside the packs and lack of O_2 in packages may minimize the oxidative deteriorative reactions and aerobic bacterial growth. Vacuum packaging can considerably extend the viable shelf life of many cooked foods. The use of vacuum packaging, in gas impermeable and heat stable materials, has many advantages, which include; no or low risks of post pasteurization contamination, ease of handling, Inhibition of growth of aerobic spoilage organisms and inhibition or slowing of deleterious oxidative reactions in the food during storage due to oxygen barrier properties of the packaging material. There are number of criteria required for the films used for vacuum packaging in large scale production methods. These requirements include: high durability, i.e., ability to withstand considerable mechanical stresses during packaging, handling, and transport, retention of flexibility even at low temperatures (–2 to 4°C) to enable satisfactory handling in the packaging and refrigeration rooms, ability to withstand heating to at least 150°C without structural damage, leaching of potentially toxic plastics or plasticizers, impermeability to liquids, including oils and fats and macromolecules, impermeability to gases, in particular oxygen, so that oxidative deterioration of the packaged food stuffs is limited or inhibited, manufactured from non-toxic, food acceptable, odorless materials and must be able to create airtight durable heat seals to close packs. Many of these criteria have been met by a range of materials mostly multilaminated plastics. Vacuum packed foods maintain their freshness and flavor 3–5 times longer than with conventional storage methods, because they don't come in contact with oxygen. Foods maintain their texture and appearance, because microorganisms such as bacteria mold and yeast cannot grow in a vacuum. Freezer burn is eliminated, because foods no longer become dehydrated from contact with cold, dry air. Moist foods won't dry out, because there's no air to absorb the moisture from the food. Dry, solid foods, won't become hard, because they don't come in contact with air and, therefore, can't absorb moisture from the air. Foods that are high in fats and oils won't become rancid, because there's no oxygen coming in contact with the fats, which causes the rancid taste and smell.

4.4.3 MODIFIED ATMOSPHERIC PACKAGING

Fresh fish is highly susceptible to spoilage from post mortem autolysis and microbial growth. The high ambient temperature of our country favors rapid growth of microorganisms. Presently ice and mechanical refrigeration are the most common means of retarding microbial and biochemical spoilage in freshly caught seafood during distribution and marketing. However, as ice melts it tends to contaminate fish accelerating spoilage and reduces shelf life. Modified atmosphere packaging, a technologically viable method has been developed as a supplement to ice or mechanical refrigeration to reduce the losses and extend the storage life of fresh seafood products. In modified atmosphere packaging air is replaced with different gas mixtures to regulate microbial activity and/or retard discoloration of the products. The proportion of each component gas is fixed when the mixture is introduced into the package; however, no control is exercised during storage. The composition of the gas mixture changes from its initial composition as a result of chemical, enzymatic, and microbial activity of the product during storage. It is primarily the enrichment of Carbon dioxide in the storage atmosphere as a means of controlling microbial growth, which results in the extension of shelf life of products. Carbon dioxide lowers the intra and extracellular pH of tissues and possibly that of microorganisms. Further, it may affect the membrane potential of microorganisms and influence on the equilibrium of decarboxylating enzymes of microorganisms. The gases normally employed are carbon dioxide, mixtures of carbon dioxide and nitrogen, carbon dioxide and oxygen and carbon dioxide, oxygen, and nitrogen with the sole objective to extend the shelf life of the product beyond that obtained in conventional refrigerated storages. Inhibition by Carbon dioxide manifests in an increased lag phase and a slower rate of growth of microorganisms during logarithmic phase. Inhibition by Carbon dioxide was found to be more effective when the product was stored at the lowest range of refrigerated temperatures. Packaging materials generally employed for this purpose are flexible films of nylon/surlyn laminates, PVC molded trays laminated with polythene, polyester/low density polythene film, etc. The use of high barrier film along with MAP that contains CO_2 effectively inhibits bacterial growth during refrigerated storage of packaged fresh fishery products.

The composition of the gas mixtures used for MAP of fresh fish varies, depending upon whether the fish in the package is lean or oily fish. For lean fish, a ratio of 30% Oxygen, 40% Carbon dioxide, 30% Nitrogen is recommended. Higher values of Carbon dioxide are used for fatty and oily fish with a comparable reduction in level of oxygen in the mixture leading to 40–60% Nitrogen. By excluding oxygen, the development of oxidative rancidity in fatty fish is slowed. On the other hand, oxygen can inhibit the growth of strictly anaerobic bacteria like *Clostridium botulinum* although there is a very wide variation in the sensitivity of anaerobes to oxygen. It is also seen that inclusion of only some oxygen with nitrogen or carbon dioxide will not prevent botulism with absolute certainty.

4.4.4 ACTIVE AND INTELLIGENT PACKAGING

4.4.4.1 ACTIVE PACKAGING (AP)

Active packaging (AP) is an innovative concept that can be defined as 'a type of packaging that changes the condition of the packaging and maintains these conditions throughout the storage period to extend shelf-life or to improve safety or sensory properties while maintaining the quality of packaged food' AP performs some desired role other than providing an inert barrier between the product and external conditions and combines advances in food technology, biotechnology, packaging, and material science, in an effort to comply with consumer demands for 'fresh like' products. This involves incorporation of certain additives into the packaging film or within packaging containers with the aim of maintaining and extending product shelf life. AP technique is either scavenging or emitting systems added to emit (e.g., N_2, CO_2, ethanol, antimicrobials, antioxidants) and/or to remove (e.g., O_2, CO_2, odor, ethylene) gases during packaging, storage, and distribution. Major AP techniques are concerned with substances that absorb oxygen, ethylene, moisture, carbon dioxide, flavors/odors, and those which release carbon dioxide, antimicrobial agents, antioxidants, and flavors. Some of the AP systems include:

- Oxygen scavenger;
- Carbon dioxide emitter;
- Moisture regulator;
- Antimicrobial packaging;
- Antioxidant release;
- Release or absorption of flavors and odors;
- Carbon dioxide scavenger;
- Active packaging systems with dual functionality (combination of oxygen scavengers with carbon dioxide and/or antimicrobial/antioxidant substances).

Other AP systems that are expected to find increased attention in the future include color containing films, light absorbing or regulating system, susceptors for microwave heating, gas permeable/breathable films, antifogging films and insect repellant package.

4.4.4.2 INTELLIGENT PACKAGING

Intelligent packaging, sometimes referred as smart packaging, senses some properties of the food it encloses or the environment in which it is kept and informs the manufacturer, retailer, and consumer of the state of these properties. Although it is distinctly different from the AP concept, features of intelligent packaging can be used to check the effectiveness and integrity of AP systems. Intelligent packaging has been defined as 'packaging systems which monitor the condition of packaged foods to provide information about the quality of the packaged food during transport and storage:

- Time temperature indicators;
- Leakage indicator;
- Freshness indicator;

- Pathogen indicator;
- Shelf life indicator.

Active and intelligent packaging systems contribute to the improvement of food safety and extend the shelf-life of the packaged foods. However, these are evolving technologies in the seafood area and many of these systems are in the developmental stage. Continued innovations in active and intelligent packaging are expected to lead to further improvements in food quality, safety, and stability.

4.5 CONCLUSION

All the above technologies have the potential to inactivate the pathogenic and spoilage causing organism. There is a need for optimizing the critical process factors to achieve the target inactivation level for specific food applications without affecting quality. In addition to this, the applicability of emerging technologies on an industrial scale needs to be compared with other technologies. Suitable packaging material and technologies along with the emerging processing technologies will further complement in enhancing safety and shelf life of food products. Systematic work is also necessary to investigate the impact on the nutritional and sensory properties of the treated foods. Most of this emerging novel technology equipment are expensive. However, research on these emerging technologies is essential to evaluate the complete functionality of these equipment with special emphasis on product quality and safety. Detailed study on emerging technologies will help in transferring these to industry to transform the way of our daily life.

KEYWORDS

- **active packaging**
- **conventional air freezing**
- **fish**
- **microwave processing**
- **novel processing techniques**
- **ohmic heating**
- **pressure shift freezing**

REFERENCES

Balachandran, K. K., (2001). *Post-Harvest Technology of Fish and Fish Products* (p. 440). Daya Publishing House, New Delhi.

Barba, F. J., Grimi, N., & Vorobiev, E., (2014). New approaches for the use of non-conventional cell disruption technologies to extract potential food additives and nutraceuticals from microalgae. *Food Engg. Rev.*, *7*(1), 45–62.

Koseki, S., & Isobe, S., (2007). Microbial control of fresh produce using electrolyzed water. *Japan Agricul. Res. Qua., 41*(4), 273–282.

Misra, N. N., Tiwari, B. K., Raghavarao, K. S. M. S., & Cullen, P. J., (2011). Nonthermal plasma in-activation of foodborne pathogens. *Food Engg. Rev., 3*(3, 4), 159–170.

CHAPTER 5

Utilization of Low-Cost Fish Through Restructured Product-Based Technologies

AMJAD K. BALANGE and R. P. DEEPITHA

Department of Post-Harvest Technology, ICAR-Central Institute of Fisheries Education, Versova, Mumbai – 400061, Maharashtra, India

ABSTRACT

Restructured fish product technology is a promising area of the fish processing sector where the science behind fish muscle protein application has tremendously contributed to the preparation of versatile products. Efficient utilization of undervalued fish species will benefit researchers and nutritionists in recommending high-quality fish protein at reasonable values across the globe. Naturally available hydrocolloids in the form of alginates from brown seaweed or carrageenan from red seaweeds are some of the best ingredients used for the preparation of restructured fish products. Endogenous enzymes like microbial transglutaminase also interact with the fish muscle proteins in enhancing the functional properties. Demand for minimal processing in the food industries has put functional properties like gelation, viscosity, emulsification, and foaming application in meat science to a diverse area. Interaction of different ingredients with the various chemical constituents of the fish muscle proteins, fats, or water molecules is the primary factor responsible for showcasing the different functional properties in restructured fish products. The majority of the restructured products are prepared from either the fish mince or surimi. The representative fish flavor, odor, or taste of the original fish is lost or reduced during the course of processing the mince or surimi, adding the desirability to introduce novel flavors to the end

Advances in Fish Processing Technologies: Preservation, Waste Utilization, and Safety Assurance.
Ranendra K. Majumder, PhD & Amjad K. Balange, PhD (Eds.)

product. Textural improvisations by adding salt, phosphates, endogenous enzymes, gums, or hydrocolloids can also valorize the fish products. The restructured product technology in fish muscle protein has remarkable scope in contributing to the global food science.

5.1 INTRODUCTION

Ever since the evolution of mankind, food became the primary need of human being, and the settlement and civilizations were endeavors thereafter. Food is a generalized term that includes all the edible commodities which is consumed by living organisms in order to maintain life.

Among the various types of food, seafood grades a remarkable position. According to the 2017–2018 Indian seafood export data, the total value of the seafood exported from India reached the all-time high of US $ 7.08 billion in 2017–2018 (Rajesh, 2018). Due to the enormous health benefits possessed, health professionals also recommend the consumption of seafood in the daily diet. Fish is a healthy food comprising all the essential nutrients necessary for human health and forms the cheapest source of easily digestible protein. The added advantage of the fish meat is the unique texture due to the lesser amount of connective tissue, especially stroma protein. The stroma content of the fish meat is reported to be less than 3%. This distinguishes the fish meat from other terrestrial meats. The protein fraction of the fish meat is also an abundant reserve of all essential amino acids which cannot be synthesized by the human body. The presence of Omega 3 fatty acids (FAs) like eicosapentaenoic acid (EPA) and docosahexaenoic acid (DHA) is another salient character of fish meat, which in turn has several health benefits.

5.2 LOW COST FISHES AND ITS UTILIZATION TREND

The extensive health advantages received by consuming seafood, pertaining to its easy digestibility, abundant reserve form of amino acids, vitamins, and minerals, has titled it as one of the top traded food commodities in the world. However, non-uniformity in the consumption pattern of fish and shellfishes has resulted in considerable pressure over certain fisheries.

This created considerable pressure on the total productivity of capture fisheries, and hence came the need for sustainability in the fisheries sector by regulating the over exploited catches and the need to focus on the underutilized resources of the ocean. Utilization of low-cost fishes landed in substantial quantities through different fishing operations and particularly as trawler by-catch has been gaining attention among the planners, researchers, and developmental agencies. Over the past few years, considerable research inputs and several technologies have come forward aiming the effective utilization of the low-cost fishes. This contributed immensely towards the field of fisheries-based value-added products and other by-products synthesis. The techno-economic feasibility of the products and technologies would help the entrepreneurs in preparation of the products at both small and large scale. Due to the abundant availability, the small size species like sardines, perches, sciaenid's, etc., fetch low price in the market. However, by understanding the textural qualities of these fish meat, it can form the perfect ingredient for several value-added fish products. In order to make efficient utilization of these underutilized catches, several efforts and technologies had been developed by researchers over the years, and restructuring technology is one among those.

5.3 RESTRUCTURED PRODUCT TECHNOLOGY

Restructuring technology usually aims in commercialization of low value fish species like the small size fishes than the commercial size (obtained in the shrimp by-catches), the trimmings from fileting of commercial fish species and non-commercial fish species that gets discarded due to its unfavorable size, appearance, flavor, color, odor or texture (Pacheco et al., 2010). The technology lets in attainment of products with high commercial value. The main process involved in the restructuring technology include grinding of fish muscle, solubilization of fish proteins with salt, structuring of fish paste and initiation of gelling phenomenon using heat (Ramirez et al., 2011). Novel products with improved mechanical and functional properties are obtained using an array of additives, thus adding value to the end products.

Recently several studies were done aiming for improvement in the mechanical and textural properties of fish meat. Additives like hydrocolloids (alginates) or enzymes (microbial transglutaminase), salts like calcium (triggers endogenous transglutaminase), sodium chloride are used in improving the overall textural properties of fish meat. Fortification using some functional ingredients to enhance the general acceptability of end products, thereby diversifying the appearance of seafood and increasing its value is the broader outcomes of restructured technology. This has introduced the science of functional foods in the fishery resources. The concept of functionality is the promoting factor behind the development of several innovative seafood products. Restructured seafood products are foods prepared from comminuted fish muscle, which are either used alone or with additional ingredients/additives, to make end product with a new appearance, texture or both (Moreno et al., 2016).

A variety of restructured fishery products are available in the markets especially seafood analogs or imitation products which are prepared from surimi. Normally fish muscle is highly perishable owing to its high nutrient content that limits its shelf life. Above all, mince which has a higher contact surface in comparison to the whole fish further limits the shelf life. Hence the products prepared from the fish flesh, requires special attention, thereby enhancing the functionality of the end-product. The unstable mince, when added with certain compounds enhance the overall textural and functional properties of the product along with an extended shelf-life. Surimi is a stable product, prepared by washing and refining the mince, added with cryoprotectants and has exceptionally good gel-forming ability and excellent frozen storage stability.

The fish mince or surimi used, are subjected to different treatments to endow the end product with improved textural properties, bringing the concept of functional properties like gelation, water holding capacity, viscosity, emulsion, and foam formation in the end product. Restructured products can serve as the base for incorporating several functional ingredients like dietary fibers which modulate the fish muscle texture.

Although there are different technologies in the restructuring process, heat induced gelation is the most widely used technique in surimi and surimi-based products. Other processes like cold gelation with the addition of different ingredients, adds new possibilities in enhancing the functional properties. Certain studies have demonstrated the use of dietary fibers in improving the functional properties of fish mince. The different technologies used in the restructured fishery products include:-

5.3.1 GELATION

In the seafood industries, gelation is the phenomenon involved in the preparation of a three dimensional viscoelastic gel from fish flesh using certain additives and processes. It is in fact one of the most commonly used techniques in the restructured fishery products preparations. Based upon the modifications in processes and kinds of additives used, the process can be induced by several factors.

5.3.1.1 HEAT-INDUCED GELATION

Nowadays, the most widely used method in the restructuring technology is the heat set gelation process for both surimi and minced muscle. The major protein fraction involved in the gelation is the myosin, a myofibrillar protein. The process of gelation is initiated with the addition of salt in the appropriate quantity that allows the myofibrillar protein to unfold, exposing the reactive surfaces facilitating intermolecular interactions. These interactions with the intermolecular bonds form a three-dimensional network resulting in a gel. The different types of bonds involved in the network formation are hydrogen bonds, ionic interactions, hydrophobic interactions, and covalent bonds (Lanier et al., 2005). Hydrophobic bonds and covalent bonds are responsible for the gels thermostability. After the unfolding process, upon addition of salt, the coils are aggregated in an ordered manner. This ordered aggregated chains when heated at temperature below 50°C, a soft gel named "suwari" is formed, and the process is called setting. The type of network formed is non-disulfide covalent bonds between the two amino acids glutamine and lysine on the unfolded proteins under the effect of calcium-dependent transglutaminase, naturally present in fish muscle (Lanier et al., 2005). The setting temperature, which is the temperature at which the protein network unfolds and forms a 3D network is species depended. Normally the cold-water species form suwari gels at 25°C whereas warm water species from the gel network at 40°C (Ramirez et al., 2011). This mechanism of suwari gel formation can also be induced using high pressure, where hydrophobic interactions play a predominant role in network formation (Perez et al., 1997). Apart from the conventional heat processing techniques, methods like ohmic heating, radio frequency (RF) or microwave has been used in inducing gelation process in fish meat. Alaska Pollock surimi prepared by ohmic heating, at a slower heating rate showed higher shear stress value, but Pacific whiting surimi which is known to be proteolytically more active showed higher shear stress values at higher heating rates (Yongsawatdigul and Park, 1996). Surimi prepared with ohmic heating from tropical species has shown increased breaking strength, reduced deformation and improved water holding capacity in comparison to water bath heating. Microwave heating is another method for heat induced gelation and this has shown improved textural properties for low-salt gels. In countries like Japan, hydrostatic pressure has been used to induce the gelation of different kinds of surimi from species like Alaska pollack, sardines, skipjack tuna, and squid.

5.3.1.2 COLD GELATION

Minimal processing has become the attention of the recent scenario due to the loss of nutritional properties of food when subjected to high temperatures like cooking. So cold gelation, gained importance in the context. In cold gelation process, the protein aggregation is mainly due to the binding agents that get active at sufficiently low temperatures inducing desirable changes in the raw product. The two most commonly used binding agents in the fish restructuration technology are sodium alginate and microbial transglutaminase and another component known as konjac glucomannan has also exhibited similar properties (Moreno et al., 2008).

5.4 HYDROCOLLOIDS USED IN THE RESTRUCTURED FISHERY PRODUCTS

The science of rheology in food system has gained attention since the recent past, and the textural modifications in the food systems using different additives is an outcome supporting the same fact. Hydrocolloids are functional carbohydrates used in the food systems to improve the quality and shelf life of the product. Several hydrocolloids are used in the food industry like starch, xanthan, guar gum, cellulose derivatives, etc. The application of hydrocolloids in seafood industries to improve the textural quality of fish meat is also gaining popularity since recent times.

5.4.1 ALGINATE

Alginate is a hydrocolloid extensively used in the food industry owing to its excellent functional properties. The most commonly used form of alginate is sodium alginate due to its high-water solubility (Nakauma et al., 2016). The gelation model of alginate is described in the form of an egg box model where divalent ions, mostly calcium ions, are trapped between the two guluronic acid blocks (Morris et al., 1978). Several studies have reported that sodium alginate has the ability to retain the water holding capacities in meat muscles thereby increasing the juiciness of the product. The inherent viscous nature of sodium alginate and the ability to form a network in the presence of a divalent cation is the reason behind the popularization of this hydrocolloid in food industry. Alginate forms thermo-irreversible gels which doesn't melt upon heating and hence have wide applications in restructured foods and ice-creams (Roopa and Bhattacharya, 2008). Alginate is a hydrocolloid gel and are commonly called the "physical gels," because unlike synthetic polymer chains where the gels are formed by covalent linkage, in hydrocolloids, the gels are formed by the physical association of polymer chains by hydrogen bonding, hydrophobic interaction and cation mediated cross-linking. Several restructured fish products with improved functional properties had been developed over the past few years by incorporating sodium alginate in fish mince.

5.4.2 CARRAGEENANS

Carrageenans are popular additives used in the meat restructure technology for enhancement of the binding properties of the muscle proteins. It is used as a thickening, stabilizing, and gelling agent. Carrageenans display low viscosities and good heat transfer mechanisms when dissolved and mostly due to this reason this hydrocolloid is used in the hot filling steps of food processing (Challen and Moorhouse, 2011). Among the different forms of carrageenan, kappa carrageenan is the widely used form in comparison to the iota and lambda forms. The wide application of carrageenan in food industries is spread over the synthesis of ice creams, whipped toppings, flan, mousse, etc.

5.4.3 AGAR

Agar is another excellent hydrocolloid used abundantly in the food industries. Due to the excellent gel-forming ability it is mainly used as a gelling agent and as a stabilizing agent for controlling the viscosity. Globally around 90% agar produced is for food applications and the remaining 10% is used for bacteriological and biotechnology works (Qin, 2018). Agar finds tremendous application in the confectionary for the preparation of jellies, marshmallows, candies, etc. In the meat industry, the same is used in the preparation of sausages to lessen the oil content from sausages.

5.4.4 GUMS

Other most commonly used gum forms in hydrocolloids in addition to carrageenan include locust bean gum, xanthan, and gum Arabic (Taheri and Jafari,

2019). Gums from aqueous dispersions and known for its excellent viscous parameters. Among the plant gums used commercially, few such as agar, carrageenan are used in food products like jams, jellies, restructured foods and bakery fillings (Saha and Bhattacharya, 2010; Soultani et al., 2014).

5.5 ENZYMES USED IN THE RESTRUCTURED FISHERY PRODUCTS

Use of enzymes in seafood is not a novel approach. Several enzymes were used by researchers in enhancing various processes. Traditionally the enzymes like proteases (bacillolysin, papain, pepsins, and trypsins) have been used for a variety of processes. Both endogenous and exogenous enzymes were used in the process of deskinning, descaling, shrimp peeling, fish sauce production, tenderization of squid meat, etc. Although different enzymes have contributory roles in the seafood industry for extending the shelf life, fermentation technology, ripening, denaturation, etc., significant contribution in the restructured fishery products is through microbial transglutaminase.

5.5.1 MICROBIAL TRANSGLUTAMINASE

Microbial transglutaminase, an endogenous enzyme plays a crucial role in enhancing the textural properties of fish muscle. It irretrievably catalyzes the covalent cross-linking of proteins by building an isopeptide bond between glutamine and lysine residues, and hence plays an integral part in setting phenomenon in fish muscle protein (Motoki and Seguro, 1998). The endogenous transglutaminase (TGase) found in different floral and faunal tissues and microbial transglutaminase (MTGase) secreted by different microorganisms like *Streptoverticillium mobaraensis* or *Streptomyces mobaraensis* are different in nature (Motoki and Seguro, 1998). Although MTGase is recommended as an additive in fish processing, especially to bind the fish muscle in restructured products, the United State Department of Commerce, and the European Food Safety Authority (EFSA) is still looking into its use in the food industry. The major benefit of this enzyme over the endogenous TGase is that the former activity is totally Ca^{2+} independent. Upon adding MTGase to any substrate, the textural characteristics of the restructured fish and other meat products, improved significantly (Kuraishi et al., 1997; Motoki and Seguro, 1998). In addition to this, MTGase is comparatively stable, catalyzes the reaction at high temperatures (Gómez-Guillen et al.,

2005), and is highly active than fish endogenous TGase (Hemung et al., 2008). MTGase is known to enhance the gel quality of dark, fatty fish, such as sardines (Karayannakidis et al., 2008), lizardfish surimi upon 10 days of ice storage (Benjakul et al., 2008) and hake muscle which otherwise reported low functional quality protein due to the longer frozen storage (Moreno et al., 2009). The optimum temperature of MTGase is around 20–50°C based on the specific enzyme and the substrate, but the activity was found even at 5°C in fish muscle (Moreno et al., 2010). This fact makes MTGase an ideal component in the preparation of restructured products at low temperature without causing changes in the end product (Moreno et al., 2008).

5.6 SURIMI AND MINCE: THE INTERMEDIATORY STAGE FOR RESTRUCTURED FISHERY PRODUCTS

With these technologies several restructured fishery products are available in the markets. Based on the method of preparation and the final appearance they are known by different names. Addition of suitable ingredients and appropriate technologies has resulted in versatile products in the market. Due to the enhanced functional properties, these products had gained much visibility in the domestic and international markets. Apart from gelation, viscosity, water holding capacity, protein solubility and emulsifying properties are other important features considered in valorizing the fishery products. As surimi and mince mostly serve as the intermediatory product, modifications can be made depending on the consumer preferences. Food processors have ample opportunities in these areas in designing versatile features that can attract consumers, in demanding for more fishery-based products. Both mince and surimi had been used in the restructuring technology, although there are differences in properties of both. Mince and surimi are primarily distinguished by the washing process, which give the characteristic white color to the latter. When compared to the mince, surimi is known for its excellent functional properties and frozen storage stability (Kim and Park, 2007). On the other hand, preparation of mince is quite simple and yield as well as nutritional property is superior than that of the surimi. However, the use of natural components like seaweed extracts or other nutritive components in the surimi mixture can resolve the issue of poor nutritional property of the surimi.

5.7 OTHER INGREDIENTS USED IN RESTRUCTURED TECHNOLOGY

5.7.1 SALT

Apart from playing a contributory role in processed meats, salts exhibit significant properties in the restructured meats. Amount of salt added varies between 0.5 to 1% (Pearson and Gillett, 1996). The functional role of salt and phosphates are in the extraction of myofibrillar proteins that enhances binding of meat particles (Schmidt, 1986). Most commonly added form of salt in improving the functionality of muscle protein is sodium chloride. The ions from the salt interact with the protein molecules and form cross-linking, which is the initiation step involved in the process like gelation. Further heating/temperature rise set the formed network into a viscoelastic sole that render the characteristic juicy texture to the final product. However, rising health issues like high blood pressure upon consumption of excess salt has made researchers to focus on other alternatives for salt in the food industry (Ruusunen and Puolanne, 2005).

5.7.2 PHOSPHATES

Phosphates in food are mainly used to enhance the effect of salt, by increasing the water-binding capacity even at lower concentrations of salt (Knipe et al., 1985). Addition of phosphates increased the biding strength in the muscle foods; it was found that with the increase in chain length of the molecule, the synergistic effect between salt and phosphates decreased linearly (Trout and Schimdt, 1986). Among all the different forms of phosphate salts, tetra-sodium pyrophosphate had the highest binding capacity followed by sodium tetra polyphosphate and sodium hexametaphosphate (Trout and Schmidt, 1984). Similar to salts the excess concentration of phosphates in the food system has certain health implications, where it influences the calcium, iron, and magnesium balance in the human body and enhances the risk of osteo-arthritis (Shahidi and Synoweicki, 1997). Due to this reason, the present research focuses on reducing the effect of this additive in the food by other natural alternatives.

5.7.3 SOY PROTEINS

Different forms of isolated protein are used in the food industries to enhance the functionality. Soy protein is an abundantly used isolated protein in the meat food system due to its binding capacity, gelling ability and emulsion stability (Renkema and Vliet, 2002) also known for its antioxidant properties and improved water absorption capacities (Pratt and Birac, 1979). Because of their high binding capacity with muscle proteins, soy proteins find tremendous applications in the restructured meat products (Tsao et al., 2002).

Other most commonly used ingredients in the restructured meat foods include wheat gluten, crude myosin extract, egg white powder, bovine, porcine, broiler plasma powder, gelatin dried apples, corn crumbs, mushrooms, rice bran oil, fiber, and walnut (Sun, 2009).

5.8 DIFFERENT RESTRUCTURED FISHERY PRODUCTS

Based on the advances in the food engineering and technologies, several interventions were developed in the meat food as well. Seafood industry used the restructured technologies to prepare diversified value added fishery products.

5.8.1 FISH SAUSAGE AND PASTE PRODUCTS

Traditional products like sausages, cakes, patties, balls, loaves, and burgers are well established products and have become the ideal vehicle for efficient utilization of minced fish. Fish mince is used in sausages, most popularly in Japan. Fish mince has also been used as an extender in preparation of frankfurters, patties, and burgers (King, 1973).

Sausage and ham are products popular in many countries nowadays, but during the initial period when the product was just introduced in 1968, it didn't gain much popularity mainly due to the poor-quality raw material, fish odor, and lack of standardized recipes and unavailability of a suitable sausage casing. But gradually with the development in the meat science and technologies, the addition of suitable ingredients, and the introduction of Ryphan (rubber hydrochloride casings) and Kurehalon (vinylidene chloride casings) the fish paste and sausage products started gaining visibility at the global level. The sausage and paste products were the improvised form of the traditional steamed paste products of Japan like Kamaboko. In India, the demand for fish sausage and paste products hasn't gained much popularity like other countries around the world.

While considering the raw material quality, preferably white meat fishes, which have strong elasticity in comparison to red meat, are chosen for the sausage preparations. Generally low-cost fishes like threadfin bream (perches), croakers, lizard fish, ribbon fish, and Lactarius are commonly used. Globally, Alaska pollack, blue marlin (*Tetrapturu samplus*), striped marlin (*Tetrapturus mitsukuri*), *Saurida* sp. are the common fishes targeted for sausage preparation.

The preparation of sausage primarily consists of grinding the fish with oil, seasonings, and starch and packing in casing followed by sealing and heat treatment. The seasoning material includes sugar, salt, sodium glutamate, coloring materials, preservatives, and condiments are separately blended and added to the silent cutter. Sodium or potassium salts of sorbic acid up to 2 g per kg of fish sausage were permitted in Japan (Ravishankar et al., 1992). The prepared sausages usually harbor thermo-tolerant and spore-forming bacteria, and hence preservatives used mainly target these micro florae. Other than sausage, several other convenient products are prepared from minced fish meat like fish fingers, fish burger, fish cutlet, salted fish mince, fish balls, fish wafers, fish soup powder, etc. In order to increase the texture and shelf life, different seasoning and binding agents are added to the product. Due to their ready-to-eat nature, these products have gained much popularity in both the domestic and international markets within a short span of time. Among the described convenient products also, some are categorized into battered and breaded products due to the additional layer of a batter and bread crumbs around the product, increasing its overall acceptability.

5.8.2 TEXTURIZED PRODUCTS

Marine beef is a type of fish protein concentrate which can be easily rehydrated for home use. The exclusive characteristic of the product is that it possesses a ground meat texture when prepared and hence has versatile applications in a variety of foods. It is prepared by concentrating fish protein without altering the amino acid balance. The fish mince is washed with water, treated with sodium bicarbonate (alkali blanching), pH adjusted and kneaded. Kneading the fish with salt breaks down the muscle fiber and forms a viscous paste. This resultant paste is mixed with alcohol to denature and coagulate the protein precipitating the muscle fiber into a new fiber structure similar to meat. After removing the fat by solvent extraction, the product is finally dried and packed as pellets.

5.8.3 SURIMI AND SURIMI-BASED PRODUCTS

Surimi is the wet concentrate of fish myofibrillar protein. Its quality depends greatly on the freshness of fish used, species, the leaching process, temperature storage period, and also the levels and types of cryoprotectants used. This intermediate product forms the base for several dishes, like the traditional kamaboko products of Japan to the shellfish substitutes famous in western countries. The unit operations involved in the manufacture of surimi includes process like preparation of mince (heading, gutting, and deboning/mincing). The mince, so prepared is subjected to washing (removes the pigment and fat), dewatering (removes excess water from the washed mince), refining, mixing (cryoprotectant addition-give frozen storage stability) and freezing (Martin et al., 2009).

Recently, seafood analogs, which are normally known as imitation products started gaining attention among the public, due to its low price. Analog products are produced from cheaply available source of fish protein or surimi and are designed to mimic high-value, luxury food items. Shellfish analogs are popular surimi seafood products which imitate shellfishes such as crab, lobster, shrimp, etc., in shape, color, appearance, flavor, and textural properties. Shrimp analog is made from surimi with added ingredients desired for creating the unique texture and taste modification (Remya et al., 2015).

Starch is one of the widely used ingredients in shellfish analogs, used primarily for enhancing the water-holding capacity and also for replacing a portion of the fish protein by maintaining desired gel properties. Starch is the main ingredient accountable for enhancing the gel strength, modifying texture, reduce cost as well as improve freeze-thaw stability in surimi (Lee, 1984). The concentration level between 4–12% starch is shown to improve gel strength and the textural properties in seafood (Okada, 1963). The commonly used starch forms in the shellfish analog products include wheat, corn, maize, tapioca, and potato.

5.9 CONCLUSION

The restructured technology has started gaining popularity since past few years. People have become more health-conscious and driving their attention towards healthy foods, like fish and its products has become the need of era. Working women and their tiring schedules are other factors that popularize the demand for ready-to-eat products in the market. The use of natural ingredients and basic food engineering technologies by researchers has contributed enormously to highlighting such products in the market. In the years to come, the technologies may improve further, thereby increasing the efficient utilization of the underutilized natural resources of the ocean. Ample scopes and opportunities lie ahead, in identifying the true properties of these oceanic resources. Global food crisis and nutritional deficiencies being the future threat for humankind, the science of restructuring technology has immense potential in modifying the properties of underutilized foods of the ocean and contributing positively towards global food security.

KEYWORDS

- **docosahexaenoic acid**
- **eicosapentaenoic acid**
- **functional properties**
- **low-cost fish**
- **processing**
- **restructured products**
- **textural properties**

REFERENCES

Benjakul, S., Phatcharat, S., Tammatinna, A., Visessanguan, W., & Kishimura, H., (2008). Improvement of gelling properties of lizardfish minces as influenced by microbial transglutaminase and fish freshness. *J. Food Sci., 73*(6), S239–S246.

Challen, I., & Moorhouse, R., (2011). Hydrocolloids in restructured foods. *Hydrocolloids in Food Processing* (pp. 165–214). Blackwell Publishing, Ltd. and Institute of Food Technologists.

Gómez-Guillén, M. C., Montero, P., Solas, M. T., & Pérez-Mateos, M., (2005). Effect of chitosan and microbial transglutaminase on the gel forming ability of horse mackerel (*Trachurus* spp.) muscle under high pressure. *Food Res. Int., 38*(1), 103–110.

Hemung, B. O., Li-Chan, E. C., & Yongsawatdigul, J., (2008). Thermal stability of fish natural actomyosin affects reactivity to cross-linking by microbial and fish transglutaminases. *Food Chem., 111*(2), 439–446.

Karayannakidis, P. D., Zotos, A., Petridis, D., & Taylor, K. A., (2008). Physicochemical changes of sardines (*Sardina pilchardus*) at–18°C and functional properties of kamaboko gels enhanced with Ca^{2+} ions and MTGase. *J. Food Process Eng., 31*(3), 372–397.

Kim, J. S., & Park, J. W., (2007). Mince from seafood processing by-product and surimi as food ingredients. In: *Maximizing the Value of Marine By-Products* (pp. 196–228). Woodhead Publishing.

King, F. J., (1973). Improving the supply of minced block for the fish stick trace, a progress report. *Mar. Fish. Rev., 38*(8), 26.
Knipe, C. L., Olson, D. G., & Rust, R. E., (1985). Effects of selected inorganic phosphates, phosphate levels and reduced-sodium chloride levels on protein solubility, stability and pH of meat emulsions. *J. Food Sci., 50*(4) 1010–1013.
Kuraishi, C., Sakamoto, J., Yamazaki, K., Susa, Y., Kuhara, C., & Soeda, T., (1997). Production of restructured meat using microbial transglutaminase without salt or cooking. *J. Food Sci., 62*(3), 488–490.
Lanier, T. C., Carvajal, P., & Yongsawatdigul, J., (2005). *Surimi Gelation Chemistry, Surimi and Surimi Seafood* (pp. 435–489). CRC Press, Canada.
Martín-Sánchez, A. M., Navarro, C., Pérez-Álvarez, J. A., & Kuri, V., (2009). Alternatives for efficient and sustainable production of surimi: A review. *Compr. Rev. Food Sci. Food Saf., 8*(4), 359–374.
Moreno, H. M., Carballo, J., & Borderías, A. J., (2008). Influence of alginate and microbial transglutaminase as binding ingredients on restructured fish muscle processed at low temperature. *J. Sci. Food Agric., 88*(9), 1529–1536.
Moreno, H. M., Carballo, J., & Borderías, A. J., (2010). Use of microbial transglutaminase and sodium alginate in the preparation of restructured fish models using cold gelation: Effect of frozen storage. *Inno. Food Sci. Emerg. Technol., 11*(2), 394–400.
Moreno, H. M., Carballo, J., & Javier, B. A., (2009). Study of two different cold restructuring processes using two different qualities of hake (*Merluccius capensis*) muscle, with addition of microbial transglutaminase. *J. Sci. Food Agric., 89*(8), 1346–1351.
Moreno, H. M., Herranz, B., Pérez-Mateos, M., Sánchez-Alonso, I., & Borderías, J. A., (2016). New alternatives in seafood restructured products. *Crit. Rev. Food Sci. Nutri., 56*(2), 237–248.
Morris, E. R., Rees, D. A., Thom, D., & Boyd, J., (1978). Chiroptical and stoichiometric evidence of a specific, primary dimerization process in alginate gelation. *Carbohydr. Res., 66*(1), 145–154.
Motoki, M., & Seguro, K., (1998). Transglutaminase and its use for food processing. *Trends Food Sci. Technol., 9*(5), 204–210.
Nakauma, M., Funami, T., Fang, Y., Nishinari, K., Draget, K. I., & Phillips, G. O., (2016). Calcium-binding and calcium-induced gelation of sodium alginate modified by low molecular-weight polyuronate. *Food Hydrocoll., 55*, 65–76.
Okada, M., (1963). Studies of elastic property of kamaboko. *Bull. Tokai Reg. Fish. Res., 36*, 21–25.
Pacheco-Aguilar, R., Ocaño-Higuera, V. M., Ezquerra-Brauer, J. M., Castillo-Yañez, F. J., García-Sánchez, G., & Marquez-Rios, E., (2010). Partial characterization of 5′-nucleotidase from giant squid (Dosidicusgigas) mantle Caracterizaciónparcial de la enzima 5′-nucleotidasa del manto de calamar gigante (Dosidicusgigas). *CyTA–Journal of Food, 8*(1), 65–71.
Pearson, A. M., & Gillett, T. A., (1996). *Processed Meats* (3rd edn., pp. 417–437). Chapman & Hall, New York, USA.
Pérez-Mateos, M., Lourenço, H., Montero, P., & Borderías, A. J., (1997). Rheological and biochemical characteristics of high-pressure-and heat-induced gels from blue whiting (*Micromesistius poutassou*) muscle proteins. *J. Agric. Food Chem., 45*(1), 44–49.
Pratt, D. E., & Paula, M. B., (1979). "Source of antioxidant activity of soybeans and soy products." *J. Food Sci., 44*(6),1720–1722.

Qin, Y., (2018). Seaweed hydrocolloids as thickening, gelling, and emulsifying agents in functional food products. In: *Bioactive Seaweeds for Food Applications* (pp. 135–152). Academic Press.

Rajesh, A., (2018). *India's Seafood Export Cross USD 7 Billion for the First Time* (p. 1). The New Indian Express, Chennai.

Ramírez, J. A., Uresti, R. M., Velazquez, G., & Vázquez, M., (2011). Food hydrocolloids as additives to improve the mechanical and functional properties of fish products: A review. *Food Hydrocoll., 25*(8), 1842–1852.

Ravishankar, C. N., Setty, T. M. R., & Shetty, T. S., (1992). Method for the preparation of sausages of acceptable quality from Indian oil sardine (*Sardinella longiceps*) and their shelf-life at different storage temperatures. *Food Control, 3*(3), 144–148.

Remya, S., Basu, S., Venkateshwarlu, G., & Mohan, C. O., (2015). Quality of shrimp analogue product as affected by addition of modified potato starch. *J. Food Sci. Technol., 52*(7), 4432–4440.

Renkema, J. M. S., & Ton, V. V., (2002). Heat-induced gel formation by soy proteins at neutral pH. *J. Agric. Food Chem., 50*(6), 1569–1573.

Roopa, B. S., & Bhattacharya, S., (2008). Alginate gels: I. Characterization of textural attributes. *J. Food Engg., 85*(1), 123–131.

Ruusunen, M., & Puolanne, E., (2005). Reducing sodium intake from meat products. *Meat Sci., 70*(3), 531–541.

Saha, D., & Bhattacharya, S., (2010). Hydrocolloids as thickening and gelling agents in food: A critical review. *J. Food Sci. Technol., 47*(6), 587–597.

Schmidt, G. R., (1986). Processing and fabrication. In: Bechtel, P. J., (ed.), *Muscle as Food* (p. 201). Academic Press, Orlando, FL.

Shahidi, F., & Synowiecki, J., (1997). Protein hydrolyzates from seal meat as phosphate alternatives in food processing applications. *Food Che., 60*(1), 29–32.

Soultani, G., Evageliou, V., Koutelidakis, A. E., Kapsokefalou, M., & Komaitis, M., (2014). The effect of pectin and other constituents on the antioxidant activity of tea. *Food Hydrocoll., 35*, 727–732.

Sun, X. D., (2009). Utilization of restructuring technology in the production of meat products: A review. *CyTA–J. Food, 7*(2), 153–162.

Taheri, A., & Jafari, S. M., (2019). Nanostructures of gums for encapsulation of food ingredients. In: *Biopolymer Nanostructures for Food Encapsulation Purposes* (pp. 521–578). Academic Press, Orlando, FL.

Trout, G. R., & Schmidt, G. R., (1984). Effect of phosphate type and concentration, salt level and method of preparation on binding in restructured beef rolls. *J. Food Sci., 49*(3), 687–694.

Trout, G. R., & Schmidt, G. R., (1986). Effect of phosphates on the functional properties of restructured beef rolls: The role of pH, ionic strength, and phosphate type. *J. Food Sci., 51*(6), 1416–1423.

Tsao, C. Y., Kao, Y. C., Hsieh, J. F., & Jiang, S. T., (2002). Use of soy protein and microbial transglutaminase as a binder in low-sodium restructured meats. *J. Food Sci., 67*(9), 3502–3506.

Yongsawatdigul, J., & Park, J. W., (1996). Linear heating rate affects gelation of Alaska pollock and Pacific whiting surimi. *J. Food Sci., 61*(1), 149–153.

CHAPTER 6

Hydrocolloids: Application in Fish Processing to Improve Mechanical and Functional Properties of Fish Products

RANENDRA K. MAJUMDER

College of Fisheries, Central Agricultural University (Imphal), Lembucherra – 799210, Tripura, India

ABSTRACT

In the present-day food processing industry, hydrocolloids are given high weightage as novel additive in the development of high-value food products. Hydrocolloids are a diverse group of long-chain polysaccharides and popularly known as gums, which are dispersible in water and form viscous dispersions and/or gels, the characteristic attribute of each hydrocolloid. Hydrocolloids alter the physical properties of the solution forming gels, and also enable thickening, emulsification, coating, and stabilization. The unique functional properties of each hydrocolloid is mainly governed by its conformation, shape, and molecular size. The properties of hydrocolloids conducive for application in fish products include solubility, viscosity, gel-forming ability, emulsifying capacity and surface activity. Protein-polysaccharides interactions play a key role in facilitating the structure and stability of many processed food products including from fish where proteins are the main ingredients. Gums are common hydrocolloids used for improving the mechanical attributes of restructured products due to their gelling, emulsifying, thickening, and stabilizing properties. Konjac is recognized as a globally accepted gel forming agent and thickener. Deacetylated konjac glucomannan in addition to form extremely stable gel, also has affinity for the salts of iron, zinc, calcium, and magnesium. Carrageenan finds successful

Advances in Fish Processing Technologies: Preservation, Waste Utilization, and Safety Assurance.
Ranendra K. Majumder, PhD & Amjad K. Balange, PhD (Eds.)

application in meat products such as cooked ham for stabilizing the injected water. Chitosan is a water-soluble polysaccharide which carry a positive charge, by, and large used as a dietary fiber as human cannot digest it. The ultimate aim for functional food development includes defense against reactive oxidative species (ROS), cardiovascular functions, gastrointestinal physiology and functions, etc., and all these expected benefits from food could be addressed by hydrocolloids.

6.1 INTRODUCTION

An increase in fish production is expected during the next several years as a consequence of a rise in demand for fisheries products with concomitant development of culture and capture fisheries as well. Numerous studies have clearly highlighted nutritional benefits of fish consumption, with significant contribution towards well-recognized health benefits attributed to fish ingredients, particularly proteins and lipids. Consequently, there was a growing interest in increasing the use of fish proteins as a good ingredient due to their functional and nutritional properties. However, the growing consumers' demand for fish and fish products is affecting the fishery resources worldwide, which is revealed by the drastic reduction of the stocks of many most valued food fishes. Whereas, at the same time, there are several underutilized fish species due to their size, flavor, odor, color, or texture, their value addition to develop new products could greatly contribute to the economy of the sector. As estimated by FAO in the year 2016, processed seafood accounts for 92% and 53% of total seafood used for food purposes in the developed and developing countries respectively. Initially, the seafood processing methods were designed to prolong shelf-life and ensure food safety, but presently, they also address the creation of new fish products through innovative ideas to make it more palatable. Currently, one of the challenges faced by the fish processing sector is the development of newer technologies to create high value fish products with superior functional properties. For example, transformation of fileting by-products from the surimi plants to high value products would significantly strengthen the economy of the sector. Although, to obtain novel products with improved mechanical and functional properties, the application of different additives plays a key role. One such novel additive is hydrocolloids, whose application in the development of high value food products is most significant in the present day food processing industry. Hydrocolloids or popularly known as gums are a diverse group of long-chain polymers (polysaccharides), which are dispersible in water forming viscous dispersions and/or gels, characteristic property of each hydrocolloid. As reported, hydrocolloids were first isolated from the exudates of trees or bushes, different plants or seaweeds, flours from seeds or grains, gummy slimes of fermented foods, and many other natural products. Presently several hydrocolloids are cataloged as permitted food additives in many countries throughout the world.

6.2 WHAT IS HYDROCOLLOID?

A colloid is a homogeneous non-crystalline substance and consists of ultramicroscopic particles of one substance dispersed through a second substance. This is an intermediate state between a solution and a suspension, such as like gels, sols, and emulsions. The term ‘hydrocolloid’ describes a group of hydrophilic polymers (in the case of food polysaccharides and to some extent, proteins, notably gelatin), which can be dispersed in water. Although, the quantity of water available decides the different states, e.g., gel or sol (liquid). There are two states of hydrocolloids, either irreversible (single-state) or reversible. Agar is an example of reversible hydrocolloid from seaweed extract as it can exist alternately in gel or solid-state with addition or elimination of heat. The dispersion ability of the hydrocolloid compounds in water is due to the presence of a large number of hydroxyl groups which noticeably enhance their affinity for water molecules so as to make them hydrophilic compounds. Further, due to the property of dispersion, an intermediate state between a true solution and a suspension, the hydrocolloid compounds behave like a colloid and exhibit the properties of it.

6.2.1 ROLE OF HYDROCOLLOIDS IN IMPROVING FOOD PROPERTIES

In foods, a wide range of functional properties are exhibited by hydrocolloids, which include significant attributes like thickening, gelling, emulsifying, stabilization, coating, etc. Several hydrocolloids are capable of producing highly viscous solutions or gels with varying textures even at low concentrations (1% or less). Hydrocolloids are also used as suspension and emulsion stabilizers, mainly due to their thickening ability as they can retard particle sedimentation and droplet creaming because of bulk viscosity effects. Recently, there has been a growing popularity of application of hydrocolloids in several food formulations and this is primarily because of their capability to modify the food system's rheology. On the basis of the states of matter in the continuous and dispersed phases, surface properties of the food systems can be categorized into four groups—sol, gel, emulsion, and foam. Sensory properties of any food are governed mainly by its texture and/or viscosity. Hydrocolloids are used as food additives to perform specific purposes and usually find applications in many food formulations, namely, soups, gravies, salad dressings, sauces, and toppings, etc., to achieve preferred viscosity and mouthfeel. Whereas, for some products like ice creams, jams, jellies, gelled desserts, cakes, and candies, they are used to achieve desired texture. Hydrocolloids have another functional property, and this is contributed by its dietary fiber aspect, as it performs several physiological functions which are beneficial for human.

6.2.2 FUNCTIONAL PROPERTIES OF HYDROCOLLOIDS

For about hundreds of years, hydrocolloids have been widely studied by mankind for the functional and nutritional perspective. However, a lot about the role of hydrocolloids in relation to the final product quality is yet to be learnt. Quality of food products specifically depends on the balance of interactions among different kinds of ingredients, including hydrocolloids. Presently, various techniques are available to assess the interactions of the food ingredients up to the molecular and supramolecular levels including phase-separated systems and gel networks. Since hydrocolloids belong to a structurally multifarious class with different levels of solubility, they can modify the physical properties of the solution such as gel formation, enable thickening, emulsification, coating, and stabilization. Gelatin, although not belonging to the polysaccharide group, but due to its excellent hydrophilicity and polydispersity; it is ascribed as hydrocolloid. Whereas, other food proteins such as whey proteins, although traditionally not classified as hydrocolloids, but are included as they exhibit aggregation and gelation behavior similar to those of polysaccharides to some extent. The modification of major properties of a food system such as texture, viscosity, and surface activity which determines appearance, taste, mouthfeelness, etc., constitutes numerous functions of hydrocolloids in foods. Each hydrocolloid has its own unique functional properties, which is governed by its chemical structure, molecular size, and shape.

6.2.2.1 SOLUBILITY

Hydrocolloids are necessarily to be dissolved completely in aqueous solution to exhibit their function in the food systems. Therefore, a knowledge on the solubility of polysaccharides is very essential for their applications. The solubility of hydrocolloids is an outcome of a continuous hydration process, which gradually converts intermolecular binding of the hydrocolloids to molecule–water binding. Complete dissolution of the hydrocolloids in the food system is influenced by the factors such as particle size, temperature, and cations. Generally, finely powdered hydrocolloids (particle size higher than 100 mesh) dissolve fast compared to coarse hydrocolloids (mesh size less than 60) as a longer time for water penetration is required in case of later. Although, occasionally fine particle size hydrocolloids may lump together to form a particle with a hydrated surface which remain as undissolved powder in the center. In case of industrial applications, such situations are solved by employing high agitation while mixing or addition of another dry ingredient such as sugar or salt.

6.2.2.2 VISCOSITY

The thickening property of hydrocolloids is contributed in the food systems due to their viscosity. Viscosity is defined as the internal friction of a liquid or, in other words, a liquid's tendency to resist flow. In the systems of colloidal suspensions, enhancement of viscosity is the result of thickening of the liquid phase and attributable to the liquid absorption with resultant swelling of the dispersed colloid. Hydrocolloids exhibit the property of viscosity due to non-specific intermolecular entanglement of polymer chains which are conformationally disordered and thereby resulting in a resistance to flow under shearing forces. Viscosity is essentially a polymer-solvent interaction. However, the thickening effect exhibited by the hydrocolloids depends on the factors like, type of hydrocolloid and its concentration, pH, and temperature of the food system and finally the kind of food systems. The factors which are responsible for achieving viscosity of the hydrocolloid system includes physical factors such as concentration, temperature, solvation, electrical charge, degree of dispersion, previous thermal and mechanical treatment; and chemical factors such as presence or absence of other lyophilic colloids and presence of both electrolytes and non-electrolytes. The viscosity is also influenced significantly by the molecular mass of the polymer. Moreover, the molecular structure of the polymer significantly influence its hydrodynamic volume in solution. Generally, linear and rigid molecules exhibit much higher viscosity compared to the highly branched and highly flexible polymers of the same molecular mass, due to the larger hydrodynamic volume of the first one.

6.2.2.3 GEL FORMING ABILITY

In food applications, one of the most desired attributes is attractive soft solid texture, and this can be achieved by using hydrocolloids due to their swollen particulate forms in gelled state. Gelling property is a phenomenon where polymer molecules cross-link, tangle, and get interconnected, resulting in a three-dimensional molecular network in an aqueous medium in a food system wherein water, solutes, and pigments remain imbibed within it. Hydrocolloids form gels through cross-linking of polymer chains which is resulted due to their physical association through hydrogen bonding, hydrophobic interaction, and cation mediated cross-linking, and are often referred to as physical gels. Whereas, in synthetic polymer gels, polymer chains are covalently cross-linked. The essential part of the gelation process is the formation of junction zones, which is a three-dimensional network formed by cross-linking of two or more polymer chains. Gels differ in their thermal behavior because of the junction zones. The extent of physical interaction of the polymers is determined by several parameters such as temperature, presence of ions, and the inherent structure of the hydrocolloid. There are three different ways by which gelation can be induced, which are ionotropic gelation, cold-set gelation, and heat-set gelation. In ionotropic gelation, the by and large negatively charged hydrocolloids form gel in the presence of cations, e.g., alginate, carrageenan, and low-methoxy pectin with Ca^{2+}. In cold-set gelation, in order to achieve full dispersion and hydration, well-powdered hydrocolloids are dissolved in warm or boiling water, enabling the formation of interchain association upon cooling, resulting in a three-dimensional network. Although, most of the hydrocolloids including agar and gelatin form gels by this mechanism. For heat-set gelation, hydrocolloids need to be heated first for inducing gel formation. Examples are, curdlan, konjac glucomannan, methylcellulose, hydroxypropyl methylcellulose, starch, and globular proteins. Heating induces unfolding of their molecular structures, followed by rearrangement into a three-dimensional network.

6.2.2.4 SURFACE ACTIVITY AND EMULSIFYING PROPERTIES

An emulsifier or emulsion stabilizer functions by retarding the process of precipitation of dispersed solid particles. As a result, the creaming rates of oil droplets and foams decrease resulting prevention of the coalescence of oil droplets. These ultimately prevent aggregation of dispersed particles with resultant prevention of syneresis of gelled systems containing oils. The mostly studied hydrocolloid as emulsifying agent is probably the gum Arabic. It has significant surface activity, promoting adsorbing onto oil-water interfaces to impart steric stabilization by creating repulsive forces. Other polysaccharides such as galactomannans, xanthan, pectin, etc., reduce surface and interfacial tensions, adsorbing onto solid surfaces resulting increase of repulsive forces leading to improved stability of oil-in-water emulsions. Recently, emulsifying property of pectin (a class of hydrocolloid) has attracted attention. The sugar beet pectin is considered to be highly surface-active compared to gum Arabic and also found effective in stabilizing fine orange oil or triglyceride oil-based emulsions at a pectin/oil ratio of 1:10. This may be due to its higher protein content.

6.2.2.5 PROPERTIES TO FORM EDIBLE FILMS (EFS) AND COATINGS

Edible films (EFs) are used as coating of food wherein it provides a barrier between food and the surrounding environment, and most significantly, they can be consumed along with the food. Casing of sausage from meat is one familiar example. Recent, there has been a trend to use many gums and their derivatives such as alginate, carrageenan, pectin, cellulose, and starch as EFs to provide a barrier against migration of moisture, gas, aroma, and lipid from food. The hydrophilic nature of the hydrocolloids has inherent moisture barrier properties. Although an inverse relationship between water vapor and oxygen permeability has been reported in some cases, such films were found effective in preventing oxidative deterioration in some lipid and other susceptible food ingredients.

6.2.2.6 HYDROCOLLOIDS AS FAT REPLACERS

Due to growing awareness about healthy diet, high value health beneficial foods can be developed by replacement of fats and oils with 'structured water' to give high fiber, low-fat, and low-calorie foods with admirable eating quality. So far numerous hydrocolloid-based products have been developed and intended for the replacement of fat in food, and thereby creating an increased demand for hydrocolloids in the food sector. In some traditional practices, starches are used as fat replacers, which form stable thermo-reversible soft and fat-like spreadable gels when dissolved in water. Further, through maximization of the synergies of hydrocolloids in combination with specific starch, almost 100% fat replacement is possible. Presently the fat replacers are occasionally referred to as fat mimetics, as these are tailored to imitate more closely to the properties of fats or oils for any particular application. Inulin, due to its particle gel characteristics functions as a fat replacer only in water-based food systems.

6.2.3 MOLECULAR STRUCTURES OF HYDROCOLLOIDS AND STRUCTURE-FUNCTION RELATIONSHIP

One of the most critical factors for the solubility, viscosity, and gel-forming ability of polysaccharides is their molecular weight as well as molecular weight distribution. Molecules with high molecular weight have a large excluded volume, results intermolecular interaction of polysaccharides leading to reduced solubility. Solubility is inversely related with the molecular weight. On the other hand, polysaccharides with high molecular weights promote higher viscosities and exhibit intermolecular associations under the same concentration. Molecular weight plays a key role in gelation process. In the gelation process, however, intermolecular associations of polysaccharides are important and these are stable only when the length of the molecular chain is long enough. On the contrary, in the case of some neutral polysaccharides such as (1→3) (1→4)-β-D-glucans, the stronger gels are formed by lower molecular weight samples compared to the higher molecular weight samples. This indicates that to some extent, the gelation rate and molecular weight of the polysaccharide is inversely proportional. Polysaccharides with low molecular weight exhibit higher mobility in the solution, probably resulting in more interactions between the molecules leading to the formation of ordered junction zones.

The polysaccharides having branches or charged groups, e.g., carboxylate, sulfate, or phosphate groups, intermolecular association is obstructed resulting higher solubility; on the contrary, structural characteristics such as linear chains, large molecular weight, and other regular conformations that support intermolecular association lead to poor solubility. The charged groups in the polysaccharides affect their solubility in two ways: (1) by increasing the affinity to water at molecular level, and (2) by producing electrostatic effects exerted by the charged group to disrupt the intermolecular associations.

The charged polysaccharides may undergo chain extension due to intramolecular repulsive activity of the charged group (e.g., carboxylic groups), may exhibit relatively higher viscosity. Normally, with the decreasing pH of the solution viscosity increase due to intermolecular association of the negatively charged polysaccharides. Due to the same reason, sometimes gel formation could also be induced.

It can be summarized that the polysaccharides' key functionalities like solubility, viscosity, and gel-forming ability is decided by the balance between the interactions such as molecule-molecule and molecule-water. The dilute solutions of low molecular weight polysaccharides exhibit low viscosity due to the dominancy of the interactions between polysaccharides and water molecules. However, as the polysaccharides achieve the critical concentration or the M_w increases, viscosity shows a sharp increase due to high intermolecular associations, and the solutions exhibit pseudoplastic flow behavior. In the case of some specific hydrocolloids, changes of temperature, pH or addition of other solutes increase the interactions between the polysaccharides resulting in ordered molecular structures, thereby inducing gelation.

6.2.4 CLASSIFICATION OF THE HYDROCOLLOIDS

On the basis of their origin and 53 methods of manufacturing, hydrocolloids are generally classified into four different groups:

- Hydrocolloids isolated from plants (without chemical modification);
- Hydrocolloids obtained by fermentation;
- Plant-derived hydrocolloids (chemically modified); and
- Hydrocolloids obtained from animals.

The common hydrocolloids are listed in Table 6.1 according to their source and chemical structure. The world hydrocolloid market is dominated by starches (~70%) followed by gelatin (~12%), pectin (~5%), carrageenan (~5%) and xanthan gum (~4%). Others include locust bean gum, alginates, CMC, guar gum, etc. Generally recognized as safe (GRAS) hydrocolloids include alginic acid (CFR No. 184.1011), agar-agar (CFR No. 184.1115), ammonium alginate (CFR No. 184.1133), calcium alginate (CFR No. 184.1187), guar gum (CFR No. 184.1339), locust bean gum (CFR No.

184.1343), pectins (CFR No. 184.1588), potassium alginate (CFR No. 184.1610) and sodium alginate (CFR No. 184.1724).

TABLE 6.1 Classification of Hydrocolloids

Basis	Class	Example
Origin	Plant	Pectin, inulin, gum Arabic, gum ghatti, gum tragacanth, gum karaya, cassia seed gum, basil seed gum, mesquite seed gum, fenugreek gum, chicle gum, oat gum, rye gum, konjac, psyllium, guar gum, locust bean gum, flaxseed gum, wattle gum, starches
	Animal	Chitin, chitosan, gelatin
	Seaweed	Agar, carrageenan, alginic acid, alginate, furcellaran, ulvan, fucoidan, red alga xylan
	Microbial	Xanthan, gellan gum, tara gum, dextran, pullulan, welan gum, curdlan, levan
	Synthetic	Methylcellulose, methyl ethyl cellulose, carboxymethylcellulose, hydroxyethylcellulose, hydroxyl propyl cellulose, hydroxypropyl methylcellulose, microcrystalline cellulose
Structure	Glucan	Starch, oat gum, barley gum, curdlan, welan gum, pullulan, dextran
	Fructan	Inulin, levan
	Xylan	Red alga xylan
	Rhamnan	Ulvan
	Galactomannan	Guar gum, locust bean gum, tara gum, cassia seed gum, basil seed gum, mesquite seed gum, fenugreek gum
	Glucomannan	Konjac, alginate
	Arabinoxylan	Psyllium, flaxseed gum (containing another galacturonan fraction), rye gum, wheat gum
	Galactan	Agar, carrageenan, fucoidan, furcellaran
	Arabinogalactan	Gum Arabic
	Galacturonan	Pectin
	Glycanorhamno-galacturonan	Gum karaya, gum tragacanth (containing another arabinogalactan fraction)
	Glycanoglucuro-nomannoglycan	Gum ghatti
	Glucosamine polymer	Chitin, chitosan
	Protein	Gelatin

Source: Li and Nie (2015).

6.2.5 *HYDROCOLLOIDS APPROPRIATE FOR FISH PROCESSING INDUSTRY*

Protein-polysaccharides interactions play a significant role in the structure and stability of many processed food products including fish where proteins are the major component. Functional properties of fish proteins, such as solubility, surface activity, structural stability, gelling capacity, emulsifying, and foaming properties are usually affected in presence of polysaccharides. Salt induced partially solubilized fish proteins and water form a continuous matrix or gel. Polysaccharides in the form of additives can be trapped into the matrix to fill the gel that influences the formation of continuous protein gel matrix during thermally induced gelation. Added polysaccharides can modify the viscosity, mobility, and other associated properties of the liquid phase of the gel, including texture, rheology, appearance, and relative volume fraction of the gel.

6.2.5.1 *CARBOHYDRATES*

Gel forming capacity of the fish muscle protein plays a significant role for development of various surimi-based or restructured fish products. Several hydrocolloids belonging to carbohydrates, such as starch, carrageenan, and konjac are normally used to obtain improved mechanical properties of surimi gels. Foodgrade gums and starches promote the formation of the continuous matrix through their interaction with proteins and water in the fish mince. Some of the hydrocolloids, through their interaction with the fish proteins, form a more structured system, while others are trapped in the cross-linked proteins gel and bind water to modify the viscosity of the gel. Moreover, the addition of carbohydrates into the formulation would significantly improve the capacity of the salt to solubilize myofibrillar proteins, thereby modifying the mechanical and functional properties of gels. Although, there are some carbohydrates which are not compatible with the muscle proteins and exhibit a disruptive effect in the gelling system.

6.2.5.2 *STARCHES*

Starch serves as the most commonly used hydrocolloid thickening agent in products based on surimi or fish mince. The reason of its popularity may be it is relatively cheap, abundant, and most significantly it does not develop any perceptible taste in the product when used at a low concentration. In surimi-based fish products, starch @ 4 to 12% can control wetness, stickiness, and thermal stability during storage. It can also increase the firmness and gel strength of surimi at lower cost. Recently, there has been growing interest to identify both native and modified starches to improve the mechanical and functional attributes of restructured and surimi-based products. It has been reported that the native wheat starch-modified the rheological properties of kamaboko gels from sardine surimi, whereas, potato starch could slightly improve the gel strength of surimi-based seafood products. However, potato starch showed a detrimental effect on the whiteness of the gel compared to the wheat starch. Further, the acetylated rice starch @ 4 to 8% improved the freeze-thaw stability of surimi paste, but reduced the gel strength and expressible moisture content of the surimi gels. Although, starch is the first choice in the case of traditional fish burgers or sausages, however, attempt is going on to improve the functionality of starch as texture modifiers by adding small amounts of proteins and non-starch hydrocolloids for development of quality products from minced fish.

6.2.5.3 GUMS

Gums are common hydrocolloids used for improving the mechanical properties of restructured products owing to their gelling, emulsifying, thickening, and stabilizing properties. They are readily available, inexpensive, and most of them are compatible with fish muscle proteins and improve yield by increasing WHC, but without any significant effect on the color. Most importantly, gums do not add calories and have beneficial effects of adding fiber on diet. Amongst all gums, konjac, and carrageenan shows high positive effect on the texture of the fish muscle. Although, other gums such as locust bean and xanthan gum have also been reported to improve the mechanical properties of surimi gel from fish.

6.2.5.3.1 Konjac Glucomannan (KGM)

Konjac is recognized as a globally accepted gelling agent and thickener. These are linear polymers of both (1→4)-linked β-D-mannose and (1→4)-linked β-D-glucose residues. The ratio of mannose to glucose molecules is on an average 1.6:1. Although this is random, but small repeats of three to five mannose blocks are frequent. Acetyl groups are randomly spread and occur every 10–19 units (3–6%) on carbon atoms C2, C3, or C6 of mannopyranose, which are responsible to promote water solubility. They are obtained through drying and pulverizing the root of the perennial herb, *Amorphophallus konjac*. The high molecular weight of konjac glucomannan (> 300 kDa) allows it to form viscous pseudoplastic solutions. Konjac flours even solubilize in ice water and its solubility can be accelerated by applying high temperature and also by raising speed of agitation. The absorption capacity of konjac flours is so high that it can absorb about 15–20 g of water or about 1–2 g of oil per g. Prolonged exposure above 80°C degrade the gum, especially in acidic media. Although konjac exhibits high viscosity at 1% solution, but its viscosity is much less compared to guar or xanthan gum at lower concentrations. Deacetylated konjac glucomannan is capable of forming random junction zones through hydrogen bonds resulting in a gel which is extremely stable and irreversible. Although glucomannan exhibits affinity for iron, zinc, calcium, and magnesium salts, however, mineral intake is reduced at higher content of glucomannan because of physical entrapment.

It is reported to be an efficient cryoprotective agent as it can decrease the denaturation/aggregation of myofibrillar proteins of fish during frozen storage. At lower concentration of about 1%, it can exhibit cryoprotective activity similar to the conventional cryoprotectants, while at 1.5 to 2% level, it can increase WHC and gel strength of the surimi, even though the color become darker. Konjac is also used to promote adhesion for coatings and bindings in the matrix of restructured meat products. It is often blended with xanthan, carrageenan, or starches to provide better result. In the intestine, as the konjac glucomannan is degraded by the enzymes such as β-D-glucanase and β-D-mannanase, there is the possibility of bacterial fermentation to occur. Konjac is capable of acting as a prebiotic in the human colon. Its low energy content is also able to reduce appetite. Moreover, amongst all gums, konjac has the ability to lower cholesterol levels more effectively. It can also regulate blood sugar level. Konjac can also act synergistically with stanols and sterols in order to reduce harmful low density lipoprotein (LDL).

6.2.5.3.2 Carrageenans

Carrageenans belong to a family of linear, negatively charged sulfated ($-SO_3^-$) galactan (repeating units of galactose and 3,6-anhydrogalactose) polysaccharides, extracted from edible red seaweeds (family Rhodophyceae). Three types of carrageenan are commercially available, namely, κ-carrageenan (kappa), λ-carrageenan (lambda), and ι-carrageenan (iota). Their structural differences are exhibited by the number and position of sulfate groups present in the repeating galactan units of the polysaccharide. Presence of sulfate groups per galactan unit of the polysaccharide determines the gel strength. λ-carrageenan with three sulfate groups per galactan unit exhibit the property of gel inhibition, and that results in good water solubility and is known as nongelling polysaccharide. Whereas, ι-carrageenan (2 sulfate groups per galactan) and κ-carrageenan (1 sulfate group per repeating unit) form soft and rigid gel, respectively. The annual global production of carrageenan is about 50,000 MT, and about 70–80% of it is used in the food industry.

Solutions of carrageenan are highly viscous, but with the increasing temperature viscosity is reduced. However, with the application of heat in neutral systems (pH 7–9), viscosity drop is reversible. At room temperature, viscosity of carrageenan is stable over a wide range of pH. However, their gelling abilities are affected by cations such as K^+, Na^+, and Ca^{2+}. κ- and ι-carrageenans is capable to form gels in the presence of K^+ and Ca^{2+} ions, and such ability forms the basis of their many applications in food. In meat products such as cooked ham, carrageenan is used successfully for stabilization of the injected water. In case of carrageenan-induced fish protein system, fish protein and the carrageenan competes for the added salt with resultant increase of water-holding capacity during heating. Moreover, the uniform distribution of κ-carrageenan concentrate the fish protein complex, resulting in a stronger protein-hydrocolloid network in the presence of KCl, leading to an increased gelling ability. The addition of κ-carrageenan has been reported to cause an increase in the whiteness of the surimi gels. Carrageenan neither have any nutritional value nor assimilated by the human body, and provide only fiber. However, it is claimed that carrageenan can cause allergies by obstructing the absorption of potassium.

6.2.5.3.3 Xanthan Gum

The extracellular polysaccharide xanthan gum is obtained through aerobic fermentation of sugar by bacterium *Xanthomonas campestris*. Structurally, xanthan is constituted by a cellulosic backbone of β-(1→4) linked D-glucose units and a trisaccharide side chain, composed of two mannose units separated by a glucuronic acid. Molecular ratio of D-glucose, D-mannose, and D-glucuronic acid in xanthan is approximately 3: 3: 2. Further, the terminal mannose units are linked to a pyruvate group, and the non-terminal residue usually carries an acetyl group. The presence of carboxyl groups on the side chains renders the anionic property of gum molecules. The average molecular weight of xanthan gum is about 2,000 kDa. Unlike most polysaccharides, there is narrow molecular-weight distribution.

Xanthan is a non-gelling polysaccharide, capable of forming highly viscous solutions at low concentrations. Xanthan can exist in solution in a rigid and ordered chain conformation. One of the reasons of huge acceptance of xanthan to the food industry is the stability of its solution which is not affected by temperature or pH. Strong synergies exist between xanthan gum and certain galactomannans (e.g., locust bean gum) or konjac glucomannan, leading to form a thermo-reversible gel. Even though in the presence of salts and/or acids, xanthan solutions also show exceptional stability while heating. With the increase of temperature, although the viscosity of the solution decreases, but complete recovery takes place upon cooling. The hydrated xanthan can tolerate up to 20–30% salt without any change in the viscosity.

It has been reported that xanthan alone have a negative effect on the mechanical properties of surimi gel as it is an anionic gum. When fish meat is processed as surimi, the pH of the myofibrillar proteins remain above the isoelectric point, and therefore, exhibit antagonistic effect due to repulsive action of both the anionic macromolecules. Although, xanthan gum exhibited the gelling of surimi upon addition of high levels of calcium (0.4%).

6.2.5.3.4 Locust Bean Gum

Locust bean gum (LBG) or Carob gum is finely ground endosperm of the seeds from the carob tree (*Ceratonia siliqua*, family Leguminosae). Chemically, it is a neutral polysaccharide having D-mannose and D-galactose in a molecular ratio on an average of 3.5: 1. Structurally, LBG is having a linear backbone of β-(1→4)-linked mannose units, in which nearly every fourth mannose molecule is linked to a galactose residue via an α-(1→6)-linkage. The average molecular weight of LBG is approx. 310 kDa. Due to lack of adequate number of side chains, LBG is partially soluble in cold water, but heating over 80°C, it forms pseudoplastic solutions and the viscosity is stable at a pH range of 3–11.

The important property of the LBG is that there occurs a synergistic interaction with κ-carrageenan and xanthan resulting gels with good cohesiveness. LBG acts as a filler ingredient when added alone in the surimi gel, where it is entrapped to form a continuous surimi gel matrix and modify the physicochemical properties of the gel's aqueous phase thus influencing texture of the final product. Synergistic effect of xanthan-LBG (ratio 0.25/0.75) with addition of calcium (0.4%) exhibited improved mechanical properties of surimi gel from Silver carp. LBG is a globally approved food additive without any negative effect and allergenic reactions in the human digestive tract.

6.2.5.3.5 Pectins

Pectin is an anionic hydrocolloid and consist of a linear backbone of α-(1→4)-linked galacturonic acid units with different degree of methoxylation. The gel-forming ability seems to be the most important property of pectin. Gels are normally formed at concentrations of 1 g/kg. The principal factor behind the gelling ability of pectin is diminished electrostatic repulsion due to the presence of hydrogen bonds between carboxyl groups (non-dissociated) and secondary alcohol groups, as well as hydrophobic interactions between the methyl ester groups. Commercially two types of pectins are available, such as HM pectins (high mol. wt.) and LM pectins (low mol. wt.). Interaction between HM pectins and proteins takes place above the isoelectric point and requires 65% sugar and pH <3.2 for gelation. The high solute content results in a low water activity, create hydrophobic environment leading to minimize the pectin-solvent interactions, which induces pectin-pectin interactions. Whereas, LM pectins are able to form gels even in the presence of calcium and over a wide range of pHs with or without sugar. It has been reported that Amidated pectins (ALM) improve mechanical properties of surimi and restructured fish products, as compared to LM pectins. It is prepared with ammonia in alcoholic suspensions of pectin. The H-bondings between ALM pectin and myofibrillar proteins improve the mechanical properties, specifically hardness and breaking force of fish gels. Although, higher concentrations of ALM pectin (20–50 g/kg) could result in detrimental effect on mechanical properties due to swelling of the hydrocolloid or changes in the native conformation of muscle proteins by pectin-protein interactions, with the presence of both hydrogen bonds and electrostatic interactions. ALM pectins have been reported to perform better under acidic conditions (pH 3.2–3.6), while non-amidated LM pectin works in the pH range of 2.8–6.5. As a soluble dietary fiber pectin serves as a natural part of the human diet, without any significant contribution to nutrition. Consumption of pectin can reduce the level of blood cholesterol. Additional favorable physiological effects exerted by pectin include improvement in mineral absorption, reduction of gastric emptying time, as well as anti-diarrhea effects.

6.2.5.3.6 Fiber

Modern diet is formulated with deliberate incorporation of dietary fiber as several health beneficial effects beyond bowel regularity such as promotion of digestive and cardiovascular health, weight management including general wellness are provided by them. It has been reported that addition of dietary fiber concentrate from white and red grape @ 20–40 g/kg to mince from horse mackerel (*Trachurus trachurus*) improved water and oil retention resulting more cooking yield and reduced lipid peroxidation during frozen storage at –20°C for 6 months without alteration of sensory acceptance. Adding wheat fiber of different particle size decreased the gel strength, cohesiveness, and water binding capacity of surimi gel, however, fiber with large particle sizes increased gel strength and hardness during freezing. Pea fiber has been added to minced fish and surimi up to 4% to produce restructured products without modifying the textural and functional properties. Further, it was also found to be compatible with carrageenan (2%) to boost the hardness of the restructured products from hake. Pea fiber with microbial transglutaminase (MTGase) @ 1% or above improved the textural properties of surimi gels from Atlantic mackerel (*Scomber scombrus*) and chub mackerel (*Scomber japonicus*).

6.2.5.3.7 Chitosan

Chitosan is obtained by deacetylation of chitin (the major component of the exoskeleton of insects and crustaceans) by treating with strong alkali. It is composed of (1→4)-linked 2-amino-2-deoxy-β-D-glucopyranosyl residues. Chitosan is a water-soluble polysaccharide, carry a positive charge and can be used in food as a dietary fiber, as it is not digested by humans. Chitosan as hydrocolloid is gaining wide popularity due mainly for its non-toxicity, biocompatibility, and biodegradability. Addition of chitosan @ 1–1.5% to improve gel strength of the surimi, showed promising results. The activity of chitosan in gel formation might be mediated by the presence of endogenous transglutaminase (TGase) in the presence of calcium ions. Although the exact mechanism by which chitosan improves gel's texture is not clear till date, however, the participation of H-bonding, hydrophobic, and electrostatic interactions might be involved to induce formation of cross-linked heavy chains of myosin through polymerization by endogenous enzymes during the setting process.

Another significant contribution of chitosan is its antioxidative property, which is comparable with the conventional antioxidants. The antioxidative function of chitosan when added to fish muscle is dependent on both of its molecular weight and concentration in the product. Chitosan with lower molecular weight (around 30 kDa) exhibits higher radical scavenging activity. The scavenging ability of chitosan is mainly due to presence of active amino and hydroxyl groups in the polymer. The compact structure of chitosan at higher molecular weight has strong intramolecular H-bonding, which is responsible for weakening the behavior of the amino and hydroxyl groups, resulting in less radical scavenging activity. The ion scavenging ability of chitosan is strongly affected by its degree of acetylation. Studies revealed that highly acetylated chitosan exerts very little chelating activity. It has also been reported that chitosan can be used effectively to extend the shelf-life of restructured fish products under chilled storage.

6.2.5.3.8 Alginates

Alginates or alginic acids are linear polyuronide, with the cations of Na, K, NH_4, or Ca and produced directly from brown seaweed (Phaeophyceae). They belong to the family of unbranched binary copolymers of (1→4)-linked β-D-mannuronic acid (M) and α-L-guluronic acid (G) residues. The M and G residues are present in the ratio of M:G=0.4–1.9: 1, which vary widely in their composition and sequences. The average molecular weight ranges from 32 and 200 kDa. During cold processing, alginates are able to form heat-stable gels. Gelling takes place in the presence of acids and all multivalent cations, except magnesium. In food industry, alginates are used to render emulsification and gelation as well as to serve as coating and thickening agents. In case of fish processing, alginates are used to develop restructured fish products by the cold binding technique. The cold binding efficiency of sodium alginate can be improved by adding $CaCl_2$ (0.1%), whereas binding capability is reduced at a higher concentration (1%). Addition of alginate to chicken myosin gels @ 0.5% resulted increase of the WHC up to 40% and that might be due to greater entrapment of water by the protein-polysaccharide matrix. In the GI tract, alginates are neither metabolized nor resorbed, however, they can minimize the rate of resorption of cationic minerals like calcium and iron. So far, many countries approved alginates (especially alginic acid, sodium, potassium, calcium, and ammonium alginate salts and propylene glycol alginate) for use in foods and are considered as GRAS.

6.3 HEALTH BENEFITS OF HYDROCOLLOIDS

Overconsumption of high fat or high calorie foods has caused an extremely alarming situation in the growth of chronic diseases, especially cardiovascular disease, diabetes mellitus and cancer. According to the report of FAO/WHO, the excessive consumption of carbohydrates, especially sugar need to be decreased and people should develop habit to increase dietary fiber consumption. Several studies on the role of a high-fiber diet on the incidence of arterial disease increased awareness and prompted to pay more attention to the nutritional values of hydrocolloids. Recently, food scientists are very much curious to reveal the nutritive and physiological effects of hydrocolloids. Hydrocolloids can be used as fibers with specific health benefits. Additional health benefits from the consumption of hydrocolloids include appetite regulation, proper bowel function, reduction of osteoporosis risk, and prevention of coronary heart diseases, type-2 diabetes mellitus and colon cancer.

The dietary fiber improves control of blood glucose by trapping ingested carbohydrates from the viscous gel formed in the intestine after digestion, which results in slow vascular absorption of glucose.

Although high-viscosity fibers help to control blood glucose levels, but they reduce the palatability of products. Insoluble fiber promotes movement of foodstuff through the GI tract, resulting in decreased absorption of nutrients (such as glucose). In addition, insoluble fiber can also reduce appetite and food intake, and moreover the short-chain fatty acids (FAs) formed during fermentation reduce the postprandial glucose response. Hydrocolloids play a significant role in the modulation of the activity of the colon. Many reports suggested that polysaccharides with high arabinose and galactose content exhibit immunomodulatory functionalities, including complement-fixing activities and/or macrophage function modulation. Hydrocolloids can modulate the passage rate of foods and enzyme digestion of starch, proteins, and lipids in the small intestine, thereby helping to reduce obesity. Carrageenan-enriched diets are shown to have beneficial effect to maintain lipid balance so as to lower the risk of cardiovascular disease. Another study has revealed that alginate is able to exert cardiovascular-beneficial effects indirectly through modulation of glucose and cholesterol uptake in the small intestine.

6.4 CONCLUSION

Although, food scientists' interest lies to improve the taste, texture or appearance of foods with longer shelf-life and also to raise efficiency in their manufacture by developing new processing technologies. But in the case of food hydrocolloids, scientists utilize the fundamental concepts of physical chemistry, macromolecular science and rheology to understand the interactions between constituent hydrocolloids and other ingredients like fats and proteins and their relative influence on the properties of the food systems, such as dispersions, emulsions, foaming, and gelling abilities. Functional and healthy food is one promising opportunity to deal with the new challenges of the modern society with high concern about food safety and health. The main targets for functional food development include defense against reactive oxidative species (ROS), cardiovascular functions, gastrointestinal physiology, functions, etc. Hydrocolloids in the diet could become one of the most probable answers for all the aforesaid life-threatening human disorders. Therefore, hydrocolloids would definitely play a noteworthy role in the field of development of functional food in the next decades to come.

KEYWORDS

- **carrageenans**
- **edible films**
- **food additives**
- **hydrocolloids**
- **konjac glucomannan**
- **surface activity**

REFERENCES

Goff, H. D., & Guo, Q., (2019). The role of hydrocolloids in the development of food structure. *Handbook of Food Structure Development* (pp. 1–28). Royal Society of Chemistry. doi: 10.1039/9781788016155-00001.

Li, J. M., & Nie, S. P., (2016). The functional and nutritional aspects of hydrocolloids in foods. *Food Hydrocoll., 53*, 46–61.

Milani, J., & Maleki, G., (2012). Hydrocolloids in food industry. In: Valdez, B., (ed.), *Food Industrial Processes: Methods and Equipment* (pp. 17–38). InTech, Croatia.

Ramirez, J. A., Barrera, M., Morales, O. G., & Vazquez, M., (2002). Effect of xanthan and locust bean gums on the gelling properties of myofibrillar proteins. *Food Hydrocoll., 16*, 11–16.

Ramirez, J. A., Uresti, R. M., Velazquez, G., & Vazquez, M., (2011). Food hydrocolloids as additives to improve the mechanical and functional properties of fish products: A review. *Food Hydrocoll., 25,* 1842–1852.

Santana, P., Huda, N., & Yang, T. A., (2013). The addition of hydrocolloids (carboxymethylcellulose, alginate and konjac) to improve the physicochemical properties and sensory characteristics of fish sausage formulated with surimi powder. *Turkish J. Fish. Aqua. Sci., 13*, 561–569.

Wustenberg, T., (2015). *Cellulose and Cellulose Derivatives in the Food Industry: Fundamentals and Applications* (1st edn.). Wiley-VCH Verlag GmbH & Co. KGaA.

CHAPTER 7

Edible Packaging for Fish Preservation: Fundamentals and Applications

P. LAYANA, R. P. DEEPITHA, K. A. MARTIN XAVIER, and AMJAD K. BALANGE

Department of Post-Harvest Technology, ICAR-Central Institute of Fisheries Education, Versova, Mumbai – 400061, Maharashtra, India

ABSTRACT

The edible packaging technology for seafood preservation has gained considerable attention in recent years due to its possibility to replace synthetic food packages in the future. The major attraction of edible packaging technology is the use of naturally occurring biodegradable polymers that may be eaten along with the food and has barrier properties similar to synthetic polymers. Polysaccharides, proteins or lipids, which may be used alone or in combination as composites, form the edible polymer matrix. An edible packaging can be converted into a bioactive package by incorporating bioactive compounds such as antioxidants and antimicrobials into the matrix in order to enhance seafood quality and safety. Such bioactive packaging can retard spoilage by scavenging free radicals responsible for oxidative rancidity or creating an unfavorable environment for bacterial growth during storage. Several kinds of active compounds such as plant essential oils, plant phenolics, organic acids, bacteriocins, probiotic bacteria are incorporated for fabricating bioactive edible packages for seafood. As the technology is still in the infant stage, it cannot entirely replace traditional packaging; however, better quality products with enhanced shelf life can be achieved by combining primary edible packaging with conventional packaging.

Advances in Fish Processing Technologies: Preservation, Waste Utilization, and Safety Assurance.
Ranendra K. Majumder, PhD & Amjad K. Balange, PhD (Eds.)

7.1 INTRODUCTION

Over the past few years, consumer focus is shifting mainly towards healthy food, which has driven considerable attention towards aquatic foods. Nutritional profiling of fish has revealed abundant nutrient reserves in its meat. However, the compositional uniqueness of fish meat with easily digestible protein, polyunsaturated fatty acids (FAs), essential vitamins and minerals create a conducive environment for bacterial growth and other spoilage reactions. Fish spoilage which initiates immediately after its death, is primarily associated with bacterial, enzymatic, and chemical reactions, categorizing fresh fish as a highly perishable commodity. In order to arrest such deleterious changes involved in limiting its shelf life, the technology of food packaging has emerged, which ensures that such perishable products are protected from external spoilage factors by forming an envelope around them. However, the rising issues associated with the enormous plastic disposal in the form of packaging materials, wrappers, straws, cutleries, and bottles into the oceans have become one of the greatest threats of the modern era. This has driven researchers to develop alternatives that reduce the usage of non-biodegradable components, especially related to food packaging.

Edible packaging technology is an eco-friendly approach and a promising tool in food packaging, ensuring the supply of safe and stable food. As the world is witnessing an increasing trend towards "Green consumerism," the demand for edible coatings (ECs) and films has improved over the past few years, owing to its biodegradability and non-toxic characteristics. Edible packaging can improve the quality and shelf life of fresh fish and fishery products by acting as a primary package. In addition to this, edible packaging technology reduces the risk of adverse environmental and health impacts associated with excessive utilization of non-biodegradable synthetic packaging materials. It can replace conventional plastic packaging materials because it is more eco-friendly, easily degradable, and safer than the latter.

7.2 HISTORY OF EDIBLE PACKAGING

The history of edible films (EFs) can be traced back to ancient kingdoms, where films derived from edible polymers were used for spying purposes. The secret information was written on the films so that when spies were captured by chance, they could easily destroy the information by eating it up. EFs and coatings, especially waxes, have been used since ancient times to prevent wilting of fruits and to impart glossy appeal on fruit surface. The use of wax coating to preserve citrus fruits in China dates back to the 12th century. The first edible film used for food preservation was made from Soya milk (Yuba) in the 15th century in Japan (Dehghani et al., 2018). Till 1967, the role of edible packaging in food applications were limited mostly to wax coatings of fruits and vegetables. However, today, the EFs and coatings are applied on a wide variety of foods in different forms to enhance their quality and safety.

7.3 WHAT IS EDIBLE PACKAGING?

An edible packaging could be defined as a primary package made from edible components which may be eaten together with or without food. They enrobe the food to enhance its quality and shelf life. Edible packaging could be done in two forms, a coating or a film.

7.3.1 EDIBLE COATINGS (ECS)

Coatings are thin layer of edible material formed directly on the food surfaces. Coatings are applied in liquid form, usually by dipping or spraying.

7.3.2 EDIBLE FILMS (EFS)

EFs are prepared separately as thin sheets and then placed over the product or between the food components. Films are first molded as solid sheets, then applied either as a wrap or used as stand-alone pouches.

Generally, the thickness of such films and coatings should be less than 0.3 mm (Pavlathand Orts, 2009). Mechanical strength, moisture, and gas barrier properties of these coatings and films mainly depend on the compatibility and the strength of the bonds formed (covalent bonds, ionic bonds, and H-bonding) between coating-forming polymer chains. The main advantage of edible packaging is that bioactive compounds such as antioxidants, antimicrobials, flavors, colorants, and probiotics can be incorporated into the polymer matrix to enhance the quality and safety of the food products. Such coatings and films can enhance the shelf-life of seafood by providing protection from mechanical damage as well as spoilage activities.

7.4 PRINCIPLES OF EDIBLE PACKAGING

An edible packaging can be employed as a barrier, carrier or enhancer to the product.

7.4.1 BARRIER

As a barrier, it regulates the gaseous and moisture transport, reduces flavor and aroma loss from the food and protect the food from exposure to microbial contaminants.

7.4.2 CARRIER

The coating act as a carrier for active compounds like antioxidants, antimicrobials, flavors, colors, nutraceuticals, etc. These compounds are released into the food to slow down chemical and microbial spoilage, enhance or maintain the product's appearance and nutritional value.

7.4.3 ENHANCER

Application of EFs or coatings improve the appearance and mechanical properties of fragile food products, especially fruits. Wax coatings are applied on some citrus fruits and apples to enhance the appearance by imparting glossiness, while possibly reducing dehydration and wilting. Chitosan coating has also been used to reduce mechanical damages on strawberries (Resende et al., 2018).

7.5 STRUCTURAL COMPONENTS OF EDIBLE PACKAGING

Edible packaging is made from biomacromolecules having the ability to form a continuous and cohesive network. Structurally, these films and coatings have at least two essential components; the matrix and additives:

- Matrix is a cohesively structured polymer network of any biomacromolecule;
- Additives are components that can be added into the material matrix, either independently or in combination, to enhance the biopolymer functionality;

- The matrix and additive components can be hydrophilic or hydrophobic in nature. To retain the edibility of packaging material, only water or ethanol could be used as a solvent to prepare edible polymer solution.

7.5.1 MATRIX MATERIALS

The matrix is usually based on hydrocolloids (polysaccharides and proteins) or lipids (waxes), which may be used alone or in combination as composites (Parreidt et al., 2018).

7.5.1.1 HYDROCOLLOID MATRIX

Polysaccharides and proteins are hydrocolloids. They are large biopolymers with many hydroxyl groups having affinity towards water. Such hydrophilic polymer coatings (especially polysaccharides such as alginates and starch) act as sacrificial moisture agents over the product. Here, the moisture evaporates from the film instead of the product surface and protects the food from dehydration (Parreidt et al., 2018). Whereas, protein-based packaging provides good mechanical stability through better polymer chain to chain interactions. Hydrocolloid films and coatings have neutral taste, colorless, oil-free appearance and can be applied to extent the shelf life of fish and fishery products by significantly reducing dehydration and oxidative rancidity (Parreidt et al., 2018).

7.5.1.1.1 Polysaccharide Based Films and Coatings

Polysaccharides are naturally occurring carbohydrate polymers where monomeric units are joined by glycosidic bonds to form long polymer chains. Polysaccharides are widely used to prepare EFs or coatings and their sources include marine algae (alginate and carrageenan), other plants (starch, pectin, and cellulose), animals (chitosan), and microorganisms (bacterial cellulose, Pullulan). Among these, chitosan and alginate are extensively studied for edible packaging of fruits, meat, and dairy products (Sanchez-Ortega et al., 2014). Generally, a polysaccharide matrix formation involves two processes:

- Disrupting the original interactions among the polymer chains during melting/dissolution process; and
- Cross-linking of adjacent chains through newly formed covalent bonds/hydrogen bonds during drying process.

The major advantages of polysaccharide films/coatings are:

- Widely available and usually cost-effective;
- Colorless, oil-free appearance and adds only minor calory to the product;
- Excellent gas barrier and mechanical properties due to the formation of well-ordered hydrogen-bonded matrix;
- Exerts barrier to lipids and oils.

However, the hydrophilic nature of these coatings exerts little or no water vapor resistance, hence functionality is wholly dependent upon the humidity conditions of storage. The major polysaccharide films evaluated for preservation of fish and fishery products are given below:

1. **Chitosan:** It is a polymer of N-acetyl-D-glucosamine. It is derived from chitin by deacetylation in the presence of strong alkali. Chitin is the major constituent of marine shrimp exoskeleton and it is the second most abundant polysaccharide found in nature. The potential of chitosan to act as an edible matrix component for the quality preservation of different fish and fishery products has been widely reported by many researchers (Fan et al., 2009; Li et al., 2012; Qiu et al., 2014; Soares et al., 2016), mostly due to their excellent structural and functional properties. The major features are the following:
   i. Allows formation of a continuous layer of coating on foods;
   ii. Films and coatings are usually crystal clear and flexible, yet they act as excellent oxygen and carbon dioxide blockers;
   iii. Inherent antimicrobial characteristics due to its cationic nature.
2. **Alginates:** These are obtained from marine brown algae. They are anionic polysaccharides composed of D-mannuronic acid (M) and L-guluronic acid (G) joined by 1–4 glycosidic bonds. The alginate edible matrix is formed when sodium alginate ionically interacts with divalent or multivalent cations. Usually, Ca^{2+} or Mg^{2+} are used for the purpose where the ions form salt bridges between G blocks from two alginate polymer chains, resulting in a three-dimensional hydrogel network. Therefore, the physical characteristics of alginate

gels/films widely depend on the proportion of guluronic (G) to mannuronic (M) acid residues present in it. Alginate films or coatings incorporated with different bioactive agents such as plant essential oils, 6-gingerol, phenolic extracts, lactoperoxidase, and other natural compounds were successfully used for the quality preservation of different fish and fishery products (Liu et al., 2016; Martinez et al., 2018; Ehsani et al., 2020).

3. **Cellulose Derivatives:** Cellulose is a natural polymer in which D-glucose units are linked by ß-1,4 glycosidic bonds. It is present in both plants and animals. Cellulose derivatives such as carboxymethylcellulose (CMC), methylcellulose, hydroxypropyl cellulose, and hydroxypropyl methylcellulose have been studied for their packaging efficiency for meat, seafood, and confectionaries. Bioactive CMC coatings were found to be effective in preserving fish such as rainbow trout during refrigerated storage and smoked eel fillets.
4. **Starch, Pectin, and Pullulan:** Starch is composed of two types of polymers, the linear chain of amylase and branched-chain of amylopectin. Films and coatings made from starch are extensively studied, especially for fruits, because they are colorless and odorless films with neutral taste and good gas barrier properties. For seafood packaging, starch-based composite films are gaining attention nowadays. A zein/potato starch/chitosan composite film incorporated with curcumin was developed as a potential bioactive packaging material for *Schizothorax prenati* filets (Xin, 2020). Starch-based film with anthocyanin extracts was developed as a bioactive-intelligent package to enable visual monitoring of spoilage process and also to enhance the shelf life of shrimp during storage (Zhang et al., 2020). Pectin is another polysaccharide composed of linear 1,4-galacturonic acids chains interspersed with branched rhamnogalacturonan (1,4-galacturonic acid to 1,2-rhamnose) chains. No major study has been carried out on the potential use of pectin and pullulan-based films and coatings for seafood packaging till date.

7.5.1.1.2 Protein-Based Films and Coatings

Protein is a polymer of amino acids linked by peptide bonds. Proteins that form EFs are derived either from plants (corn zein, soy protein, wheat gluten) or animals (collagen, gelatin, whey protein, casein, and egg albumin). Among these, gelatin, whey protein, corn zein and wheat gluten have been evaluated for their films/coatings forming properties for fruits, meat, dairy, and seafood preservation.

The initial step in protein film formation involves denaturation of protein accomplished by the application of heat, solvents, or a pH shift. The matrix is formed when the denatured polypeptide chains associate with each other through new intermolecular interactions (covalent, ionic, hydrogen, and van der Waals). The mechanical, barrier, and other functional properties of such films largely depend upon the degree of polymer chain interactions. Higher degree of interactions results in films with better water vapor and gas barrier properties (Hassan et al., 2018). The major advantage of protein-based films or coatings are its effective mechanical resistance and oxygen blocking capacity even at low relative humidity. However, like polysaccharide films, protein films also act as weak moisture barrier, i.e., the moisture loss from the products is significantly high (two to four orders of magnitude) in comparison to conventional packaging materials (Dehghani et al., 2018).

1. **Gelatin:** Collagen is the main component of bone, skin, and connective tissue. Hydrolysis of collagen yields gelatin. The presence of the amino acids glycine, proline, and hydroxyl-proline in a very well-defined order is the specific feature of gelatin. The films or coatings produced from gelatin exhibit good transparency, mechanical, and barrier properties and can be manufactured by simple casting or extrusion processes. Usually, gelatin-coating formulation contains 20–30% gelatin, 10–30% plasticizer (glycerin or sorbitol) and 40–70% water (Hassan et al., 2018). Antimicrobial and antioxidant gelatin films were fabricated and successfully used for preservation of seafood (Gomez-Jiang et al., 2019). Edible bag made out of fish gelatin bilayer films laminated with polylactic acid was successfully used to store fried salmon skin, indicated by a significant reduction in lipid oxidation rate (Nilsuwan et al., 2020). Gelatin pouches incorporated with phenolic compounds were also found suitable to reduce lipid oxidation in tuna oil during storage (Le et al., 2020).
2. **Whey Protein:** Whey protein-based films and coatings have been assessed for their potential in seafood preservation due to its excellent gas barrier properties. Antioxidant films for Asian sea bass (Chaijan et al., 2020), frozen Atlantic salmon (Rodriguez-Turienzo et al., 2012) and antimicrobial films for smoked salmon slices (Boyaci et al., 2016), fabricated from whey protein were found to be efficient in shelf-life extension of these products. Whey protein coatings also

prevent the loss of aromatic compounds present in food. However, as whey protein is hydrophilic in nature, the films possess weak moisture barrier properties. Incorporation of different kinds of lipids has been tried to enhance the moisture blocking properties of these films (Hassan et al., 2018).

3. **Corn Zein and Wheat Gluten:** Zein protein is a prolamin protein obtained from corn. Zein is hydrophobic in nature and can be dissolved in 70–80% ethanol. Zein films or coatings are very good moisture blockers than whey protein films. Bioactive zein coatings were fabricated to preserve the quality of fish balls during storage (Lin et al., 2009, 2011). Wheat gluten is hydrophobic globular protein obtained from wheat flour. Wheat gluten has two components, i.e., prolamin and glutelin. Gluten exerts elasticity and cohesiveness and thus facilitate the process of film formation. The composite protein film prepared from wheat gluten, corn zein and whey protein was found as a promising tool for storage of hot smoked rainbow trout with 2–3 weeks extension in shelf-life (Dursun and Erkan, 2014).

7.5.1.2 LIPID-BASED FILMS AND COATINGS

Unlike hydrocolloids, lipids are not biopolymers and they do not form self-supporting film structures. The application of lipids as such is limited to the formation of ECs. However, lipids reincorporated into polysaccharide and/or protein films in order to make composites with better moisture barrier properties, owing to its low polarity. Waxes and resins are the commonly used lipid matrices. Beeswax and carnauba wax are commonly used for coating fruits such as oranges and apples. The major advantages of lipid matrix are as follows:

- Can be used to add gloss as well as protection to the product;
- Lipids reduce moisture transmission due to their hydrophobicity;
- Addition of lipids into hydrocolloids can improve cohesiveness and moisture barrier property of the composite films/coatings.

The main disadvantages are:

- Opaqueness, waxy taste and texture, fragility, and chances of rancidity;
- They do not form self-supporting cohesive structures, hence cannot be used as such for film formation;
- Lipid-based coatings are extra brittle due to their hydrophobicity.

Different types of lipids used for edible packaging are listed below:

1. **Waxes and Paraffin:** Waxes are esters of acids with alcohols. Waxes include paraffin wax (obtained from crude petroleum), carnauba wax (palm tree leaves), beeswax (honeybees), candelilla wax (candelilla plant). Wax coatings, only if applied as a thin layer, are considered fit to be eaten. Wax coatings are usually applied for cheese, uncooked fruits, and vegetables.
2. **Fatty Acids (FAs):** Saturated FAs are better moisture blocker than unsaturated FAs. The presence of double bonds in the structure and branching of acyl chains (less efficient lateral packing of acyl chains) cause increased water vapor permeability in unsaturated FAs matrix compared to saturated ones (Stuchell and Krochta, 1995).
3. **Aceto-Glycerides:** Acetic acid esters of monoglycerides are called acetylated monoglycerides (AMG). AMGs can be easily solidified into waxlike form from its molten state. AMGs can be extended up to 800% of its length whereas most of the lipids can be extended up to only 102% of their actual length (Hassan et al., 2018). Films and coatings made with AMG are usually applied on meat products to inhibit dehydration during storage.
4. **Shellac Resins:** These are secretions of insect *Lacciferlacca*. These resins aren't generally recognized as safe (GRAS) but have been approved as a secondary chemical in ECs. The major use of shellac-based films and coatings are in the pharmaceutical field, and very few studies have been done for its use in food products.

7.5.1.3 COMPOSITE FILMS OR COATINGS

Composite matrices are films or coatings with enhanced mechanical and barrier properties. As discussed above, the hydrocolloid coatings exhibit excellent mechanical and structural properties, but possess poor moisture barrier properties. Whereas, lipids can't make stand-alone structures but exhibit good moisture barrier capacity due to its low polarity. To overcome these problems, these materials can be used in combination. Here, the structural integrity of the hydrocolloids and hydrophobic behavior of the lipid could be complemented in order to enhance the functional properties of the resulting films. Such films/coatings are implemented in the form of suspension, emulsion or as bilayers:

1. **Emulsion:** Lipid is dispersed in the biopolymer matrix.
2. **Bi-Layer Composites:** Lipid forms the second layer over the polysaccharide or protein layer.
3. **Multi-Layer Composites:** Usually consist of two or more layers of film-forming polymers.

7.5.2 ADDITIVES

The commonly used additives in ECs/films are described in subsections.

7.5.2.1 PLASTICIZER

These are non-volatile, low molecular weight substances used to reduce the inherent brittleness in most biopolymers. Plasticizers reduce the polymer chain-to-chain interaction by positioning themselves between polymer molecules and thus increasing the stretchability of the film. The addition of plasticizer makes the edible film soft and flexible with increased toughness and impact resistance. In order to make the EFs with better structural properties, different kinds of food-grade polyols (glycerol and sorbitol), glycols (polyethylene glycol), sugars (glucose, fructose, and sucrose) and lipids and their derivatives (phospholipids, FAs and surfactants) are used as plasticizers. Among these, glycerol is the most commonly used plasticizer.

7.5.2.2 CROSSLINKING AGENTS

They enhance intermolecular interaction between polymer chains. Transglutaminase or genipin are used to improve crosslinking between protein-polymer chains and citric or tannic acid are used for polysaccharides.

7.5.2.3 REINFORCEMENTS

Micro or nanosized particles are incorporated into the polymer matrix to improve the mechanical and functional properties of films such as gas and moisture barrier properties, tensile strength, etc.; for example, cellulose nanofibers, starch nanocrystals, and chitosan nanowhiskers.

7.5.2.4 *EMULSIFIERS*

They are surface-active compounds having both hydrophilic and hydrophobic properties capable of preventing phase separation in otherwise immiscible system. An emulsifier stabilizes composite coatings and improves its adhesion. Common food-grade emulsifiers or surfactants in use are tweens, spans, lecithin, sodium lauryl sulfate, and glycerol monostearate. The amount usually incorporated is less than 1% of the biopolymer weight (Rossman, 2009).

7.5.2.5 *BIOACTIVE COMPOUNDS*

Various components such as antioxidants, antimicrobials, nutraceuticals, flavor/color agents can be incorporated into the polymer matrix to improve the functionality of the package. Such films or coatings formulated with bioactive compounds are known as bioactive films/coatings.

7.6 BIOACTIVE EDIBLE PACKAGING SYSTEM FOR FISH AND FISHERY PRODUCTS

The application of ECs/films based on polysaccharides, proteins, and lipids has recently gained interest as bioactive edible packaging systems, a measure to extend the shelf-life of seafood by reducing the rate of chemical and microbial spoilage. Incorporation of antimicrobial and antioxidant compounds into the matrix has the advantage of allowing slower release of these compounds into the products (Dehghani et al., 2018). The preservative properties of coatings may largely relate to the presence/release of these compounds and their interaction with the food. Such bioactive films/coatings can retard fish spoilage by scavenging/reduce free radicals responsible for oxidative rancidity in fish or by creating an unfavorable environment for the action for autolytic enzymes and bacteria. Currently, several kinds of active compounds such as plant essential oils (cinnamon, oregano, thyme, and lemongrass), plant phenolics, organic acids (benzoic, propionic, acetic acid, and lactic), bacteriocins (peroxidase and nisin), probiotic bacteria are incorporated for fabricating bioactive edible package for seafoods, each of which has different effects on lipid oxidation, and proteolytic and microbial spoilage (Sanchez-Ortega et al., 2014). Chitosan has been the widely used matrix component for this purpose due to its cationicity and inherent antimicrobial properties.

Among the natural bioactive compounds, phenolics, and essential oils extracted from plants and spices have been the most studied to preserve seafoods as they were reported to possess both antioxidant and antimicrobial potential. Alginate coatings incorporated with tea polyphenols could extend the shelf life and quality of refrigerated bream (Song et al., 2011). Jiang et al. (2011) found that chitosan-based EFs and coatings containing sodium lactate, sodium diacetate and potassium sorbate could efficiently inhibit the growth of *Listeria monocytogenes* on cold-smoked salmon during refrigerated storage. Antimicrobial films have also been prepared by incorporating natural preservatives (grape seed extract and licorice extract) into a chitosan matrix (Li et al., 2013). Remya et al. (2016) reported an eight day-extension of shelf life of Barracuda steaks when coated with chitosan containing ginger essential oil. A study conducted by Li et al. (2020) found that chitosan-sodium alginate coatings containing ε-polylysine could efficiently retard the dehydration and maintain the quality of pufferfish during refrigerated storage. Such findings indicate that the bioactive edible packaging can successfully replace the illegal use of harmful chemical preservatives (formaldehyde and benzoate) for the shelf-life extension of fish and fishery products.

7.6.1 MASS TRANSFER INVOLVED IN BIOACTIVE EDIBLE PACKAGING

The interactions that can occur between the package, food, and its environment are governed by different mass transfer processes (Salgado et al., 2015). They are as given in subsections.

7.6.1.1 MIGRATION

The bioactive compounds incorporated into the edible matrix can migrate/release into the product or headspace and pursue their protective action.

7.6.1.2 ADSORPTION

Components of food or headspace adsorbed into the inner surface of the film or coating.

7.6.1.3 ABSORPTION

Components of food or headspace are finally incorporated into the polymer matrix.

7.6.1.4 PERMEATION

Moisture or gas molecules diffuses out/in through the polymer matrix.

7.6.2 FILM/COATING FORMING TECHNIQUES

An edible film is a dried network of biopolymer, whereas coatings are applied in wet form. The solvents used for the process should be food grade. Usually, water or ethanol are used for dissolving the components.

7.6.2.1 FILM FORMING TECHNIQUES

The process involves obtaining a well-ordered network of solid phase from a liquid phase. The film-forming methods can be generally classified into two:

1. **Wet Process:** This involves solubilization/dispersion of all ingredients into water or ethanol and subsequent drying to form the film structure. It is regarded as high energy-consuming process as the solvent needs to be evaporated.
2. **Dry Process:** It is based on the thermoplastic properties of the edible polymer. Here, the heat is applied to increase the temperatures above the melting point of the film forming polymer and cause them to flow (e.g., extrusion process). The use of solvent is avoided in dry process, therefore do not require an evaporation step.

7.6.2.1.1 Hydrocolloid Films

This could be achieved by any of the techniques like simple coacervation, complex coacervation, solvent casting, thermal gelation and extrusion (Parreidt et al., 2018):

1. **Simple Coacervation:** The hydrocolloid which is dispersed in water can be precipitated by different techniques such as simple drying (evaporation of solvent), addition of water-soluble non-electrolyte in which the hydrocolloid is not soluble (e.g., alginate precipitates in the presence of ethanol) or change in pH with the addition of an electrolyte (salting out or cross-linking will take place).
2. **Complex Coacervation:** Polymer complex is obtained by mixing two oppositely charged hydrocolloid solutions.
3. **Gelation or Thermal Coagulation:** Gelation is achieved either by simple heating of the macromolecule (e.g., Ovalbumin) or by subsequent cooling of the heated solution (e.g., Gelatin, Agar).

7.6.2.1.2 Lipid/Composite Films

Lipid films can be formed by different methods such as initial melting and subsequent solidification, dissolving in organic solvent, or as an emulsion in water, followed by drying. The composite films are formed either an emulsion or bilayers:

1. **Emulsion:** In emulsion-based films, lipid along with the surfactant are mixed with the hydrocolloid polymer solution to form a stable emulsion. The solution is heated at temperature above the melting point of the lipid, then homogenized and cast to form films (Krochta, 2002).
2. **Bilayer:** Films can be prepared by two different methods:
   i. **Lamination:** Molten/solubilized lipid is laminated on a hydrocolloid film surface.
   ii. **Emulsion:** Lipid is dispersed in hydrocolloid solution to form an unstable emulsion. Bilayer is formed upon drying due to phase separation.

7.6.2.2 COATING TECHNIQUES

Coatings can be obtained in various ways such as dipping, brushing or spraying, followed by drying.

7.6.2.2.1 *Hand Spreading*

Involves covering the product surface with coating solution using brush or roller. Then allowed to solidify at ambient temperature or by heating.

7.6.2.2.2 *Spray Coating*

Ensures uniform coating over the product. Electro-spraying is a recent technology which produces thin and uniform coating over the product.

7.6.2.2.3 *Dipping*

The final coating will be thicker, and less uniform compared to other methods. The method is suitable for products with irregular shapes and sizes.

7.6.3 *IMPORTANT REQUIREMENTS OF EDIBLE FILMS (EFS)/ COATINGS*

- All components used for edible package must attain GRAS status.
- The films should possess sufficient mechanical strength to protect the food from the external loads. Tensile strength, Young's modulus, and elongation at break of the edible package should be on par with that of synthetic packaging materials.
- For preventing transmission of liquids/gases between the contained food and the environment, sufficient gas/moisture barrier properties must be present.
- The material should be heat-sealable for making an ideal food packaging system.
- Sensory properties such as appearance, color, texture, taste, and odor are the important criteria chosen for evaluating EFs with respect to consumer acceptance.
- The film properties must remain constant with respect to time, to protect and preserve the confined food.

7.6.4 *ADVANTAGES AND DISADVANTAGES OF EDIBLE PACKAGING*

7.6.4.1 *ADVANTAGES*

- Greater acceptability due to the clear and transparent nature of plasticized EFs.

- Tensile strength and Youngs Modulus of most EFs are comparable to synthetic ones.
- Excellent heat sealability and gas barrier properties comparable to those of plastics.
- Waste disposal problem is almost zero.
- Landfill requirement is reduced considerably due to the faster biodegradable nature of the film in comparison to the other synthetic films.
- As the components are non-toxic, transmission of packaging molecules into the food may not be detrimental to consumer health.
- Transmission of packaging molecules into food does not create any health issues.

7.6.4.2 DISADVANTAGES

- EFs have poor elongation, poor resistance to gases and liquids in comparison to synthetic plastic materials.
- As most of the edible polymers are hydrophilic in nature, the films derived from them exhibit dimensional instability and lead to microbial growth due to moisture absorption during storage.
- The bilayer composites may delaminate with the passage of time and develops cracks and pinholes.
- A thick coating of edible material may inhibit oxygen exchange, causing anaerobic environment and further off-flavor development.
- EFs production is still in laboratory scale. Hence, it could be employed as a primary package only.

7.6.5 CONSIDERATION FOR INDUSTRIAL SCALING UP OF EDIBLE PACKAGING SYSTEM

- Unplasticized EFs have inferior film properties as they are brittle than plastic packaging material. However, good heat sealability and gas barrier properties make it comparable to plastics. Further research is still required to produce films/coatings with superior properties.
- At present, laboratory scale production has certain limitations such as difficulties in creating continuous films, extended drying time (2–3 days) and restrictions in thickness control and making of large sized films (> 25 cm). High cost of production and energy consumption

are other factors limiting the upgradation. Development of continuous films, less production time and cost are targets to be achieved in scaling up the production.

- Science of nanotechnology can be used as an ideal tool to improve the functional properties. But, research on nanocomposite EFs is at infant phase and further studies in the field are required to investigate the toxicological effects of nanomaterials in the film.
- An In-depth investigation is required in the aspects like edibility, biodegradability, and safety w.r.t. the use of bioactive compounds in edible packaging system.
- Future edible film would be multifunctional and compatible to the modern packaging technologies.
- Addressing the cost issues is another focus area to be attended. The total cost of the packaging should not exceed 10% of the product cost to justify the adaptation of EFs.
- Lack of proper knowledge on edibility, biodegradability, sensory aspects, lack of awareness and cultural issues, insufficient marketing strategies, etc., can affect the consumer acceptance of EFs. These aspects need to be focused for the commercialization of edible packaging systems (Jeya et al., 2020).

7.7 CONCLUSION

Quality food for the future generation is still remaining the greatest challenge for present-day researchers and food processors. Though technologies and tools are developed by scientists to inhibit the upcoming global food crisis, serious attention needs to be given to retaining the quality of food produced. Biodegradable edible packaging technology is the most promising way to address the current issues associated with quality preservation of highly perishable food commodities, plastic pollution, and its removal could be a positive step to move towards a greener globe. But the field still being in its nascent stage requires further research and skills to reach the full potential. At present, EFs and coatings cannot entirely replace traditional packaging, however, better quality and shelf life of the produce can be achieved by combining primary edible packaging with conventional packaging. Further triumph in the field can evolve with researches, to make this budding technology to reach greater heights.

KEYWORDS

- **bioactive compounds**
- **biodegradable packing**
- **chitosan**
- **edible packaging**
- **plant phenolics**
- **plastic replacer**
- **seafood quality**

REFERENCES

Boyaci, D., Korel, F., & Yemenicioglu, A., (2016). Development of activate-at-home-type edible antimicrobial films: An example pH-triggering mechanism formed for smoked salmon slices using lysozyme in whey protein films. *Food Hydrocoll., 60*, 170–178.

Chaijan, S., Panpipat, W., Panya, A., Cheong, L. Z., & Chaijan, M., (2020). Preservation of chilled Asian sea bass (*Lates calcarifer*) steak by whey protein isolate coating containing polyphenol extract from ginger, lemongrass, or green tea. *Food Control*, 107400.

Dehghani, S., Seyed, V. H., & Joe, M. R., (2018). "Edible films and coatings in seafood preservation: A review". *Food Chem., 240*, 505–513.

Dursun, S., & Erkan, N., (2014). The effect of edible coating on the quality of smoked fish. *Italian Journal of Food Science, 26*(4), 370.

Ehsani, A., Hashemi, M., Afshari, A., Aminzare, M., Raeisi, M., & Zeinali, T., (2020). Effect of different types of active biodegradable films containing lactoperoxidase system or sage essential oil on the shelf life of fish burger during refrigerated storage. *LWT, 117*, 108633.

Fan, W., Sun, J., Chen, Y., Qiu, J., Zhang, Y., & Chi, Y., (2009). Effects of chitosan coating on quality and shelf life of silver carp during frozen storage. *Food Chem., 115*(1), 66–70.

Gomez-Estaca, J., López De, L. A., Gómez-Guillén, M. C., López-Caballero, M. E., & Montero, P., (2009). Antimicrobial activity of composite edible films based on fish gelatin and chitosan incorporated with clove essential oil. *J. Aquat. Food Prod. Technol., 18*(1, 2), 46–52.

Hassan, B., Chatha, S. A. S., Hussain, A. I., Zia, K. M., & Akhtar, N., (2018). Recent advances on polysaccharides, lipids and protein based edible films and coatings: A review. *Int. J. Biol. Macromol.,109*, 1095–1107.

Jeya, J., Chandrasekaran, M., Venkatesan, S. P., Sriram, V., Britto, J. G., Mageshwaran, G., & Durairaj, R. B., (2020). Scaling up difficulties and commercial aspects of edible films for food packaging: A review. *Trends Food Sci. Technol., 100*, 210–222.

Jiang, W., Hu, S., Li, S., & Liu, Y., (2019). Evaluation of the preservation effect of gelatin-water soluble chitosan film incorporated with Maillard peptides on bluefin tuna (*Thunnus thynnus*) slices packaging. *LWT, 113*, 108294.

Jiang, Z., Neetoo, H., & Chen, H., (2011). Control of *Listeria monocytogenes* on cold-smoked salmon using chitosan-based antimicrobial coatings and films. *J. Food Sci., 76*(1), M22–M26.

Krochta, J. M., (2002). Proteins as raw materials for films and coatings: Definitions, current status, and opportunities. *Protein-Based Films and Coatings, 1*, 1–40.

Le, T., Takahashi, K., Okazaki, E., & Osako, K., (2020). Mitigation of lipid oxidation in tuna oil using gelatin pouches derived from horse mackerel (*Trachurus japonicus*) scales and incorporating phenolic compounds. *LWT*, 109533.

Li, P., Zhou, Q., Chu, Y., Lan, W., Mei, J., & Xie, J., (2020). Effects of chitosan and sodium alginate active coatings containing ε-polysine on qualities of cultured pufferfish (*Takifugu obscurus*) during cold storage. *International Journal of Biological Macromolecules*.

Li, T., Hu, W., Li, J., Zhang, X., Zhu, J., & Li, X., (2012). Coating effects of tea polyphenol and rosemary extract combined with chitosan on the storage quality of large yellow croaker (*Pseudosciaena crocea*). *Food Control, 25*(1), 101–106.

Li, T., Li, J., Hu, W., & Li, X., (2013). Quality enhancement in refrigerated red drum (*Sciaenops ocellatus*) fillets using chitosan coatings containing natural preservatives. *Food Chem., 138*(2, 3), 821–826.

Lin, L. S., Wang, B. J., & Weng, Y. M., (2009). Preservation of commercial fish ball quality with edible antioxidant-incorporated zein coatings. *J. Food Process Pres., 33*(5), 605–617.

Lin, L. S., Wang, B. J., & Weng, Y. M., (2011). Quality preservation of commercial fish balls with antimicrobial zein coatings. *J. Food Qual., 34*(2), 81–87.

Liu, X., Jia, Y., Hu, Y., Xia, X., Li, Y., Zhou, J., & Liu, Y., (2016). Effect of *Citrus wilsonii* Tanaka extract combined with alginate-calcium coating on quality maintenance of white shrimps (*Litopenaeus vannamei* Boone). *Food Control, 68*, 83–91.

Martinez, O., Salmerón, J., Epelde, L., Vicente, M. S., & De Vega, C., (2018). Quality enhancement of smoked sea bass (*Dicentrarchus labrax*) fillets by adding resveratrol and coating with chitosan and alginate edible films. *Food Control, 85*, 168–176.

Nilsuwan, K., Guerrero, P., De La Caba, K., Benjakul, S., & Prodpran, T., (2020). Properties and application of bilayer films based on poly (lactic acid) and fish gelatin containing epigallocatechin gallate fabricated by thermo-compression molding. *Food Hydrocoll.*, 105792.

Parreidt, S. T., Müller, K., & Schmid, M., (2018). Alginate-based edible films and coatings for food packaging applications. *Foods, 7*(10), 170.

Pavlath, A. E., & Orts, W., (2009). Edible films and coatings: Why, what, and how? In: *Edible Films and Coatings for Food Applications* (pp. 1–23). Springer, New York, NY.

Qiu, X., Chen, S., Liu, G., & Yang, Q., (2014). Quality enhancement in the Japanese sea bass (*Lateolabrax japonicas*) fillets stored at 4 C by chitosan coating incorporated with citric acid or licorice extract. *Food Chem., 162*, 156–160.

Remya, S., Mohan, C. O., Bindu, J., Sivaraman, G. K., Venkateshwarlu, G., & Ravishankar, C. N., (2016). Effect of chitosan based active packaging film on the keeping quality of chilled stored barracuda fish. *J. Food Sci. Tech., 53*(1), 685–693.

Resende, N. S., Gonçalves, G. A. S., Reis, K. C., Tonoli, G. H. D., & Boas, E. V. B. V., (2018). Chitosan/cellulose nanofibril nanocomposite and its effect on quality of coated strawberries. *J Food Qual., 2018*.

Rodriguez-Turienzo, L., Cobos, A., & Diaz, O., (2012). Effects of edible coatings based on ultrasound-treated whey proteins in quality attributes of frozen Atlantic salmon (*Salmo salar*). *Innov. Food Sci. Emerg. Technol., 14*, 92–98.

Rossman, J. M., (2009). Edible films and coatings for food applications. In: Embuscado, M. E., & Huber, K. C., (eds.), *Films and Coatings for Food Applications* (pp. 367–390). Springer Science Business Media, LLC, New York, NY.

Salgado, P. R., Ortiz, C. M., Musso, Y. S., Di Giorgio, L., & Mauri, A. N., (2015). Edible films and coatings containing bioactives. *Curr. Opin. Food Sci., 5*, 86–92.

Sanchez-Ortega, I., García-Almendárez, B. E., Santos-López, E. M., Amaro-Reyes, A., Barboza-Corona, J. E., & Regalado, C., (2014). Antimicrobial edible films and coatings for meat and meat products preservation. *Sci. World J., 2014.*

Soares, N. M., Fernandes, T. A., & Vicente, A. A., (2016). Effect of variables on the thickness of an edible coating applied on frozen fish–Establishment of the concept of safe dipping time. *J. Food Eng., 171*, 111–118.

Song, Y., Liu, L., Shen, H., You, J., & Luo, Y., (2011). Effect of sodium alginate-based edible coating containing different anti-oxidants on quality and shelf life of refrigerated bream (*Megalobramaamblycephala*). *Food Control, 22*(3, 4), 608–615.

Stuchell, Y. M., & Krochta, J. M., (1995). Edible coatings on frozen king salmon: Effect of whey protein isolate and acetylated monoglycerides on moisture loss and lipid oxidation. *J. Food Sci., 60*(1), 28–31.

Xin, S., Xiao, L., Dong, X., Li, X., Wang, Y., Hu, X., Sameen, D. E., Qin, W., & Zhu, B., (2020). Preparation of chitosan/curcumin nanoparticles based zein and potato starch composite films for *Schizothoraxprenati* fillet preservation. *I. J. Biol. Macromol.*

Zhang, K., Huang, T. S., Yan, H., Hu, X., & Ren, T., (2020). Novel pH-sensitive films based on starch/polyvinyl alcohol and food anthocyanins as a visual indicator of shrimp deterioration. *I. J. Biol. Macromol., 145*, 768–776.

CHAPTER 8

Disruptive Technologies: Its Scope in Seafood Processing and Traceability of Fish Products

AJIT KUMAR ROY,[1] B. M. YADAV,[2] and S. A. JOSHI[3]

[1]*Ex. Pr. Scientist and Head, Social Sciences, ICAR-CIFA, Bhubaneswar – 751002, Odisha, India*

[2]*Associate Professor, College of Fisheries, Dr. B. S. K. K. V., Dapoli, Ratmagiri – 415629, Maharashtra, India*

[3]*Assistant Professor, College of Fishery Science, MAFSU, Nagpur – 440006, Maharashtra, India*

ABSTRACT

The United Nations has estimated that the global population may reach 9.7 billion by 2050. The need for higher protein and meat are increasing the global food demand due to dietary changes and rising incomes in developing countries. An additional 27 million tons of fish should be produced by 2030 as estimated by the Food and Agriculture Organization (FAO) for maintaining the present level of fish consumption. Technological interventions have played a pivotal role in fish processing to support innovation, productivity, waste reduction and recovery as well as utilization, increasing shelf-life of the fish, improving food safety standards and to facilitate exports. At present, the demand for good quality ingredients, transparency in product, sustainable, and responsible sourcing has a huge impact on the development of value-added products dominating the food industry. The public has over the past years broadly accepted "disruptive technologies" as a term which refers to Blockchain, Internet

Advances in Fish Processing Technologies: Preservation, Waste Utilization, and Safety Assurance.
Ranendra K. Majumder, PhD & Amjad K. Balange, PhD (Eds.)

of Things (IoT), artificial intelligence (AI), drones, 3D printing, machine learning, biometrics, virtual, augmented, and mixed reality and other cutting-edge technologies, etc. In this communication, it is highlighted that how AI, the Internet of Things (IoT), 3-D printing, and Blockchains have the huge potential to change the types of trade, traders, and trading commodity. Some of the startup driven innovations are highlighted to update the impactful 'Trends and Innovations' of promising startups and emerging companies in the areas of Artificial Proteins using 3D food printers that allow for the personalized, exact, and reproducible nutrition containing the optimum quantity of nutrients necessary for a specific sex, way of lifestyle and the health conditions. Food robotics improves the fish quality; enhance output and decreases expenses in food and liquid processing. Notable disruptive innovations in fish processing technology in frozen seafood, Octopus farming, and shrimp cultivation are gaining base and contributing to the expansion of the processed fish seafood and equipment market. The data which is high volume and complex which require cost effective, innovative information processing for enhanced insight and decision making as "the fourth paradigm of sciences" also discussed.

8.1 INTRODUCTION

This chapter gives an overview of new and promising technologies while discussing the potential of innovative monitoring technologies to better manage fish processing technologies and related areas. The chapter also discusses the latest development in big data, artificial intelligence (AI), IoT, and blockchain technologies along with their applications in fisheries and aquaculture sector, particularly in fish processing for managing the fisheries resources in a sustainable manner right from harvesting to consumer's plate.

8.2 GLOBAL FISHERIES SCENARIO

The depletion in fish stocks from their biologically sustainable levels is reported to be 90% in 1974 to 65.8% in 2017. The 78.7% of present landings comes from biologically sustainable stocks. Of the total stocks, the under-fished stocks and the maximally sustainably fished stocks were found to be 6.2% and 59.6%, respectively. The improved implementation of management measures has increased since 1989. The total global capture fisheries production recorded was 96.4 million tons with an upsurge of 5.4%. In 2018, World aquaculture production reached to 114.5 million tons, with farm-gate auction value of 263.6 billion USD. This production is comprised of 82.1 million and

32.4 million tons of aquatic animals and aquatic algae, respectively along with 26,000 tons of ornamental seashells and pearls. Finfish contributed with volume of 54.3 million tons in aquatic animal farming. Food fish aquaculture production (57 million tons) has outpaced that of the non-food fish subsector in world aquaculture. World production of farmed aquatic animals raised on average at 5.3% per year during 2001–2018. Fisheries and aquaculture sector provided employment to 59.51 million people involving 14% female population. In the entire employment, aquaculture sector contributed employment of 20.53 million population while fisheries sector gave opportunities to 38.98 million. The global total of fishing vessels was calculable to be 4.56 million and with 3.1 million vessels, Asia, lasts to have 68% share.

In terms of fish use and processing, about 88% of the total fish production of 179 million tons is used for direct human consumption; the remaining 12% is used for non-food purposes. The live, fresh or chilled fish contributed 44% of fish applied for direct human consumption. It has been observed that a substantial percentage of world fisheries production is processed into fishmeal and fish oil. Fish oil being a most nutritious and digestible component is still widely used for aquaculture feed. The proportion of fish oil and fishmeal used in pelleted fish feed has shown decreasing trend. The fish oil contains long-chain polyunsaturated fatty acids (PUFAs) which is proven to be good for health. About 25 to 35% of fish meal and fish oil is produced from processing waste of fish industries.

8.2.1 LOSS AND WASTAGE OF HARVEST

From the world fish production, it is found that 35% of the worldwide production is either lost or wasted per year. To reduce the waste and losses in the fisheries sector, it is necessary to implement the appropriate policies, regulatory frameworks, enhance capacity building, and develop services and infrastructure and physical access to markets. For improving stock sustainability and food security as well as reducing pressure on fish stocks, fish loss and waste reduction is foremost important.

8.2.2 FISH CONSUMPTION

The average yearly development of total food fish utilization expanded at 3.1%, outpacing the yearly population development rate (1.6%) between 1960 to 2017. The evaluations for per-capita fish utilization in 2018 presently remain at 20.5 kg. In 2017, fish represented around 17% of total animal protein, 7% of all proteins, devoured globally. Fish provided around 20% of their average per capita utilization of animal protein. Since 2016, at the worldwide level, aquaculture has been the primary source of fish available for human utilization. In 2018, this offer was 52%, a figure that can be relied upon to keep on expanding in the long term. Thus, aquaculture has expanded fish availability promoting further developed nourishment and food security.

8.2.3 FISH TRADE AND PRODUCTS

In 2018, 67 million tons of fish were traded internationally, which accounted for 38% of total fish caught and cultured throughout the world. In 2018, the total export of USD 164 billion contributed 11% to agricultural products. From 1976 to 2018, the worth of worldwide exports of fish and fish products expanded at the yearly rate of 8% in ostensible terms and of 4% in genuine terms. There is a negative impact on trade among exporters and importers due to the upsurge of coronavirus disease (COVID-19). In 2018, fish imports by developing nations addressed 31% of the worldwide total by esteem and 49% by quantity (live weight). In 2018, processed products (i.e., excluding fresh whole and live fish) contributed more than 90% of the quantity (live weight equivalent) of trade in fish and fish products, from which frozen products representing the noteworthy share. For human consumption, 78% of the fish products were exported.

8.2.4 FUTURE DEMAND OF FISH

FAO of the United Nations (UN) expects that an additional 27 million tons of fish production will be required just to keep with the current degree of utilization in 2030. It should not surprise that additional ag-tech investments in aquaculture will be at record levels. The eventual fate of fish farming looks more sustainable, more recognizable, and more profitable as well as more productive (FAO, 2015).

8.3 NOTION ABOUT POST-HARVEST FISHERIES AND ITS IMPORTANCE

Post-harvest fisheries comprise the activities that occur from the time of fish landed or harvested until it is devoured. Post-harvest fisheries embrace the social, ecological, monetary, institutional, technical, and promoting parts of the inventory, preservation, processing, and distribution of fish and fish products. It is widely accepted that disruptive technology can affect and drive industry transformation. 'Disruptive technology' was defined as techniques, skills, methods, and processes that accomplish a target better compared to what went before it, yielding huge advantages to its key stakeholders, and has the potential to be carried out and scaled up amazingly.

8.3.1 WORLD FISH UTILIZATION

As per a report, the fish utilization globally is highest for direct human consumption (45%), followed by dried, salted, smoked, or cured forms (12%), prepared, and preserved form (13%), and frozen form (30%) as on 2014.

8.3.2 SCENARIO OF GLOBAL SEAFOOD PROCESSING SECTOR

The global seafood processing sector is considered the most intricate part of all food sectors. It is based on about 1,000 commercial species and has a more extensive scope in processing technologies as compared to other food sectors. With an increasing worldwide interest for high-quality seafood items, two significant patterns in seafood supply have arisen during the last two decades. From wild capture fisheries, total seafood production has declined, although the worth has expanded, and aquaculture production has risen consequently by about 10% per year. Today wild-caught seafood represents approximately half of all seafood consumed globally whereas aquaculture supplies the other approximately 50%.

8.3.3 RISING GLOBAL DEMAND FOR FOOD

As indicated by the latest estimate by the UN, we may reach 9.7 billion individuals by 2050. This development, along with rising incomes in developing countries bringing about dietary changes, for example, consuming more protein and meat is driving up global food demand. There will be an increment from 59% to 98% in food demand by 2050 that will shape agricultural markets in such a way that we have not seen previously. Farmers worldwide should increase crop production, either by expanding the amount of agricultural land to develop crops or by enhancing the productivity on existing agricultural lands through fertilizer, manure, and irrigation and acquiring new strategies like precision farming (UN, 2015). It is believed the seafood industry is on the brink of dramatic change, driven by environmental concerns and market moves such as a developing and more health-conscious global middle class and a boom in aquaculture. Overall, from 2016 to 2030, the world's fish consumption is predicted to increase by 20%, driven by global population growth, the expansion of the middle class, and more noteworthy urbanization.

8.3.4 FOOD PROCESSING AND PACKAGING

Food processing and packaging are the most important aspects of the food industry. More processed and packaged food is utilized as a proportion of the total in better-off, urbanizing, and industrializing economies (Kearney, 2010). Purchaser interest in the sustainable production of food varieties and food-related issues are relied upon to be an increasing pattern, and legislation is beginning to build up this trend towards "socially responsible items" (FAO, 2015). In a general sense, any food packaging should contain, illuminate, preserve, inform, and provide convenience while recognizing the imperatives placed upon their utilization from both legal and environmental points of view (Kim and Seo, 2018). Additionally, packaging technologies need to address customer assumptions for product quality. In such a manner, this highly inter-disciplinary review looks at how a circular economy principle can be applied to the seafood industry by using food processing by-products in environmentally friendly active packaging (AP) solutions for the reduction of food spoilage, post-processing, and to extend the shelf life. In recent years, there are many new food processing technologies have been extensively examined in food processing research. These technologies share key benefits for enhancing the preservation and quality of conventional foods, for combating the growing challenges posed by globalization, increased competitive pressures, and various consumer demands.

8.3.5 SEAFOOD PROCESSING TECHNOLOGY

This area is continually evolving to meet the needs of the seafood sector as it seeks to respond to many factors such as environmental, regulatory, and market factors. Innovative technologies, as well as the innovative application of older technologies, have brought forward various innovative seafood products like MAP live mussels, HPP lobster meat, and farmed seafood products. In addition, technologies are being applied to extract more value from seafood processing by-products and create valuable products such as biomedical chitosan and calcium supplements. The main goal of these technologies is to allow the seafood sector to produce high value-low-cost products and thereby remain competitive in global markets.

8.3.6 THE SEAFOOD SECTOR: THE MOST WASTEFUL FOOD SECTOR

According to the United States Department of Agriculture, of the 70 million metric tons of fish harvested globally per year, over half ends up as a by-product or waste. As compared to today's market, seafood processors have to produce high value-low-cost products. The role of technology in seafood processing has evolved rapidly over the last decade in response to the above drivers to support innovation, reduction, recovery, and utilization of waste, productivity, increase shelf-life; improve food safety, and facilitate exports. Various inventive processing technologies have recently developed as a result. The seafood industry is increasingly moving towards new product development and also acquiring new processing methods that allow us to either do things we were unable to do before or to improve and do more efficiently than we did before. Perhaps one of the most innovative advancements to date is high-pressure processing (HPP). High-pressure processing is a cold pasteurization technique in which the food products are subjected to an exceptionally significant degree of hydrostatic pressure (e.g., up to 87,000 psi) for a couple of seconds to a few minutes. The impact of the HPP treatment is to inactivate the vegetative microorganisms which enhance product shelf-life, yet there are insignificant changes to product texture, flavor, and nutritional value. High-pressure processing subsequently can be used to retain the quality and freshness of seafood, improve food safety (e.g., inactivation of *Escherichia coli, Salmonella,* and *Listeria*), and increase the shelf-life of the product (Heather et al., 2011; Priyadarshini et al., 2019). So far usage of fish waste by-products are concerned, it is expected that in 2025, fish meal accounts for around 38% of the world's fish meal production which will produce from fish waste, compared with 29% for 2013 to 2015 average level (Omar, 2015).

8.4 SURVEY ON AREAS OF DISRUPTION IN THE FOOD INDUSTRY THROUGH STARTUP DRIVEN INNOVATION

Financial pressure, consumer preference, and technology are three important elements to bring changes in the food industry in terms of substantial changes and disturbances in the food manufacturing systems. The food industry these days is dominated by influential effects of demand of nutritious components, transparency of product and environmentally-conscious sourcing for development of products. World food retail market is projected to attain a value of 7,210 billion euro by the end of 2020. New outlooks and developing solutions laterally with startups are the important forces of the imminent future. Some of the startup-driven innovations are highlighted as follows *to* update the impactful 'Trends and Innovations' of promising startups and emerging companies.

8.4.1 IMPOSSIBLE FOODS

One illustration of a startup exploiting these propensities is *Impossible Foods.* Those companies make the plant-based *Impossible Burger*, which is described by some as "the game-changer in the food industry." The burger is made with soy leghemoglobin which carries heme, mimicking the taste of real meat.

8.4.2 ARTIFICIAL PROTEINS

3D food printers utilize various sorts of liquid or powdered food material to bring additional nutritional value to our plates and also provide extra income to companies. For a particular gender, lifestyle, or medical condition, 3D food printers allows personalized, precise, and reproducible nutrition containing the right percentage of nutrients which is taken into considerations. *Natural Machines*, the startup behind the 3D food printer *Foodini*, enhances cooking with fresh ingredients and this idea isolates the 3D food printer from others. The machine utilizes open capsules that users load up with foodstuffs of their choice to create savory or sweet cuisine. Thus, *Foodini* deals with the troublesome and tedious pieces of food preparation.

8.4.3 FOOD ROBOTICS

Robotics has not neglected to hit the food industry with an expected worth of 2.1 billion Euro by 2022. Food robotics enhances quality, output, and reduces expenses, especially in food and fluid processing. Though safety guidelines recognize food robotics from other automation, more companies utilize robotics to propel product consistency as well as overall efficiency. The need and interest for therapeutic and health benefit diets, nutritional value, as well as food inclinations, fluctuate widely between consumers. Thus, the *formation of personalized nutrition, or personalized meals*, has effectively influenced the food industry. *Personalized nutrition* is the following enormous development opportunity for organizations in the industry as consumers increasingly turn to individually-tailored diets.

Momentum Machines develops a fully autonomous burger-bot that allows you to cut toppings, bake patties and assemble and package burgers without human intervention. Headquartered in London, Nutrifix (https://nutrifix.co/) meets these demands by indorsing meals that are cooked, purchased or delivered to meet the nutritional needs of consumers. Age, height, weight, and health goals are taken into consideration by an app to calculate a personal recommended daily allowance and guides users toward their body's needs. *Too Good To Go* (https://toogoodtogo.co.uk/en-gb), a UK-based anti-food waste social startup, acts as a link between eateries with excess food and customers who can place orders via an app to collect it at a discounted price. The take-away administration is committed to cutting the amount of edible food thrown away by restaurants and is right now active in the UK, Denmark, France, and Norway.

8.4.4 IOT TECHNOLOGIES FOR FOOD WASTE

Some advantages of the application of 'IoT technologies in the food industry incorporate diminished maintenance costs, higher productivity, the transparent tracking of the supply chain as well as predictive maintenance. One application region for IoT in the food industry approaching a fast pace is the *smart kitchen*. To completely accomplish this achievement, devices, and equipment with built-in sensors are grouped into one network of data that continuously exchanges information.

Innit presents a universal platform for all kitchen gadgets that are equipped for analyzing various items in a refrigerator, monitoring their expiration date, suggesting recipes, and even preparing meals. Additionally, the in-assembled acknowledgment framework system supports consumers in keeping an eye on their fridge via smartphone or tablet. TellSpec is developing comprehensive solutions for cloud-based spectroscopy. A real-time cloud analytics engine powered by patented AI helps to monitor events of food fraud as well as events of food contamination locally and precise areas. In the food industry, aiming for Food Safety, delivery companies started focusing on AI solutions paired with automation strategies such as automated guided vehicles (AGVs), which are perceived as the ground equivalent to drones.

8.4.5 DISPATCH

Making a platform for local delivery powered by a fleet of autonomous vehicles designed for sidewalks and pedestrian spaces. The founders plan on relieving the cost of delivery due to savings in labor costs by bringing packages right to a customer's door and making their service available to the masses.

8.4.6 DELIVERY

The advancement areas are lead innovations that set the vibe for the overall market and furnish companies in the food industry with the right ingredients to remain on top (StartUS Insights, 2010).

8.5 DISRUPTIVE TECHNOLOGY

Disruptive technology is new expertise that is completely changing the way things are performed. Even though we cannot be certain which technologies will accomplish this in the future, the public has over the past years broadly accepted "disruptive technologies" as a term which refers to Blockchain, the Internet of Things, AI, drones, 3D printing, machine learning, biometrics, virtual, augmented, and mixed reality and other cutting-edge technologies, etc.

According to the 2018 World Trade Report (WTO), we are entering a new period where a series of inventions leveraging the Internet will have a significant impact on trade costs and international trade. IoT, AI, and 3D printed Blockchains have the potential to significantly change the way we transact, transact, and account. What need to be taken into consideration are the capacity building programs that can support a more balanced spread of adoption across the globe. This can be observed through the development of supply chains and the manner in which information is created and stored, which can be seen through the emergence of Blockchain technology and the way it has caught the imagination of many. There is a myriad of technologies on the horizon that have the potential to revolutionize not only the consumer markets in which they will operate, but also, for customs, the way its work is done (WCO, 2019).

8.5.1 DISRUPTIVE INNOVATION IN FISH PROCESSING TECHNOLOGY

The processing techniques used can vary significantly and are reliant on:

- Customer taste;
- Obtainability and prices of the raw material;
- Technical acquaintance;
- Time required for processing;
- Price of the finished product;
- Storing amenities;
- Marketability and periodic variations.

Generally, the fish is processed as the fish is not able to be sold in fresh condition, cold storage facilities are not sufficient, or the product is intended for distant markets. Domestic activities such as salting, smoking, drying, and fermenting are undertaken by women at home or on open grounds. Women prefers such activities because these functions could be performed along with their regular household works. Economic pressure to shift from existence to commodity production has recurrently resulted in reforming women's roles. Modern processing units tend to take over small-scale industries. The majority of fish processing plants focus on export potential rather than local consumption. The various types of fishes obtainable from domestic markets and the marketing prospects for women have altered as a result of the advancement of industrial processing. Food processing technology is constantly evolving. Frozen seafood, octopus farming, and shrimp farming are all gaining traction, contributing to the expansion of processed seafood and seafood processing equipment. With the advancement in food processing technologies, frozen seafood, octopus, and shrimp farming are also gaining momentum leading to huge market available for processing equipment.

8.6 STARTUPS ON AQUACULTURE AS THE PROCESSED SEAFOOD

With the acceleration of urbanization and the diversification of food choices, aquaculture is becoming the fastest growing form of food production in the world, contributing to the fastest growing animal husbandry industry. As the market for processed seafood and seafood processing equipment has grown exponentially over time, some start-up companies and established companies are willing to invest in seafood and aquaculture. The growing idea of clean meat is not limited to terrestrial animals, as various organizations are investing in research and development departments to compound the idea of clean seafood. These include Finless Foods, which are making efforts for the use biotechnology to process shellfish in the laboratory.

High-pressure processing (HPP) is the modern technology that the food manufacturers are aiming towards, which improves the shelf-life of a product by many folds. This non-thermal processing technique is welcomed by many seafood processors thereby accelerating seafood processing equipment market throughout the world.

Maxwell Chase's innovative seafood packaging line aims to accelerate the growth of the processed seafood market: As processors make groundbreaking changes in the outlook for the processed seafood market, Maxwell Chase announced in 2018 the creation of a unique packaging line that will extend the shelf life of fresh and frozen seafood. The integrated absorption technology implemented at the bottom of the tray absorbs excess liquid, thus increasing the shelf life.

With the estimate of CAGR of 3.5%–5% during the period 2019–2025, The global frozen food market reached between USD 225–228 billion in 2018. Therefore, the real market Marine products are expected to continue to thrive in the coming years.

Disruption in market for processed seafood and equipment for seafood processing by establishing of processing facilities around the world. Manufacturers are inclined towards this food processing technology because it can produce a variety of fish filets, such as straight cut, diagonal cut, contour cut and round cut. Various types of fish cutting equipment can be found, including manual, automatic, semi-automatic, and electric. These Fish cutting equipment are the most sought-after seafood processing equipment.

In cooperation with IMB, *Aquacloud* is all set to become global leader in the seafood processing equipment market and have designed the Seafood Innovation Cluster. To avoid and envisage sea lice epidemic, the company collects information from different salmon cultivators of Norway and installs machine learning expertise.

This ground-breaking innovation will attract the attention of various fish farmers and processors, thereby increasing the market leader in seafood processing equipment. Till 2018, the world seafood processing and seafood processing equipment market is estimated at approximately US $ 2.3 billion, and is anticipated to grow at a compound annual growth rate (CAGR) of 4.8% during the estimated period 2019–2025 (Industry ARC, 2019).

8.7 TRANSPARENCY AND TRACEABILITY OF SEAFOOD

A greater number of food manufacturers have digital systems for tracking and traceability, quality assurance and management of supply chain. Digital systems collect a lot of information which is used to gain a modest advantage. Find out how this data can be used to become more crystal clear for consumers, coordinate supply chains and decrease food waste from harvesting to consumer.

FAO defines Blockchain as an information technology that acts as a common record to digitally store and track data related to a product or service, from the stage of raw production to the point of delivery, i.e., consumers in real-time. The technology now being trialed in fisheries and in the food safety sector that holds considerable potential to improve market access, especially for small-scale fishers and fish farmers, FAO said. The difficulty of damaging information from the blockchain enhances the traceability of fish foodstuffs along the value chain, with more fisheries, farms, and fish processing facilities in the country of origin and plant hygiene standards of many countries. We will be able to meet such import requirements. Increased traceability can meet the demands of increasing buyers of legal and responsibly procured fish. Some fisheries and farms help meet certification requirements (Neil, 2018).

Currently, seafood consumers are thinking that they want to know exactly where their fish come from, how they were cultured, if they are safe to eat, and whether they got to their grocery seafood case at doorstep. We know that top suppliers are indispensable for doing business in the 21st century. Some say that while product quality is the most important thing on the consumer's mind, sustainability, and tracking are being pulled. Casey Marion's concept of tracking relates to the safety and quality of food and specific lot or shipment numbers. Transparency often involves mapping the supply chain and truly considerate everything contained in a particular distribution product. Today's industry requires full traceability and transparency throughout the supply chain.

David Pilat, Vice President of Business Development at Blue Circle Foods in Washington, DC, said the first step in developing a first-class traceability and transparency program would be knowing exactly where the fish came from. Patrick McBride, director of sales at McKinley Town, California, Wild Planet Foods, said: Traceability is "the ability to accurately trace the source of the seafood we supply without resorting to the use of physical resources measures designed to deceive." Transparency is "our attempt to interconnect and make this data available to all customers, users, and purchasers of Wild Planet products," he added. Traceability gives consumers the tools to make well-versed decisions about their seafood consumption. Traceability and transparency not only transfers power to consumers, but also embraces seafood suppliers responsible for the procedures and strategies they use to source seafood from the ocean. The cost of ensuring the sustainability, traceability, and transparency of seafood is high. The reputation of traceability and transparency has grown exponentially in recent years. Blue Circle Foods is at the forefront of traceability. Traceability gives consumers the assurance that they are eating fish sold in the market and that they have not been traded for other fish. This removes the risk of seafood fraud.

8.8 SEAFOOD FRAUD

Food fraud has grown commonplace in the sale of seafood items with false deceptive labels, descriptions, or promises. It not only jeopardizes the profits of reputable fishermen and seafood traders, but it also jeopardizes the progress made by sustainable fisheries and illegal and unregulated fishing to go unnoticed. According to a recent global analysis, around 30 of seafood across the world is falsely labeled or mislabeled. Each year, an estimated 11–26 million tons of IUU fish are landed, affecting the fishing sector between $10 billion and $23.5 billion. Despite the obvious dangers posed by seafood fraud, there is reason to be hopeful. The Marine Stewardship Council (MSC) has created its Chain of Custody Standard in 1999 to ensure that every distributor, processor, and retailer engaging in MSC-certified sustainable seafood has effective traceability systems in place. At present, throughout the world there are over 3,000 MSC certified companies (Lucy, 2016).

8.8.1 DNA TESTING PROCESS

During the processing of seafood products, the color, shape, and texture of fish and other marine species are frequently changed beyond recognition. As a result, identifying the species of fish present in a product just by sight is nearly impossible. However, regardless of how a seafood product is preserved (fresh, canned, frozen), or what form it is in (filet, eggs, fins, or processed product), even if it is fresh, canned, or frozen, it is still a seafood product. However, regardless matter how a marine product is stored (fresh, canned, frozen), or what form it is in (filet, eggs, fins, or processed product), even the smallest fragment has its own genetic code. Scientists can identify which species is present in a sample by comparing a certain segment of DNA to a reference library containing the genetic codes of most fish species, as well as for some species with very distinct populations they can identify where in the world that species was caught, Over the last decade, the use of DNA testing has revolutionized seafood traceability, and research and development is ongoing to increase efficiency and accuracy for an ever-growing number of species. It is a valuable tool for verifying and determining the authenticity of seafood products, preventing seafood fraud, and preventing the commercialization of endangered and vulnerable fish species (Lucy, 2016). The MSC undertook DNA research in 2016 to see if MSC-certified seafoods are meeting their traceability standards. The findings revealed that MSC branded seafood had a low rate of mislabeling (less than 1%) (Andy, 2019).

8.8.2 IMPORTANCE OF FOOD TRACEABILITY DURING COVID-19 PANDEMIC

Consumers, supermarket stores, and their supplier partners are becoming increasingly interested in tracking the progress of a food product from farm to fork. With the coronavirus, the trend has only accelerated, as industry members up and down the supply chain work to reassure consumers that, while they may have to fear interaction with their fellow citizens, they can at least rely on a safe food supply. Even before the pandemic, the US Food and Drug Administration (FDA) have put a greater emphasis on traceability. According to the FDA, the agency's Modern Era of Smarter Food Safety initiative builds on existing efforts to implement the Food Safety Modernization Act (FSMA) by using new technology to create a "more digitized, traceable, and safer system to help protect consumers from contaminated food."

The agency's latest initiative promotes traceability as well as transparency. The FDA hopes to release a strategy blueprint for implementation of the New Era of Smarter Food Safety this year, after stakeholder feedback that began last year. When it comes to food traceability, the effort's proposed use of new and emerging digital technologies will be critical in tracing the origin of contaminated food to its source in minutes, or in seconds, rather than days or weeks. The technologies that will lead the way are a Blockchain-based supply tracking system, distributed ledgers, sensors, the internet of things, and AI.

8.8.3 LOWERING RESPONSE TIMES

According to Thomas Burke, a food safety and traceability scientist at the Chicago-based Institute of Food Technologists (IFT), traceability is still focused on enhancing grocery stores' and suppliers' responsiveness to food safety incidents, particularly for fresh produce. "There has been a lot of excitement over the last two years on the application of Blockchain to automate supply chains," he said. While there is ongoing interest in Blockchain, pilots, and limited implementations among retailers and their suppliers have demonstrated that this technology is not required to meet their traceability demands.

Standardization and data collecting mechanisms for traceability, for example, are key to enabling external traceability. COVID, according to Burke, has generated increased interest in digital technologies, such as those that improve traceability, because it has made traditional third-party audits and verification of food safety measures more challenging. Burke noted that after COVID, there may be a lasting shift in consumer behavior when it comes to food consumption. Many consumers were dining more outside than inside in the months prior to the pandemic. Now they've improved their home cooking skills, and shops have increased their capacity to provide food straight to customers, he said (Nelson Andy, 2020).

A Blockchain-based traceability mechanism called Fishcoin: Singapore-based Eachmile Technologies has created Fishcoin, a Blockchain-based traceability system claims will improve the fish industry's stability and security. The white-colored white farming industries of Fishcoin say that the data is utilized to increase "efficiency and innovation," and this concept and the development of this seafood are based on this notion.

8.8.4 TRANSPARENCY AND TRACEABILITY OF OPERATIONS

The CEOs of eight global seafood companies signed a major ocean sustainability pledge to improve traceability and transparency in their businesses. The pledge also aims to reduce antibiotic use, participate in scientific research and development to improve aquaculture, reduce plastic usage and greenhouse gas emissions, and more.

Ecotrust Canada is launching 'ThisFish,' the world's first fish traceability program. Consumers can input a code online and learn information about their fish, who caught it, how, and where they caught it, and can even see a video clip. Consumers are becoming more interested in values-based companies, whether they are organic, eco-certified, Fair Trade, locally created, or sustainably sourced. Companies began to recognize that by improving their practices, they could improve social conditions, and even consumers responded favorably.

8.8.4.1 IMPORTANCE OF TRACEABILITY FOR THE SEAFOOD INDUSTRY

Companies can limit their risk exposure by using traceability. They can avoid seafood that comes from illegal fishing, for example. It could also be caught legally but mislabeled at the store. Illegal fishing, seafood fraud, and mislabeling are all risks, but human rights violations have also gained attention. Human slavery in fish supply chains has been reported in the New York Times, the Associated Press, and The Guardian. Of course, major brands are absolutely opposed to this. They can avoid dangerous supply chains and bad operators via traceability. The problem is that supply chains have become so complex that tracking a product's origins isn't always as simple as following a shipment from point A to point B.

Seafood for the future—with support from the European Union's (EU) Horizon 2020 program, a project called seafood future was launched to provide high-quality, balanced, and healthful seafood free of environmental toxins.

The project's goal is to create new, sustainable solutions to improve the safety and nutritional value of seafood throughout Europe. The project will create new information to develop commercially viable eco-innovative solutions for improving the socio-economic and environmental sustainability of European r645 seafood production and the processing industry, in order to fulfill a growing market demand for safe and sustainable seafood.

The project's knowledge management structures will ensure that innovative solutions in the seafood production and processing industries are transferred to end-users. The following are the expected outcomes:

- Producing seafood of market-driven, tailor-made, and eco-innovative with Better quality, traceability, authenticity, and safety of market-driven, tailor-made, eco-innovative.
- Production of novel, commercially feasible seafood production and processing methods that reduce production risks and damage to the environment.
- Preventing or eradicating contaminants such as norovirus, Listeria, and marine toxins from seafood with new innovative strategies.
- New technologies and procedures are being developed to help with traceability, authentication, labeling, and benchmarking of EU seafood products.
- To increase consumer confidence and trust in European seafood, a benchmark for seafood quality and traceability certification schemes has been established (Seafood Tomorrow, 2020).

8.9 BIG DATA ANALYTICS

Big data analytics mentions to tools and practices aimed at transforming large amounts of raw data to help understanding of human behavior. Most importantly, Bigdata being, a large, fast, and versatile information asset that requires information processing in the form of cost-effective innovations to enhance insight and decision-making, in addition to the ability to reveal stereotyped facts from large data sets as the "fourth paradigm of science." It is data (Gray et al., 2009). If data is the new oil, it is argued that knowing how to refine it into actionable intellect is the key to unlocking its potential. Big data sources typically include share some or all of these characteristics like i) Digitally generated ii) Passively produced – a byproduct of our everyday routine or dealings with digital amenities iii) Automatically collected iv) Geographically or temporally traceable vi) Continuously analyzed in real-time.

8.9.1 USES OF BIG DATA

So far, much work has been described on tools, software, platforms, analytics, etc., practical to big data (Roy, 2017; Dirk and Balietti, 2011;

Jeremy et al., 2008). These are considered as new industrial revolution. It also includes big data analytics to address the challenges of the fisheries and aquaculture.

8.9.2 IMPACT AND USE CASES OF CUTTING-EDGE TECHNOLOGIES

Using the state-of-the-art technology described above, petabyte raw information can provide clues to everything from reducing medical costs to TB prevention. Big data has the potential to innovate everything (Jeremy et al., 2008; Eysenbach, 2006). Big data is reported to monitor the epidemic and quality of water in Aquaculture. Based on Google queries in 2008 a tool is launched for flue like symptoms called Google Flu Trends. As per the U.S. Centers for Disease Control and Prevention (CDC), vast amounts of health-related online data can be mined to detect disease outbreaks "before diagnostic testing or laboratory confirmation." Google Dengue Trends shows similar possibilities that work equally well. Reports are available on how AI, IoT robots can be used to prevent disease and increase production value to improve labor efficiency.

8.9.3 BLOCKCHAIN IN AQUACULTURE AND FISHERIES IN THE SEAFOOD INDUSTRY

According to FAO, world fish production is reached to 171 million tons in 2016, 88% is used for human consumption. FAO estimates that by 2030, world fish consumption will increase by 20% from 2016 levels. Fish shine and many other renowned scientific institutions confirm that this could lead to one of the most valued industries in the world. It falls under the weight of its own growth. Therefore, lending is an important partner in collaboration with Fishcoin. This will accelerate the move towards integration, the SDG2 advocacy hub of the Unified World Food Program, for more sustainable seafood production. The Fishcoin ingenuity targets to solve the issue of traceability of the first mile for sustainable seafood production. Eachmile Technologies, based in Singapore, has developed Fishcoin, a Blockchain-based traceability system that it says can upsurge the stability and security of the seafood industry. Fishcoin claims that the data are used to increase "efficiency and innovation" behind this concept and the development of seafood (Yu, 2019; Jaya, 2018).

8.9.4 *COMBATING WITH ILLEGAL, UNREPORTED, AND UNREGULATED (IUU) FISHING*

However, one has to bear in mind that IUU fishers can use shared data to plan illegal activities or escape monitoring, if data sharing is done without proper control. In spite of significant improvements, around 20% of global fisheries production still comes from illegal, unreported, and unregulated (IUU) fishing. The IUU fishery costs the global economy up to $ 23 billion annually, threatens the sustainability of fisheries in our oceans and contributes to human rights abuses. Significant progress has been made in electronic surveillance of marine fisheries through AI, machine learning and digital image recognition. For some vessels, data is now transmitted in real-time over the Internet, even in remote fishing grounds.

8.9.5 *FAO NOTES DISRUPTIVE TECH COULD AID OVERFISHING*

In the freshly released "Status of aquaculture in the world" (SOFIA) report, FAO clarified that there are few drawbacks for the initiation of "disruptive technologies" in the fisheries field. It cites Blockchain technology, sensors, and automatic identification systems as largely positive developments, but does note there is room for abusing these. New technologies in the fisheries sector offer opportunities for improved fishing practices (e.g., more selective targeting of species or reduced losses of fishing gear). However, if abused, they can also be used to facilitate IUU [illegal, unregulated or unreported] fishing or If not focused in fisheries management, it can generally increase fishing capacity and lead to overfishing of resources. In the case of Blockchains, for example, they can make it possible to collect more data and to use it more resourcefully and successfully, thus increasing predictive fishing capacity. FAO also added that some new technologies have also created barriers to fisheries that lack capacity and funds. These factors underscore the importance of ensuring effective management in order to improve the fishery's sustainability using new technologies without compromising its sustainability. It is a need of an hour to understand the problems faced by fishers and fish farmers to use these new innovations, and accordingly for the best use of disruptive technologies, training and demonstrations need to be undertaken. It provides immeasurable opportunities to increase technological and financial efficiency, create new jobs, improve food security and livelihoods, and contribute to the 2030 Agenda, particularly the SDGs

[Sustainable Development Goals], in well-managed and disruptive technology sectors.

Jayson Berryhill, expert in mobile technology and partner in seafood Blockchain initiative Fishcoin told Undercurrent News that responsible safeguards should be in place to mitigate the risks of technology. The FAO describes a Blockchain as an information technology that acts as a shared ledger for digital storage and tracking of data associated with a product or service, from the raw material to the consumer's hand in real-time. FAO stated that the technology now being trialed in fisheries and in the food safety sector holds considerable potential to improve market access, particularly for moderate fishers and aquaculturists.

KEYWORDS

- **artificial intelligence**
- **disruptive technologies**
- **fish trade**
- **food traceability**
- **internet of things**
- **seafood processing**

REFERENCES

Andy, N., (2020). *Food Traceability Becomes Even More Important in Times of COVID-19.* Supermarket perimeter. https://www.supermarketperimeter.com/articles/5490-food-traceability-becomes-even-more-important-in-times-of-covid-19 (accessed on 09 March 2022).

Dirk, H., & Balietti, S., (2011). *From Social Data Mining to Forecasting Socio-Economic Crises.* http://arxiv.org/pdf/1012.0178v5.pdf (accessed on 09 March 2022).

Eysenbach, G., (2006). Infodemiology: Tracking flu-related searches on the web for syndromic surveillance. In: *AMIA Annual Symposium Proceedings* (Vol. 2006, p. 244). American Medical Informatics Association.

FAO, IFAD, and WFP, (2015). *The State of Food Insecurity in the World 2015.* Meeting the 2015 international hunger targets: Taking stock of uneven progress. Rome, FAO.

Gray, J., Tansley, S., & Tolle, K., (2009). *The Fourth Paradigm: Data-Intensive Scientific Discovery.* Microsoft Research. Redmond, Washington. http://research.microsoft.com/en-us/collaboration/fourthparadigm/contents.aspx (accessed on 09 March 2022).

IndustryARC, (2019). *Processed Seafood and Seafood Processing Equipment Market – Forecast (2020–2025).* Report Code: FBR 76622. https://www.industryarc.com/Research/Processed-Seafood-Seafood-Processing-Equipment-Market-Research-504486 (accessed on 09 March 2022).

Jaya, N., (2018). *Blockchain in Aquaculture and Fisheries of Seafood Industry.* http://www.techiversy.com/Blockchain/Blockchain-in-aquaculture-and-fisheries-of-seafood-industry/ (accessed on 09 March 2022).

Jeremy, G., Mohebbi, M. H., Patel, R. S., Brammer, L., Smolinski, M. S., & Brilliant, L., (2008). Detecting influenza epidemics using search engine query data. *Nature*, 1012–1014.

Kim, D., & Seo, J., (2018). A review: Breathable films for packaging applications. *Trends Food Sci. Technol., 76*, 15–27.

Lucy, A., (2016). How DNA testing helps to ensure traceable, sustainable seafood? *Traceability: From Ocean to Plate*. https://msc20-live-cd2.msc.org/docs/default-source/default-document-library/what-we-are-doing/msc-from-ocean-to-plate-traceability-and-dna-report-2016.pdf?sfvrsn=fbce98a4_4 (accessed on 05 April 2022).

Manuel, H., Thompson, M., & Carroll, K., (2011). Emerging trends in seafood processing technologies. Putting the 'fish' in efficiency, *J. Ocean Technol., 6*(4), 15–20.

Neil, R., (2018). *FAO Notes Disruptive Tech Could Aid Overfishing*. Undercurrent News, seafood business news from bench to the surface. https://www.undercurrentnews.com/2018/08/01/fao-notes-disruptive-tech-could-aid-overfishing/ (accessed on 09 March 2022).

Omar, R., (2015). *Fish Waste Management: Turning Fish Waste into Healthy Feed.* FAO, United Nations, Rome, Italy.

Priyadarshini, A., Rajauria, G., O'Donnell, C. P., & Tiwari, B. K., (2018). Emerging food processing technologies and factors impacting their industrial adoption. *Crit. Rev. Food Sci. Nutri.* doi: 10.1080/10408398.2018.1483890.

Roy, A. K., (2017). *Applied Big Data Analytics: Business Analytics, Healthcare, Capital Market & Life Sciences (Bigdata Book 1).* Amazon Kindle Edition. https://www.amazon.in/Applied-Big-Data-Analytics-Healthcare-ebook/dp/B010IOJ12U (accessed on 09 March 2022).

StartUS insights, (2010). *Disrupting The Food Industry: A Breakdown on Startup Driven Innovation.* https://www.startus-insights.com/innovators-guide/disrupting-food-industry-breakdown-startup-driven-innovation/ (accessed on 09 March 2022).

United Nations, (2015). Department of economic and social affairs, population division. *World Population Prospects: The 2015 Revision, Key Findings and Advance Tables.* Working Paper No. ESA/P/WP.241.

World Customs Organization (WCO), (2019). *Study Report on Disruptive Technologies*. Rue du Marché, 30 B-1210 Brussels, Belgium.

Yu, L. G., (2019). *Fishcoin initiative Aims to Solve Traceability Challenges in First Mile for Sustainable Seafood Production.* https://www.foodnavigator-asia.com/Article/2019/11/28/Fishcoin-initiative-aims-to-solve-traceability-challenges-in-first-mile-for-sustainable-seafood-production? (accessed on 09 March 2022).

CHAPTER 9

Secondary Raw Material and Low-Cost Fish for Fish Meal Production and Its Implications in the Animal Feed Industry

L. NARASIMHA MURTHY,[1] GIRIJA G. PHADKE,[1] A. JEYAKUMARI,[1] and S. VIJAYAKUMAR[2]

[1]*Mumbai Research Center of ICAR-Central Institute of Fisheries Technology, CIDCO Admin Building, Sector I, Vashi, Navi Mumbai – 400703, Maharashtra, India*

[2]*Fisheries Research Station and Information Center, Buthnal, KAFSU, Bidar, Karnataka, India*

ABSTRACT

In order to cope up with the change in demand and market for fish meal, there is a need to enhance knowledge on fish meal in regard to increase its nutritional value and economic value in feed industry. Fish meal, raw material used, quality, and regulations, nutritional aspects, processing, and preservation, importance, applications of fish meal have been explained in the present book chapter. There is a need to explore new areas for source of material which will enable industry-driven approach. The use of alternative to fish meal in aquafeed industry and search for such alternative options are current trends in the industry.

9.1 INTRODUCTION

Fish is a significant and important wellspring of proteins. Processing of fish create huge waste which represents disposal problems if not managed

Advances in Fish Processing Technologies: Preservation, Waste Utilization, and Safety Assurance.
Ranendra K. Majumder, PhD & Amjad K. Balange, PhD (Eds.)

appropriately. Usage of fish handling waste for arrangement of fish meal diminishes the threat and also leads to produce fish feeds (Murthy et al., 2013). Apart from whole fish, fish byproducts are likewise used in fish feed processing. Production of fish over the globe has been estimated at 179 million tons in 2018 wherein, direct human consumption share accounts for 156.4 million tons with 22.2 million tons being utilized for non-food uses. Global per capita consumption of fish is 20.5 kg (FAO, 2020). Small fish species including low cost and trash fish form the bulk when a fishery is concerned. In total, around 15–18% of the fish landings is utilized in fish meal and fish oil manufacture globally. About 50% of the total landings are usually consumed as food.

The fisheries sector plays an important role in socio-economic development with significant contributions to food security, livelihood option, and diversified agriculture. Fish and fishery products processing is one of the rising sectors in the Indian economy for earning foreign exchange through worldwide export of substantial amount of fish and fishery products (Shinoj et al., 2009).

9.2 FISH MEAL MANUFACTURE FROM FISH WASTE

Processing industry in fisheries lead to generation of high quantity of raw material as waste which may be termed as secondary raw material for production of high value byproducts and fish meal. The quantity of the waste generated ranges from 45 to 55% depending on the product manufactured. Waste from fish processing includes head, fins, skin, visceral organs, fileting frames, etc., which, if left unattended lead to environmental pollution and a few other problems as well. Conversion of these secondary raw materials into fish meal and oil may lead to overcome the problem. Aquafeeds are reliant on fish meal and fish oil as major ingredients. Increased aquaculture activities demand for more quantity of aquafeeds to be produced. Potential of byproducts and rest raw materials need to be considered to ensure proteins for aquaculture industries (Hua et al., 2019). Filleting operations in the processing of fish generate large proportions which forms a valuable source as a raw material for production of fish meal and fish oil which is at the level of 30% at present.

9.2.1 FISH MEAL: HISTORICAL PERSPECTIVE

Fish meal which was utilized distinctly as manure since 1910 is also utilized in animal feed. Fish meal manufacture started in Europe and America path back to 18th century. Idea behind fish meal and oil production was using it as a fertilizer. The use of secondary raw material from fish processing as an animal feed started as early as 800 AD with the use of herring in Norway. After oil pressing, remains were utilized as fish feed. Fish meal primitive form was explained in 14th century itself. British and Japanese used fish meal as fertilizer in the past which now have great demand in animal feed industry due to its nutritional importance.

9.2.2 *WHAT IS FISH MEAL?*

Fish meal is obtained by crushing fish and other fish parts and expulsion of water and all/some oil (Ruiter, 1995). Fish meal is a product in dry with brown-gray color in a coarse to fine powdered form. Fish meal is a brown powder which normally contains a high level of protein and appreciable quantities of lipid and minerals (Barlow, 2003). Approximately 90% of the fatty fish species are used in fish meal and fish oil production (Barlow, 2003). It is generally obtained by evacuating the majority of the water nearly (90 to 95%) and a few or the entirety of the oil from fish or fishery waste or secondary raw materials. In general, pelagic species of fish having higher content of oil are used in the manufacture of fish meal. Fish meal is commonly sold as a powder, and is utilized for the most part in compound nourishments for poultry, pigs, and cultivated fish; it is unreasonably significant to be utilized as a fertilizer. Fish meal is viewed as profoundly gathered nutritious enhancement in terms of proteins, vitamins, minerals, and so forth. Essential mineral compounds namely calcium, iron, and phosphorus are found in higher proportions in fish meal. General fish meal manufacture involves cooking, squeezing, drying, and pounding the fish and fishery waste. Fish meal production leads to a byproduct known as fish oil which usually contains fatty acids (FAs) in the form of triglycerides with phospholipid, ethers of glycerol and esters of wax in variable proportions. Carbon atoms ranging between 14 to 22 and a higher degree of unsaturation up to six double bonds are the important characteristics of fish oil (FAO, 1986).

9.2.3 *RAW MATERIAL USED IN PRODUCTION OF FISH MEAL*

Raw material utilized for fish meal processing could be any fish, bycatch, waste generated from canning industry and fileting operation, secondary raw materials generated from fish processing, etc. Fish meal is specially water-soluble vitamins. It is viewed as a wellspring of vitamin D. Fish meal manufacture process is subject to the raw material to be utilized, fat content and protein content. High protein content in fish meal permits its utilization in manufacture of aquatic and animal feed. Fish meal production includes dry and wet rendering out of which dry rendering is applicable for lean assortments of fish while wet rendering is appropriate for fish species having high lipid content.

When global fish meal scenario is concerned, anchoveta is among the main group of fish species used in fish meal and fish oil production. Manhadens from Gulf and Atlantic, Norway pout which are not marketed as food are utilized in fish meal manufacture. Forage fish form the food-grade category include Anchovies from Peru, Japan, Europe, and other countries Capelin, blue whiting, and sprat from Europe which have small demand, are also used as raw material in fish meal production. Fish which are rejected in food markets also go in fish meal manufacture. Mackerel, sardines, and herrings are few species utilized in fish meal production (FAO, 1986).

9.2.4 QUALITY OF RAW MATERIAL USED IN FISH MEAL MANUFACTURE

Optimal raw material is necessary for producing quality fish meal and fish oil as it will lead to optimal nutritional value. Raw material for fish meal production is highly perishable, leading to spoilage. Total volatile base nitrogen content of raw material is used as an important parameter when fish meal quality is considered. European Food Safety Authority (EFSA) also suggests checking sensory parameters of the raw material along with biochemical quality parameters.

Significant portion of fish production is converted to fish meal and oil which is further utilized in animal and aquafeeds indirectly making contribution to human consumption. Production of fish meal and oil is generally carried out from sustainably managed fish stocks which are having less demand for direct consumption. Fish meal quality can be high quality, low temperature fish meal and prime quality fish meal which has different application field's right from small units of aquaculture to animal feed industries. Indian oil sardine (*Sardinella longiceps*) is widely used in the manufacture of fish meal and oil in India. Freshness of raw material is first and foremost important point to be considered in determining quality of fish meals. It has an impact on nutritional quality characteristics and cost of the final produce. Levels of TVB-N, proteins, fats, and biogenic amines are important parameters to be considered. Microbial and enzymatic actions reduce the protein quality of the final produce. Oxidation of fatty compounds due to non-enzymatic reactions may lead to oxidation of final produce.

9.2.5 RAW MATERIAL SOURCES USED IN FISH MEAL PRODUCTION

In general, sources of raw material are categorized into three categories by FAO (1986) as: (i) raw material, i.e., fishes caught exclusively for production of fish meal and oil from different regions such as Peru, Norway, USA; (ii) raw material caught as bycatch; and (iii) secondary raw material produced as an output of processing such as fish frames, fish cuts, etc. There is a wide variety of fish species used in the production of fish meal and fish oil. Species of fish that are smaller in size and fatty in nature are used in fish meal and fish oil production. Secondary raw material generated from fish processing such as offal's, trimmings, fileting waste and cannery waste is also used in manufacture of fish meal and fish oil. Product yield depends on raw material quality and compositional factor (FAO, 1986). Fishes having protein content in the range of 14 to 18%, minerals in the range of 1–4%, fat in the variable range between 2–31% and moisture content between 60–80% are utilized in fish meal production.

Seasonal Changes in fish composition, diet, zooplankton diversity and availability, size of fish, reproductive stage are changing factors in relation to season which determines nutritional composition of fish especially FA composition of fishes. Due to the varied activity of endogenous proteases, variation in level of spoilage is observed (Table 9.1).

TABLE 9.1 Raw Material Utilized in Fish Meal and Fish Oil Production in Various Countries

Country	Raw Material
Peru	Peruvian anchovy
Chile	Anchovy and Chilean mackerel
China	Anchovy, sardine, mackerel, variety of species
Denmark	Eels, sprats, pouts
Norway	Herrings, whiting
South Africa	Pilchard
Japan	Mackerel, sardine, pilchard
India	Indian oil sardine, Indian mackerel, anchovies
Thailand	Variety of species ranging from sardine, mackerel to anchovy
USA	Pollock, menhaden

9.2.6 SECONDARY RAW MATERIALS

Filleting and canning operations generate a large amount of secondary raw materials accounting for up to 70% of whole fish. Responsible sourcing of raw material is essential to ensure the quality and safety of fish meal.

9.2.7 FISH MEAL PROCESSING

Fishmeal is the crude flour obtained after milling and drying fish or fish parts, while fish oil is usually a clear brown/yellow liquid obtained through the pressing of the cooked fish. The main purpose in fish meal production is reducing moisture content of raw material to the level below 10% which means 90 to 95% of moisture forms the fish and fishery waste is removed wherein oil should not exceed 10%. Hence, oil content from fish need to be removed by 80 to 90% during the manufacture of fish meal. The yield varies depending on raw material utilized but it is approximately 20% and 5% for fish meal and fish oil from raw material.

In general, fish meal production is carried out by two methods which are:

- Dry rendering process suitable for raw material having low fat content; and
- Wet rendering process suitable for raw material having low fat and high fat content.

The wet rendering process is used more when compared with dry rendering method in fish meal production. Dry rendering or dry reduction method is suitable for lean fish containing less than 2–3% oil. It is not a continuous process.

9.2.8 FISH MEAL AND OIL MANUFACTURE FROM THE FISH AND FISH PROCESSING WASTE

In wet rendering process of fish meal manufacture fish meal and fish oil are produced in a single unit. Main flow steps in the fish meal and oil manufacture are heating to coagulate the proteins and rupture depots of fat and liberate oil and bound water, pressing to remove large amount of liquid portions from the solid mass, separation of water (stick water) and lipids which is excluded if the raw material contains lesser proportion of fat than 3%, evaporation of stick water to fish soluble, drying of the solid mass with soluble added to form a stable fish meal by removal of water, grinding/pulverizing dried material to get fish meal of desired fineness.

9.2.8.1 RAW MATERIAL RECEIVING

It is the first and foremost step from where fish meal production process starts. In general, standardized processing of fish meal contains coagulation by means of heat and separation of fat by mechanical means followed by dewatering by means of heat. Raw material is minced when fish and secondary raw materials are utilized (FAO, 1986). Knife hasher type of mincing equipment is utilized which consists of rotor and knives (FAO, 1986).

9.2.8.2 COOKING

Fish and fish processing waste is cooked at temperatures in the range of 95°C to 100°C for a period of about 15–20 minutes using cookers as a source of heating the fish and fishery waste quickly. Cooking is generally carried out with supply of steam in indirect means. Cooking leads to coagulation or denaturation of proteins in fish and fishery waste and also ruptures the cell wall of fish tissues leading to separation of fat and water in fish. Heating process also lead to inactivate microbial load and parasites present in the raw material (Ruiter, 1995). Temperature varies from 70°C to 100°C as targeted temperature. In order to reduce energy cost, precookers are also used which are powered by evaporator generated steam for efficient energy utilization. The cookers are steam jacketed with middle screw heated for material feeding for active heat transfer. Cookers are attached to strainers to remove the released water and oil during the cooking step.

9.2.8.3 PRESSING

It separates solid and liquid phase, i.e., press cake and press liquor. Screw presses either single screw press or double screw press is utilized in pressing operation. Double screw pressing is preferred in order to achieve removal of maximum quantity of oil and water from the cooked raw material. Press cake usually has about 45 to 55% water with nearly 3% fat. Liquor is usually used further to extract oil. High moisture levels may cause higher energy consumption (FAO, 1986).

9.2.8.4 DECANTING

It separates solids and liquids and solids are returned to the press cake. Decanting may be two-phase or three phases wherein two-phase decanter segregates solids and liquids whereas three-phase decanter segregates solids, liquids, and oils (FAO, 1986).

9.2.8.5 FLUFFING

Large lumps of press cake from twin-screw press are converted to small pieces of approximately 1 cm size. Fluffing operation increases drying efficiency and reduce drying time. Due to advancement in screw press machinery, this step is not essential as press cake disintegration takes place in screw conveyors itself during its movement to dryer.

9.2.8.6 DRYING

Drying reduces moisture content to less than 10%. Drying can be direct or indirect. In direct type of drying, flame is used in direct mode to dry the press cake whereas in indirect type of drying, steam is used to dry the material. Drying operation usually needs about 20 minutes to achieve a moisture content of less than 10%. In order to remove steel-like contaminants, magnetic separator is fixed at the end of the drying line before cooling. Drying uses two-step functioning with removal of moisture from surface in the first step and removal of moisture within the material in the second step. Dryers of indirect steam type are preferred due to simplicity, lower investments needed with processing of large quantity at a time.

9.2.8.7 COOLING

Coolers are utilized to cool the fish meal dried. These are similar to dryers wherein the only difference is cool water is passed instead of steam as in the case of dryers which enables cooling of fish meal to ambient temperatures (FAO, 1986).

9.2.8.8 SIEVING

Press cake after drying is subjected to a screen which is of vibratory type to segregate extraneous matter such as shells, hooks, wood, cloths, etc., before milling.

9.2.8.9 MILLING

Fine fish meal particle size is obtained using milling averaging around No. 40 mesh Tyler screen. The small-sized particles thus produced pass through the sieve fixed at the bottom of the chamber (FAO, 1986).

9.2.8.10 PACKAGING, LABELING, AND STORAGE OF FINAL PRODUCE

Fish meal after milling is packed in jute bags or paper bags or HDPE bags which are usually polyethylene (PE) lined. Proper labeling and storage at ambient temperature are carried out till further use. Ventilation is necessary in the area where fish meal is stored. Strict control of temperature and humidity levels is essentials. Fishmeal's are normally packaged into bags of different nature and different materials (FAO, 1986). Bulk storage is the preferred method of storing fish meal.

9.2.8.11 TECHNIQUES FOR FISH AND FISH MEAL PRESERVATION

Cold chain management during transportation and storage of raw material to fish meal plant is essential. Chilling of fish is essential to get quality raw material. Due to the sensitive nature of fish meal and fish oil, oxidation of fat from fish meal and fish oil is common. Antioxidants such as ethoxyquin are utilized in order to prevent oxidation of fats. Addition of synthetic antioxidants such as butylated hydroxytoluene (BHT) is also common (FAO, 1986). However, raised public concerns have regulated use of such antioxidants in fish meal. Schematic representation of steps involved in fish meal and fish oil production is given in Figure 9.1.

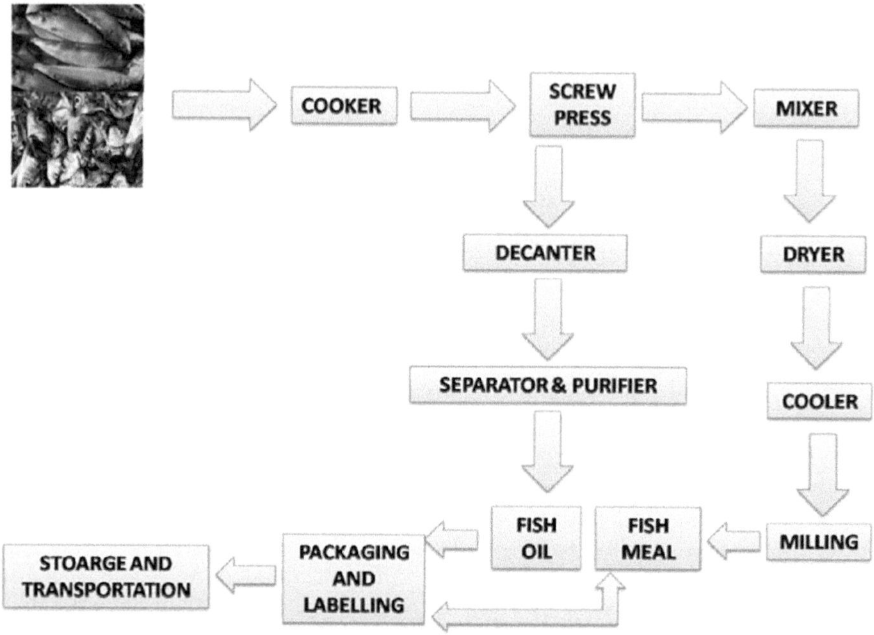

FIGURE 9.1 Fish meal and oil manufacture: Steps in production.

9.3 NOVEL METHODS IN PROCESSING OF FISH MEAL

Standardization of conventional methods of fish meal processing lead to enhanced energy-efficient operations yields of oil and protein. Enzymatic protein hydrolysates and fish protein concentrates are the options alternative to fish meal which find applications in a range of products from food, feed, fertilizers to nutraceuticals. Protein concentrates are usually manufactured from fish processing waste. Mincing of raw material followed by addition of organic acids to lower down pH to less than 5 which produces silage is heated using decanter centrifugation to obtain fish protein concentrates which are categorized into three as type A with fat less than 0.75%, type B with fat content up to 3% and type C which is hygienically produced fish meal, as per FAO. When compared with fish meal, it has higher levels of protein and lower levels of minerals. Diet supplementation with FPC can fulfill deficiency in proteins. Hydrolyzed fish proteins are produced by conversion of proteins to low molecular weight peptides using enzymes. It is possible to use crude visceral enzymatic protein hydrolysates from fish proteins which have a wide range of applications. These are highly valued bioactive protein hydrolysates and can be produced from frame waste of fish as well.

9.4 QUALITY OF FISH MEAL

Quality of fish meal solely depends on the raw material utilized in the production of fish meal and fish oil as it is directly related to nutritional quality and price of the final produce. Species of fish, conditions such as temperature, level of pollution, feed utilized to grow fish, seasonal conditions, preservation techniques are important parameters that determine the quality of fish meal, its yield and nutritional value. Quality and fresh raw material is important factor to be considered in fish meal production. A higher value of total volatile basis nitrogen is unacceptable if the intended use is in shrimp feed industries. If fish waste is to be used, it is essential to determine quality parameters for the waste as well. Drawback in the use of waste as raw material is lower proteins and higher ash content. In order to produce fish meal of grade I, use of only waste as a raw material is not suggested. Microbial and enzymatic actions degrade the quality of the final produce.

Fish meal and squid meal produced in the Gujarat region exhibited higher cadmium content, especially in the squid meal. There is a need for reduced contamination from heavy metals by use of raw material with high quality and by adoption of good manufacturing practices and hazard analysis critical control point related measures (Murthy et al., 2013). There is certain non-enzymatic reactions cause oxidation affecting nutritional quality of lipids.

Important physical properties of fish meal includes color which is usually range from light brown to dark brown free from burning, odor-free from rancid odor, fine granulous to powdered texture, free from visible molds and contaminants, etc. Chemical components of fish meal show variation in proteins from 50–70%, crude fats in the range between 5–10%, ash contents ranged between 12–33% and moisture content lies between 6–10%. Water-soluble vitamins of B group are present with average values of riboflavin–7.3 mg/100 g of fish meal, niacin 126 mg/100 g of fish meal, pantothenic acid–30.60 mg/100 g of fish meal, Vitamin B12–0.25 mg/100 g of fish meal, pyridoxine–5.7 mg/100 g and choline–4,000 mg/100 g of fish meal. Residual oil in fish meal makes it rich in vitamin D as well (5,000 IU/kg of fish meal). Fish meal of Indian origin exhibits higher proportions of phosphorus to calcium 1:1 against 1:2 proportions in other fish meals and mineral composition includes zinc–70 mg/kg, iodine–70 mg/kg, iron–250 mg/kg, copper–7 mg/kg, manganese–4 mg/kg, cobalt–0.1 mg/kg with proportions of fat which oxidizes on storage (Cho and Kim, 2011).

Along with biochemical quality, microbiological quality standards for evaluation of fish meal available are Indian Standards and European Union (EU) standards which are applicable for Indian markets and European markets, respectively. Indian standards for fish meal were published in 1967 with revision in 1983, i.e., IS: 4307–1983, which are followed in India which recommends no growth of visible fungi and free material from *Salmonella* (Murthy et al., 2019).

9.5 FISH MEAL: ANALYSIS OF QUALITY

A range of methods right from traditional methods to sophisticated methods like GC-MS are used in analyzing the quality of fish meal. In general, crude protein content, water-soluble proteins, moisture content, fat content, mineral content, salt content, total volatile base nitrogen content, histamine level, are important quality parameters analyzed during evaluation of fish meal quality (Murthy et al., 2019). Crude protein level is estimated in terms of total nitrogen content using the Kjeldahl method. Moisture and ash level using standard ISO methods, salt content using Volhards titrimetric method, peroxide value and free FA content using chloroform extraction and titrimetric method, thiobarbituric acid substances using spectrophotometric method for malondialdehyde produced, total volatile base nitrogen using Conway micro diffusion analysis are used in quality evaluation. HPLC size exclusion chromatography and UV adsorption methods are used to determine the size of peptide fractions. Various physical quality parameters such as flow properties, bulk density, are also evaluated using various standardized methods. Sophisticated method such as solid phase microextraction and gas chromatography are used. ISO methods for *Salmonella* are used to detect the presence of *Salmonella*.

9.6 IMPORTANCE OF FISH MEAL

Major fish meal manufacturing countries include Peru, Chile, Asian, and southeast Asian Countries, Denmark, and the USA. Fish meal is an important byproduct from fisheries industry. Its application in aquaculture industry and feed industry provides essential omega 3 FAs such as EPA and DHA to human beings through consumption of aquacultured fish. These omega three FAs importance in cell membrane functioning, cardiovascular functioning, brain development, immunity functioning is well documented. The importance of fish meal and fish oil also lies as a feed for farmed and aquatic animals due to its nutritional characteristics and digestibility. In order to support physiological functions and growth of fish and other animals consuming fish meal based feeds, micronutrients in the form of amino acids, vitamins, and minerals are provided. Peculiar characteristics of fish meal includes high content of proteins ranging from 60% to even more than 70%, presence of long-chain omega 3 FAs, presence of calcium, phosphorus, potassium, selenium like minerals, presence of B group vitamins like B1, B2, B6, and B12, its higher digestibility values, etc. Fish meal proteins are not replaceable as it contains lysine, which is deficient in plant proteins. In addition to proteins, fat-soluble vitamins are also found in higher quantities in fish meal. The fat content of fish meal aids in fish growth and weight gain, and it also functions as a source of energy. Fish meal higher in mineral content aids in the formation of bones in fishes fed with fishmeal-based diets. Calcium, iron, trace elements such as copper, iodine are required in the physiological functioning of animals. Lower fiber content enables ease for absorption in fish and poultry. Growth factors present in fish meal makes it more nutritive when utilized in feed industry. At the end of this decade, about $1/6^{th}$ of the captured fish landed will be utilized for the production of fish meal with an estimated level of production of fish meal would be 5.3 million tons. More than one-third of the fish waste obtained from seafood processing would be used in fish meal production.

9.7 FISH MEAL AND ITS APPLICATIONS

The peculiar characteristic of fish oil makes it useful in reducing the level of cholesterol in blood with a positive effect on metabolic and physiological body regulations. Apart from its use in human food, fish oil also possess several industrial applications such as drying oils in varnish and paint manufacture, as a filling medium in fish canning industries, in production of margarine and as a carrier of fat-soluble vitamins and it is applicable manufacture of linoleum, detergents industries, rubber manufacture, ink manufacture, lubricant manufacturing industries, soap making units, etc. (FAO, 1986). Fish meal depending on raw material is divided into 4 categories as: (a) herring type; (b) anchovy/pilchard; (c) menhaden; and (d) white-fish

(Barlow, 2003). High biological value of fish meal is useful in animal diets. Rich amino acid profile especially by presence of lysine and sulfur amino acids is important in animal feeds. Fish meal possess high digestibility of nitrogen which is 85% for poultry and 89% for piggery (Barlow, 2003). In the poultry industry, broilers are fed with fish meal based diets followed by birds and layers with levels equivalent to 2 to 8% (Barlow, 2003).

9.7.1 APPLICATION IN ANIMAL FEED INDUSTRY

Fish meal is a major source of proteins in the manufacture of animal feeds and commercial aquafeeds. Fish meal cost depends on various parameters such as availability of raw material used in formulation of fish meal, site, climatic conditions, composition, export potential, stability, palatability, and protein content. In the global view, fish meal and fish oil production is an important industry. However, static production is recorded in the last few years. Fish meal price declined almost by 50% from 2014 to 2017 in just 3 years from 2,400 USD to 1,100 USD per ton with the enhanced use of soya meal and other ingredients as substitutes for fish meal in the manufacture of fish feed. Optimization of process parameters due to increased level of competition from fish meal and oil substitutes are a need of an hour to make it beneficial for improved value chain management in fish meal industry.

9.7.2 APPLICATION OF FISH MEAL IN AQUACULTURE SECTOR AND FEED INDUSTRY

Aquafarming is one of the fastest-growing food sectors in the globe wherein fish meal formed a major source of proteins in feeds. With the limited supply of raw material for fish meal production in recent years, the level of fish meal addition to feeds have been observed to be drastically reduced (Olsen and Hasan, 2012). Fish meal was cheap raw material to be used in animal feed manufacture. Rapid growth of aquaculture sector has led to increased demand and price of fish meal (Olsen and Hasan, 2012). Fish meal produced in South Africa is widely used as an ingredient in Chicken feed at 3 to 10% proportions due to its nutritional composition. Fish meal production from South Africa accounts for 1.5% of fish meal production around the globe. Culture of finfish and shellfish depends on fish meal as a source of dietary ingredient. Aquaculture is highly dependent on it especially in shrimp culture and fish culture. Costs of fish meal are increasing globally thus feed manufacturers are trying to reduce levels of inclusion in aquafeed (Tacon and Metian, 2008).

Fish feeds are usually manufactured using extrusion processing. Fish meal is added during the extrusion process as one of the major ingredients in fish feed production. Fish meal has unique nature which is different from plant-based proteins demanding higher moistures and energy levels for pellet making process. Increasing aquaculture production is demanding for higher quantity of fish feed wherein fish meal is an important ingredient. About 88% of the total catch is utilized for human consumption whereas; 12% of the catch is utilized for non-food uses such as fish meal (FAO, 2020). In addition 30% of secondary raw materials are converted to fish meal. With the contribution of over 60% on feed costs in aquaculture, alternatives for fish meal are also utilized due to higher cost of fish meal.

9.7.3 ALTERNATIVE OPTIONS FOR FISH MEAL IN FEED INDUSTRY

For growing aquaculture industry, there is a need of high protein diets which are usually obtained from fish meal. Due to unsustainable use of forage fish in aquafeed production, additionally protein source alternative options are a need of an hour. Alternatives can be from plant sources or aquaculture byproducts, insect meals, etc., which have a great potential through bioconversion in order to ensure environmental sustainability (Hua et al., 2019). Replacing fish meal in aquafeed is complex as fish meal exhibits excellent amino acid composition along with other features such as nutrient digestibility and lack of anti-nutritional compounds in fish meal. There are several studies on fish meal replacement with other ingredients.

No successful attempts and research is there on replacing fish meal completely with other protein sources due to anti-nutritional compounds and other reasons related to nutrition. There are comparative studies on amino acid profile of fish meal and plant based meal indicating need of removing anti-nutritional factors such as trypsin inhibitors, tannins, etc., from plant based meals. Less phosphorus content in plant meal is also a barrier in replacing fish meal with other ingredients. Form of phosphorus in plant, i.e., phytase phosphorus is not bioavailable to fishes and its use in organic farming is restricted. Further, several ingredients like microbes, fungi, yeasts, microalgae, blood meal, worm meal, insects, gastropods are also suitable as partially replacement options for fish meal as an ingredient in aquaculture feed. Microbes are candidate replacement in carnivore diets and salmonid diets. Microalgae being the part of primary production have also gained importance in fish feed industry wherein fermentation and photo bioconversion techniques are adopted for making them palatable. Animal based proteins and utilization of insects are also one of the potential alternative options due to their cheaper price. House flies pupae, silkworm pupae have gained importance in carp, tilapia, and catfish culture.

9.7.4 *QUALITY OF FISH FEED*

Fish meal is one of the most important ingredients in fish meal. Feed quality is a direct impacting factor on aquaculture practices and economic returns as well. Commercially produced fish feeds from India indicated contamination of fish feeds with microbes. Microbiological quality analyzes indicated TPC levels up to 4 log values with the presence of *E. coli*, Enterobacteria, yeasts, and molds in a few feed samples. Standardized quality of feeds is essential as many of feed manufacturers are not following regulations in India.

9.7.4.1 HISTAMINE

Fishmeal industry faces problems of high levels of histamine. Study on histamine levels in fishmeal and shrimp feed exhibited values equivalent to 138 $\mu g.g^{-1}$ and 191 $\mu g.g^{-1}$ respectively (Kennedy et al., 2004). A study of fish meal quality from Taiwan revealed higher TVB-N content in 87% fish samples and higher histamine levels in 10% fish meal samples.

Biogenic amines are produced from various microbial actions causing enzymatic hydrolysis. Conversion of histidine to histamine, lysine to cadaverine, arginine to putrescine and tyrosine to tyramine are few examples of biogenic amines. These are important from a safety point of view. Higher levels of these biogenic amines are risky to human health. These are stable during heating but is water-soluble making them solubilized in stick water. Histamine being the major biogenic amine, limits is set to ensure the quality of the fish products. Fish species from Scombridae, Clupeidae, Pomatomidae, and Engraulidae have been associated with higher levels of histamine. If stale raw material is utilized in preparation of fish meal, it contains higher proportion of biogenic amines which affects protein digestibility. For warm water tropical species of fish such as mackerel, sardine, and anchovy, histamine is utilized as criteria to determine fish meal quality. Histamine has an impact on growth. Poor quality and spoiled raw material lead to higher histamine levels. The recommended quantity of biogenic amines for a high quality fishmeal should be less than 1,000 ppm for histamine, and the total sum of all four of the main biogenic amines (cadaverine, putrescine, tyramine, and histamine) should be less than 2,000 ppm.

9.7.5 *DIOXINS*

Anthropogenic activities and environmental pollution lead to dioxins which are found to be in higher proportions in fish meal and fish oils however; these are reduced during oil refining. Supercritical Carbon dioxide extraction and activated carbon extraction remove dioxins to 95%.

9.7.6 *USE OF ETHOXYQUIN*

Protection of animal feed against lipid oxidation is carried out by extensive use of ethoxyquin (E324) which will stabilize long-chain polyunsaturated FAs which may be carried out from farmed fish to fish feed. EQ is an approved fish feed additive in EU with an upper limit of 150 mg/kg alone or in combination with BHT and butylated hydroxyanisole (BHA) as per EC 1831/2003. Study on levels of ethoxyquin in fish feed and shrimp feeds manufactured in Andhra Pradesh in India indicated ethoxyquin level below 1 ppm which is within the permissible range of utilization with fish feed containing more ethoxyquin compared to shrimp feed.

The areas where research is needed are evaluation of suitability of TVB-N as spoilage parameter, establishing freshness parameters for fish meal raw material, and developing quality index methods suitable for enhanced fish meal quality. In countries like India, fish meal and fish oil production plays an important role in the economy. However, static phase is a barrier.

KEYWORDS

- **butylated hydroxytoluene**
- **fish meal**
- **fish oil**
- **polyethylene**
- **raw material**

REFERENCES

Barlow, S. M., (2003). Fish meal. In: Benjamin, C., Finglas, P., & Toldra, F., (eds.), *Encyclopedia of Food Science, Food Technology and Nutrition* (2nd edn., pp. 2486–2491). Academic Press.

Cho, J. H., & Kim, I. H., (2011). Fish meal–nutritive value, *J. Animal Physiol. Animal Nutr., 95*, 685–692.

FAO, (1986). *The Production of Fishmeal and Oil*. FAO Fisheries Technical Paper, 142, Food and Agriculture Organization of the United Nations, Rome, Italy. ISBN 92-5-102464-2.

FAO, (2020). *The State of World Fisheries and Aquaculture (SOFIA)-2020* (p. 244). Food and Agriculture Organization, Rome, Italy.

Hua, K., Cobcroft, J. M., Cole, A., Condon, K., Jerry, D. R., Mangott, A., Praeger, C., et al., (2019). The future of aquatic protein: Implications for protein sources in aquaculture diets. *One Earth*, 1. https://doi.org/10.1016/j.oneear.2019.10.018.

Kennedy, B., Karunsagar, I., & Karunasagar, I., (2004). Histamine level in fishmeal and shrimp feed marketed in India. *Asian Fish. Sci., 17*, 9–19.

Miles, R. D., & Chapman, F. A., (2006). *The Benefits of Fishmeal in Aquaculture Diets*. University of Florida, IFAS Extension. Retrieved from: https:/edis.ifas.ufl.edu/pdffiles/FA/FA12200.pdf (accessed on 09 March 2022).

Murthy, L. N., Jeyakumari, A., Laly, S. J., & Phadke, G. G., (2019). *Training Manual on Quality Assurance and Biochemical Evaluation* (p. 117). Mumbai Research Centre of ICAR-Central Institute of Fisheries Technology, 2019, CIDCO Admin Building, Sector I, Vashi, Navi Mumbai, Maharashtra – 400 703.

Murthy, L. N., Mohan, C. O., Ravishankar, C. N., & Badonia, R., (2013). Biochemical quality and heavy metal content of fish meal and squid meal produced in Veraval, Gujarat. *Ind. J. Fish., 60*(3), 113–117.

Olsen, R. L., & Hasan, M. R., (2012). A limited supply of fishmeal: Impact on future increases in global aquaculture production. *Trends Food Sci. Technol., 27*(2), 120–128.

Ruiter, A., (1995). Fish and fishery products composition, nutritive properties and stability. In: Schmidtdorff, W., (ed.), *Fish Meal and Fish Oil-Not Only By-Products* (pp. 347–376). Biddles Limited: United Kingdom.

Tacon, A. G. J., & Metian, M., (2008). Global overview on the use of fish meal and fish oil in industrially compounded aquafeeds: Trends and future prospects. *Aquacult., 285*(1–4), 146–158.

Part II

Seafood Composition and Scope of Developing High-Value Products

CHAPTER 10

Marine Fish Fatty Acids: Variation in Fatty Acid Profiles of Pelagic and Demersal Fishes

R. RAMESH[1] and G. VENKATESHWARLU[2]

[1]*Regional Research Center of ICAR-Central Institute of Freshwater Aquaculture, Fish Seed Farm, Penamaluru, Poranki P. O., Vijayawada – 521137, Andhra Pradesh, India*

[2]*Education Division, Indian Council of Agricultural Research, KAB-II, New Delhi – 110012, India*

ABSTRACT

The world's total marine catch was 79.3 million tons in 2016, representing a decrease of almost 2 million tons from the 81.2 million tons in 2015. Among various sources, the marine fishes are nutritionally significant in human diet due to their high levels of omega-3 (n-3) PUFA, which have been reported to reduce the risk of coronary heart diseases, lower blood pressure and plasma triacylglycerol levels and several other health benefits such as prevention of human breast cancer growth, diabetes mellitus, asthma, Ulcerative colitis and Crohn's Disease and inflammatory disorders of the immune system. The fatty acid (FA) profiles of seafood varies considerably with temperature, salinity, and the type and availability of food, factors which, in turn, are affected by geographical area and by season. Commercially available pelagic species at global level display the widest range of energy intensities of fisheries for human consumption. Both pelagic and demersal fishes have higher concentrations of n-3 and n-6 PUFA for apparent positive effects on human health. Multivariate statistical analysis has been successfully applied

Advances in Fish Processing Technologies: Preservation, Waste Utilization, and Safety Assurance.
Ranendra K. Majumder, PhD & Amjad K. Balange, PhD (Eds.)

in lipid research especially for FAs with respect to species, size, and seasonal changes. Principal Component Analysis (PCA) is an unsupervised method which acts to reduce the dimensionality of multivariate data while preserving most of the variance within samples. Pelagic fishes caught in post-monsoon season were rich in n-3 FAs such as EPA, 18:4n-3 and DHA. The pelagic and demersal fishes with different sizes captured during different seasons could be differentiated from one group to other group based on FA composition.

10.1 INTRODUCTION

Fish is one of the most important food sources in the human diet due to its high nutritional quality and health beneficial characteristics. It provides macronutrients, micronutrients such as vitamins and minerals and lipids encompassing essential fatty acids (FAs) and cardioprotective polyunsaturated FAs. Amount and type of lipids present in fish body parts vary from species to species. Proximate composition of fish exhibits strong inter-species and intra-species variations depending on different factors such as sex, age, season, and fishing area. Lean fish, like cod, generally have between zero and 2% fat content whereas fatty fish, such as sardine, anchovies have 16% fat. Fish contains largely polyunsaturated FAs, which are popularly known as PUFA. As far as human health is concerned; the long-chain polyunsaturated FAs play an important role especially eicosapentaenoic acid (EPA) as well as docosahexaenoic acid (DHA) belonging to omega-3 class.

Among various sources, the marine fishes are nutritionally significant in human diet due to their high levels of omega-3 (n-3) PUFA, which have been reported to reduce the risk of coronary heart diseases (Erkkila et al., 2003) and lower blood pressure and plasma triacylglycerol levels (Dallongeville et al., 2003). These FAs have been reported to have several health benefits such as prevention of human breast cancer growth, diabetes mellitus, asthma, Ulcerative colitis and Crohn's Disease and inflammatory disorders of the immune system (Miles and Calder, 2012; Spite, 2013).

In view of these health benefits, there are numerous recommendations for the intake of omega-3 PUFA for the prevention of deficiency in essential FAs and to decrease the risk of developing cardiovascular disease. The World Health Organization (WHO) recommends for the prevention of CVD and ischemic stroke, the intake of omega-3 PUFA in the diet should represent 6 to 10% of total energy intake (5–8% n-6 PUFA and 1–2% n-3 PUFA), and a regular intake of one to two servings of fatty fish per week, equivalent to

200 to 500 mg/day of EPA and DHA (Crawford, 2007). Moreover, the EPA + DHA recommendation daily intake (RDI) by the European Food Safety Authority (EFSA) based on cardiovascular risk considerations for European adults is between 250 and 500 mg/day, and the American Heart Association (AHA) recommends a daily intake of 400–500 mg of EPA + DHA. The U.S. Department of Agriculture (USDA) recommends that Americans eat at least eight ounces of seafood per week, which provides 250 milligrams (mg) of DHA and EPA. This level of seafood consumption correlates with decreased rates of cardiac arrest, reduced risk of cardiovascular disease, and vitamin D deficiency as well as with improved infant health outcomes when consumed by pregnant and breastfeeding mothers (HHS and USDA, 2015).

10.2 MARINE FISH FATTY ACID (FA) PROFILES

Marine fish are a rich source of lipids, such as n-3 polyunsaturated FAs (n-3 PUFAs), particularly, the very important n-3 PUFA, eicosapentaenoic (EPA) and docosahexaenoic (DHA) acids (EFSA, 2012). The abundance of the FAs in seafood products (fish, crustaceans, cephalopods, bivalves, etc.), varies widely. It is well known that seafood FA composition depends on several factors, such as ecological, environmental, and nutritional parameters (species, sex, sexual maturity degree, size, location, water temperature, type of feeding, and season, etc.). The variation in the FA composition is explained by fluctuations in the quality and amount of available food, especially phytoplankton (Soccol and Oetterer, 2003).

The FA profiles of seafood are quite well documented; this profile varies considerably with temperature, salinity, and the type and availability of food, factors which, in turn, are affected by geographical area and by season (Hayashi and Takagi, 1978). It was reported that a decrease in temperature leads to a higher PUFA content due to an enhanced activity of desaturases, which contribute to the biosynthesis of more unsaturated FAs (with lower melting point) in order to keep their physiological role. Nonetheless, it must be stressed that FA profile in seafood results from a balance between FAs from diet and those formed by biosynthesis (Cowey and Sargent, 1972). But, whenever food is rich enough in the necessary FAs, fish, and other aquatic organisms do not spend energy in FA biosynthesis (Worthington and Lovell, 1973).

Seafood products contain a wide variety of lipid classes, ranging from TAG, free fatty acids (FFAs), and sterols (cholesterol) to phospholipids (PL).

These lipids are composed of different components, including a FA moiety. Among these FAs, n-3 PUFAs represent a large share (Medale, 2003). Particularly, marine fish are rich sources of EPA (20:5 n-3) and DHA (22:6 n-3) (Ackman, 1988). In wild fish, EPA, and DHA may represent as much as 90% of the total PUFA. EPA and DHA are considered to be essential FAs in that they cannot be biosynthesized with a level of efficiency sufficient to meet human needs (Arts et al., 2001). In fact, there have been several studies pointing to a very limited extent of α-linolenic (ALA, 18:3 n-3) conversion to EPA and DHA (Ackman, 1992).

Most of the marine fishes contain less than 6% lipid by weight. Total lipids of the marine fishes contain high levels of unsaturated FAs. The composition of FAs showed that total ω-3 polyunsaturated fatty acids (n-3 PUFA; 29.7–48.4%) were the highest, n-6 PUFA (11.0–20.0%), saturated fatty acids (3.63–11.4%), and finally, monounsaturated fatty acids (MUFA; 1.37–9.12%). All marine fishes have higher content of n-3 PUFA when compared to cultivable fishes. Most of them have higher n-3/n-6 ratio than the freshwater fishes.

Regarding saturated fatty acids (SFAs), palmitic acid (16:0) is the main SFA in fish, being found in all lipid classes (TAG, PL, others). 16:0 is a key metabolite and its content is not significantly affected by fish feeding. Myristic acid (14:0) and stearic acid (18:0) have lower percentages, and the first of these is preferably in TAGs. The SAFAs with longer chain such as 20:0, 22:0, and 24:0 are typically found at levels below 1%. FAs with an odd number of carbon atoms have also been detected and are probably derived by oxidation of the alcohol present in copepods or microorganisms. Indeed, it has been claimed that these organisms biosynthesize FAs via propionyl-coenzyme A leading to less usual FAs.

The most abundant monounsaturated fatty acid (MUFA) in seafood lipids are usually palmitoleic acid (16:1 n-7) and oleic acid (18:1 n-9). With regard to MUFAs with a long chain, such as 20:1 and 22:1, they appear to have an exogenous origin, possibly from the marine food chain. The marine fishes are having good FA profile due to abundant supply of dietary essential PUFA through natural feeding on phytoplankton and zooplankton and other marine algae. The PUFA and HUFA are the main FAs predominate in marine food web and carnivorous marine fish species. Carnivorous marine fishes consume the small fishes, shrimps, and algae which are rich in 20:5n-3 and 22:6n-3.

In the background of several recommendations for intake of omega-3 PUFA, it is very important to know the levels of these FAs in various fishes.

Further, the information on variations with respect to type, size, and season also assume significance. Hence, an attempt has been made to bring out the details with respect to these health beneficial FAs in various pelagic and demersal fishes.

10.3 PELAGIC FINFISH RESOURCES AND THEIR FATTY ACID (FA) PROFILES

Pelagic fish harvested for human consumption encompass a diverse variety of species. They range from the small, densely schooling, largely planktivorous herrings, to large, top-level carnivores including various tuna and billfish species. Commercially available pelagic species at global level which display the widest range of energy intensities of fisheries for human consumption listed in Table 10.1. The 350 species of pelagic fishes occur in ocean surface waters to a depth of 200 m. This habitat type can be divided into 10 different regions based on faunal differences, with more joining of Southern Hemisphere areas because of the relative lack of large landmasses. These regions are Arctic and Antarctic polar, North Pacific cold temperate, North Pacific warm-temperate, tropical Indo-Pacific, North Atlantic boreal, North Atlantic warm-temperate, Atlantic tropical, southern warm-temperate and southern cold temperate.

Many of the world's most important fisheries species occur in pelagic regions, including numerous sharks, sardines, herrings, salmons, codfishes, pollocks, hakes, haddocks, sauries, mackerels, and tunas. In addition, about 100 species of mostly pelagic fishes have a worldwide distribution. This group includes pelagic sharks (White, Whale, Tiger, and perhaps Megamouth), Swordfish, and ocean sunfishes.

The epipelagic layer of the ocean extends from the surface down to about 660 feet (200 meters). Light penetrating the water at this level allows the growth of plankton, algae, and floating seaweed. Small foraging fish that feed off plankton are common in this zone, including herring, anchovy, scad, sprat, sardines, small mackerels and blue whiting. These foraging fish live in the coastal waters above the continental shelf. Larger coastal fish such as salmon, large mackerel, billfish, and dolphinfish feed off the smaller fish. Apex predators such as tuna, large rays, Bonita, pomfrets, and ocean sharks are able to spend longer periods of time in the deeper waters beyond the continental shelf.

TABLE 10.1 Top Pelagic Fish Species Production (tons) around the World

Fish	Scientific Name	Average Production
Peruvian anchovy	*Engraulis ringens*	3,192,476
Skipjack tuna	*Katsuwonus pelamis*	2,829,929
Sardines	*Sardinella* spp.	2,289,830
Jack and horse mackerels	*Trachurus* spp.	1,743,917
Atlantic herring	*Clupea harengus*	1,639,760
Pacific chub mackerel	*Scomber japonicus*	1,598,950
Yellowfin tuna	*Thunnus albacares*	1,462,540
Japanese anchovy	*Engraulis japonicus*	1,304,484
Scads nei	*Decapterus* spp.	1,298,914
European pilchard	*Sardina pilchardus*	1,281,391
Large head hairtail	*Trichiurus lepturus*	1,280,214
Blue whiting	*Micromesistius poutassou*	1,190,282
Atlantic mackerel	*Scomber scombrus*	1,138,053
Seer fishes nei	*Scomberomorus* spp.	918,967
Gulf menhaden	*Brevoortia patronus*	618,719
European sprat	*Sprattus sprattus*	584,577
Japanese pilchard	*Sardinops melanostictus*	531,466
Atlantic chub mackerel	*Scomber colias*	511,618
Indian mackerel	*Rastrelliger kanagurta*	499,474

Source: FAO (2018).

The total FA profile of commercially important pelagic fishes has been reported in the world mentioned (Tables 10.2 and 10.3). Pelagic fishes have a substantial amount of health beneficial FAs viz. Arachidonic Acid (20:4, n-6), EPA (20:5, n-3) and DHA (22:6, n-3). Some fishes had the highest amount of MUFA and SAFA levels due to the high amount of 16:1 n-7, 18:1 n-9, 20:1 n-9 and 14:0, 16:0, 18:0, and 20:0 respectively. A wide range of investigations have shown that the composition of FAs in fish is influenced by the composition of FAs in the diet. The richer a tissue is in triglycerides, the closer is the resemblance of its FA composition with that of the diet (Viga and Grahl-Nielson, 1990). It has also been established that the FA compositions of tissue lipids of fish is are readily influenced by the FA composition of dietary lipid (Torstensen et al., 2000; Bell et al., 2001, 2002; Sargent et al., 2002). Highest levels of EPA/DHA were reported within small pelagic fish species (average of EPA/DHA = 778/966 mg/100 g, including Atlantic mackerel, Pacific herring, Atlantic herring, and European anchovy) (Tacon, 2013).

Pelagic fishes are more active due to their feeding behavior and contain more amount of fat as they consume phytoplankton and zooplankton, which are rich in highly unsaturated FAs abundant in the pelagic waters than compared to demersal waters. The pelagic fishes had more red meat than demersal fishes which contain more fat compared to light muscle. Further, pelagic fishes need more energy for their movement compared to the sluggish nature of demersal fishes. Most of the pelagic and demersal fishes are carnivorous in nature. They feed mainly on small worms, fishes, crabs, and shrimps, and some fishes like Bombay duck, horse mackerel and lizard fishes feed on their own young ones. Due to this cannibalistic nature, the FAs like EPA and DHA were directly accumulated in their body.

TABLE 10.2 Total Fatty Acid Profiles (%) of Pelagic Fish Species of Different Parts of the World

Locations	Fish Species	Total SAFA	Total MUFA	Total n-3 PUFA	References
Canada	*Lampanyctus* spp.	12.0	40.5	40.0	Ozdemir et al. (2019)
	Chaulidos sloani	20.2	41.0	35.6	
	Serrivomer beanii	15.33	25.0	42.5	
China	*Coilia mystus*	45.6	19.3	26.4	Zhang et al. (2020)
	Ilisha elongata	33.5	37.3	17.6	
	Thryssa kammalensis	37.9	28.7	21.4	
Black sea	*Belone belone*	34.7	26.4	35.8	Tanakol et al. (1999)
Egypt	*Pomatomus saltatrix*	28.78	48.32	22.81	Yazeed (2013)
Philippines	*Decapterus macarellus*	66.5	23.0	17.0	Metillo et al. (2014)
	Anchovy	68.7	15.7	15.0	
	Sardine	66.23	20.02	10.53	
Panama	*Scomberomorus sierra*	39.11	32.64	28.13	Murillo et al. (2014)
	Caranx caballus	37.77	28.66	33.66	
India	*Harpodon nehereus*	35.8	23.9	33.5	Ramesh et al. (2013)
	Ribbon fishes	38.3	20.21	40.6	Ramesh et al. (2015)

TABLE 10.3 Major Fatty Acid Composition (% of Total Fatty Acids) of Commercially Important Pelagic Fishes

Fatty Acids	Sardines	Mackerels	Carangids	Tunas	Bombay Duck	Horse Mackerel	Ribbon Fishes	Herrings
16:0	20–25	16–20	34–39	14–19	14–19	12–17	13–15	12–14
18:0	5–6	11–12	10–12	6–9	6–8	11–16	8–10	10–13
18:1n-9	8–10	8–10	18–20	7–14	10–14	9–13	7–10	10–12
18:2n-6	1–2	1–2	1–2	1–1.5	0.5–1.5	1.0–2.0	1–1.5	1–1.5
18:3n-3	0.5–1	1–2	0.5–1	2–8	0.5–1	0–1.5	0–1.0	0–0.5
20:4n-6	2–3	3–5	2–4	4–10	3–5	5–8	3–5	2–4
20:5n-3	11–13	0.5–1	1–2	6–9	6–10	4–9	7–9	6–8
22:6n-3	15–20	34–40	11–14	24–45	18–20	10–15	28–33	15–20

Source: Gopakumar (1997); Ramesh (2007).

Ramesh et al. (2013, 2015) identified total 37 FAs in muscle of pelagic fishes, i.e., Bombay duck and Ribbon fishes including such as EPA and DHA. These pelagic fish edible portion contained high levels of n-3 PUFA, which are good sources of human health-promoting FAs. The most common SAFA were palmitic acid accounts for 14.45–46.55% (Tables 10.5–10.7). Among MUFA, oleic acid (18:1n-9) was the most abundant comprising of 7.57–14.45% of the total FAs. Arachidonic acid (AA, 20:4n-6) was the principal n-6 PUFA at a level of 2.65–9.42% of total FAs. EPA (20:5n-3) and DHA (22:6n-3) were the major n-3 PUFA identified, accounting 5.77–10.07% and 16.37–25.21%, respectively.

10.4 DEMERSAL FINFISH RESOURCES AND THEIR FATTY ACID (FA) PROFILES

Demersal fish of economic importance are concentrated on the continental shelf and on the continental slope down to about 500 m. They live on or near the bottom of the ocean. A regular landing from a deep-sea trawling session would include catfish, croakers, soles, groupers, shiny nose, snappers, and megalops (Table 10.4). Other commercially valuable and contribute substantially to the global economy from marine capture fisheries such as flounder, sole, turbot, plaice, and halibut also deep fishes like important are sharks, rays, threadfin breams, pomfrets, cod, hake, redfish, haddock, bass, and chimeras. World catches of demersal species increased rapidly during the first three-quarters of the 20th century. Since then the catches from some stocks have declined, due to overfishing, while other previously underexploited stocks have increased their yields.

World total marine catch was 79.3 million tons in 2016, representing a decrease of almost 2 million tons from the 81.2 million tons in 2015 (FAO, 2018). Alaska pollock again surpassed anchoveta as the top species in 2016, with the highest catches since 1998. However, anchovies remain as the top species among pelagic group followed by Skipjack tuna and sardines occupied consecutive rankings, respectively. The Northwest Pacific continues to be by far the most productive fishing area, with catches in 2016 of 22.4 million tons, slightly higher than in 2015. All other temperate areas have shown decreasing trends for several years, with the sole exception of the Northeast Pacific, where catches in 2016 were higher than the average for 2005–2014 because of catches of Alaska pollock, Pacific cod, and North Pacific hake. In contrast to the temperate areas, and the upwelling areas, which are characterized by high annual variability in catches, tropical areas have experienced a continuously rising trend in production as catches of large (mostly tuna) and small pelagic species continue to increase.

TABLE 10.4 Top Demersal Fish Species Production (tons) around the World

Fish	Scientific Name	Average Production
Alaska pollock	*Theragra chalcogramma*	3,476,149
Pollock	*Pollachius virens*	385,227
Atlantic cod	*Gadu smorhua*	1,329,450
Pacific cod	*Gadus macrocephalus*	425,467
Threadfin breams	*Nemipterus* spp.	683,213
Sharks, skates, rays	*Elasmobranchii*	337,819
Croakers	*Sciaenidae*	492,528
Flatfishes nei	*Pleuronectiformes*	289,551
Argentine hake	*Merluccius hubbsi*	526,573
Cape hakes	*Merluccius capensis*	266,854
South Pacific hake	*Merluccius gayi*	197,911
Haddock	*Melanogrammus aeglefinus*	273,459
Blue grenadier	*Macruronus novaezelandiae*	255,421
Patagonian grenadier	*Macruronus magellanicus*	230,221
Norway pout	*Trisopterus esmarkii*	299,145
Atlantic redfishes	*Sebastes* spp.	318,383
Sea catfish nei	*Ariidae* spp.	242,815
Sand eels	*Ammodytes* spp.	1,003,343
Filefish	*Cantherhines* spp.	234,446

Of the total world catch of over 80 million tons of fish, roughly 36 million tons are accounted for by groups of typically pelagic fish (Khedkar, 2003). Of the total fish catch, approximately 25 million tons are used for reduction purposes, and it is likely that a very high proportion of this comes from pelagic fish landed in bulk. The production of fish meal from oily fish is around 6 million tons, with an associated 1.4 million tons of oil, implying a conversion ratio of about 3.4:1. Allowing for approximations, these figures imply that about 70% of tropical pelagic fish caught are used for purposes other than direct human consumption. In addition to meal and oil, over 1.7 million tons of canned pelagic fish products is produced.

FAs in demersal fishes predominant by DHA which accounts for 9.0–33% followed by palmitic acid and C18:1n-9 in decreasing order as 12–32% and 7–19%, respectively. The concentrations of EPA and AA were the range of 2–10% and 1–6%, respectively (Tables 10.5 and 10.6). The most common SAFA were palmitic acid (16:0) (15.22–23.1% and stearic acid (18:1n-9) (7.4–17.35%) were recorded in threadfin breams (Ramesh et al., 2016). However, oleic acid (18:1n-9) was the most abundant comprising of 7.4–17.35% of the total FAs found among MUFA. Arachidonic acid (AA, 20:4n-6) was the principal n-6 PUFA at a level of 3–9% of total FAs. EPA (20:5n-3), and DHA (22:6n-3) were the major n–3 PUFA identified, accounting 2.19–6.96% and 14.56–25.98%, respectively.

TABLE 10.5 Total Fatty Acid Profiles of Different Demersal Fish Species of Different Parts of the World

Locations	Fish Species	Total SAFA	Total MUFA	Total n-3 PUFA	References
Canada northwest Atlantic Ocean	*Rajella fyllae*	22.5	40.8	37.8	Ozdemir et al. (2019)
	Alepocephalus bairdii	24.6	20.0-	50.9	
	Borostomias antarcticus	21.6	18.5	48.0	
China	*Collichthys lucidus*	38.6	38.7	15.2	Zhang et al. (2020)
	Arius sinensis	44.0	35.8	14.7	
	Cynoglossus lida	36.4	31.6	23.4	
Panama	*Cynoscionphoxo cephalus*	42.5	33.47	23.59	Murillo et al. (2014)
India	*Nepiterus* sp.	36.2	18.11	39.0	Ramesh et al. (2016)

10.5 VARIATION IN THE FATTY ACID (FA) PROFILES OF PELAGIC AND DEMERSAL FISHES

FA compositions of selected pelagic and demersal fish species are summarized (Tables 10.7–10.9). Wide variations in the FA composition were observed among different fish species. FAs were categorized into SAFA, which accounted for 12% to 68% of total FAs; MUFA, which constituted 15% to 48%; and PUFA, which comprised 10% to 50. In general, most of the pelagic and demersal species contained a higher proportion of SAFA than that of MUFA and PUFA. The marine fish species were typically characterized by high percentage share of long-chain FAs. However, the levels of PUFA reported in pelagic and demersal fish species were significant. It is well known that dietary composition, especially the diet's FA profile, exerts a qualitative and quantitative impact on the FA composition of fish lipids. The difference in the FA signature pattern might be influenced by the diet variations.

TABLE 10.6 Major Fatty Acid Composition (% of Total Fatty Acids) of Commercially Important Demersal Fishes

Fatty Acids	Pomfrets	Perches	Catfishes	Sharks	Sciaenids	Lizard Fishes	Thread Fin Breams	Groupers
16:0	22–24	17–21	26–28	30–31	30–32	15–19	12–20	15–17
18:0	14–15	7–14	10–18	17–20	6–7	10–12	9–12	9–15
18:1n-9	13–19	13–15	9–10	12–16	10–12	9–10	7–9	9–11
18:2n-6	3–6	1–2	0.5–1	0.5–1	1–2	0.5–1.5	1–1.5	1–1.5
18:3n-3	2–3	0.8–1	0.5–1	0.5–1	0.5–1	0.2–0.5	0.5–1.0	0–0.5
20:4n-6	1–2	5–6	5–6	6–7	3–6	4–6	4–9	4–6
20:5n-3	8–10	4–7	4–7	1–2	4–5	3–5	3–6	3–4
22:6n-3	30–33	24–26	21–24	9–16	14–16	20–30	19–25	18–20

The lipid content of fishes is highly variable between and within species, depending upon different biotic and abiotic factors. In the wild, this variability in fishes reflects the trophic level of the species, the location of the catch (with seasonal changes in natural productivity and hence the availability of food), and the size and physiological status of the fishes. The FA composition not only reflects the dietary resources available but also the 'bioconversion' capacities of the species concerned (Sargent, 1995). Although marine fishes have comparatively high levels of DHA in their lipids, it is generally known that their FA composition has seasonal variations and that the DHA levels in marine fish is often influenced by their prey lipids because of their lack of Δ5 and Δ6 desaturases (Kanazawa et al., 1979). Bergstrom (1989) observed that the FA composition of the prey, as represented by the stomach contents, influenced the FA composition of the fish and both were subject to seasonal variations.

It has traditionally been accepted that 18:3 n-3 is converted to 22:6 n-3 by a pathway combining the sequential action of Δ6, Δ5, and Δ4 desaturases with chain elongation reactions (Henderson and Tocher, 1987). But in the marine environment, DHA or its structural precursors are synthesized by marine algae, and then pass through along with the food chain. It has also been reported that lipid content and FA composition changes according to species, diet, geographical origin and season (Hatano, 1990). In addition to the seasonal variations, body size, location, and sexual maturity also have been reported to influence the FA composition (Ackman, 1982).

However, FA distribution of fish is known to be affected by the season, water temperature, and nature of fats in the diet, maturity, and gender (Ackman, 1982). Changes in tissue FA composition are under various metabolic influences such as FA desaturation and elongation processes (Henderson et al., 1995), preferential incorporation and β-oxidation (Henderson and Sargent, 1985), or lipogenic activity. The FA composition may also be affected by the size or age of animals (Kiessling et al., 2001), affecting metabolic activity.

Seasonal variations and localities have been observed to have an effect on some marine fish and shellfishes, but the differences generally had little effect on oil quality and the processed fish. The FA composition varied little among areas and body sizes (Osako et al., 2003). Within the general patterns, the FA composition is known to be affected by diet (Olsen and Skjervold, 1995), food deprivation (De silva et al., 1997b) and non-dietary factors including environmental temperature (Satoh et al., 1984).

10.5.1 SEASONAL CHANGES IN FATTY PROFILES OF PELAGIC AND DEMERSAL FISHES

The actual nutritional content of fish varies substantially by species and species group (Bogard et al., 2015). Variation also occurs according to a number of factors such as seasonality, habitat, trophic level, and diet (Sushchik et al., 2017). However, the composition of FA in fish fluctuates widely according to several factors, mainly species and seasonal factors (Saito et al., 1997). The amounts of DHA in lipids greatly change within species as well as during climatic seasons (Greenberg et al., 2008). The changes in seasons having an effect on lipid and FA composition were reported by Mourente and Tocher (2009). Further, it was noticed that the seasonal accumulation of sufficient reserves may be a necessary for maturity in some fishes and particularly in migrating tuna. The lipid storage allows tuna to swim across the ocean for a long time even during that time the food is limited (Mourente et al., 2002).

FA profiles of edible tissue of pelagic and demersal fishes with respect to different seasons (Monsoon (Season-I), post-monsoon (Season-II) and pre-monsoon (Season-III) were presented in Tables 10.7–10.9. The total PUFA varied seasonally due to the presence of high amounts of DHA, EPA, and AA. DHA was the dominant n-3 PUFA in both the groups, the contribution of DHA and EPA decreased from pre-monsoon to post-monsoon. Saturated FAs (16:0 and 18:0) were highest in season-III followed by season-II and I in decreasing order of III>II>I. The total MUFA showed equal quantities in both pelagic and demersal fishes. However, Oleic acid was found highest in season-III and compared to other seasons. It was derived that difference between the pelagic and demersal was attributed due to mainly dietary constitution of the fishes (Ramesh et al., 2013, 2016).

Bombay duck caught in summer season were reported rich in DHA, stearic acid and 18:1n-7 (Ramesh et al., 2013). This might be due to environmental temperature may affect the FA composition of the tissue in that the degree of unsaturation increases with decreasing temperature (Morris and Culkin, 1989). PUFA content increased during summer in Bombay duck, maybe due to its cannibalistic feeding habits and due to the unavailability of marine algae in the food chain during summer (Govindan, 1972). It was reported that the though the lipid content is low Bombay duck, the concentration of EPA and DHA which are considered as human health beneficial FAs (Joydeep et al., 1999), observed were very high.

PUFA content in fish and shellfish varies inversely with water temperature, while the SAFA content changes positively with water temperature (Dunstan et al., 1999). Habitat temperature is known to affect the FA composition of poikilothermic fish in a way that the degree of unsaturation increases with decreasing temperature (Morris and Culkin, 1989). In agreement with the earlier studies, PUFA content increased during the post-monsoon and winter seasons in ribbonfish. This could be due to its cannibalistic feeding habits and availability of marine algae in the food chain during monsoon and winter seasons (James et al., 1986).

10.5.2 EFFECT OF SIZE ON FATTY ACID (FA) VARIATION

Muscle lipid levels are reported to increase with fish size (Shearer et al., 1994; Hemre and Sandnes, 1999) and this makes it difficult to quantify the effects on filet lipid content. Despite this, some studies indicate increased fat deposition during autumn, independent of increasing fish size (Morkore and Rorvik, 2001). In addition to the seasonal variations, size, and location, the sexual maturity also has been reported to influence the FA composition (Ackman, 1982).

In pelagic fishes, total SAFA were reported higher in small size of horse mackerel (48.41%) caught in pre-monsoon season. Total n-6 PUFA in wolf herring (5.69–11.35%) and n-3 PUFA in ribbon fishes (23.37–41.14%) were significantly higher in muscle of the pelagic fishes (Tables 10.7–10.9). The DHA was the most dominant FA found in pelagic fishes which was found highest in medium size (35.98%) herring captured in post-monsoon season (Ramesh, 2007). Similarly, a higher amount of EPA (10.07%) was observed in Bombay duck and AA (6.35%) was observed in wolf herring in the same season.

Small size of Bombay duck had lowest lipid content in irrespective of the seasons, possessed relatively high proportion of nutritionally important PUFA and emphasizes excellent quality of edible meat for human consumption. In the light of the diversified roles of n-3 and n-6 PUFA, the right balance between these FA groups is recommended. n-3 PUFA have been reported to modulate the undesirable effects of n-6 PUFA. Similarly, small size ribbon fish had lowest lipid contents compared to the other size groups during all the seasons. However, all the size groups of ribbonfish harvested during the different seasons possessed relatively high proportion of nutritionally important PUFA with highly beneficial essential FAs.

The EPA and DHA content in demersal species reported with different size and seasonal variations were shown in Tables 10.7–10.9. Small fish had n-3 FAs EPA and DHA concentration greater than that of other sizes and fishes caught monsoon season had lower levels than those caught in other seasons. From the individual FAs, palmitic acid found higher (24.86%) in lizardfish and higher n-3 PUFA levels (25.53–36.33%) was found in *N. japonicas* due to significant contribution of EPA and DHA.

10.6 CHEMOMETRIC METHOD FOR DETERMINING FATTY ACID (FA) COMPOSITIONS

Multivariate statistical analysis has been successfully applied in lipid research especially for FAs with respect to species, size, and seasonal changes (De Silva et al., 1997a). Chemometry of tissue FA profiles of fishes has been utilized in discrimination at the species level, stock level and individual level. Hove and Grahl-Nielsen (1991) developed a Chemometric method for determining FA compositions of tissue of marine animals. Application of these statistic methods to chromatographic data treatments became popular in the last two decades.

Principal Component Analysis (PCA) is an unsupervised method which acts to reduce the dimensionality of multivariate data while preserving most of the variance within samples (Eriksson et al., 2001). Multivariate statistical analyzes were applied for species and season classifications using in particular FAs. Such methods were successfully applied in lipid research (Kakela et al., 2010; Ramesh et al., 2013, 2015, 2016). PCA has been applied to summarize the relative differences among the samples in relation to their overall FAs profiles and also to determine the contribution of individual FAs.

The PCA offered a better understanding of the FA composition of pelagic and demersal fishes. The "score" plot revealed information about patterns in the samples and it was found useful to interpret differences and similarities among the fishes. The plot of PC_1 and PC_2 was used in this study since these two components summarized more variation between 60–90% in the data than any other pair of components.

The PCA analyzes of pelagic and demersal fishes were clearly distinguished based on FA profiles (Kakela et al., 2010). The FA composition of the fatty pelagic sardine (4% lipid) differed in many respects from the lean demersal Atlantic spotted flounder (1% lipid).

TABLE 10.7 Fatty Acid Profiles (% of Total Fatty Acids) of Pelagic and Demersal Fishes Caught During the Season-I (Monsoon)

Fishes Size		16:00	Total SAFA	18:1n-9	Total MUFA	18:2n-6	20:4n-6	Total n-6 PUFA	18:3n-3	20:5n-3	22:6n-3	Total n-3 PUFA
Bombay duck	a	19.1	37.54	12.24	27.46	0.95	4.34	7.33	0.28	6.26	18.35	27.67
	b	14.59	33.9	10.23	22.71	1.12	4.73	8.36	0.51	10.07	20.95	35.03
	c	18.14	35.84	13.82	27.34	1.2	4.26	7.22	0.18	8.74	18.61	29.6
Ribbonfish	a	15.34	32.05	9.61	22.44	1.39	4.51	8.03	0.38	7.15	28.28	37.48
	b	15.02	33.82	8.98	18.97	1.37	4.63	7.81	0.58	9.42	26.94	39.4
	c	15.48	32.29	7.57	15.89	1.28	4.69	8.58	0.41	7.49	33.16	43.14
Wolf herring	a	12.28	33.31	11.77	22.9	1.5	4.34	8.4	0.26	8.49	21.57	35.39
	b	12.95	35.6	12.37	22.17	1.28	4.67	8.67	0.18	6.14	22.98	33.56
	c	14.24	36.51	12.84	23.12	1.47	4.31	8	0.17	5.48	23.25	32.37
Horse mackerel	a	17.84	45.26	12.93	22.47	1.24	5.84	8.94	0.15	4.02	15.6	23.33
	b	12.73	35.45	11.52	23.59	1.74	5.34	9.37	0.15	9.33	15	31.59
	c	12.02	36.04	9.54	23.48	1.89	5.66	9.5	0.15	8.74	15.21	30.98
Lizardfish	a	15.21	33.13	9.1	16.79	0.13	6.68	11.59	0.12	4.67	31.58	38.49
	b	19.41	41.62	10.6	20.06	1.35	6.6	11.32	0.26	3.89	21.31	27.0
	c	15.03	36.47	9.34	17.78	1.41	5.72	11.57	0.23	4.41	27.09	34.18
N. Mesoprion	a	17.25	33.79	7.4	16.67	1.22	9	14.65	0.3	6.49	25.83	34.89
	b	15.22	32.83	7.99	16.43	1.35	8.32	14.47	0.35	6.96	25.9	36.27
	c	23.74	46.15	9.25	18.87	1.72	4.55	9.96	0.88	3.51	19.12	25.02

TABLE 10.7 *(Continued)*

Fishes Size		16:00	Total SAFA	18:1n-9	Total MUFA	18:2n-6	20:4n-6	Total n-6 PUFA	18:3n-3	20:5n-3	22:6n-3	Total n-3 PUFA
N. Japonicus	a	12.96	32.29	7.52	16.97	1.39	8.47	14.63	0.7	8.08	23.05	36.11
	b	15.8	36.23	9.29	16.48	1.1	7.63	14.19	0.33	5.72	23.18	33.1
	c	12.71	33.86	9.75	18.99	1.37	6.41	13.19	0.68	6.32	22.72	33.96
Grouper	a	17.71	38.95	11.51	20.9	1.23	6.46	12.03	0.17	5.03	19.43	28.12
	b	15.71	37.06	10.37	21.96	1.37	4.36	10.6	0.34	4.26	20.63	30.38
	c	16.82	36.82	9.67	20.53	1.72	5.71	11.31	0.32	6.14	20.24	31.34

Note: a: Small; b: Medium; c: Large.

Source: Ramesh (2007).

TABLE 10.8 Fatty Acid Profiles (% of Total Fatty Acids) of Pelagic and Demersal Fishes Caught During the Season-II (Post-Monsoon)

Fishes Size		16:00	Total SAFA	18:1n-9	Total MUFA	18:2n-6	20:4n-6	Total n-6 PUFA	18:3n-3	20:5n-3	22:6n-3	Total n-3 PUFA
Bombay duck	a	19.48	40.62	9.93	23.98	1.4	5.08	8.83	0.67	6.82	16.37	26.57
	b	17.55	38.9	11.3	27.9	0.86	3.76	6.57	0.32	5.77	17.51	26.63
	c	17.86	37.95	11.95	28.02	0.95	4.03	6.73	0.24	6.04	18.49	27.3
Ribbonfish	a	15.94	38.54	11.83	22.19	1.52	4.46	8.58	0.71	4.05	20.96	30.69
	b	16.02	38.66	12.06	20.21	1.28	4.51	8.87	0.4	4.03	24.16	32.26
	c	17.53	37.69	12.51	19.38	1.03	4.7	8.66	0.36	5.31	25.82	34.27

TABLE 10.8 *(Continued)*

Fishes	Size	16:00	Total SAFA	18:1n-9	Total MUFA	18:2n-6	20:4n-6	Total n-6 PUFA	18:3n-3	20:5n-3	22:6n-3	Total n-3 PUFA
Wolf herring	a	23.6	39.52	5.76	11.35	1.25	5.33	9.19	0.3	3.9	32.44	37.94
	b	22.56	38.5	5.67	10.48	1.32	5.36	11.05	0.37	2.81	35.98	39.97
	c	22.5	39.49	5.36	10.86	1.29	6.35	10.5	0.31	4.31	33.21	39.15
Horse mackerel	a	15.36	36.46	12.55	26.93	1.46	3.79	9.11	0.42	5.67	17.81	27.5
	b	20.79	37.84	9.67	22.65	1.16	6.08	10.57	0.42	5.56	20.49	28.94
	c	22.97	44.63	13.5	21.93	1.19	5.49	11	0.52	5.02	13.88	22.44
Lizardfish	a	18.82	39.94	9.92	19.65	0.64	5.82	9.77	0.59	6.73	20.41	30.64
	b	14.92	35	8.83	18.57	1.17	5.61	10.9	0.23	10.08	22.44	35.53
	c	13.97	34.35	7.4	22.67	1.41	5.02	8.84	0.36	12.01	17.56	34.14
N. Mesoprion	a	16.44	35.84	9	18.68	2.62	6.56	13.6	0.34	6.54	21.96	31.88
	b	16.85	37.68	8.91	18.62	1.13	7.08	13.28	0.12	6.53	20.64	30.42
	c	18.21	38.75	14.08	24.97	0.94	5.9	10.8	0.32	6.83	15.36	25.48
N. Japonicus	a	12.56	34.73	9.99	18.77	1.39	5.33	10.99	0.36	4.53	25.6	35.51
	b	16.09	39.62	8.44	15.87	1.1	5.64	10.45	0.19	3.25	26.63	34.06
	c	14.91	34.58	10.78	18.11	1.25	5.21	10.98	0.3	3.69	28.45	36.33
Grouper	a	17.54	36.43	16.16	32.07	0.48	3.8	8.34	0.12	4.11	15.01	23.16
	b	17.62	36.31	14.5	28.69	0.73	4.48	9.28	0.13	4.66	16.53	25.72
	c	17.08	38.03	13.64	28.87	0.83	4.12	8.86	0.18	4.57	14.74	24.24

Note: a: Small; b: Medium; c: Large.

Source: Ramesh (2007).

TABLE 10.9 Fatty Acid Profiles (% of Total Fatty Acids) of Pelagic and Demersal Fishes Caught During the Season-III (Pre-Monsoon)

Fishes Size		16:00	Total SAFA	18:1n-9	Total MUFA	18:2n-6	20:4n-6	Total n-6 PUFA	18:3n-3	20:5n-3	22:6n-3	Total n-3 PUFA
Bombay duck	a	14.45	32.93	13.43	26.43	1.31	4.52	7.14	0.24	6.14	25.21	33.5
	b	18.77	36.38	10.81	21.81	1.14	5.88	8.68	0.43	8.07	22.32	33.13
	c	17.06	37.21	12.35	22.34	1.2	5.31	7.89	0.28	6.78	23.86	32.56
Ribbonfish	a	17	37.25	13.57	21.65	1.25	5.43	10.14	0.41	4.68	20.19	30.96
	b	29.26	57.55	14.45	23.18	1.44	3.9	6.9	0.33	2.65	7.27	12.37
	c	22.45	45.23	16.81	25.47	1.51	4.34	8.16	0.32	3.89	13.45	21.14
Wolf herring	a	26.54	45.22	9.55	18.33	1.21	3.24	6.07	0.54	5.16	23.64	30.38
	b	32.99	50.93	5.87	13.06	1.13	2.6	5.69	0.36	5.25	22.87	30.32
	c	29.22	45.59	7.21	16.06	1.12	3.56	6.19	0.4	4.92	25.09	32.16
Horse mackerel	a	27.45	48.41	11.05	19.75	1.42	3.12	6.1	0.29	3.81	19.2	25.9
	b	24.43	47.33	12.09	22.14	1.16	3.8	6.95	0.35	3.77	15.99	23.58
	c	25.23	47.34	10.41	20.05	1.25	3.21	6.21	0.44	4.51	18.71	26.4
Lizardfish	a	24.76	43.19	13.51	23.38	1.21	4.26	8.73	0.33	3.75	18.77	24.7
	b	24.86	43.62	15.94	26.34	1.13	4.59	8.2	0.38	3.58	15.86	21.84
	c	22.45	41.59	14.34	24.94	1.05	4.21	8.22	0.37	2.45	20.25	25.25

TABLE 10.9 *(Continued)*

Fishes Size		16:00	Total SAFA	18:1n-9	Total MUFA	18:2n-6	20:4n-6	Total n-6 PUFA	18:3n-3	20:5n-3	22:6n-3	Total n-3 PUFA
N. Mesoprion	a	27.69	46.63	15.89	33.69	0.76	3.09	5.95	0.26	2.19	9.27	13.73
	b	23.1	42.97	15.06	26.96	1.33	3.31	7.37	0.21	3.25	16.81	22.7
	c	23.05	43.07	17.35	27.27	1.8	4.05	9.06	0.32	3.12	14.56	20.6
N. Japonicus	a	21.84	40.71	8.66	16.1	1.22	6.95	13.27	0.41	4.69	21.85	29.92
	b	16.8	39.35	13.19	24.85	1.45	4.46	8.99	0.61	6.34	16.75	26.81
	c	19.16	39.02	12.71	24.37	1.52	5.89	11.08	0.54	4.62	17.41	25.53
Grouper	a	18.82	39.92	17.46	33.39	0.84	3.42	7.24	0.47	5.95	8.17	19.45
	b	17.97	42.97	11.76	28.17	1.1	3.54	7.05	0.37	7.54	8.19	21.81
	c	18.5	43.32	12.06	28.25	1.5	3.42	7.52	0.33	6.15	9.21	20.91

a: Small; b: Medium; c: Large.

Source: Ramesh (2007).

It was found that compared to the sardine (19%); the Atlantic flounder (27.1%) contained in general more monounsaturated fatty acids (MUFAs); The flounder (6.8%) was also rich in n-6 polyunsaturated fatty acids (PUFAs) were twice those of the sardine (3.9%) due higher levels of 20:4n-6, 22:4n-6, and 22:5n-6, but 18:2n-6 was slightly less abundant in flounder than in sardine. In contrast, n-3 PUFAs contained less in the flounder (27.3%) than in the sardine (37%) due to 18:3n-3, 18:4n-3, 20:5n-3 and 22:6n-3. Total 60% of variance explained by PC_1 and PC_2 based the FA composition.

Further, the pelagic and demersal fishes with different sizes captured during different seasons could be differentiated from one group to other group based on FA composition. The "loading" plot clearly showed that the variables such as SAFA (14:0, 16:0, 18:0) MUFA (16:1n-7, 18:1n-9), n-6 PUFA (18:2n-6, 20:4n-6) and n-3 PUFA (18:3n-3, 22:5n-3 and 22:6n-3) which were responsible to bring out the differences among the samples. The Bi-plots presented "score" plots (samples are marked in blue color) together with the corresponding "loading" (variables are marked in red color) plots, for the first two components. Bi plot of PCA can present variation among the samples and also gives mutual differences among whole data and also PCA can explain the relationship among the different samples based on score plot and loading plots (Ramesh et al., 2007).

10.6.1 DISCRIMINATION OF PELAGIC AND DEMERSAL FISHES BASED ON FATTY ACID (FA) PROFILES

The PCA was carried out on the data set of the FA composition of the pelagic and demersal fishes. PC_1 explained 51% of the original information of the data set. The loading plots indicated that there was a significant contribution of total MUFA, total n-6 PUFA, AA, EPA, DHA variables with positive loadings and total SAFA variables with negative loadings. The PC_1 clearly distinguishes the pelagic and demersal fishes with total MUFA and total n-3 PUFA and EPA levels. All pelagic fishes except mesopelagic fishes like horse mackerel (small) were positively correlated. In demersal fishes, groupers, and *N. mesoprion* (large) and lizardfish (medium) were negatively correlated. They were high in SAFA level, lower in EPA and total n-3 PUFA levels (Ramesh et al., 2007).

The PC_2 described 22% of the total variance of the data and presented with high positive loadings of MUFA and n-3PUFA, EPA variables.

Moreover, SAFA variable were presented with high negative loading in PC_2. It was verified that the first two PCs were accounted for almost 73% of the total variance of the data set. The graphical representation of the biplot of the first two principle components (PC_1 vs PC_2) was presented in Figure 10.1, with scores and loading plots. The PCA has revealed clearly the correlation among samples and variables based on FA composition. 14:0, 18:0, 16:1n-7, 18:1n-9, 22:5n-6, DPA, and DHA are major FAs which contributing to discriminate pelagic and demersal fishes (Figure 10.2).

Biplot of distribution of muscle FAs in pelagic and demersal fishes showed that the pelagic group is clearly separated from demersal by PC_2. The discriminating FAs are distinguishably shown in correlation loading plot of PC_1 and PC_2 (Figure 10.2). PC_1 explained 53% of the data, which describes DHA and 22:5n-6 with positive loadings and DPA was the negative loadings in correlation plot. PC_2 explained 18% of the data which describes 14:0, 16:0, and 16:1n-7, 18:1n-7 with positive loadings and 18:0, 19:0 with negative loadings. Total 71% of the data is explained by these two principle components. The PCA analysis could reveal the discrimination of pelagic fishes and demersal fishes based on higher contribution of MUFA and n-3 PUFA and summarized that PCA found to be a useful tool in differentiating pelagic and demersal fishes based on their FA profiles (Ramesh et al., 2007).

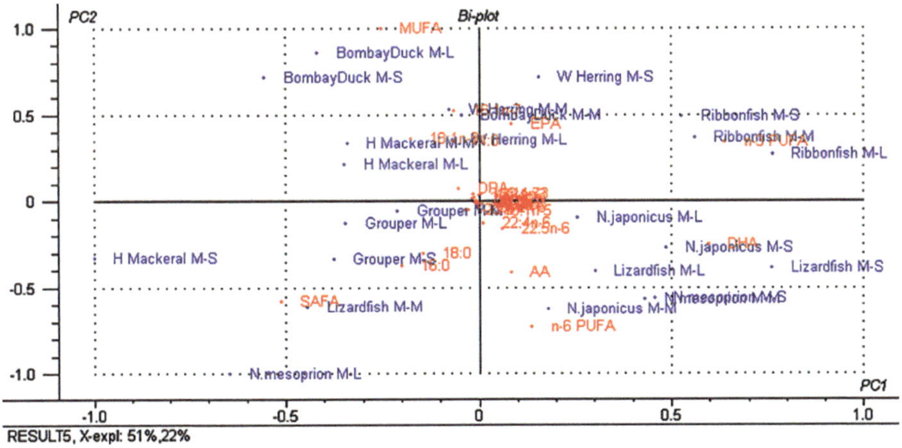

FIGURE 10.1 Distribution of muscle fatty acids in pelagic and demersal fishes with respect to size.

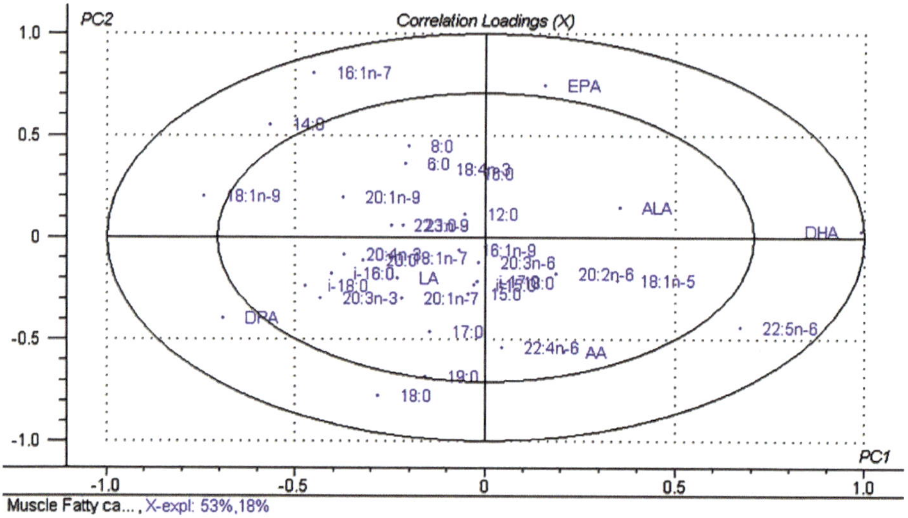

FIGURE 10.2 Correlations of muscle fatty acids in different pelagic and demersal fishes.

10.6.2 VARIATION IN FATTY ACID (FA) PROFILES IN DIFFERENT SIZE OF PELAGIC AND DEMERSAL FISHES

Of the tissues and FAs investigated in the present study, muscle tissue was found the most suitable for discrimination between pelagic and demersal species based on differences in FA composition. Joensen and Grahl-Nielsen (2000) reported the discrimination of different stocks of *Sebastes* species from Faroe Islands based on FA composition. The total SAFA content were less percentages in all the pelagic fishes compared to demersal fishes' percentages. However, total MUFA were found higher in pelagic compared than demersal fishes. Both n-3 PUFA and n-6 PUFA were showing similar pattern in pelagic and demersal fishes.

The relationships among different sizes of pelagic fishes with respect to FAs, a graphic representation of the projection of variables and samples onto the first two principal components is given in Figure 10.3 in the form of bi-plot. About 59% of the total information of the data was explained by the two principle components PC_1 and PC_2. PC_1, which explained 43% of the total information in the data, has clearly separated post-monsoon from other seasons. Pelagic fishes were segregated into clusters mainly based on the season rather than size, demonstrating the profound effect of season on FA profile compared to the size of fish.

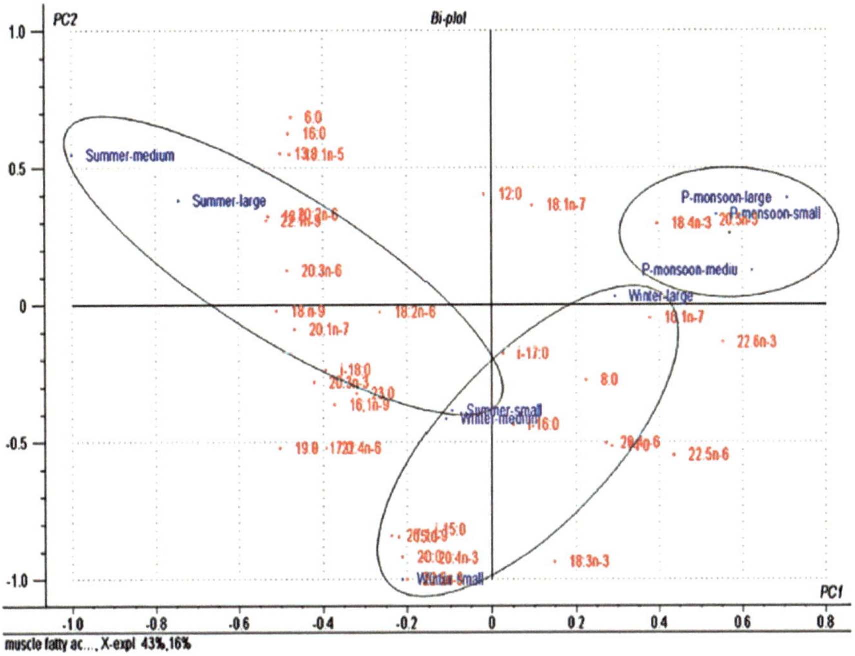

FIGURE 10.3 Bi-plot shown fatty acid distribution in different sizes of pelagic fishes based on sizes.

10.6.3 SEASONAL VARIATION OF FATTY ACID (FA) PROFILES OF PELAGIC AND DEMERSAL FISHES

The PCA could determine the seasonal changes of pelagic and demersal fishes based on FA composition with respect to different sizes (Ramesh, 2007). Score plots of the PCA clearly showed the differences among different seasons which were given in Figures 10.4 to 10.6. PC_1 clearly discriminated the fishes according to seasonal patterns. In case of small size, the response of these species to seasonal influence is strong as reflected in Figure 10.4. Some species like Bombay duck, ribbon fishes, grouper, and *N. japonicus* were remained partially overlapped. Season-II is at the center of the plot indicating no such variation differentiating the fishes. The better nutritive value of fishes with respect to health beneficial FAs (DHA) was found during season-I.

Medium size pelagic and demersal fishes (Figure 10.5) could be categorized based on PC_1 (48% of the total information), but did not resolve in the total separation of season-I and season-II groups. PC_2 (31% of the total information) negatively correlated to EPA and positively to 16:0. However, the large size gave a good separation among seasons (Figure 10.6). The main FAs positively loaded on the PC_1 (51% of the total information) are Arachidonic acid (20:4n-6), DHA, and on the negative part, 18:1n-9 and 16:1n-7. The PC_2 (26% of the total information) is positively loaded with 16:0 and negatively with EPA, successfully separated the three seasons. Thus, the FA distributions of the pelagic and demersal fishes based on the seasons with respect to sizes were explained by PC_1 and PC_2. In all the size groups, the fishes were distributed more or less in a similar manner.

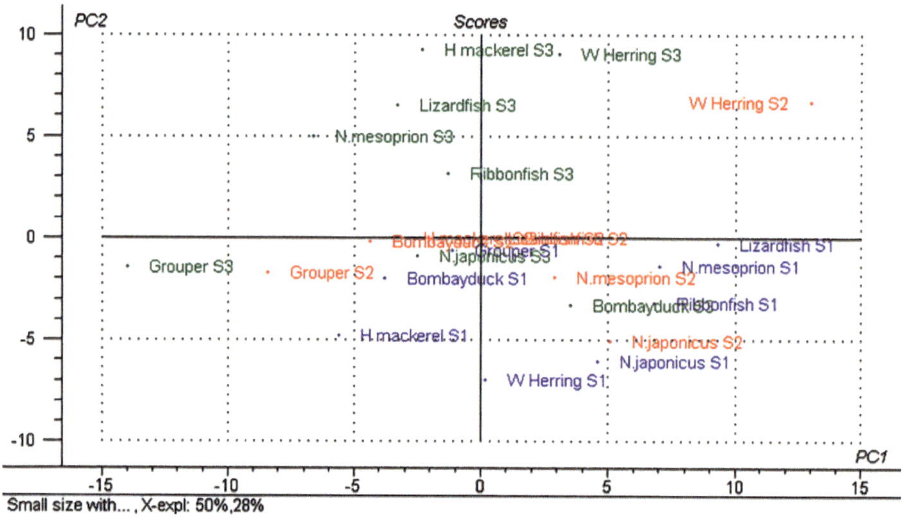

FIGURE 10.4 Distribution of pelagic and demersal fishes (small) with respect to seasons based on fatty acid composition.

Significant levels of seasonal variations were observed in the composition of SAFA, MUFA, and PUFA. The PCA analysis has revealed that there is a clear correlation between samples and variables based on FA composition. The distributions of fishes were similar during all the seasons with different size groups were given in Figures 10.4 to 10.6. PC_1 contains 48–51% of the original information of the data set. The second PC described 26–31% of the total variance. It was observed that the first two PCs accounted for almost 77–79% of the total variance of the data set in all the seasons. Further, it is clear from the bi-plot that, pelagic fishes caught in post-monsoon season were rich in n-3 FAs such as EPA, 18:4n-3 and DHA. The fish caught in pre-monsoon season were found rich in SAFA namely palmitic (16:0) and stearic acid (18:0).

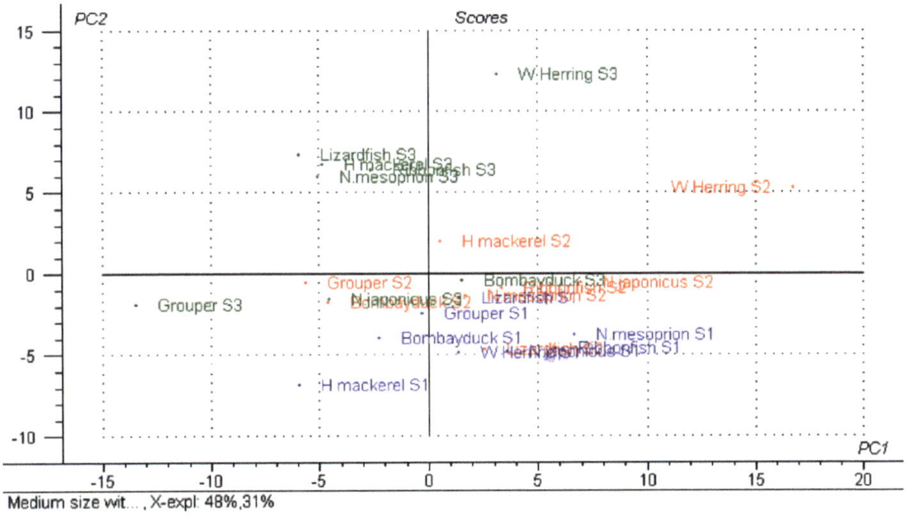

FIGURE 10.5 Distribution of pelagic and demersal fishes (medium) with respect to seasons based on fatty acid composition.

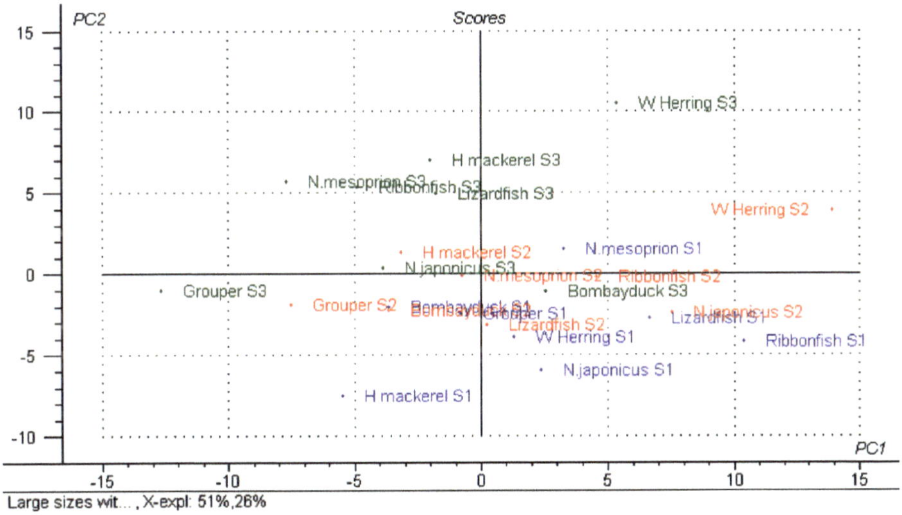

FIGURE 10.6 Distribution of pelagic and demersal fishes (large) with respect to seasons based on fatty acid composition.

10.7 CONCLUSIONS

Marine fishes are rich sources for human health-promoting n-3 polyunsaturated FAs such as EPA and DHA, which have been reported to have several health benefits including reducing the risk of cardiovascular diseases. Both pelagic and demersal fishes have higher concentrations of n-3 and n-6 PUFA for apparent positive effects on human health. These beneficial FAs were also reflected by different seasons, particularly with respect to sizes. The small marine fishes, which are available at cheaper price, have higher concentrations of EPA and DHA compared to those of other sizes. The fishes caught during monsoon season have lower levels than those caught in other seasons. The principle component analysis (PCA) is found to be an important tool to investigate the relationships of vast data of FAs present in pelagic and demersal fishes of different sizes caught in different seasons. The pelagic and demersal fishes could be discriminated by PCA based on FA profile in terms of their size and season.

KEYWORDS

- **demersal fish**
- **docosahexaenoic acid**
- **eicosapentaenoic acid**
- **fatty acid profiles**
- **health benefit of fish**
- **pelagic fish**
- **principle component analysis**

REFERENCES

Ackman, R. G., & Ratnayake, W. M. N., (1992). Non-enzymatic oxidation of seafood lipids. In: Flick, G. J., & Martin, R. E., (eds.), *Advances in Seafood Biochemistry. Composition and Quality* (pp. 245–267). Technomic Publishing Co., Inc.: Basel, Switzerland.

Ackman, R. G., (1982). Origin of marine fatty acids in nutritional evaluation of long-chain fatty acids. In: Barlow, S. M., & Stansby, M. E., (eds.), *Fish Oil* (pp. 25–88). Academic Press, London.

Ackman, R. G., Ratnayake, W. M. N., & Olsson, B., (1988). The "basic" fatty acid composition of Atlantic fish oils: Potential similarities useful for enrichment of polyunsaturated fatty acids by urea complexation. *J. Am. Oil Chem. Soc., 65*, 136–138.

Bell, J. G., Henderson, R. J., Tocher, D. R., McGhee, F., Dick, J. R., Porter, A., Smullen, R., & Sargent, J. R., (2002). Substituting fish oil with crude palm oil in the diet of Atlantic salmon (*Salmo salar*) affects tissue fatty acid compositions and hepatic fatty acid metabolism. *J. Nutr., 132*, 222–230.

Bell, M. V., Dick, J. R., & Porter, A. E. A., (2001). Biosynthesis and tissue deposition of docosahexaenoic acid (22:6n-3) in rainbow trout (*Oncorhynchus mykiss*). *Lipids, 36*, 1153–1159.

Bergstrom, E., (1989). Effect of natural and artificial diets on seasonal changes in fatty acid composition and total body lipid content of wild and hatchery-reared Atlantic salmon (*Salmo salar* L.) parr-smolt. *Aquaculture, 82*, 205–217.

Bogard, J. R., Thilsted, S. H., Marks, G. C., Wahab, M. A., Hossain, M. A., Akobsen, J., & Stangoulis, J., (2015). Nutrient composition of important fish species in Bangladesh and potential contribution to recommended nutrient intakes. *J. Food Com. Anal., 42*, 120–133.

Cowey, C. B., & Sargent, J. R., (1972). Fish nutrition. *Adv. Mar. Biol., 10*, 383–492.

Crawford, R. J. M., (2007). Food, fishing and seabirds in the Benguela upwelling system. *J. Ornithol., 148*(Suppl. 2), S253–260. doi:10.1007/s10336-007-0228-z.

Dallongeville, J., Yarnell, J., Ducimetiere, P., Arveiler, D., Ferriers, J., Montaye, M., Luc, G., et al., (2003). Fish consumption is associated with lower heart rates. *Circulation, 108*(7), 820–825.

De Silva, S. S., Gunasekera, R. M., & Austin, C. M., (1997b). Change in the fatty acid profiles of hybrid red tilapia, *Oreochromis mossambicus* x *O. niloticus*, subjected to short-term starvation, and a comparison with changes in seawater raised fish. *Aquaculture, 153*, 273–290.

De Silva, S. S., Gunasekera, R. M., Austin, C. M., Ingram, B. A., & Collins, R., (1997a). Changes in the fatty acid profile of the Australian shortfin eel in relation to development. *J. Fish Bio., 50*, 992–998.

Dunstan, G. A., Olley, J., & Ratkowsky, D. A., (1999). Major environmental and biological factors influencing the fatty acid composition of seafood from Indo-Pacific to Antarctic waters. *Rec. Res. Dev. Lipid Res., 3*,63–86.

EFSA, (2012). EFSA panel on dietetic products, nutrition and allergies (NDA). *Scientific Opinion Related to the Tolerable Upper Intake Level of Eicosapentaenoic Acid (EPA), Docosahexaenoic Acid (DHA) and Docosapentaenoic Acid (DPA)*. URL: http://www.efsa.europa.eu/en/efsajournal/doc/2815.pdf (accessed on 09 March 2022).

Eriksson, L., Johansson, E., Kettaneh-Wold, N., & Wold, S., (2001). *Multi and Megavariate Data Analysis*. Umetrics Academy, Umea, Sweden.

Erkkila, A. T., Lehto, S., Pyorala, K., & Uusitupa, M. I., (2003). n-3 fatty acids and 5-y risks of death and cardiovascular disease events in patients with coronary artery disease. *Amer. J. Clinical Nutr., 78*, 65–71.

FAO, (2018). *The State of World Fisheries and Aquaculture 2018-Meeting the Sustainable Development Goals*. Rome.

Gopakumar, K., (1997). *Indian Food Fishes: Biochemical Composition* (pp. 22–28). Central Institute of Fisheries Technology. Indian Council of Agricultural Research, Cochin, India.

Govindan, T. K., (1972). *Harpodon nehereus. Sea Food Export J., 6*, 21–24.

Greenberg, J. A., Bell, S. J., & Ausdal, W. V., (2008). Omega-3 fatty acid supplementation during pregnancy. *Rev. Obe. Gyn., 1*, 162–169.

Hatano, M., (1990). Lipids. In: Takeuchi, M., (ed.), *The Nutritive Components of Fish and New Fish Utilization Technology* (pp. 34–43). Kouseisha Kouseikaku, Tokyo.

Hayashi, K., & Takagi, T., (1978). Seasonal variation in lipids and fatty acids of Japanese anchovy *Engraulis japonica. Bull. Fac. Fish Hokkaido University, 29*, 38–47.

Hemre, G. I., & Sandnes, L., (1999). Effect of dietary lipid on muscle composition in Atlantic salmon *Salmo salar. Aqua. Nutri.*, 5, 9–16.

Henderson, R. J., & Sargent, J. R., (1985). Chain length specificities of mitochondrial and peroxisomal beta-oxidation of fatty acids in livers of rainbow trout. *Comp. Biochem. and Physiol., 82B*, 79–85.

Henderson, R. J., & Tocher, D. R., (1987). The lipid composition and biochemistry of freshwater fish. *Prog. Lipid Res., 26*, 281–347.

Henderson, R. J., Park, M. T., & Sargent, J. R., (1995). The desaturation and elongation of ^{14}C-labelled polyunsaturated fatty acids by pike *Esoxlucius* L. *in vivo. Fish Physiol. Biochem., 114*, 223–235.

HHS (U.S. Health and Human Services) and USDA (U.S. Department of Agriculture), (2015). *2015–2020 Dietary Guidelines for Americans.* 8th Edition. Washington, D.C.: U.S. Department of Health and Human Services and U.S. Department of Agriculture.

Hove, H. T., & Grahl – Nielsen, O., (1991). Fatty acid composition of start–feeding salmon (*Salmo salar*) larvae. *Aquaculture, 96*, 305–319.

James, P. S. B. R., Narasimham, K. A., Meenakshisunderam, P. T., & Appana, S. Y., (1986). *The Present Status of Ribbonfish Fishery in India* (Vol. 24, pp. 1–4). CMFRI special publication, Cochin, India.

Joensen, H., & Grahl-Nielsen, O., (2000). Discrimination of *Sebastes viviparus, S. marinus* and *S. mentella* from Faroe Islands by chemometry of the fatty acid profile in heart and gill tissues and in the skull oil. *Com. Biochem. Physiol. Part B, 126*, 69–79.

Joydeep, D., Debasish, P., Bhattacharya, A., Chakraborty, J., Banerjee, D., & Ghosh, A., (1999). Comparative study on fatty acids and either lipids of fresh and sun-dried Bombay duck (*Harpadon nehereus*). *J. Food Sci. Technol., 36*(6), 506–510.

Kakela, R., Kakela, A., Martínez, A. A., Sarzo, B., Louzao, M., Gerique, C., Villuendas, E., et al., (2010). Fatty acid signature analysis confirms foraging resources of a globally endangered Mediterranean seabird species: Calibration test and application to the wild. *Mar. Ecol. Prog. Ser., 398*, 245–258. doi: 10.3354/meps08291.

Kanazawa, A., Teshima, S. I., & Ono, K., (1979). Relationship between essential fatty acid requirements of aquatic animals and the capacity of bioconversion of linoleic acid to highly unsaturated fatty acids. *Comp. Biochem. Physiol., 63B*, 295–298.

Khedkar, G. D., & Jadhao, B. V., (2003). FISH: Pelagic species of tropical climates. In: Benjamin, C., (ed.), *Encyclopedia of Food Science and Nutrition* (pp. 2442–2447). Pelagic species of tropical climates. Publisher: Elsevier Science Ltd.

Kiessling, A., Pickova, J., Johansson, L., Asgard, T., Storebakken, T., & Kiessling, K. H., (2001). Changes in fatty acid composition in muscle and adipose tissue of farmed rainbow trout (*Oncorhynchus mykiss*) in relation to ration and age. *Food Chemistry., 73*, 271–284.

Medale, F., Lefevre, F., & Corraze, G., (2003). *Nutritional and Health Promoting Value of Fish: Main Factors Affecting Flesh Components, 38*(1), 37–44.

Metillo, E. B., & Eya, A. A. A., (2014). Fatty acids in six small pelagic fish species and their crustacean prey from the Mindanao Sea, Southern Philippines. *Trop. Life Sci. Res., 25*(1), 105–115.

Miles, E. A., & Calder, P. C., (2012). Influence of marine *n*-3 polyunsaturated fatty acids on immune function and a systematic review of their effects on clinical outcomes in rheumatoid arthritis. *British J. Nutr., 107*, S171–S184. doi: 10.1017/S0007114512001560.

Morkore, T., & Rorvik, K. A., (2001). Seasonal variations in growth, feed utilization and product quality of farmed Atlantic salmon (*Salmo salar* L.) transferred to seawater as 0+ smolts or 1+smolts. *Aquaculture,199*, 145–158.

Morris, R. J., & Culkin, F., (1989). Fish. In: Ackman, R. G., (ed.), *Marine Biogenic Lipids, Fats and Oils* (pp. 145–178). CRC Press, Boca Raton, Florida.
Mourente, G., & Tocher, D. R., (2009). Tuna nutrition and feeds: Current status and future perspectives. *Rev. Fish. Sci., 17*, 374–391.
Mourente, G., Megina, C., & Diaz-Salvago, E., (2002). Lipid in female Northern bluefin tuna (*Thunnus thynnus* L.) during sexual maturation. *Fish Physiol. Biochem., 24*, 351–363.
Murillo, E., Rao, K., & Durant, A. A., (2014). The lipid content and fatty acid composition of four eastern central Pacific native fish species. *J. Food Compos. Anal., 33*(1), 1–5.
Olesen, Y., & Skjervold, H., (1995). Variation in content of ω-3 fatty acids in farmed Atlantic salmon, with special emphasis on effects on non-dietary factors. *Aqua. Inter., 3*, 22–35.
Osako, K., Yamaguchi, A., Kurokawa, T., Kuwahara, K., Saito, H., & Nozaki, Y., (2003). Seasonal variation in docosahexaenoic acid content in horse mackerel caught in the East China Sea. *Fish. Sci., 69*(13), 589–596.
Ramesh, R., (2007). *Total Lipid Content and Fatty Acid Profiles of Important Pelagic and Demersal Fishes off Mumbai Coast*. M.F.Sc. Dissertation, CIFE, Mumbai.
Ramesh, R., Chakraborty, S. K., & Venkateshwarlu, G., (2015). Seasonal changes in total lipid content and fatty acid profiles of ribbon fish, *Lepturacanthus savala* (Cuvier, 1829). *Fishery Technology, 52*, 95–100.
Ramesh, R., Chakraborty, S. K., & Venkateshwarlu, G., (2016). Seasonal variation in fatty acid profiles of threadfin breams, *N. Mesoprion* and *N. Japonicus* (Bloch, 1791) with respect to their size. *Indian J. Geo-Marine Sci., 45*(5), 696–702.
Ramesh, R., Chakraborty, S. K., Biradar, R. S., & Venkateshwarlu, G., (2007). Differentiation of pelagic and demersal fishes by chemometry of the fatty acid profiles. *Book of Abstracts, 8th Asian Fisheries Forum* (p. 194). Kochi, India.
Ramesh, R., Pal, A. K., Chakraborty, S. K., & Venkateshwarlu, G., (2013). Variation in total lipid content and fatty acid composition in the muscle of Bombay duck *Harpodon nehereus* (Hamilton, 1922) with respect to size and season. *Indian J. Fish., 60*(1), 111–116.
Saito, H., Ishihara, K., & Murase, T., (1997). The fatty acid composition in tuna bonito (*Euthynnus pelamis*) caught at three different localities from tropics to temperate. *J. Sci. Fd. Agric., 73*, 53–59.
Sargent, J. R., (1995). (n-3) polyunsaturated fatty acids and farmed fish. In: Hamilton, R. J., & Rice, R. D., (eds.), *Fish Oil: Technology, Nutrition and Marketing* (pp. 67–94). London: Society of Chemical Industry.
Sargent, J. R., Henderson, R. J., & Tocher, D. R., (2002). The lipids. In: Halver, J., & Hardy, E., (eds.), *Fish Nutrition* (pp. 181–257). Academic Press, Elsevier, San Diego, California, USA.
Satoh, S., Takeuchi, T., & Watanabe, T., (1984). Effect of starvation and environmental temperature on proximate and fatty acid composition of *Tilapia niloticus*. *Bull. Jap. Soc. Sci. Fish., 50*, 79–84.
SenO¨zdemir, N., Parrish, C. C., Parzanini, C., & Mercier, A., (2019). Neutral and polar lipid fatty acids in five families of demersal and pelagic fish from the deep Northwest Atlantic. *ICES J. Mar. Sci.*, 1–9. doi: 10.1093/icesjms/fsz054.
Shearer, K. D., Asgard, T., Andorsdottir, G., & Aas, G. H., (1994). Whole-body elemental and proximal composition of Atlantic salmon *(Salmo salar*) during the life cycle. *J. Fish Biol., 44*, 785–797.

Soccol, M. C. H., & Oetterer, M., (2003). Seafood as functional food. *Braz. Arch. Biol. Technol., 46*(3), 443–454.

Spite, M., (2013). Deciphering the role of *n*-3 polyunsaturated fatty acid-derived lipid mediators in health and disease. *Proc. Nutr. Soc., 72*, 441–450. doi: 10.1017/S0029665113003030.

Sushchik, N. N., Rudchenko, A. E., & Gladyshev, M. I., (2017). Effect of season and trophic level on fatty acid composition and content of four commercial fish species from Krasnoyarsk reservoir (Siberia, Russia). *Fish. Res., 187*, 178–187.

Tacon, A. G. J., & Metian, M., (2013). Fish matters: Importance of aquatic foods in human nutrition and global food supply. *Rev. Fish. Sci., 21*(1), 22–38.

Tanakol, R., Yazıcı, Z., Sener, E., & Sencer, E., (1999). Fatty acid composition of 19 species of fish from the black sea and the Marmara Sea. Paper No. L7859. *Lipids, 34*, 291–297.

Torstensen, B. E., Li, O., & Froyland, L., (2000). Lipid metabolism and tissue composition in Atlantic salmon (*Salmo salar L.*) effects of capelin, palm and oleic acid-enriched sunflower oil as dietary lipid sources. *Lipids, 35*, 653–664.

Viga, A., & Grahl-Nielsen, O., (1990). Genotypic and phenotypic fatty acid composition in the tissues of salmon, *Salmo salar. Comp. Biochem. and Physiol. Part B, 96*(4), 721–727.

Worthington, R. E., & Lovell, R. T., (1973). Fatty acids of channel catfish *(Ictalurus punctatus*): Variance components related to diet, replications within diets, and variability among fish. *J. Fish. Res. Board Can., 30*, 1604–1608.

Yazeed, A. A. M., (2013). Fatty acids profile of some marine water and freshwater fish. *J. Arabian Aqua. Soc., 8*(2), 283–292.

Zhang, X., Ning, X., He, X., Sun, X., Yu, X., & Cheng, Y., (2020). Fatty acid composition analyses of commercially important fish species from the pearl river estuary, China. *Plus One, 15*(1), e0228276. https://doi.org/10.1371/journal.pone.0228276.

CHAPTER 11

Endogenous Enzymes: Their Roles in Quality of Fish/Shellfish and Their Products

SOOTTAWAT BENJAKUL,[1] AVTAR SINGH,[1] THANASAK SAE-LEAW,[1] and AMJAD K. BALANGE[2]

[1]*International Center of Excellence in Seafood Science and Innovation, Faculty of Agro-Industry, Prince of Songkla University, Hat Yai, Songkhla – 90110, Thailand*

[2]*Department of Post-Harvest Technology, ICAR-Central Institute of Fisheries Education, Versova, Mumbai – 400061, Maharashtra, India*

ABSTRACT

Endogenous enzymes, especially proteases, lipolytic enzymes, polyphenol oxidase (PPO), lipoxygenase (LOX), transglutaminase (TGase), and trimethylamine-*N*-oxide demethylase (TMAOase), contribute to the overall quality of seafoods during postmortem handling, storage, and processing. Lipolytic enzymes, including lipases and phospholipases, play a major role in the hydrolysis of lipids in seafoods, thus liberating free fatty acids (FFAs). FFAs are prone to lipid oxidation, producing off-flavor compounds. In addition, FFAs can interact with proteins, which may modify the texture of seafoods. Enzymatic browning caused by endogenous PPO is one of the most important reactions, which detrimentally affects the quality and acceptability of shellfish or crustaceans. Enzymatic lipid oxidation catalyzed by LOX is another cause of quality deterioration, especially rancidity or off-odor/off-flavor in seafoods. The textural loss in fish/shellfish and their products is mainly associated with endogenous protease and TMAOase,

Advances in Fish Processing Technologies: Preservation, Waste Utilization, and Safety Assurance.
Ranendra K. Majumder, PhD & Amjad K. Balange, PhD (Eds.)

which can lead to softening and hardening of the products, respectively. However, TGase is responsible for enhancing the gelling properties of surimi gels via augmentation of setting phenomenon. Despite the negative impact of proteases, they get involved in the manufacturing of fermented salted fish/shellfish products. Therefore, eating quality and sensory characteristics of seafoods are directly governed by endogenous enzymes.

11.1 INTRODUCTION

Fish and shellfish have been known to possess high nutritional value and have become popular globally due to their delicacy. Freshness of seafood and their products is considered as prime factor for quality determination in market. Due to high contents of amino acids, lipids, and moisture, the loss in freshness or deterioration of seafood occur rapidly after capture and slaughter, leading to the loss in quality and acceptability (Nielsen and Nielsen, 2012). These changes are not only related to microbial spoilage, but also due to the endogenous enzymes such as proteases, lipases, polyphenol oxidase (PPO), etc.

Moreover, during refrigerated/frozen storage, enzyme such as trimethylamine-*N*-oxide demethylase (TMAOase) results in the loss of quality of certain fish via muscle hardening (Leelapongwattana et al., 2008). Lean fish has been consumed widely as raw, cooked or their processed counterpart. Because of overexploitation of lean fish, industrial demand on dark-fleshed fish has been increased. However, lipid, myoglobin, and sarcoplasmic protein contents, heat-activated proteases and lipolytic enzymes (lipases and phospholipases) are present at high levels in the meat of those species, causing rapid deterioration, off-flavor/odor, destruction of vitamins and also showing the adverse effects on functional properties (Toyomizu et al., 1981; Singh and Benjakul, 2018). Similarly, shellfish such as shrimp, prawn, crab, and lobster have the limited shelf-life, depending on pretreatment, condition, and period of storage. Melanosis or black spot, which is triggered by PPO, reduces the organoleptic properties and market value. Although the melanosis is harmless to consumers, it drastically reduces the overall quality of crustaceans and their products (Sae-Leaw and Benjakul, 2019b).

On the other hand, some endogenous enzymes in fish/shellfish are of demand for processing, such as autolysis mediated by proteases, etc. Those enzymes responsible for fish silage or fish sauce production and for ripening of salted-cured fish, etc. Particular enzyme is important for obtaining the desired texture of gel-type products such as endogenous transglutaminase

(TGase) (Nielsen and Nielsen, 2012). Therefore, to maintain the freshness and quality of fish and shellfish, the activities of various undesirable endogenous enzymes have been controlled via application of several inhibitors (Sae-Leaw and Benjakul, 2019b; Singh and Benjakul, 2018; Sriket, 2014). Conversely, the conditions rendering optimal activities of desired enzymes must be optimized, in which those enzymes can be fully exploited. Hence, this chapter will provide the information related to the enzymes responsible for the loss of quality and organoleptic properties of fish or shellfish and their possible solutions are reviewed. Additionally, endogenous enzymes, which can improve the quality of seafood and their products, are also revisited (Figure 11.1).

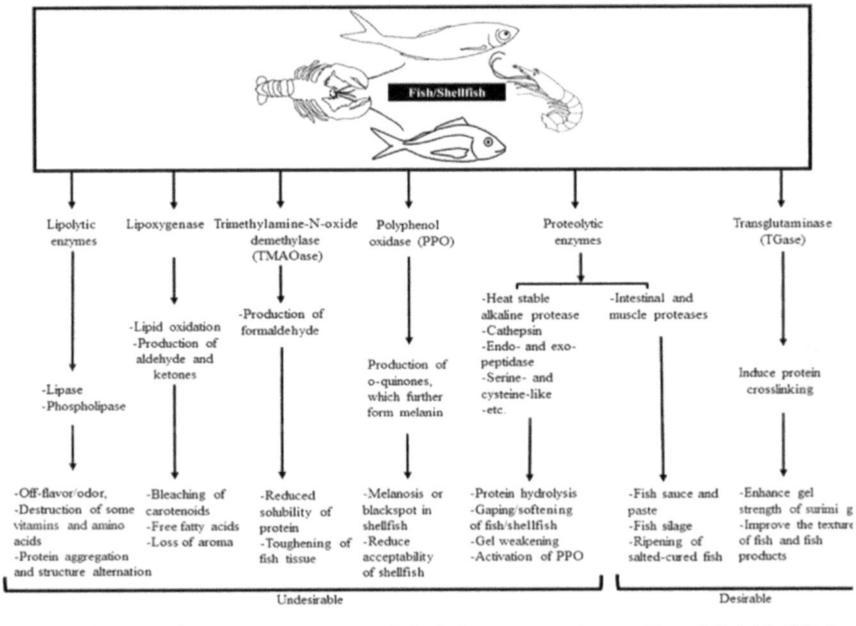

FIGURE 11.1 Endogenous enzymes and their impacts on the quality of fish/shellfish and their products.

11.2 LIPOLYTIC ENZYMES

Lipolytic enzymes play a major role in hydrolysis of lipids in fish and fish products. The major lipolytic enzymes in seafoods are lipases and phospholipases, which induce the cleavage of ester bonds of triacylglycerols (TAGs) and phospholipids, respectively, thus liberating FFAs. FFAs are able to interact with proteins to form insoluble protein-lipid complexes, which may modify the texture of seafoods. Additionally, FFAs are more susceptible to lipid oxidation, producing off-flavor compounds (Erickson, 2000). Lipases can also induce hydrolysis of ester bonds in other substrates such as cholesteryl esters, and vitamin esters, etc. (Wong and Schotz, 2002). Although lipases are water-soluble proteins, their substrates are insoluble in a purely aqueous system. Within the above classification, the so-called "true lipase" (EC 3.1.1.3 triacylglycerol lipase) is distinguished from other lipolytic enzymes such as phospholipases, sterol esterase, and retinyl-palmitate esterase. Seafood lipases have an important physiological role in hydrolysis of water-insoluble TAG to more polar diacylglycerol (DAG), monoacylglycerol (MAG), FFA, and glycerol for absorption and transportation through the membrane (Iijima et al., 1998). The reactivity of seafood digestive lipases toward lipids differs from that of digestive lipases from other sources. Fish lipases are highly reactive toward lipid-containing long-chain polyunsaturated fatty acids (PUFAs), whereas mammalian digestive lipases most actively cleave ester bonds of lipids having fatty acids with ≤20 carbons and a low degree of unsaturation (Kurtovic et al., 2009). Phospholipases are another group of lipolytic enzymes found in seafoods. Phospholipase hydrolyzes phospholipid substrate at specific ester bonds and produces acyl-lysophospholipids and FFAs (Richmond and Smith, 2011). In general, lipase activity is higher in dark muscle than in white muscle of the same fish species. Hydrolysis of acylglycerols predominates in dark muscle tissue, mainly due to the higher lipid content. In lean fish, phospholipase shows the marked impact on hydrolysis, since most of the lipid in this type of tissue is phospholipid (Shewfelt, 1981).

The optimal pH and temperature of most lipases are between 7–9 and 35–60°C, respectively. Lipases are active over a wide temperature range of –20 to 65°C, but the usual range is 30–45°C (Shahidi and Janak Kamil, 2001). Sae-Leaw and Benjakul (2018) extracted crude lipases from Asian seabass digestive organs, including stomach, liver, pyloric caeca and intestine. Lipase from liver had the highest activity, compared to those from other organs. Lipase from liver of seabass, with a molecular weight (MW) of 60 kDa, was purified to homogeneity using ammonium sulfate precipitation and a series of chromatographies, including diethylaminoethyl sepharose (DEAE) and Sephadex G-75 size exclusion columns. The optimal pH and temperature were 8.0 and 50°C, respectively (Sae-Leaw and Benjakul, 2018). Nayak et al. (2003) prepared crude lipase extracts by centrifuging tissue homogenates from rohu (*Labeo rohita*), oil sardine (*Sardinella longiceps*), mullet (*Liza subviridis*) and Indian mackerel (*Rastrelliger kanagurta*). Rohu had the highest lipolytic activity in all the tissues (intestine, stomach/pyloric ceca, liver/hepatopancreas and red muscle), compared to the other fish. In general, the highest lipase activities were observed in the intestine and stomach/pyloric caeca. Lipolytic activity in the muscle was the lowest among all the tissues for all fish, except mackerel, in which the highest activity was found in the muscle. A lipase was extracted from the hepatopancreas of red sea bream and purified using a combination of anion exchange chromatography, hydrophobic column, and gel filtration (Iijima et al., 1998). The MW of the lipase was 64 kDa and the enzyme exhibited the maximal activity at pH 7–9. The activity of the red sea bream lipolytic enzyme is Ca^{2+} independent. Cherif and Gargouri (2009) purified digestive lipase from crab hepatopancreas by heat treatment, DEAE–cellulose column, ammonium sulfate precipitation, S-200, Mono-Q, and S-200 columns. The enzyme had a MW of 65 kDa. The maximum activity of crab lipase was documented at pH 8. Furthermore, the hydrolysis was found to be specifically dependent on the presence of Ca^{2+} to trigger the hydrolysis of tributyrin emulsion (Cherif and Gargouri, 2009).

11.2.1 ROLE OF LIPOLYTIC ENZYMES ON QUALITY OF SEAFOODS

Lipolysis or hydrolysis of lipids has been associated with quality deterioration of seafoods during postmortem handling, chilling, and frozen storage. Lipid hydrolysis can be catalyzed by endogenous lipases, phospholipases, and bacterial enzymes, thus producing FFAs as the products. Accumulation of non-esterified FFAs in seafoods during postmortem handling and storage contributes to quality loss. FFAs may be responsible for the formation of off-flavor/off-odor, the destruction of some vitamins and amino acids and the changes in texture and water-holding capacity of muscle proteins (Toyomizu et al., 1981). The latter two deteriorations generally occur in frozen fish and are related to the propensity of fatty acids to bind, denature, and cross-link myofibrillar proteins (Haard, 1992a).

The relationship between quality deterioration and the liberation of FFAs mediated by lipolytic activity during frozen storage of fish was reported (Toyomizu et al., 1981). Refsgaard et al. (2000) also observed a marked increase in FFAs in Atlantic salmon stored at –10 and –20°C. This increase in FFAs content was in accordance with the lowered sensory property, suggesting that enzymatic hydrolysis of neutral lipids played a significant role in the deterioration of the quality of salmon during frozen storage.

Oxidative rancidity of mullet was related to the presence of phospholipid-FFAs (Audley et al., 2006). During the refrigerated storage of horse mackerel after freeze-thawing process, both enzymatic lipolysis and lipid oxidation occurred accompanied by quality loss (Kuizumi et al., 1988).

11.2.2 PREVENTION OF QUALITY LOSS CAUSED BY LIPOLYTIC ENZYMES IN SEAFOODS

Endogenous lipolytic enzymes are relatively resistant to low temperatures and retain their activity in the frozen state (De Koning and Mol, 1990; Jeong et al., 1990). Therefore, inactivation of lipolytic enzymes can be another means to prevent quality loss of seafoods. Lipases are generally inhibited by certain organophosphorus compounds (e.g., diethyl *p*-nitrophenyl phosphate) through irreversible phosphorylation of active site serine. A variety of chemical compounds, including anionic surfactants (e.g., sodium dodecyl sulfate, SDS), cationic surfactants (e.g., quaternary ammonium salts), certain proteins, metal ions, certain carbamates, β-lactones, and diisopropylfluorophosphate, were shown to inhibit lipases (Patkar and Bjorkling, 1994). Other lipase inhibitors are metal ions such as Fe^{2+}/Fe^{3+}, iodine, boronic acids, sulfhydryl reagents and mercury derivatives. Calcium and zinc ions tend to activate lipases or have no effect (Patkar and Bjorkling, 1994; Gargouri et al., 1997), depending on the individual lipases. Nevertheless, these chemicals are not applied commercially to seafood products due to their toxicity. Other chemicals have been shown to lower lipolytic activity. Takiguchi (1989) found that high level of NaCl inhibited lipid hydrolysis in sardine. Immersion of fish samples in $MgSO_4$ or sodium hypochlorite also led to a significant inhibition in FFA formation (Jonas, 2011; Ke and Ackman, 2011). Magnesium may compete with endogenous calcium, which is known to stimulate lipase-catalyzed hydrolytic activity, while hypochlorite may directly oxidize lipolytic enzymes, resulting in the inactivation of lipase (Weete, 1998). Recently, Gulzar and Benjakul (2019) also reported that tannic acid could inhibit lipase from Pacific white shrimp cephalothorax in a dose-dependent manner.

The handling and pretreatment conditions prior to freezing have the influence on lipolysis in frozen seafoods. Holding of silver hake for 5 days in refrigerated seawater prior to freezing led to the decreased lipolysis during frozen storage, compared to those held on ice (Hiltz et al., 1976). Icing method has been reported to affect lipolysis in fish. A lower rate of FFAs formation was observed in horse mackerel kept in slurry ice than in those stored in traditional flake ice (Losada et al., 2005). Losada et al. (2007) also found that FFAs content induced by lipolytic enzyme was reduced when a small pelagic fatty fish (*Sardina pilchardus*) was subjected to chilling in slurry ice prior to freezing.

Freezing method and frozen storage condition also have a profound role in lipolysis in fish. Slow freezing and fluctuations of storage temperature from –12 to –35°C could enhance the release of lysosomal lipase in rainbow trout muscles (Geromel and Montgomery, 1980). During frozen storage, the release of lipase was maximal during the first month. The rate of release of the enzyme was enhanced at higher frozen storage temperature. However, intermediate, and fast freezing do not trigger lipase release from lysosomes. This was in line with the results of Partman (1975) who found that temperature reduction to the freezing range should be performed within 30 min in order to minimize quality alterations caused by lipolytic enzymes.

High hydrostatic pressure or high-pressure processing (HPP) has recently been applied in food processing and several commercial food products have already been marketed. High hydrostatic pressure results in protein denaturation, inhibition of some inherent enzymatic activities and some microorganisms. With pressurization of cod (*Gadus macrocephalus*) and mackerel (*Scomber japonicus*) muscles at 400 MPa for more than 15 min, FFA accumulation was reduced, while phospholipid levels were not significantly affected, indicating that the enzymatic degradation of phospholipids was inhibited by HPP (Ohshima et al., 1993). Preheating is another means to inactivate lipase in fish. FFAs formation was inhibited by preheating of fillet and minced flesh of silver hake at 60°C (Lall et al., 1975).

11.3 LIPOXYGENASE (LOX)

LOX (E.C. 1.13.11.12, linoleate: oxygen oxidoreductase) is an iron-containing enzyme that catalyzes the incorporation of oxygen into the moiety of cis, cis-1,4 pentadiene in polyunsaturated fatty acids (PUFAs) (e.g., arachidonic, linoleic, and linolenic acid) via pentadienyl radical intermediate after H abstraction. The pentadienyl radical may react with oxygen to give peroxyl radical isomers, followed by fatty acid hydroperoxides as the final products (Kanner and Rosenthal, 1992). LOXs exist in different tissues of many higher plants, e.g., peas, soybean, corn, rice, tomato, and cucumber. LOXs are also present in seafoods. In fish, LOXs are typically found in skin, muscle, and gill. LOXs have been detected in several fish tissues, including skin of seabass (Sae-Leaw and Benjakul 2014b), skin of trout (German et al., 1985), skin, and muscle of herring (Medina et al., 1999), muscle, fin, roe, skin, milt, and gill of Baltic herring (Stodolnik and Samson, 2000), skin of sardine (Mohri et al., 1990) and skin of Nile tilapia (Sae-Leaw et al., 2013), etc. The highest LOX activity in lake herring was reported in light muscle, followed by dark muscle and skin, respectively (Wang et al., 1991). In tiger shrimp, LOX activity was only found in the hemolymph but not in the muscle, branchiae, midgut gland (hepatopancreas) eye, brain, or spinal cord (Kuo et al., 1994). Gill and skin of rainbow trout had the highest LOX activity, followed by brain, ovary, muscle, eye, liver, and spleen, respectively (Knight et al., 1995).

11.3.1 ROLE OF LOX ON QUALITY LOSS OF SEAFOODS

Lipid oxidation plays an important role in the development of aroma in seafoods. PUFAs in seafoods are extremely prone to autoxidation and LOX-catalyzed lipid oxidation and the hydroperoxides formed can be readily decomposed during handling, storage, processing, and cooking. LOX present in fish tissues can oxidize PUFAs into position-specific hydroperoxides as the primary products. These unstable hydroperoxides formed can be decomposed by homolytic cleavage and β-scission to produce various low molecular weight secondary products, such as aldehydes, alcohols, ketones, and hydrocarbons. These compounds directly influence the quality, especially flavor and palatability of seafoods and their products (Hsieh and Kinsella, 1989). The action of LOX on PUFAs can produce free radicals that further increase the rate of lipid oxidation (German et al., 1985). During storage, lipids in seafoods develop strong oxidized odors known as unpleasant rancid, stale, fishy, and cod liver oil-like (Karahadian and Lindsay, 1989; Sae-Leaw and Benjakul, 2014a). Sae-Leaw and Benjakul (2014b); and Sae-Leaw et al. (2013) reported that the development of fishy odor in seabass and Nile tilapia skins was in accordance with the marked increase in LOX activity during iced storage. The use of fresh fish skin for production of collagen, gelatin or protein hydrolysate is recommended to prevent undesirable fishy odor induced by LOX (Sae-Leaw and Benjakul, 2015).

The presence of a complex mixture of unsaturated carbonyl compounds is primarily responsible for the off-flavors of oxidized seafood products. The contribution of LOX-initiated oxidation toward PUFAs, in which specific oxidative volatile carbonyl compounds are generated, has been reported. Josephson and Lindsay (1987) demonstrated that off-odor compounds, e.g., (Z)-4-heptenal can be formed from (E,Z)-2,6-nonadienal, which primarily arises due to LOX action in fresh fish. Occurrence of (Z)-3-hexenal, (Z,Z)-3,6-nonadienal in the aroma profiles of boiled trout (Milo and Grosch, 1993) and salmon and cod (Milo and Grosch, 1996) indicated the initial involvement of LOX in the formation of the site-specific hydroperoxides. LOX present in fish tissues generates 12-hydroperoxy eicosatetraenoic acid (12-HPETE), and 14-hydroperoxydocosahexaenoic acid (14-HPDHE) (Hsieh et al., 1988a). The alpha-cleavage at either side of the carbon atoms bearing the hydroperoxy groups can result in the formation of volatile compounds responsible for the typical oxidative fishy odor, such as 2-nonenal from 12-HPETE and 3,6-nonadienal from 14-HPDHE (Josephson and Lindsay, 1986).

The attractive red or pink or orange color of flesh from some fish species such as salmon is attributed to the presence of carotenoids. The fading of the skin color during handling and processing also poses a serious problem with respect to the acceptability of these fish. Tsukuda and Amano (1966) examined these pigments in the skin of 18 species of fish and reported the major carotenoids to be astaxanthin and tunaxanthin. Any discoloration in these fish was attributed to deleterious changes in these pigments. Haard (1992b) reported that LOX present in fish tissue was involved in the discoloration or fading the pink or red color of skin or flesh of various fish at chilling temperature. Endogenous LOX has been known to catalyze the oxidation of PUFAs to form corresponding hydroperoxides, which subsequently induce discoloration of carotenoids in fish (Tsukuda and Amano, 1968). Astaxanthin, tunaxanthin, and β-carotene in fish can be degraded to colorless compounds, e.g., carbonyl compounds at refrigeration temperature (Simpson, 1982). However, pigmentation of fish can be achieved by supplementation of carotenoids in the diets. Rahman et al. (2016) investigated the effects of dietary astaxanthin levels on muscle pigmentation in juvenile rainbow trout. With increasing astaxanthin in the diet, astaxanthin content and redness of muscle increased.

11.3.2 PREVENTION OF LOX-CATALYZED LIPID OXIDATION IN SEAFOODS

Various antioxidants have been employed to inhibit LOX-catalyzed oxidation in seafoods. Plant polyphenols have been reported to show LOX inhibitory activity in several fish. Esculetin, a plant phenolic, was used for inhibition of LOX in comparison with vitamin E, vitamin C and synthetic butylated hydroxytoluene (BHT) (Saeed and Howell, 2001). Esculetin (5 mM) completely inhibited the formation of a product of lipid oxidation catalyzed by LOX, hydroxyeicosatetraenoic acid (HETE). Partial inhibition was observed in the presence of vitamin C (500 ppm), vitamin E (500 ppm) and BHT (200 ppm). The LOX inhibitory activity may be attributed to the structural similarities of the compounds tested and the accessibility to the catalytic site in LOX where the enzyme-radical complex is formed. This inhibitory mechanism was also likely due to the ability of esculetin to bind iron present in LOX (Saeed and Howell, 2001). Hsieh et al. (1988b) studied the relative efficacy of 10 flavonoid compounds and some common antioxidants in inhibiting LOX of rainbow trout gill. Fisetin and quercetin were the most potent LOX inhibitors, with IC_{50} values of 0.25 and 0.4 μM, respectively.

Green tea extract (GTE) was also able to inhibit lipid oxidation catalyzed by LOX from the gills of gray mullet and tilapia (Liu and Pan, 2004). Addition of GTE at 34 mg/L could inhibit the activities of 15-, 12-, and 5-LOX of tilapia gill by 69.7, 28.7, and 55.1%, respectively. Banerjee (2006) documented that commercially available green tea polyphenols were shown to effectively inhibit the LOX activity of mackerel muscle. Epigallocatechin gallate (EGCG) exhibited the most effective inhibitor (IC_{50} of 0.13 nM), followed by epicatechin gallate (ECG) (IC_{50} of 0.8 nM), epicatechin (EC) (IC_{50} of 6.0 nM), epigallocatechin (EGC) (IC_{50} of 9.0 nM) and catechin (C) (IC_{50} of 22.4 nM), respectively. All the tea catechins showed a non-competitive type inhibition kinetic. The presence of a gallate group in the C_3 position in C ring is crucial for efficient inhibitory activity towards LOX. Mohri et al. (1999) examined the retardation of LOX-catalyzed lipid oxidation in several fish skin extracts by different types of antioxidants. The inhibitory effects of catechin, EGC, EGCG, caffeic acid and Polyphenon G, commercially available tea catechin extracts, were stronger than those of LOX inhibitors (eicosatetraenoic acid) and ferrous iron-chelating agents, while chlorogenic acid, an ester of caffeic acid, showed lower efficacy than other compounds. The primary mechanism of LOX inhibition by polyphenols was associated with the reduction of ferric ion in LOX or via complex formation with the ferric iron at the active site of the LOX. In addition, green tea polyphenols have potent free radical quenching and antioxidant activities and have structural features that may specifically interfere with the arachidonic acid cascade, including the LOX pathway (Hong and Yang, 2003). Thus, the use of polyphenols or plant extracts can be another controlling method to prevent lipid oxidation induced by LOX.

11.4 TRIMETHYLAMINE-N-OXIDE DEMETHYLASE (TMAOASE)

TMAOase (EC 4.1.2.32) is generally found in muscle tissues and visceral organs of fish and invertebrates, which catalyzes the breakdown of trimethylamine oxide (TMAO) to equimolar quantities of formaldehyde and dimethylamine (DMA) (Leelapongwattana et al., 2010a). TMAOase was found in white muscle of blue ling, cod, and saithe, whereas no TMAOase was detected in hake, haddock, and whiting muscle (Rehbein, 1988). Kidney of lizardfish contained a high amount of TMAOase, which can contaminate into fish muscle during processing, e.g., meat separation (Benjakul et al., 2004a). Kimura et al. (2000) extracted TMAOase from the walleye pollack muscle with 1 M NaCl at pH 4.5, which was further separated by DEAE-cellulose and gel filtration chromatographies. The MW of the enzyme was estimated to be 400 kDa by the gel filtration chromatography and to be 25 kDa on sodium dodecyl sulfate-polyacrylamide gel electrophoresis. On the other hand, Benjakul et al. (2003a) estimated MW of TMAOase from lizardfish kidney to be 128 kDa based on activity staining. The reducing agents such as $FeCl_2$, ascorbate, dithiothreitol, and cysteine are the cofactors required for TMAOase activity, which helps to maintain the active form of iron as Fe^{2+} (Benjakul et al., 2003a; Kimura et al., 2000). Moreover, the formation of DMA and formaldehyde is depending on the amount of substrate, the amount of dark muscle, cofactors, temperature, and the period of frozen storage (Leelapongwattana et al., 2005, 2010a). The absence of oxygen has also been shown to accelerate the rate of DMA and formaldehyde production (Benjakul et al., 2004a; Tolstorebrov et al., 2016).

11.4.1 ROLE OF TMAOASE

During the refrigerated storage, TMAOase has been known to lower the quality of gadoid fish associated with the crosslinking of muscle proteins and hardening of tissue (Leelapongwattana et al., 2005). This was more likely due to the formation of formaldehyde, which could cause the crosslinking of proteins via exposing hydrophobic groups of myosin chain (Leelapongwattana et al., 2010a). Formaldehyde is also able to induce the cross-linking via methylene bridge (Lu et al., 2010). Ang and Hultin (1989) reported loss in solubility and ATPase activity of partially purified cod myosin stored at –25°C or –80°C due to the formation of formaldehyde induced by TMAOase. Similar results were obtained from lizardfish proteins as reported by Benjakul et al. (2004a). Parkin and Hultin (1982) reported the lower DMA and formaldehyde formation during iced storage of red hake packed in oxygen-permeable film. Moreover, the amount and reactivity of formaldehyde were enhanced in minced muscle by the rupture of cellular integrity in the muscle structure. In general, the reducing conditions enhance, whereas oxidizing conditions inhibit the enzymatic degradation of TMAO to DMA and formaldehyde (Harada, 1975).

11.4.2 INHIBITION OF TMAOASE

Several compounds such as sodium citrate, pyrophosphate, H_2O_2, sodium alginate, xanthan gum, or carrageenan were used to lower the toughness of fish muscle via inhibiting TMAOase activity in frozen fish/shellfish (Ponte et al., 1986; Leelapongwattana et al., 2008, 2010a). This is more likely due to the chelating effect on ferrous ion, which is located in the active site of the enzyme and is also required for TMAOase activity (Leelapongwattana et al., 2008, 2010a). Racicot et al. (1984) treated the red hake muscle with various oxidizing and reducing agents and found that H_2O_2 was more promising than NaOCl or $KBrO_3$ to decrease the production of DMA. Conversely, all reducing agents such as sodium erythorbate and ascorbic acid enhanced the DMA production. When lizardfish mince was stored at –20°C for 24 weeks, the reduction in TMAOase activity and formation of formaldehyde was noticed with the addition of 0.5% sodium alginate and 0.3% pyrophosphate in combination with 4% sucrose and 4% sorbitol as the cryoprotectants, compared with the control (Leelapongwattana et al., 2008). Leelapongwattana et al. (2010b) also observed reduction in TMAOase activity in Haddock mince with the addition of 0.3% pyrophosphate and 0.5% alginate. However, the increase in TMAOase activity was noticed with the addition of antioxidants such as ascorbic acid and tocopherol in haddock mince, during frozen storage for 6 weeks. This was more likely due to the reducing ability of antioxidants used, which might provide iron to function through a cyclic mechanism between the ferrous and ferric states (Kimura et al., 2000). Nonetheless, Gómez-Estaca et al. (2020) reported that the addition of sodium citrate, tea polyphenolic extracts, and phytic acid inhibited *in vitro* activity of TMAO-demethylase enzyme, which was in accordance with *in vivo* study during frozen storage (–12°C/150 days) of hake mince.

11.5 POLYPHENOLOXIDASE (PPO)

PPO (1,2-benzenediol; oxygen oxidoreductase, EC1.10.3.1) has been known as tyrosinase, polyphenolase, phenoloxidase, phenolase, cresolase, catecholase, and catechol oxidase. PPO is a copper-containing enzyme and its active site consists of two copper atoms, each liganded with three conserved histidine. Those two aforementioned copper ions are involved in the interaction of PPO with its phenolic substrates and molecular oxygen (Kim et al., 2000). PPO plays a crucial role in physiological functions, especially for the primary immune response, self-recognition, healing of injuries and cuticle sclerotization of crustaceans. PPO oxidizes phenolic substrates, especially tyrosine in crustaceans, to *o*-quinones. These highly reactive products are able to cross-link with side groups of chitins and cuticular proteins, resulting in sclerotization or shell hardening. PPO is mostly distributed in cuticle as well as hemolymph of crustacean blood as zymogen, that requires an activation process (Sae-Leaw and Benjakul, 2019b).

For melanosis, PPO catalyzes the oxidation of phenolic substrates in crustaceans, principally tyrosine, to corresponding *o*-quinones, which further undergo several reactions, producing a wide variety of compounds. Those compounds react non-enzymatically and polymerize to form insoluble high molecular weight brown pigments or melanins (Sae-Leaw and Benjakul, 2019a). PPO is located in various tissues or anatomical parts of crustaceans with different activities (Sae-Leaw and Benjakul, 2019a). PPOs in crustaceans are mainly localized in the cephalothorax, caudal zone, and cuticle of the abdominal segments, where the cuticles and pleopods are attached (Ogawa et al., 1984). Sae-Leaw and Benjakul (2019a) reported that cephalothorax of Pacific white shrimp had the highest PPO activity, followed by pereopods, telson, pleopods, carapace, cuticle, and muscle, respectively.

11.5.1 ROLE OF PPO ON QUALITY LOSS OF SEAFOODS

Crustaceans, including shrimp, prawn, crab, and lobsters are among the most marketable seafoods, having a great economic value because of their delicacy. Generally, crustaceans have a limited shelf-life, primarily caused by melanosis (black spot discoloration) together with microbial and chemical deteriorations. Melanosis reduces organoleptic properties of crustaceans and their products, thus contributing to the loss in commercial value and rejection by consumers (Montero et al., 2001). Melanosis in crustaceans is basically triggered by PPO. It usually takes place rapidly within a few hours after harvesting. Melanosis normally appears at the cephalothorax and subsequently expands to the tail. Melanosis not only occurs in the raw crustaceans, but also in pre-cooked counterpart. This is due to the insufficient heat, which cannot fully inactivate PPO. The remaining PPO is still active and induces melanosis in pre-cooked shrimp stored under the refrigerated condition (Sae-Leaw et al., 2018).

Degree of melanosis developed varies with species, harvesting, handling methods and physiological conditions. Traumatic circumstances and rough handling of the catch can stimulate the defense mechanism in crustaceans via synthesis of PPO, resulting ultimately in an increased melanosis development in crustaceans after death or during the storage (Sae-Leaw and Benjakul, 2019b). Benjakul et al. (2005) reported that the melanosis intensity in different crustaceans is governed by different concentrations of substrate and PPO. Storage temperature is another prime factor determining PPO activity and melanosis in crustaceans. Cold storage is necessary to slow down the rate of melanosis in crustaceans throughout distribution and retailing as it can decrease PPO activity (Gonçalves and de Oliveira, 2016). Nevertheless, refrigeration at low temperature does not prevent or stop undesirable melanosis reaction completely because PPO still remains active under this condition. During freezing, PPO activity in crustaceans becomes lowered. After thawing of crustaceans, PPOs localized in the digestive glands and hemolymph are easily liberated and activated, resulting in the enhanced melanosis (Sae-Leaw and Benjakul, 2019b). Fluctuation of temperature during cold chain transportation or storage can also induce the spoilage or deterioration caused by enzymes and microorganisms. Melanosis generally develops at a faster rate than microbial spoilage (Thepnuan et al., 2008). Thus, it can be used as the index for quality loss of crustacean prior to microbial deterioration, in which the crustacean cannot be consumed safely.

11.5.2 INHIBITION OF PPO AND MELANOSIS IN CRUSTACEANS

Anti-melanosis agents have been extensively employed to reduce melanosis and prolong the shelf-life of crustaceans and their products. Sulfite-based compounds and their derivatives are commonly used as PPO inhibitors in crustaceans by reacting with intermediate quinones and inhibiting the subsequent polymerization. However, sulfite-based compounds cause the serious allergic reactions (Montero et al., 2001). As a consequence, safer alternatives from natural sources such as polyphenols or plant extracts with anti-melanosis properties have gained attention (Gokoglu and Yerlikaya, 2008; Sae-Leaw et al., 2018; Nirmal and Benjakul, 2009). Additionally, different preservation methods such as modified atmosphere packaging (MAP) (Nirmal and Benjakul, 2011a), ultrasound treatment (Sae-Leaw et al., 2018), uses of slurry ice (Aubourg et al., 2007) and edible coating (Yuan et al., 2016a) have been applied in conjunction with anti-melanosis agents to augment the efficacy in controlling of melanosis.

Plant polyphenols and extracts have been demonstrated to show PPO inhibitory activity and anti-melanosis in crustaceans via several mechanisms or kinetics. Certain polyphenols are able to interact with the active site of PPO or compete with PPO because of their similar structure to substrates (Sae-Leaw and Benjakul, 2019b). Additionally, polyphenols could also interact with enzymes or protein via either hydrogen bonding or hydrophobic interaction (Cirkovic and Stanic-Vucinic, 2018). This may bring about the loss in PPO activity. Moreover, the hydroxyl groups of polyphenols may be involved in electron donation to intermediate quinone, in which further oxidation is terminated. Also, polyphenols are able to chelate metal ions, especially Cu^{2+} in the active site of PPO. They also form hydrogen bonding with the active site of PPO, or form a Schiff's base through its aldehyde group, thus lowering PPO activity (Kubo and Kinst-Hori, 1998).

Among all PPO inhibitors, 4-Hexylresorcinol (4-HR) has been documented to inhibit PPO and control enzymatic browning in crustaceans effectively (Galvão et al., 2017). 4-HR has numerous advantages over sulfiting agents, including its lower effective level, higher specificity, chemical stability and negligible beaching impact toward treated crustaceans. 4-HR has been recognized as generally recognized as safe (GRAS) in the United States and is accepted by the Food and Drug Administration (FDA) for the use in shrimp. Moreover, the use of 4-HR has no effect on texture, taste, or color of the treated shrimp when its residuals in shrimp meat are less than 1 ppm (Iyengar et al., 1991). However, 4-HR is still expensive, compared to other anti-melanosis agents. Therefore, other plant polyphenols or extracts have gained increasing interest as the alternative anti-melanosis agents.

Polyphenols from different plants are proven to have the potential as PPO inhibitor. Tea generally has a high amount of polyphenols, especially catechin and derivatives with a wide variety of bioactivities. Tea catechins, including catechin (C), EGC, EC, ECG, and EGCG were able to inhibit PPO from Pacific white shrimp cephalothorax (Sae-Leaw et al., 2017). EGCG showed the greatest PPO inhibitory activity, followed by C, EC, ECG, and

EGC, respectively (Sae-Leaw et al., 2017). Hydroxyl group at 5′ position of B ring is mainly involved in PPO inhibition. Gallate moiety of EGCG more likely contributes to the improved PPO inhibitory activity. Pacific white shrimps dipped in 0.5 or 0.75% EGCG solution had the lower melanosis scores than the control and those treated in 1.25% sodium metabisulfite (SMS) during the storage at 4°C of 10 days (Sae-Leaw et al., 2017). Nirmal and Benjakul (2011b) reported that shrimps soaked in 0.5% ethanolic extract from de-chlorophyllized green tea showed the lower melanosis than the control and exhibited similar melanosis score to those treated with 1.25% SMS. Kojic acid has also been proved as an effective anti-melanosis agent in many crustaceans. Chen et al. (1991) demonstrated that kojic acid (20–200 μg/mL) showed a concentration-dependent inhibitory effect on PPO from various crustaceans, including Florida spiny lobster, grass prawn and white shrimp. DL-DOPA and catechol oxidized by lobster PPO was inhibited by 80 and 70%, respectively, when kojic acid (20 μg/mL) was implemented. Kojic acid was generally more effective in inhibiting lobster PPO than white shrimp and grass prawn PPO. Both ferulic acid (FA) and oxygenated FA (OFA) at various concentrations (0.1–2%, w/v) were able to inhibit PPO from Pacific white shrimp (Nirmal and Benjakul, 2009). FA demonstrated a higher PPO inhibitory activity than did OFA counterpart, when the same level was used. Quinone in OFA plausibly lost the reducing power. Thus, DOPA-quinone formed as induced by PPO was not reduced to DOPA.

Plant extracts have been documented to contain a wide variety of polyphenols. Thai indigenous leaf extracts (Liang, Mun-poo, cashew, and lead) are the potential sources of PPO inhibitors. Those extracts rich in polyphenols could lower the activity of PPO from cephalothorax of Pacific white shrimp (Sae-Leaw and Benjakul, 2019c). Cashew leaf extract possessed the greatest PPO inhibitory activity at all concentrations (0.025–0.5%) tested, followed by lead, Liang, and Mun-poo leaf extracts, respectively. Soaking of Pacific white shrimp in 1% cashew leaf extract for 30 min was able to retard the melanosis during refrigerated. Polyphenols in cashew leaf extract might contribute to PPO inhibition, thus lowering melanosis in Pacific white shrimp kept at 4°C for 12 days. Chamuang (*Garcinia cowa* Roxb.) leaf extract (CLE) was employed to inhibit PPO and black spot discoloration in Pacific white shrimp during refrigerated storage (Shiekh et al., 2019). CLE was rich in polyphenolic glycosides, in which chrysoeriol 6-C-glucoside-8-C-arabinopyranoside and 2-feruloyl-L-sinapoyl-gentiobiose were dominant. Polyphenols present in CLE might inactivate PPO by interaction with PPO active sites, chelating copper ion or reduction of DOPA-quinone to

DOPA via its ability to donate electrons to intermediate quinone. Several extracts from different parts of plants, including litchi seed (Prasad et al., 2009), pomegranate peel (Basiri et al., 2015), grape seed (Sun et al., 2014; Gokoglu and Yerlikaya, 2008), acerola pulp (Gonçalves et al., 2015), avocado seed and peel (Dao et al., 2016) and saffron tepal (Abbasvali et al., 2016), have also been documented to show PPO inhibitory activity and can be used as natural anti-melanosis agents.

Another potential method to allow PPO inhibitor to penetrate to the target PPO localized underneath the shell or other organs is a promising approach to increase the efficacy in melanosis prevention of crustaceans. Sae-Leaw et al. (2018) demonstrated that pre-cooked Pacific white shrimps soaked in 0.25% EGCG solution along with ultrasonication had lower melanosis scores, compared to those dipped in EGCG before pre-cooking. Generally, EGCG was able to penetrate into pre-cooked shrimp at higher extent than into raw counterpart. Proteins and connective tissues underwent denaturation when cooking at high temperature was applied. As a result, the tissues were destroyed and became looser. Ultrasonication also helps PPO inhibitor to migrate into the shrimp via cavitation effect, favoring the increased penetration of inhibitor to the target PPO (Chemat et al., 2011).

Edible coating is another means to minimize undesirable changes in seafood products. Edible coatings (ECs) have been implemented for controlling the moisture loss and formation of oxygen barrier (Cuppett, 1994). Furthermore, they can hold additives or preservatives on the surface of seafoods, thus maintaining their eating quality. The use of an edible coating in conjunction with PPO inhibitor to control melanosis in crustaceans has been reported. Chitosan coating containing GTE was used for treatment of Pacific white shrimp (Yuan et al., 2016b). Melanosis scores of the samples were in the ascending order: the sample treated with 1% chitosan in conjunction with 1% GTE, those treated with 1% GTE alone, 1% chitosan alone and the control. The retardation of melanosis in shrimp treated with chitosan coating was plausibly due to reducing power and oxygen barrier properties of chitosan. Yuan et al. (2016a) studied the impact of chitosan coating containing pomegranate peel extract (PPE) on melanosis of Pacific white shrimp during 10 days of iced storage. The shrimp coated with chitosan in the presence of 1.5% PPE demonstrated lower melanosis score than those treated with 1.5% PPE or chitosan coating alone, indicating the synergistic effect between chitosan coating and PPE.

Modified atmosphere packaging (MAP) has been introduced as the powerful means for shelf-life extension of seafoods and their products

(Nirmal and Benjakul, 2011a). Atmosphere with low oxygen was able to control melanosis of crustacean by lowering hydroxylation in blackening process. MAP has been implemented in combination with plant polyphenols or extracts to retard melanosis in crustaceans during prolonged storage. Nirmal and Benjakul (2011a) studied the impact of GTE without and with ascorbic acid in combination with MAP (50% CO_2; 5% O_2; 45% N_2) on melanosis formation of Pacific white shrimp during refrigerated storage of 10 days. Shrimp treated with GTE prior to storage under MAP exhibited the lower melanosis score than those stored under MAP alone (without GTE treatment) or the control, irrespective of ascorbic acid incorporated. Basiri et al. (2014) investigated the impact of methanolic pomegranate peel extract (PPE) and vacuum packaging on melanosis of Pacific white shrimp stored for 10 days under refrigeration conditions. Shrimps treated with 2% PPE and packed under vacuum showed the lowest melanosis score, followed by that treated with 1% PPE and packed under vacuum, that packed under vacuum alone and the control, respectively. PPE containing high amount of polyphenols could act as PPO inhibitor, whereas vacuum packaging with limited oxygen level decreased the hydroxylation. As a whole, melanosis can be lowered via the combined effect.

11.6 PROTEASES

Proteases are hydrolytic enzymes that cleave peptide bonds of proteins. Proteases are generally divided into 2 subclasses, depending on the site of cleavage: endopeptidases and exopeptidases. Endopeptidases cleave peptide bonds at the internal peptide bonds, whereas exopeptidases hydrolyze the peptide bonds at the amino or carboxyl ends (Singh and Benjakul, 2018). Proteases can also be classified as trypsin-like, chymotrypsin-like, chymosin-like, or cathepsin-like, etc. (Simpson et al., 2012). Moreover, proteases can be further distinguished by their active site. Aspartic, cysteine, serine, and metalloproteases have been proposed by the Enzyme Commission of the International Union of Biochemists (Singh and Benjakul, 2018). Fish and shellfish are known to possess proteases with different molecular properties, which are influenced by species, organs, as well as environmental factors (Sriket, 2014). Those proteases have a wide range of MWs, optimal temperature, and pH, and their location in fish/shellfish can be varied.

During the post-mortem handling and storage, fish, and shellfish undergo various changes including rigor mortis, resolution of rigor, autolysis, and bacterial spoilage (Singh and Benjakul, 2018; Teklemariam et al., 2015). The pH and high moisture content in fish and shellfish accelerate microbial and enzymatic degradation (Lougovois and Kyrana, 2005). The enzymatic degradation is mainly related to the proteases present in the digestive tract, which can be liberated via rupturing of digestive organs along with release of blood and water during storage and transportation, thus enhancing the extensive autolysis (Singh and Singh, 2020; Singh and Benjakul, 2018; Pushparajan et al., 2013). Additionally, cathepsins and calpains are associated with softening of fish muscles, caused by the degradation of muscle proteins, especially myosin heavy chain (MHC). Myofibrillar associated protease plays a role in the degradation of various fish muscle (Benjakul et al., 1997; Ladrat et al., 2003; Ho et al., 2000). The cathepsins degraded MHC and α-actinin, whereas tropomyosin and actin were susceptible to the action of cathepsin L. On the other hand, actin is resistant to proteolysis, and troponin T is tolerant to hydrolysis induced by cathepsin D. Desmin was degraded by cathepsins B and L (Benjakul et al., 1997; Ladrat et al., 2003). Ho et al. (2000) reported that the cathepsins B, L, and L-like present in mackerel muscle have the highest activity at pH 5.5–6.5 when incubated at 55°C. Compared to cathepsin, calpains are less responsible for cleavage of myofibrillar proteins in fish muscle. However, they are known to increase the susceptibility of other proteases to degrade the myofibrillar proteins (Singh and Benjakul 2018).

11.6.1 ROLE OF PROTEASES IN FISH AND SHELLFISH

11.6.1.1 SOFTENING/MUSHINESS AND GAPING

After the death of fish, proteases majorly cause the softening and gaping of fish muscle. Generally, fish muscles are supported by the collagenous tissue such as perimysium and endomysium connective tissues, which undergo degradation during the storage (Singh and Singh, 2020; Yoshida et al., 2009). Moreover, proteins localized in the Z-line and H-zones were also cleaved primarily by the action of proteases (Singh and Benjakul, 2018; Sriket, 2014). Shigemura et al. (2004) reported the collapse of collagen fibrils of yellowtail within 24 of storage at 4°C. Histological study of Pacific bluefin tuna muscle during storage revealed disintegration of the pericellular connective tissue (Nakamura et al., 2005). For shellfish, Sriket et al. (2011a) stated that the leakage of protease from hepatopancreas to the first segment or the second segment of the abdomen was more likely responsible for the softening of freshwater prawn meat during extended iced storage. The presence of aminopeptidase, carboxypeptidases A and B, cathepsin C, chymotrypsin, and collagenases having peptidase and protease activity in the hepatopancreas was documented (Garcia-Carreno et al., 1994; Sriket et al., 2011a).

Generally, fish muscles are organized as tissue blocks of myofibers running in parallel and attached to sheaths of connective tissue so-called myocommata. The endogenous proteases hydrolyzed the collagenous fibrils, which connect the myocommata of the fish muscles (Ofstad et al., 2006; Bao et al., 2020). This leads to the separation of myocommata and causes gaping in the fish muscles during the prolonged chilled storage (Singh and Benjakul, 2018; Bao et al., 2020). Serine proteases, such as trypsin, and chymotrypsin can cleave connective tissues, thus inducing the gaping. Additionally, matrix serine proteases (MSPs) and matrix metalloproteases (MMPs) are also known to involve in gaping. MSPs were detected in wolffish and cod muscle, while salmon muscle contained only MMP (Lodemel and Olsen, 2003). Nalinanon et al. (2008) reported that fish skin from bigeye snapper contain proteases capable of degrading the β-, α_1-, and α_2-chains of collagen. Bao et al. (2020) observed the weakening and breakage of myofibrils, as well as the increased inter-myofibrillar and inter-myofiber spaces of myofibrils of black carp stored at different temperatures (4 and –0.5), which was mainly related to gelatinolytic and caseinolytic proteases. Ofstad et al. (2006) concluded that the increased gaping during storage was due to degradation of proteoglycans and glycoproteins, which are important for the spatial organization of the collagen fibers and in anchoring cells in the extracellular matrix to the collagen network.

11.6.1.2 AUTOLYSIS DURING THERMAL PROCESSING OF FISH OR SHELLFISH AND THEIR PRODUCTS

The mushiness or liquefaction of fish or shellfish during cooking is governed by the parasite Myxosporea (protozoa: myxozoa) associated proteases, which were active at high temperature (50–60°C) (Tsuyuki, Williscroft, Kabata, and Whitpiker, 1982). Additionally, endogenous proteases mainly cause gel weakening or modori phenomenon of surimi gels (Singh and Benjakul, 2017a; Liu et al., 2020). Proteases in muscle tissue display a high activity at temperatures around 50 to 60°C and trigger rapid and severe degradation of myofibrillar proteins, particularly MHC, whereas actin is relatively stable against proteolysis (Singh and Benjakul, 2018; Sriket, 2014). There are two types of major proteases responsible for the proteolysis of myofibrillar proteins: cathepsin and heat-stable alkaline proteases (Singh and Singh, 2020). Gel degradation inducing factor (GIF) and modori inducing proteases (MIPs) were reported to induce modori in threadfin bream muscles (Toyohara et al., 1990). MIP were classified into four classes: sarcoplasmic-50°C-MIP (Sp-50-MIP); sarcoplasmic-60°C-MIP (Sp-60-MIP); myofibrillar-50°C-MIP (Mf-50-MIP), and myofibrillar-60°C-MIP (Mf-60-MIP) (Toyohara et al., 1990). Yongsawatdigul and Piyadhammaviboon (2004) found that autolysis of minced increased with increasing incubation time and the maximum autolysis of mince and surimi from lizardfish was obtained at pH 6 and 7, respectively, when heated at 65°C.

11.6.1.3 ACTIVATION OF POLYPHENOL OXIDASE (PPO) IN SHRIMP

Proteases from hepatopancreas were found to activate PPO, which in turn increased the melanosis. Trypsin, which is the dominant protease in hepatopancreas, was able to activate the latent form of PPO (pro-PPO) (Manheem et al., 2012). Generally, invertebrate PPO is synthesized as a pro PPO, which is activated by a protease cascade triggered by microbial compounds (carbohydrates and lipopolysaccharides) and a series of other proteins (Wang et al., 2006). This is the primary immune response for the development of cuticle sclerotization and healing of injuries in crustaceans. Moreover, hydrolyzed protein via proteolytic enzymes such as free tyrosine might act as a substrate for active PPO (Ali et al., 1994). Manheem et al. (2012) observed increased PPO activity when proteases from hepatopancreas of Pacific white shrimp were incorporated. In the presence of proteases from hepatopancreas, the relative activity of 116% was obtained in comparison with the control (without proteases from hepatopancreas).

11.6.2 PREVENTION OF PROTEOLYSIS IN FISH AND SHELLFISH MUSCLE AND THEIR PRODUCTS

The control of protease-mediated muscle degradation is necessary to maintain fish meat quality for prolonged durations (Singh and Singh, 2020). Therefore, various processes such as employment of protease inhibitors, washing of fish mince, etc., have been exploited widely. The washed mince of lizardfish was less prone to the autolysis, which was more likely due to the removal of sarcoplasmic proteases during the washing process (Benjakul et al., 2003b). Moreover, Yarnpakdee et al. (2009) studied the effect of washing and skin removal on autolysis of goatfish and observed that the autolysis of mince was more pronounced than that of washed mince. In the presence of skin, a higher autolysis in the mince was obtained. For protease inhibitors, whey protein, egg white, chicken, and porcine blood plasma, legume seeds, etc., have been used to inhibit proteolysis of fish muscle and their products (Singh and Benjakul, 2018; Singh and Singh, 2020). Softening or mushiness of fish and shellfish is reduced by soaking or injection of protease inhibitors into muscles or target organs containing proteases in fish or shellfish. Wang et al. (2011) treated Atlantic cod filets with various protease inhibitors, phenylmethanesulfonyl fluoride, ethylenediaminetetraacetic acid, phenanthroline, pepstatin A, antipain, and E-64, which showed the inhibition against cathepsins B, D, and L. However, due to consumer safety and ignorance towards synthetic additives, food-grade protease inhibitors are commonly used. Crude recombinant soy cystatin and allicin from soy and ginger were used to inhibit proteolysis of Arrowtooth flounder and common carp, respectively (Ge et al., 2016; Kang and Lanier, 2005). For surimi gel weakening or softening, whey protein concentrate and egg white powder are the most widely used protease inhibitors. However, porcine or chicken or salmon blood plasma, fish roes, squid ovary, etc., have also been reported to improve gelling properties of surimi from Pacific whiting, Alaska pollock, Indian mackerel, sardine, bigeye snapper and lizardfish surimi, mainly via inhibition of endogenous proteases (Kato et al., 2010; Benjakul et al., 2004b; Rawdkuen et al., 2007; Fowler and Park, 2015; Singh and Benjakul, 2017a, b, 2018). Additionally, legume seed protease inhibitors from soybean, Bambara groundnuts, cowpeas, adzuki bean, etc., have been known to enhance gel properties via inhibiting proteases associated with myofibrillar proteins (Singh and Benjakul, 2018). Injection of legume seed extract directly to the hepatopancreas of whole freshwater prawn could lower softening of abdomen during iced storage (Sriket et al., 2011b). This indicated the major role of released digestive protease in hydrolysis of muscle protein, particularly abdomen close to cephalothorax, where hepatopancreas is located.

11.6.3 ROLE OF PROTEASES IN DEVELOPMENT OF SEAFOOD PRODUCTS

Although proteases resulted in the loss of the quality of fish/shellfish. However, during seafood processing such as salting, heating, fermentation, etc., intestinal, and muscle proteases could be used to provide desired taste and texture. The endogenous proteases in pickled products that can degrade proteins into polypeptides and free amino acids are cathepsins, calpains, dipeptidases, and aminopeptidases (Toldra et al., 2000). Wu and Cao (2018) reported that the formation of flavoring substances during the processing of dry-salted fish (*Decapterus maruadsi*) was due to the action of endogenous proteases. Similarly, fish sauce or paste have been prepared via prolonged hydrolysis of heavily salted small pelagic fish mediated by intestinal proteases (Nielsen and Nielsen, 2012), in which fish sauce was prepared as liquefied protein fraction, and fish paste is the "solid" protein fraction. Fish silage is a liquefied product generated from whole fish or parts of fish using endogenous proteases, in the presence of acid. The cod viscera having two active forms of pepsins (I and II) could be used for the production of fish silage by adding protease-rich cod viscera (Gildberg and Almas, 1986).

11.7 TRANSGLUTAMINASE (TGASE)

TGase (EC 2.3.2.13), also called protein-glutamine γ-glutamyltransferase or glue of proteins, belongs to the class of transferase. It catalyzes the formation of an isopeptide bond between the carboxamide (-(C=O)NH_2) group of glutamine residues (donor) and ε-amine groups (-NH_2) of lysine side chain, resulting in the release of ammonia (NH_3) (Kieliszek and Misiewicz, 2014). It transfers an acyl onto a lysine residue of polypeptide chain and induces cross-linking, i.e., the formation of inter- or intramolecular ε-(γ-Glu)Lys cross-links (Zhu et al., 2011). TGase has been widely extracted, purified, and characterized from various parts of animal tissues and body fluids, plants, fish muscle, and microorganisms (Razavian et al., 2020). Recently, Razavian et al. (2020) purified the TGase from the bovine liver extract by gel filtration on Sephacryl S-200 HR column with a yield of 36.7%. TGase with MW of 74 kDa was stable at temperatures below 55°C and in a narrow pH range of 6.5–8.0. Due to the lower cost production and availability of microbes, microbial transglutaminase (MTGase) has been produced and used widely. MTGase has been approved by FDA and considered as GRAS since 1998 (Mostafa, 2020). MTGase has been commercially produced and incorporated in various foods, leather industry, tissue engineering, etc. (Mostafa, 2020; Zhu and Tramper, 2008).

11.7.1 ROLE OF TGASE/MTGASE IN IMPROVEMENT OF SEAFOOD PRODUCT QUALITY

Endogenous TGase and MTGase played a vital role in the quality development of fish and shellfish and their products. The formation of ε-(g-Glu) Lys linkages increases the gel strength of the surimi gel obtained from various fish (Singh et al., 2020b; Benjakul et al., 2008). Generally, "setting" or "suwari" phenomenon is induced by the endogenous TGase during the incubation of surimi paste at 40°C (Benjakul et al., 2008; Singh and Benjakul, 2017b). During setting, myosin heavy chain (MHC) undergoes polymerization via formation of non-disulfide covalent cross-links induced by endogenous TGase. This phenomenon is varied among the different species of fish, which is more likely associated with varying amount and activity of TGase as well as the sufficient availability of MHC and confirmation of actomyosin (Benjakul et al., 2008; Wan et al., 1995; Araki and Seki, 1993). Fish TGase is a Ca^{2+}-dependent enzyme, however, the requirement varies among fish species. The addition of calcium compounds to surimi enhanced endogenous TGase-mediated setting, resulting in stronger gels (Benjakul et al., 2008). Moreover, setting temperature and times are other factors determined the activity of TGase and gel properties of surimi. Gel strengthening of surimi from various dark and white-fleshed fish including mackerel, lizardfish, bigeye snapper, etc., via incorporation of MTGase into surimi paste have been documented (Cardoso et al., 2009; Chanarat and Benjakul, 2013). Benjakul et al. (2008) studied the effect of different setting conditions, in which setting either 25 or 40°C, prior to heating at 90°C resulted in the increase in both breaking force and deformation of surimi from both species of bigeye snapper (*P. tayenus* and *P. macracanthus*), particularly when setting time was augmented. Chanarat et al. (2012) studied the effect of MTGase at various concentrations on protein cross-linking and gel property of surimi from various fish such as threadfin bream, Indian mackerel and sardine in the presence and absence of endogenous TGase. The EDTA was used to inhibit the activity of endogenous TGase. It was noted that the gel added with MTGase had higher breaking force, regardless of EDTA addition. In the presence of EDTA, the percentage of increase in breaking force was lower, suggesting the combined effect of endogenous TGase with MTGase on surimi gel properties. With the addition of MTGase, the gel with the highest increase in breaking force showed the highest decrease in MHC band intensity. Moreover, the highest decreasing rate in ε-amino group content of natural actomyosin was observed, which indicated higher cross-linking associated with the higher activity of MTGase. Additionally, properties of gels from squid mantle surimi, e.g., from giant squid were improved by the addition of MTGase (Singh et al., 2020a; Moreno et al., 2009). The higher cross-linking activity of MTGase resulted into the decreased digestibility and hence reduced the bioavailability of proteins or peptides for the absorption into the body. Recently, Singh et al. (2020b) used MTGase in combination with ethanolic coconut husk extract (ECHE) that not only enhanced the gel properties of spotted golden goatfish (*Parupeneus heptacanthus*) surimi, but also increased the antioxidant potential of *in vitro* digest, which was more likely attributed to phenolic compounds in ECHE.

11.8 SUMMARY

Seafoods can serve as an excellent source of nutrients, meeting the increasing public awareness of the significance of diet to health. However, loss in quality of seafoods is of major concern since deterioration or spoilage takes place rapidly. Endogenous enzymes play a major role in quality loss, while some enzymes can help improve the quality or act as the processing aid. Therefore, the condition related to their activities, such as pH, temperature, inhibitor, activator, etc., need to be investigated in which the negative impact on the fish/shellfish and their products can be well controlled, whereas the activity of desired enzymes should be maximized. Moreover, understanding of mechanism of action of endogenous enzymes could bring about the better exploitation or prevention of particular enzymes in seafoods.

KEYWORDS

- **lipolytic enzymes**
- **lipoxygenase**
- **polyphenol oxidase**
- **proteases**
- **quality**
- **seafood**
- **transglutaminase**
- **trimethylamine-n-oxide demethylase**

REFERENCES

Abbasvali, M., Ranaei, A., Shekarforoush, S. S., & Moshtaghi, H., (2016). The effects of aqueous and alcoholic saffron (*Crocus sativus*) tepal extracts on quality and shelf-life of Pacific white shrimp (*Litopeneous vannamei*) during iced storage. *J. Food Qual., 39*, 732–742.

Ali, M. T., Gleeson, R. A., Wei, C. I., & Marshall, M. R., (1994). Activation mechanisms of pro-phenoloxidase on melanosis development in Florida spiny lobster (*Panulirus argus*) cuticle. *J. Food Sci., 59*, 1024–1030.

Ang, J. F., & Hultin, H. O., (1989). Denaturation of cod myosin during freezing after modification with formaldehyde. *J. Food Sci., 54,* 814–818.

Araki, H., & Seki, N., (1993). Comparison of reactivity of transglutaminase to various fish actomyosins. *Nippon Suisan Gakkai Shi., 59*, 711–716.

Aubourg, S. P., Losada, V., Prado, M., Miranda, J. M., & Barros-Velázquez, J., (2007). Improvement of the commercial quality of chilled Norway lobster (*Nephrops norvegicus*) stored in slurry ice: Effects of a preliminary treatment with an antimelanosis agent on enzymatic browning. *Food Chem., 103*, 741–748.

Audley, M., Shetty, K., & Kinsella, J., (2006). Isolation and properties of phospholipase A from Pollock muscle. *J. Food Sci., 43*, 1771–1775.

Banerjee, S., (2006). Inhibition of mackerel (*Scomber scombrus*) muscle lipoxygenase by green tea polyphenols. *Food Res. Int., 39*, 486–491.

Bao, Y., Wang, K., Yang, H., Regenstein, J. M., Ertbjerg, & Zhou, P., (2020). Protein degradation of black carp (*Mylopharyngodon piceus*) muscle during cold storage. *Food Chem., 308*, 125576.

Basiri, S., Shekarforoush, S. S., Aminlari, & Akbari, S., (2015). The effect of pomegranate peel extract (PPE) on the polyphenol oxidase (PPO) and quality of Pacific white shrimp (*Litopenaeus vannamei*) during refrigerated storage. *LWT–Food Sci. Technol., 60*, 1025–1033.

Basiri, S., Shekarforoush, S. S., Aminlari, M., Abhari, K., & Berizi, E., (2014). Influence of combined vacuum packaging and pomegranate peel extract on shelf life and overall quality of Pacific white shrimp (*Penaeus vannamei*) during refrigerated storage. *Iran. J. Vet. Res., 15*, 23–29.

Benjakul, S., Phatcharat, S., Tammatinna, A., Visessanguan, W., & Kishimura, H., (2008). Improvement of gelling properties of lizardfish mince as influenced by microbial transglutaminase and fish freshness. *J. Food Sci., 73*, S239–S246.

Benjakul, S., Seymour, T. A., Morrissey, M. T., & An, H., (1997). Physicochemical changes in Pacific whiting muscle proteins during iced storage. *J. Food Sci., 62*, 729–733.

Benjakul, S., Visessanguan, W., & Tanaka, M., (2003a). Partial purification and characterization of trimethylamine-N-oxide demethylase from lizardfish kidney. *Comp. Biochem. Physiol. B: Biochem. Mol. Biol., 135*, 359–371.

Benjakul, S., Visessanguan, W., & Tanaka, M., (2004a). Induced formation of dimethylamine and formaldehyde by lizardfish (*Saurida micropectoralis*) kidney trimethylamine-N-oxide demethylase. *Food Chem., 84*, 297–305.

Benjakul, S., Visessanguan, W., & Tanaka, M., (2005). Properties of phenoloxidase isolated from the cephalothorax of kuruma prawn (*Penaeus japonicus*). *J. Food Biochem., 29*, 470–485.

Benjakul, S., Visessanguan, W., & Tueksuban, J., (2003b). Heat-activated proteolysis in lizardfish (*Saurida tumbil*) muscle. *Food Res. Int., 36*, 1021–1028.

Benjakul, S., Visessanguan, W., Tueksuban, J., & Tanaka, M., (2004b). Effect of some protein additives on proteolysis and gel-forming ability of lizardfish (*Saurida tumbil*). *Food Hydrocoll., 18*, 395–401.

Cardoso, C., Mendes, R., Vaz-Pires, P., & Nunes, M. L., (2009). Effect of dietary fibre and MTGase on the quality of mackerel surimi gels. *J. Sci. Food Agric., 89*, 1648–1658.

Chanarat, S., & Benjakul, S., (2013). Effect of formaldehyde on protein cross-linking and gel-forming ability of surimi from lizardfish induced by microbial transglutaminase. *Food Hydrocoll., 30*, 704–711.

Chanarat, S., Benjakul, S., & H-Kittikun, A., (2012). Comparative study on protein cross-linking and gel enhancing effect of microbial transglutaminase on surimi from different fish. *J. Sci. Food Agric., 92*, 844–852.

Chemat, F., Zill-e, H., & Khan, M. K., (2011). Applications of ultrasound in food technology: Processing, preservation and extraction. *Ultrason Sonochem., 18*, 813–835.

Chen, J. S., Wei, C. I., Rolle, R. S., Otwell, W. S., Balaban, M. O., & Marshall, M. R., (1991). Inhibitory effect of kojic acid on some plant and crustacean polyphenol oxidases. *J. Agric. Food Chem., 39*, 1396–1401.

Cherif, S., & Gargouri, Y., (2009). Hermoactivity and effects of organic solvents on digestive lipase from hepatopancreas of the green crab. *Food Chem., 116*, 82–86.

Cirkovic, V. T. D., & Stanic-Vucinic, D. J., (2018). The role of dietary phenolic compounds in protein digestion and processing technologies to improve their antinutritive properties. *Comp. Rev. Food Sci. Food Saf., 17*, 82–103.

Cuppett, S. L., (1994). Edible coatings as carriers of food additives, fungicides, and natural antagonists. In: Krochta, J. M., Baldwin, E. A., & Nisperos-Carriedo, M. O., (eds.), *Edible Coatings and Films to Improve Food Quality* (pp. 121–137). Lancaster/Basel: Technomic Publishing.

Dao, P. T. A., An, N. T. H., Thuy, N. T. T., Tuyet, N. T. A., & Truc, T. T. M., (2016). Screening on antioxidant activities of by-products from vegetables and fruits in Tay Nguyen region and applying for shrimp cold storage. In: *2016 3rd International Conference on Green Technology and Sustainable Development* (pp. 93–97). Kaohsiung, Taiwan.

De Koning, A. J., & Mol, T. H., (1990). Rates of free fatty acid formation from phospholipids and neutral lipids in frozen cape hake (*Merluccius* spp) mince at various temperatures. *J. Sci. Food Agric., 50*, 391–398.

Erickson, M., (2000). Chemical treatments to control enzymes in fishery products. In: Haard, N. F., & Simpson, B. K., (eds.), *Seafood Enzymes: Utilization and Influence on Postharvest Seafood Quality* (pp. 489–504). New York, USA: Marcel Dekker, Inc.

Fowler, M. R., & Park, J. W., (2015). Salmon blood plasma: Effective inhibitor of protease-laden Pacific whiting surimi and salmon mince. *Food Chem., 176*, 448–454.

Galvão, J. A., Vázquez-Sánchez, D., Yokoyama, V. A., Savay-Da-Silva, L. K., Brazaca, S. G. C., & Oetterer, M., (2017). Effect of 4-hexylresorcinol and sodium metabisulphite on spoilage and melanosis inhibition in *Xiphopenaeus kroyeri* shrimps. *J. Food Process. Preserv., 41*, e12943.

Garcia-Carreno, F. L., Hernandez-Cortes, M. P., & Haard, N. F., (1994). Enzymes with peptidase and proteinase activity from the digestive systems of a freshwater and a marine decapod. *J. Agric. Food Chem., 42*, 1456–1461.

Gargouri, Y., Ransac, S., & Verger, R., (1997). Covalent inhibition of digestive lipases: An *in vitro* study. *Biochim. Biophys. Acta, Lipids Lipid Metab., 1344*, 6–37.

Ge, L., Xu, Y., Jiang, X., Xia, W., & Jiang, Q., (2016). Broad-spectrum inhibition of proteolytic enzymes by allicin and application in mitigating textural deterioration of ice-stored grass carp (*Ctenopharyngodon idella*) fillets. *Int. J. Food Sci. Technol., 51*, 902–910.

German, J. B., Chen, S. E., & Kinsella, J. E., (1985). Lipid oxidation in fish tissue. Enzymic initiation via lipoxygenase. *J. Agric. Food Chem., 33*, 680–683.

Geromel, E. J., & Montgomery, M. W., (1980). Lipase release from lysosomes of rainbow trout (*Salmo gairdneri*) muscle subjected to low temperatures. *J. Food Sci., 45*, 412–415.

Gildberg, A., & Almas, K., (1986). Utilization of fish viscera. In: Maguer, M., & Jelen, P., (eds.), *Food Engineering and Process Applications 2, Unit Operations* (pp. 425–435). London: Elsevier Applied Science.

Gokoglu, N., & Yerlikaya, P., (2008). Inhibition effects of grape seed extracts on melanosis formation in shrimp (*Parapenaeus longirostris*). *Int. J. Food Sci. Technol., 43*, 1004–1008.

Gómez-Estaca, J., Gómez-Guillen, M. C., Marín-Peñalver, D., & Montero, M. P., (2020). Functional aptitude of hake minces with added TMAO-demethylase inhibitors during frozen storage. *Food Chem., 309*, 125683.

Gonçalves, A. A., & De Oliveira, A. R. M., (2016). Melanosis in crustaceans: A review. *LWT–Food Sci. Technol., 65*, 791–799.

Gonçalves, A. A., Oliveira, A. R. M. D., & Abrantes, M. R., (2015). Acerola fruit as a possible antimelanotic agent in white shrimp (*Litopenaeus vannamei*). *J. Food Process, 2015*, 1–9.

Gulzar, S., & Benjakul, S., (2019). Effect of pre-treatments on yield and properties of lipid extracted from cephalothorax of Pacific white shrimp (*Litopenaeus vannamei*) by ultrasonic-assisted process. *LWT–Food Sci. Technol., 100*, 106–113.

Haard, N. F., (1992a). Biochemical reactions in fish muscle during frozen storage. In: Bligh, G., (ed.), *Seafood Science and Technology* (pp. 176–209). Oxford: Fishing News Books.

Haard, N. F., (1992b). Biochemistry and chemistry of color and color change in seafoods. In: Flick, G. J., & Martin, R. E., (eds.), *Advances in Seafood Biochemistry, Composition and Quality* (pp. 305–360). Lancaster, UK: Technomic Publishing Co.

Harada, K., (1975). Studies on enzymes forming formaldehyde and dimethylamine in fish and shellfish. *J. Shimoneseki Univ. Fish., 23*, 161–241.

Hiltz, D. F., Lall, B. S., Lemon, D. W., & Dyer, W. J., (1976). Deteriorative changes during frozen storage in fillets and minced flesh of silver hake (*Merluccius bilinearis*) processed from round fish held in ice and refrigerated seawater. *J. Fish. Res. Board. Can., 33*, 2560–2567.

Ho, M. L., Chen, G. H., & Jiang, S. T., (2000). Effects of mackerel cathepsins L and L like, and calpain on the degradation of mackerel surimi. *Fish. Sci., 66*, 558–568.

Hong, J., & Yang, C. S., (2003). Effects of tea polyphenols on arachidonic acid metabolism in human colon. In: Shahidi, F., Ho, C. T., Watanabe, S., & Osawa, T., (eds.), *Food Factors in Health Promotion and Disease Prevention* (pp. 27–38). Washington, DC: American Chemical Society.

Hsieh, R. J., & Kinsella, J. E., (1989). Lipoxygenase generation of specific volatile flavor carbonyl compounds in fish tissues. *J. Agric. Food Chem., 37*, 279–286.

Hsieh, R. J., German, J. B., & Kinsella, J. E., (1988a). Lipoxygenase in fish tissue: Some properties of the 12-lipoxygenase from trout gill. *J. Agric. Food Chem., 36*, 680–685.

Hsieh, R. J., German, J. B., & Kinsella, J. E., (1988b). Relative inhibitory potencies of flavonoids on 12-lipoxygenase of fish gill. *Lipids, 23*, 322–326.

Iijima, N., Tanaka, S., & Ota, Y., (1998). Purification and characterization of bile salt-activated lipase from the hepatopancreas of red sea bream, *Pagrus major*. *Fish Physiol. Biochem., 18*, 59–69.
Iyengar, R., Bohmont, C. W., & McEvily, A. J., (1991). 4-hexylresorcinol and prevention of shrimp blackspot: Residual analyses. *J Food Compost Anal., 4*, 148–157.
Jeong, B. Y., Ohshima, T., Koizumi, C., & Kanou, Y., (1990). Lipid deterioration and its inhibition of Japanese oyster *Crassostrea gigas* during frozen storage. *Nippon Suisan Gakkaishi, 56*, 2083–2091.
Jonas, R., (2011). Effect of magnesium sulphate and cryoprotective agents on free fatty acid formation in cold-stored fish muscle. *J. Fish. Res. Board. Can., 26*, 2237–2240.
Josephson, D. B., & Lindsay, R. C., (1986). Enzymatic generation of volatile aroma compounds from fresh fish. In: Parliment, T. H., & Croteau, R., (eds.), *Biogeneration of Aromas* (pp. 201–221). Washington, D.C.: American Chemical Society.
Josephson, D. B., & Lindsay, R. C., (1987). Retro-aldol degradations of unsaturated aldehydes: Role in the formation of c4-heptenal from t2, c6-nonadienal in fish, oyster and other flavors. *J. Am. Oil Chem. Soc., 64*, 132–138.
Kang, I., & Lanier, T., (2005). Inhibition of protease in intact fish fillets by soaking in or injection of recombinant soy cystatin or bovine plasma. *J. Agric. Food Chem., 53*, 9795–9799.
Kanner, J., & Rosenthal, I., (1992). An assessment of lipid oxidation in foods. In: *Pure and Applied Chemistry* (pp. 1959–1964).
Karahadian, C., & Lindsay, R. C., (1989). Evaluation of compounds contributing characterizing fishy flavors in fish oils. *J. Am. Oil Chem. Soc., 66*, 953–960.
Kato, N., Suzuki, Y., Kunimoto, M., Kitakami, S., Murakami, Y., & Arai, K., (2010). Comparison of effects of plasma and albumen powders on the physical property of heated gels formed from salt-ground of three species fish frozen surimi. *J. Jap. Soc. Food Sci. Technol., 57*, 26–31.
Ke, P., & Ackman, R., (2011). Inhibition of formation of free fatty acids in cod liver with sodium hypochlorite solution (Javex). *J. Fish. Res. Board. Can., 32*, 297–300.
Kieliszek, M., & Misiewicz, A., (2014). Microbial transglutaminase and its application in the food industry: A review. *Folia Microbiol., 59*, 241–250.
Kim, J., Marshall, M. R., & Wei, C., (2000). Polyphenoloxidase. In: Haard, N. F., & Simpson, B. K., (eds.), *Seafood enzyme: Utilization and Influence on Post Harvest Seafood Quality* (pp. 271–315). New York: Marcel Dekker.
Kimura, M., Seki, N., & Kimura, I., (2000). Purification and characterization of trimethylamine-N-oxide demethylase from walleye pollack muscle. *Fish. Sci., 66*, 967–973.
Knight, J., Holland, J. W., Bowden, L. A., Halliday, K., & Rowley, A. F., (1995). Eicosanoid generating capacities of different tissues from the rainbow trout, *Oncorhynchus mykiss*. *Lipids, 30*, 451–458.
Kubo, I., & Kinst-Hori, I., (1998). Tyrosinase Inhibitors from cumin. *J. Agric. Food Chem., 46*, 5338–5341.
Kuizumi, C., Chang, C. M., Ohshima, T., & Wada, S., (1988). Influences of freeze-thawing process on the quality of sardine and horse mackerel during refrigerated storage. *Nippon Suisan Gakkai Shi., 54*, 2203–2210.
Kuo, J. M., Pan, B. S., Zhang, H., & German, J. B., (1994). Identification of 12-lipoxygenase in the hemolymph of tiger shrimp (*Penaeus japonicus* bate). *J. Agric. Food Chem., 42*, 1620–1623.

Kurtovic, I., Marshall, S. N., Zhao, X., & Simpson, B. K., (2009). Lipases from Mammals and Fishes. *Rev. Fish. Sci, 17*, 18–40.

Ladrat, C., Verrez-Bagnis, V., Noël, J., & Fleurence, J., (2003). *In vitro* proteolysis of myofibrillar and sarcoplasmic proteins of white muscle of sea bass (*Dicentrarchus labrax* L.): Effects of cathepsins B, D and L. *Food Chem., 81*, 517–525.

Lall, B. S., Manzer, A. R., & Hiltz, D. F., (1975). Preheat treatment for improvement of frozen storage stability at −10 C in fillets and minced flesh of silver hake (*Merluccius bilinearis*). *J. Fish. Res. Board. Can., 32*, 1450–1454.

Leelapongwattana, K., Benjakul, S., Visessanguan, W., & Howell, N. K., (2005). Physicochemical and biochemical changes during frozen storage of minced flesh of lizardfish (*Saurida micropectoralis*). *Food Chem., 90*, 141–150.

Leelapongwattana, K., Benjakul, S., Visessanguan, W., & Howell, N. K., (2008). Effect of some additives on the inhibition of lizardfish trimethylamine-N-oxide demethylase and frozen storage stability of minced flesh. *Int. J. Food Sci. Technol., 43*, 448–455.

Leelapongwattana, K., Benjakul, S., Visessanguan, W., & Howell, N. K., (2010a). Effects of trimethylamine-n-oxide demethylase (TMAOase) inhibitors and antioxidants on physicochemical and biochemical changes of haddock muscle induced by lizardfish TMAOase during frozen storage. *J. Food Biochem., 34*, 1032–1048.

Leelapongwattana, K., Benjakul, S., Visessanguan, W., & Howell, N. K., (2010b). Effects of trimethylamine-N-oxide demethylase (TMAOase) inhibitors and antioxidants on physicochemical and biochemical changes of haddock muscle induced by lizardfish TMAOase during frozen storage. *J. Food Biochem., 34*, 1032–1048.

Liu, J. Y., Yoshida, A., Gao, Y. L., Shirota, K., Shiina, Y., & Osatomi, K., (2020). Identification of a modori-inducing proteinase in the threadfin bream: Molecular cloning, tissue distribution and proteinase leakage from viscera during ice storage. *Food Chem., 330*, 127246.

Liu, Y. J., & Pan, B. S., (2004). Inhibition of fish gill lipoxygenase and blood-thinning effects of green tea extract. *J. Agric. Food Chem., 52*, 4860–4864.

Lødemel, J. B., & Olsen, R. L., (2003). Gelatinolytic activities in muscle of Atlantic cod (*Gadus morhua*), spotted wolffish (*Anarhichas minor*) and Atlantic salmon (*Salmo salar*). *J. Sci. Food Agric., 83*, 1031–1036.

Losada, V., Barros-Velázquez, J., & Aubourg, S. P., (2007). Rancidity development in frozen pelagic fish: Influence of slurry ice as preliminary chilling treatment. *LWT–Food Sci. Technol., 40*, 991–999.

Losada, V., Piñeiro, C., Barros-Velázquez, J., & Aubourg, S. P., (2005). Inhibition of chemical changes related to freshness loss during storage of horse mackerel (*Trachurus trachurus*) in slurry ice. *Food Chem., 93*, 619–625.

Lougovois, V., & Kyrana, V., (2005). Freshness quality and spoilage of chill-stored fish. *Food Policy. Control. Res., 1*, 35–86.

Lu, K., Ye, W., Zhou, L., Collins, L. B., Chen, X., Gold, A., Ball, L. M., & Swenberg, J. A., (2010). Structural characterization of formaldehyde-induced cross-links between amino acids and deoxynucleosides and their oligomers. *J. Am. Chem. Soc., 132*, 3388–3399.

Manheem, K., Benjakul, S., Kijroongrojana, K., & Visessanguan, W., (2012). The effect of heating conditions on polyphenol oxidase, proteases and melanosis in pre-cooked Pacific white shrimp during refrigerated storage. *Food Chem., 131*, 1370–1375.

Medina, I., Saeed, S., & Howell, N., (1999). Enzymatic oxidative activity in sardine (*Sardina pilchardus*) and herring (*Clupea harengus*) during chilling and correlation with quality. *Eur. Food Res. Technol., 210*, 34–38.

Milo, C., & Grosch, W., (1993). Changes in the odorants of boiled trout (*Salmo fario*) as affected by the storage of the raw material. *J. Agric. Food Chem., 41*, 2076–2081.
Milo, C., & Grosch, W., (1996). Changes in the odorants of boiled salmon and cod as affected by the storage of the raw material. *J. Agric. Food Chem., 44*, 2366–2371.
Mohri, S., Cho, S. Y., Endo, Y., & Fujimoto, K., (1990). Lipoxygenase activity in sardine skin. *Agric. Biol. Chem., 54*, 1889–1891.
Mohri, S., Tokuori, K., Endo, Y., & Fujimoto, K., (1999). Prooxidant activities in fish skin extracts and effects of some antioxidants and inhibitors on their activities. *Fish. Sci., 65*, 269–273.
Montero, P., Lopez-Caballero, M. E., & Perez-Mateos, M., (2001). The effect of inhibitors and high pressure treatment to prevent melanosis and microbial growth on chilled prawns (*Penaeus japonicus*). *J. Food Sci., 66*, 1201–1206.
Moreno, H. M., Cardoso, C., Solas, M. T., & Borderías, A. J., (2009). Improvement of cold and thermally induced gelation of giant squid (*Dosidicus gigas*) surimi. *J. Aquat. Food Prod. Technol., 18*, 312–330.
Mostafa, H. S., (2020). Microbial transglutaminase: An overview of recent applications in food and packaging. *Biocatal. Biotransformation., 38*, 161–177.
Nakamura, Y. N., Ando, M., Seoka, M., Kawasaki, K. I., & Tsukamasa, Y., (2005). Comparison of the proximate compositions, breaking strength and histological structure by the muscle positions of the full-cycle cultured Pacific bluefin tuna *Thunnus orientalis*. *Fish. Sci., 71*, 605–611.
Nalinanon, S., Benjakul, S., Visessanguan, W., & Kishimura, H., (2008). Improvement of gelatin extraction from bigeye snapper skin using pepsin-aided process in combination with protease inhibitor. *Food Hydrocoll., 22*, 615–622.
Nayak, J., Viswanathan, N. P., Ammu, K., & Mathew, S., (2003). Lipase activity in different tissues of four species of fish: Rohu (*Labeo rohita* Hamilton), oil sardine (*Sardinella longiceps* Linnaeus), mullet (*Liza subviridis* Valenciennes) and Indian mackerel (*Rastrelliger kanagurta* Cuvier). *J. Sci. Food Agric., 83*, 1139–1142.
Nielsen, M., & Nielsen, H., (2012). Seafood enzymes. In: Benjamin, K. S., (ed.), *Food Biochemistry and Food Processing* (pp. 247–262). John Wiley and Sons, Inc. Hoboken, NJ, United States.
Nirmal, N. P., & Benjakul, S., (2009). Effect of ferulic acid on inhibition of polyphenoloxidase and quality changes of Pacific white shrimp (*Litopenaeus vannamei*) during iced storage. *Food Chem., 116*, 323–331.
Nirmal, N. P., & Benjakul, S., (2011a). Retardation of quality changes of Pacific white shrimp by green tea extract treatment and modified atmosphere packaging during refrigerated storage. *Int J. Food Microbiol., 149*, 247–253.
Nirmal, N. P., & Benjakul, S., (2011b). Use of tea extracts for inhibition of polyphenoloxidase and retardation of quality loss of Pacific white shrimp during iced storage. *LWT–Food Sci. Technol., 44*, 924–932.
Ofstad, R., Olsen, R., Taylor, R., & Hannesson, K., (2006). Breakdown of intramuscular connective tissue in cod (*Gadus morhua* L.) and spotted wolffish (*Anarhichas minor* O.) related to gaping. *LWT-Food Sci. Technol., 39*, 1143–1154.
Ogawa, M., Perdigao, N. B., Asntiago, M. E. D., & Kozima, T. T., (1984). On physiological aspects of black spot appearance in shrimp. *Nippon Suisan Gakkaishi, 50*, 1763–1769.
Ohshima, T., Ushio, H., & Koizumi, C., (1993). High-pressure processing of fish and fish products. *Trends. Food Sci. Technol., 4*, 370–375.

Parkin, K. L., & Hultin, H. O., (1982). Some factors influencing the production of dimethylamine and formaldehyde in minced and intact red hake muscle. *J. Food Process. Preserv., 6*, 73–97.

Partman, W., (1975). The effects of freezing and thawing on food quality. In: Duckworth, R. G., (ed.), *Water Relations in Food* (p. 505). London: Academic Press.

Patkar, S., & Bjorkling, F., (1994). Lipase inhibitors. In: Woolley, P., & Petersen, S. B., (eds.), *Lipases* (pp. 207–224). Cambridge, MA: Cambridge University Press.

Ponte, D. D., Roozen, J., & Pilnik, W., (1986). Effects of additions, irradiation and heating on the activity of trimethylamine oxide-ase in frozen stored minced fillets of whiting. *Int. J. Food Sci. Technol., 21*, 33–43.

Prasad, K. N., Yang, B., Yang, S., Chen, Y., Zhao, M., Ashraf, M., & Jiang, Y., (2009). Identification of phenolic compounds and appraisal of antioxidant and antityrosinase activities from litchi (*Litchi chinensis* Sonn.) seeds. *Food Chem., 116*, 1–7.

Pushparajan, N., Varadharajan, D., & Soundarapandian, P., (2013). Microbial studies of prepared curries stored in tin free steel cans. *J. Food Process. Technol., 4*, 1000238.

Racicot, L. D., Lundstrom, R. C., Wilhelm, K. A., Ravesi, E. M., & Licciardello, J. J., (1984). Effect of oxidizing and reducing agents on trimethylamine N-oxide demethylase activity in red hake muscle. *J. Agric. Food Chem., 32*, 459–464.

Rahman, M. M., Khosravi, S., Chang, K. H., & Lee, S. M., (2016). Effects of dietary inclusion of astaxanthin on growth, muscle pigmentation and antioxidant capacity of juvenile rainbow trout (*Oncorhynchus mykiss*). *Prev. Nutr. Food Sci., 21*, 281–288.

Rawdkuen, S., Benjakul, S., Visessanguan, W., & Lanier, T. C., (2007). Effect of chicken plasma protein and some protein additives on proteolysis and gel-forming ability of sardine (*Sardinella gibbosa*) surimi. *J. Food Process. Preserv., 31*, 492–516.

Razavian, S. M. H., Kashfi, A., & Khoshraftar, Z., (2020). Purification of bovine liver transglutaminase by gel filtration. *Appl. Biol. Chem., 63*, 6.

Refsgaard, H. H. F., Brockhoff, P. M. B., & Jensen, B., (2000). Free polyunsaturated fatty acids cause taste deterioration of salmon during frozen storage. *J. Agric. Food Chem., 48*, 3280–3285.

Rehbein, H., (1988). Relevance of trimethylamine oxide demethylase activity and haemoglobin content to formaldehyde production and texture deterioration in frozen stored minced fish muscle. *J. Sci. Food Agric., 43*, 261–276.

Richmond, G. S., & Smith, T. K., (2011). Phospholipases A_1. *Int. J. Mol. Sci., 12*, 588–612.

Saeed, S., & Howell, N. K., (2001). 12-Lipoxygenase activity in the muscle tissue of Atlantic mackerel (*Scomber scombrus*) and its prevention by antioxidants. *J. Sci. Food Agric., 81*, 745–750.

Sae-Leaw, T., & Benjakul, S., (2014a). Fatty acid composition, lipid oxidation, and fishy odor development in seabass (*Lates calcarifer*) skin during iced storage. *Eur. J. Lipid Sci. Technol., 116*, 885–894.

Sae-Leaw, T., & Benjakul, S., (2014b). Fatty acid composition, lipid oxidation, and fishy odor development in seabass (*Lates calcarifer*) skin during iced storage. *Eur. J. Lipid Sci. Technol., 116*, 885–894.

Sae-Leaw, T., & Benjakul, S., (2015). Physico-chemical properties and fishy odor of gelatin from seabass (*Lates calcarifer*) skin stored in ice. *Food Biosci., 10*, 59–68.

Sae-Leaw, T., & Benjakul, S., (2018). Lipase from liver of seabass (*Lates calcarifer*): Characteristics and the use for defatting of fish skin. *Food Chem., 240*, 9–15.

Sae-Leaw, T., & Benjakul, S., (2019a). Distribution and characteristics of polyphenoloxidase from Pacific white shrimp (*Litopenaeus vannamei*). *J. Food Sci., 84*, 1078–1086.
Sae-Leaw, T., & Benjakul, S., (2019b). Prevention of melanosis in crustaceans by plant polyphenols: A review. *Trends in Food Sci. Technol., 85*, 1–9.
Sae-Leaw, T., & Benjakul, S., (2019c). Prevention of quality loss and melanosis of Pacific white shrimp by cashew leaf extracts. *Food Control, 95*, 257–266.
Sae-Leaw, T., Benjakul, S., & Simpson, B. K., (2017). Effect of catechin and its derivatives on inhibition of polyphenoloxidase and melanosis of Pacific white shrimp. *J. Food Sci. Technol.*, 54, 1098–1107.
Sae-Leaw, T., Benjakul, S., & Vongkamjan, K., (2018). Retardation of melanosis and quality loss of pre-cooked Pacific white shrimp using epigallocatechin gallate with the aid of ultrasound. *Food Control, 84*, 75–82.
Sae-Leaw, T., Benjakul, S., Gokoglu, N., & Nalinanon, S., (2013). Changes in lipids and fishy odor development in skin from Nile tilapia (*Oreochromis niloticus*) stored in ice. *Food Chem., 141*, 2466–2472.
Shahidi, F., & Janak, K. Y. V. A., (2001). Enzymes from fish and aquatic invertebrates and their application in the food industry. *Trends. Food Sci. Technol., 12*, 435–464.
Shewfelt, R. L., (1981). Fish muscle lipolysis—A review. *J. Food Biochem., 5*, 79–100.
Shiekh, K. A., Benjakul, S., & Sae-Leaw, T., (2019). Effect of chamuang (*Garcinia cowa* Roxb.) leaf extract on inhibition of melanosis and quality changes of Pacific white shrimp during refrigerated storage. *Food Chem., 270*, 554–561.
Shigemura, Y., Ando, M., Harada, K., & Tsukamasa, Y., (2004). Possible degradation of type I collagen in relation to yellowtail muscle softening during chilled storage. *Fish. Sci., 70*, 703–709.
Simpson, B. K., Nollet, L. M., Toldrá, F., Benjakul, S., Paliyath, G., & Hui, Y., (2012). *Food Biochemistry and Food Processing.* John Wiley and Sons.
Simpson, K. L., (1982). Carotenoid pigments in seafood. In: Martin, R. E., Flick, G. E., Hebard, C. E., & Ward, D. W., (eds.), *Chemistry and Biochemistry of Marine Food Products* (pp. 115–136). Westport, CT: AVI.
Singh, A., & Benjakul, S., (2017a). Effect of serine protease inhibitor from squid ovary on gel properties of surimi from Indian mackerel. *J. Texture Stud., 48,* 541–549.
Singh, A., & Benjakul, S., (2017b). Serine protease inhibitors from squid ovary: Extraction and its effect on proteolysis and gel properties of surimi. *J. Food Sci. Technol., 54*, 267–275.
Singh, A., & Benjakul, S., (2018). Proteolysis and its control using protease inhibitors in fish and fish products: A review. *Compr. Rev. Food Sci. Food Saf., 17*, 496–509.
Singh, A., Mittal, A., & Benjakul, S., (2020a). Full utilization of squid meat and its processing by-products: Revisit. *Food Rev. Int.*, 1–25. 10.1080/87559129.2020.1734611.
Singh, A., Prabowo, F. F., Benjakul, S., Pranoto, Y., & Chantakun, K., (2020b). Combined effect of microbial transglutaminase and ethanolic coconut husk extract on the gel properties and in-vitro digestibility of spotted golden goatfish (*Parupeneus heptacanthus*) surimi gel. *Food Hydrocoll.*, 106107.
Singh, J., & Singh, B., (2020). Inhibition of post-mortem fish muscle softening and degradation using legume seed proteinase inhibitors. *J. Food Sci. Technol., 57*, 1–11.
Sriket, C., (2014). Proteases in fish and shellfish: Role on muscle softening and prevention. *Int. Food Res. J., 21*, 433.

Sriket, C., Benjakul, S., Visessanguan, W., & Hara, K., (2011b). Effect of legume seed extracts on the inhibition of proteolytic activity and muscle degradation of fresh water prawn (*Macrobrachium rosenbergii*). *Food Chem., 129*, 1093–1099.

Sriket, C., Benjakul, S., Visessanguan, W., & Kishimura, H., (2011a). Collagenolytic serine protease in fresh water prawn (*Macrobrachium rosenbergii*): Characteristics and its impact on muscle during iced storage. *Food Chem., 124*, 29–35.

Stodolnik, L., & Samson, E., (2000). Lipoxygenase activity of selected tissues and organs of Baltic herring. *Acta Ichthyol. Piscat., 2*, 47–57.

Sun, H., Lv, H., Yuan, G., & Fang, X., (2014). Effect of grape seed extracts on the melanosis and quality of pacific white shrimp (*Litopenaeus vannamei*) during iced storage. *Food Sci. Technol. Res., 20*, 671–677.

Takiguchi, A., (1989). Effect of NaCl on the oxidation and hydrolysis of lipids in salted sardine fillets during storage. *Nippon Suisan Gakkai Shi., 55*, 1649–1654.

Teklemariam, A. D., Tessema, F., & Abayneh, T., (2015). Review on evaluation of safety of fish and fish products. *International J. Fish. Aquat. Stud., 3*, 111–117.

Thepnuan, R., Benjakul, S., & Visessanguan, W., (2008). Effect of pyrophosphate and 4-hexylresorcinol pretreatment on quality of refrigerated white shrimp (*Litopenaeus vannamei*) kept under modified atmosphere packaging. *J. Food Sci., 73*, S124–S133.

Toldrá, F., Aristoy, M. C., & Flores, M., (2000). Contribution of muscle aminopeptidases to flavor development in dry-cured ham. *Food Res. Int., 33*, 181–185.

Tolstorebrov, I., Eikevik, T. M., & Bantle, M., (2016). Effect of low and ultra-low temperature applications during freezing and frozen storage on quality parameters for fish. *Int. J. Refrig., 63*, 37–47.

Toyohara, H., Kinoshita, M., & Shimizu, Y., (1990). Proteolytic degradation of threadfin-bream meat gel: A research note. *J. Food Sci., 55*, 259–260.

Toyomizu, M., Hanaoka, K. I., & Yamaguchi, K., (1981). Effect of release of free fatty acids by enzymatic hydrolysis of phospholipids on lipid oxidation during storage of fish muscle at –5°C. *Nippon Suisan Gakkai Shi., 47*, 615–620.

Tsukuda, N., & Amano, K., (1966). Studies on the discoloration of red fishes. I. Content of carotenoid pigments in eighteen species of red fishes. *Nippon Suisan Gakkaishi, 32*, 334–345.

Tsukuda, N., & Amano, K., (1968). Studies on the discoloration of red fishes-V: Enzyme involved in the discoloration of carotenoid pigments in fish skin tissues. *Nippon Suisan Gakkaishi, 34*, 633–639.

Tsuyuki, H., Williscroft, S. N., Kabata, Z., & Whitpiker, D. J., (1982). *The Relationship between Acid and Neutral Protease Activities and the Incidence of Soft Cooked Texture in the Muscle Tissue of Pacific Hake (Merluccius productus) Infected with Kudoapaniformis and/or K. thyrsitis, and Held for Varying Times Under Different Pre-Freeze Chilled Storage Conditions* (pp. 39), Fish. Aquat. Sci. Can. Tech. Rep. No. 1130.

Wan, J., Kimura, I., Satake, M., & Seki, N., (1995). Causes of inferior gel-forming ability of salmon surimi paste. *Fish. Sci., 61*, 711–715.

Wang, P. A., Vang, B., Pedersen, A. M., Martinez, I., & Olsen, R. L., (2011). Post-mortem degradation of myosin heavy chain in intact fish muscle: Effects of pH and enzyme inhibitors. *Food Chem., 124*, 1090–1095.

Wang, Y. C., Chang, P. S., & Chen, H. Y., (2006). Tissue distribution of prophenoloxidase transcript in the Pacific white shrimp *Litopenaeus vannamei*. *Fish. Shellfish Immunol., 20*, 414–418.

Wang, Y. J., Miller, L. A., & Addis, P. B., (1991). Effect of heat inactivation of lipoxygenase on lipid oxidation in lake herring (*Coregonus artedii*). *J. Am. Oil Chem. Soc., 68*, 752–757.

Weete, J. D., (1998). Microbial lipases. In: Akoh, C. C., & Min, D. B., (eds.), *Food Lipids. Chemistry, Nutrition, and Biotechnology* (pp. 641–664). New York, USA: Marcel Dekker.

Wong, H., & Schotz, M. C., (2002). The lipase gene family. *J. Lipid Res., 43*, 993–999.

Wu, Y., & Cao, S. M., (2018). Study on endogenous protease and protein degradation of dry-salted *Decapterus maruadsi*. *CyTA – J. Food, 16*, 350–356.

Yarnpakdee, S., Benjakul, S., Visessanguan, W., & Kijroongrjana, K., (2009). Autolysis of goatfish (*Mulloidichthys martinicus*) mince: Characterization and effect of washing and skin inclusion. *Food Chem., 114*, 1339–1344.

Yongsawatdigul, J., & Piyadhammaviboon, P., (2004). Inhibition of autolytic activity of lizardfish surimi by proteinase inhibitors. *Food Chem., 87*, 447–455.

Yoshida, A., Bae, I., Sonoda, H., Masuo, R., Oda, S., Cao, M. J., Osatomi, K., & Hara, K., (2009). Characterization of gelatinolytic enzymes in the skeletal muscle of red sea bream pagrus major. *Fish. Sci., 75*, 1317–1322.

Yuan, G., Lv, H., Tang, W., Zhang, X., & Sun, H., (2016a). Effect of chitosan coating combined with pomegranate peel extract on the quality of Pacific white shrimp during iced storage. *Food Control, 59*, 818–823.

Yuan, G., Zhang, X., Tang, W., & Sun, H., (2016b). Effect of chitosan coating combined with green tea extract on the melanosis and quality of Pacific white shrimp during storage in ice. *CyTA – J. Food, 14*, 35–40.

Zhu, D., Wu, Q., & Wang, N., (2011). 3.02-industrial enzymes. In: Moo-Young, M., (ed.), *Comprehensive Biotechnology* (2nd edn., pp. 3–13). Burlington: Academic Press.

Zhu, Y., & Tramper, J., (2008). Novel applications for microbial transglutaminase beyond food processing. *Trends. Biotechnol., 26*, 559–565.

CHAPTER 12

Green Technologies for Developing High-Value Novel Marine Nutraceuticals: Scope and Innovations

LEKSHMI R. G. KUMAR, NILADRI S. CHATTERJEE, and SUSEELA MATHEW

Biochemistry and Nutrition Division, Central Institute of Fisheries Technology, Cochin – 29, Ernakulam, Kerala, India

ABSTRACT

Marine organisms serve as rich source of biologically active substances such as polyunsaturated fatty acids, proteins, peptides, pigments, sterols, etc., having immense therapeutic benefits. The increased awareness regarding the health benefits of marine bioactive among the general public has further fostered its extraction for diverse applications. Conventional solvent-based methods are often employed for extraction of bioactive compounds from marine sources. However, such methods are time consuming, require a longer duration and need further purification/concentration step. In addition to this, there is an increasing global concern on the use of organic solvents for extraction as they are not eco-friendly methods. Hence, innovative and sustainable extraction technologies such as supercritical fluid extraction (SFE) were introduced as an alternative to conventional separation methods. SFE helps in the efficient extraction of thermolabile compounds as the process is often carried out at lower temperature. Apart from SFE, other greener technologies that are currently employed for extraction of marine bioactives include ultrasound-assisted extraction (UAE), microwave-assisted extraction (MAE), eutectic solvents, deep eutectic solvents (DES), switchable solvents (SS), etc. The present review highlights the sustainable

Advances in Fish Processing Technologies: Preservation, Waste Utilization, and Safety Assurance.
Ranendra K. Majumder, PhD & Amjad K. Balange, PhD (Eds.)

extraction methods used in isolation of marine bioactive compounds with special emphasis on the most commonly employed techniques, SFE, UAE, and MAE.

12.1 INTRODUCTION

Bioactive compounds from natural sources have garnered significant attention as potential candidates in the field of nutraceuticals and biomedical industries. Marine biome is endowed with wide diversity of organisms that serve as potential repositories of bioactive compound possessing significant bioactivity. Polyunsaturated fatty acids, essential amino acids, unique proteins, vitamins, minerals, sulfated polysaccharides, pigments, etc., are present in different quantities in different organisms. Many of such bioactive compounds are usually acquired as an adaptation to the extreme conditions of the marine environment. Tapping such compounds for excellent biodiscoveries, especially as biodrugs is one of the recently developing research areas, especially in the wake up of pandemics such as COVID-19. Apart from the marine organisms, it has been reported that the polluting by-products from fish processing industry can also serve as an excellent source of biomolecules, if properly extracted. There has been mounting evidence that the consumption of marine foods and biomolecules can reduce the incidence of many ailments (Lordan et al., 2011). The success of such biomolecules application in various realms depends to a greater extent on how effectively such molecules are being extracted, how pure such compounds are, and also their physicochemical characterizations to endorse further applications. Conventional extraction based on solvents is the most commonly employed method for extraction of biomolecules. However, such methods have serious setbacks in terms of the enormous quantity of solvents employed, the storage and disposal of the solvents after use, the solvent residues, and most importantly, the yield and bioactivity of such extracts. In lieu of conventional-based extraction, researchers have tried to develop effective alternative methods for the isolation of biomolecules. Several novel extraction approaches such as supercritical fluid extraction, pressurized liquid extraction (PLE), microwave-assisted extraction (MAE), ultrasound-assisted extraction (UAE), extraction using deep eutectic solvents (DES), etc., have been developed for the extraction of biomolecules. Some of the extraction methods most commonly employed in the isolation of marine biomolecules are discussed in detail in this chapter.

12.2 SUPERCRITICAL FLUID EXTRACTION (SFE)

Seafood, especially, fish is considered as one of the richest nutrient sources of proteins and other micronutrients for millions of people worldwide. The increased fish consumption along with processing for producing diverse fish products has in turn created a large quantum of processing discards. Such processing discards fish has posed serious impacts on the environment as well as economy-related to its disposal. As per the 2018 FAO report, about 20 million tons of seafood processing discards are being generated globally. Further research has shown that such seafood processing discards are rich in terms of their bioactive nutrient content. Hence, by devising a proper extraction protocol conversion of such processing wastes into useful bioactive molecules can be effectively achieved. Conventional practice of extraction using acid/alkaline/solvent have serious setbacks in terms of the type and amount of solvents employed, their storage and disposal, quality, and quantity of biomolecules being extracted. In this regard, several novel and sustainable extraction practices have been developed for isolating marine biomolecules. Of late, supercritical fluid extraction has emerged as one of the most feasible, economic, and environmentally friendly methods for extraction of commercially important bioactive compounds from natural resources. Recently, a substantial amount of research has been carried out to establish the effectiveness of SFE in extracting biomolecules from marine sources/fish-processing discards.

Briefly, supercritical fluid extraction can be defined as a green eco-friendly technology wherein a supercritical fluid (SCF) is used as solvent for extraction of natural biomolecules without affecting their bioactivity. Supercritical fluid can be any substance whose temperature and pressure should be above its critical point (CP) to form a homogenous phase possessing properties of both liquid and gas, commonly referred to as the mesophase. In simpler terms, supercritical fluid can be defined as that form of matter in which the liquid and gaseous phases are indistinguishable. The properties of SCFs are listed below:

- Highly compressed gas having properties of both gas and liquids;
- Have low viscosity, high diffusivity and hence enhanced transport properties;
- Can fine tune the solvating properties, density, and solubility by changing the pressure/temperature accordingly.

Basically, the qualitative and quantitative aspects of superfluid extraction are observed to be affected by various factors such as the SCF solvent employed, pressure, and temperature conditions, the flow rate of solvents, type of solvents used, particle size, rate of diffusion, etc. The most commonly used solvent in SFE is carbon dioxide (CO_2) as its critical conditions (critical pressure, P_c – 74 bar, critical temperature, T_c – 31.1°C) are easily achievable at ambient conditions. Non-toxicity, safety, non-flammable, inexpensive, availability, inertness, high diffusion coefficient, lower density, increased mass transfer makes it as an excellent candidate for SFE. Furthermore, such lower temperatures make it favorable for extraction of many thermolabile compounds. Being non-polar in nature, supercritical CO_2 (SC-CO_2) can extract only non-polar compounds or substances of lower polarity. For this reason, certain solubility enhancers known as co-solvents/modifiers can be added to increase the polarity of SC-CO_2. Methanol, ethanol, chloroform, ethane, etc., are some of the co-solvents being used in SFE among others. Among the available co-solvents, ethanol is the most recommended as it is non-toxic, miscible in CO_2 and food-grade (Liza et al., 2010).

SFE have several advantages over the conventional methods and some are listed below:

- Lower viscosity and high diffusion coefficient attributes SCFs enhanced transport properties, facilitating efficient extraction;
- Solvent strength of SCF can be modified by changing the extraction pressure and temperature;
- SFE leaves no chemical residue making the process sustainable;
- Lower energy requirements;
- Highly efficient process in terms of increasing yield and lower extraction times;
- No degradative changes to the bioactive compounds extracted using SFE.

However, the technology has its own limitations such as high cost of the equipment in the production scale. The sample for SFE should be dried in such a way that its moisture content should be below 20%. In case of thermolabile substances, freeze-drying has to be done which will in turn increase the operating cost (Rubio-Rodríguez et al., 2008).

12.2.1 SFE FOR EXTRACTION OF MARINE BIOACTIVE COMPOUNDS

Since SCF possess liquid-like solvating properties and gas like transport properties, it makes more effective in the selective extraction of bioactive compounds such as polyunsaturated fatty acids rich fish oil, astaxanthin, carrageenan, fucoxanthin, etc. (Singh et al., 2017). Létisse et al. (2006) has reported that SFE is an effective process for extraction and fractionation of high quality fish oils. A comparative study on the extraction performance of conventional methods such as cold extraction, wet reduction, enzymatic extraction with that of SFE under moderate conditions have shown that SFE was effective in obtaining fish quality in terms of the oxidation aspects (Rubio-Rodríguez et al., 2012). Sahena et al. (2010) have studied the various methods of supercritical fluid extraction such as continuous, cosolvent, soaking, and pressure swing for obtaining oil from skin of mackerel and found out that pressure swing is the most effective technique in terms of extractability of polyunsaturated fatty acids such as EPA, DHA, CO_2 consumption. Another study has suggested the effectiveness of SFE in concentrating carotenoids from shrimp processing waste (Mezzomo et al., 2013). Sánchez-Camargo et al. (2011) have reported the utility of SFE in valorizing Brazilian red-spotted shrimp waste. It was observed that pressure and temperature had less significant effect on lipid yield whereas it had significant effects on astaxanthin yield. Sánchez-Camargo et al. (2012) have reported that the use of ethanol as a co-solvent along with supercritical CO_2 has been found to significantly improve the extraction yield of astaxanthin and major fatty acids. Similarly, Saravana et al. (2017) have studied the effects of various co-solvents on yield of seaweed phenolics and found that sunflower oil performed better as a co-solvent giving better fucoxanthin and phlorotannin content. Michalak et al. (2016) has reported that algal extracts obtained by SFE was found to have growth stimulant properties as evident by the improved chlorophyll and carotenoid content in plant shoots. Some of the recent SFE based approaches for extraction of marine biomolecules has been given in Table 12.1.

12.3 ULTRASOUND-ASSISTED EXTRACTION (UAE)

Ultrasound extraction is often considered as a rapid, non-thermal, eco-friendly, and less energy consumption technique used for isolation of bioactive compounds. The technology employs ultrasonic waves for the extraction process. In general, it has been reported that ultrasonic waves in the frequency range of 20–1,000 KHz are often employed for the extraction purpose (Ojha et al., 2020). The main principle behind ultrasonic extraction is the acoustic cavitation. The ultrasonic waves propagate through the samples via compression and rarefaction. The process of compression and rarefaction induces the formation of bubbles followed by its growth and collapse resulting in the formation of cavitation. The shear forces generated during the process of bubble collapse further facilitates in the disruption of matrices and release of bioactive compounds.

TABLE 12.1 SFE for Extraction of Marine Bioactive Compounds

Marine Biomolecules	Source for Extraction	References
Bioactive Lipids/*Omega* 3 Fatty Acids		
EPA and DHA	Northern shrimp (*Pandelus borealis Kreyer*) by-products	Amiguet et al. (2012)
PUFAs	Trout (*Oncorhynchus mykiss*) by-products	Fiori et al. (2012)
Fish oil	Common carp (*Cyprinus carpio*)	Kuvendziev et al. (2018)
Phospholipids	Atlantic salmon frame bone	Haq and Chun (2018)
PUFAs	Sardine (*Sardina pilchardus*) waste	Melgosa et al. (2020)
Bioactive lipids	Microalga *Crypthecodinium cohnii*	Couto et al. (2010)
Bioactive lipids	Microalga (*Scenedesmus obliquus, Chlorella protothecoides* and *Nannochloropsis salina*)	Solana et al. (2014)
Bioactive lipids	Microalga (*Scenedesmus obliquus* and *Scenedesmus obtusiusculus*)	Lorenzen et al. (2017)
Omega-3FAs, carotenoids	Microalgae (*Phaeodactylum tricornutum, Nannochloropsis oculata,* and *Porphyridium cruentum)*	Feller et al. (2018)
Omega-3 FAs	Microalgae (*Nannochloropsis* sp.)	Leone et al. (2019)
Astaxanthin		
Astaxanthin	Crayfish	López et al. (2004)
Astaxanthin and lipids	Brazilian red-spotted shrimp waste (*Farfantepenaeus paulensis*)	Sánchez-Camargo et al. (2011)

TABLE 12.1 *(Continued)*

Marine Biomolecules	Source for Extraction	References
Astaxanthin and *Omega*-3 FAs	Brazilian red-spotted shrimp waste (*Farfantepenaeus paulensis*)	Sánchez-Camargo et al. (2012)
Astaxanthin	Microalgae (*Haematococcus pluvialis*)	Wang et al. (2012)
Astaxanthin and triglycerides	Microalgae (*Haematococcus pluvialis*)	Kwan et al. (2018)
Astaxanthin (ASX) rich oil	Shrimp waste (*Peneaus monodon*) waste and mackerel skin (*Scomberomorus niphonius*) skin	Roy et al. (2020)
Astaxanthin, lutein, and FAs	Crayfish shells	Sanzo et al. (2018)
Seaweed based Biomolecules		
Fucoxanthin and phenols	Brown seaweed (*Undaria pinnatifida*)	Roh et al. (2008)
Fucoxanthin, phenolics, and FAs	Brown seaweed (*Sargassum muticum*)	Conde et al. (2015)
Fucoxanthin	Brown seaweeds (*Saccharina japonica* and *Sargassum horneri*)	Sivagnanam et al. (2015)
Fucoxanthin and astacene	Freshwater alga	Fabrowska et al. (2016)
Seaweed extracts	*Gracilaria mammillaris*	Ospina et al. (2017)
Fucoxanthin and phlorotannin	Brown seaweed *(Saccharina japonica)*	Saravana et al. (2017)

The extraction efficiency of the process can depend on a number of factors such as the ultrasound intensity, time, solvent to substrate ratio, power, temperature, propagation cycle, etc. (Medina-Torres et al., 2017). When compared to the modern extraction techniques which are used currently for extraction of bioactive compounds, the equipment cost is less. The lesser extraction time, lower solvent usage and lower operating temperature makes it a favorable process for extraction for many biologically sensitive and thermal labile compounds. The utility of UAE in isolating biomolecules from a wide range of matrices has been already reported. However, rather than being employed as a main extraction protocol, UAE finds application in as a pre-treatment technique. The ability to disrupt the cell wall combined with the shorter extraction duration makes it a favorable pre-treatment process in many plants based matrices.

12.3.1 UAE IN EXTRACTION OF MARINE BIOACTIVE COMPOUNDS

Ultrasound assisted extraction is being used for the isolation of bioactive compounds from fish/fish by-products. However, unlike SFE, UAE finds lesser application in extraction of bioactive compounds from fish matrices. Instead, it is reported to have wider applications in extraction biomolecules from micro and macroalgae as the latter have a rigid cell wall compared to the former. The potential of UAE in improving the functional attributes of biomolecules and extraction yield has been studied and reported recently. Zou et al. (2017) have reported that collagen extracted from soft-shelled turtles using ultrasonic assisted extraction have better physicochemical and functional attributes. The extraction yield and thermal stability was found to be better in collagen using UAE than the conventional treatments. The application of UAE as a pre-treatment method in the production of collagen from soft-shelled turtle (*Pelodiscus sinensis*) has been studied and reported by Zou et al. (2017). The application of UAE has been found to improve the yield, solubility, and antioxidant properties suggesting its feasibility for producing collagen for biomedical applications. Sinthusamran et al. (2018) have explored the feasibility of UAE in valorization of solid residue obtained after shrimp protein hydrolysis. UAE was found to be successful in extracting polyunsaturated fatty acids like docosahexaenoic acid (DHA), eicosapentaenoic acid (EPA) and carotenoids such as astaxanthin, astaxanthin monoester, astaxanthin diester and canthaxanthin. A recent study by Siewe et al. (2020) has suggested that UAE has promising potential in valorization of fish products for producing umami flavor providing better yield, umami concentration, taste, and degree of hydrolysis (DH) and has outperformed the conventional extraction techniques.

UAE has found wider applications in seaweeds for isolation of carotenoids like fucoxanthin and phenolic compounds such as phlorotannin, pigments such as phycoerythrin, phycocyanin, and biopolymers such as carrageenan, laminarin, alginate, etc. Youssouf et al. (2017) have reported that the use of UAE for alginate and carrageenan extraction has not affected their chemical structure and molar mass distribution. Zhu et al. (2017) have reported the positive effect of combining UAE with ultrafiltration for effective recovery of pigments from seaweeds. Table 12.2 shows the application of UAE in obtaining biomolecules from seaweeds.

TABLE 12.2 Ultrasonic Assisted Extraction of Seaweed Biomolecules

Biomolecules Extracted	Source of Extraction	References
Total phenolics, fucose, and uronic acid	*Ascophyllum nodosum*	Kadam et al. (2015)
R-phycoerythrin	*Grateloupia turuturu*	Le Guillard et al. (2015)
Phycobiliproteins	*Gelidium pusillum*	Mittal et al. (2017)
Seaweed pigments	*Undaria pinnatifida*	Zhu et al. (2017)
Fucoxanthin	*Padina tetrastromatica*	Raguraman et al. (2018)
Alginate and carrageenan	*Sargassum binderi* and *Turbinaria ornata*	Youssouf et al. (2017)
Alginate	*Sargassum muticum*	Flórez-Fernández et al. (2019)

12.4 MICROWAVE-ASSISTED EXTRACTION (MAE)

Microwave assisted extraction is a green extraction technique used for obtaining bioactive compounds from natural products using microwave energy. Microwaves are non-ionizing electromagnetic radiations in the frequency range of 300 MHz to 300 GHz. The efficiency of the extraction process depends on the ability of a material to absorb the microwave energy and its subsequent conversion to thermal energy. Briefly, when a material is exposed to microwave material, it absorbs the microwave radiation and causes the vibration/oscillation of polar molecules. This leads to inter and intramolecular friction and thereby collision of the charged ions causing heating of the matrices. The pressure build up during this process will help in cell wall rupture favoring the extraction of bioactives by interaction with the solvents. The changes in temperature and pressure generated during the process of microwave heating plays a key role in extraction efficiency (Grosso et al., 2015). Apart from pressure and temperature, the duration of the process, microwave power, frequency, nature of solvent, dielectric constant of the solvent, nature of the sample, moisture content, etc., are some of the factors that have significant effect on the extraction efficiency (Bruno et al., 2019). The shorter duration, less solvents and energy requirements make MAE a favorable extraction process when compared to the traditional process.

12.4.1 *MAE FOR EXTRACTION OF MARINE BIOACTIVE COMPOUNDS*

MAE has been extensively used for extraction of biomolecules from fish, fish by-products, and seaweeds A range of biomolecules such as chitin, chitosan, hydroxyapatite, fucoidan, etc., were isolated using MAE. El-Knidri et al. (2016) has reported a very efficient, eco-friendly, and time saving extraction protocol for obtaining chitin and chitosan from shrimp shell waste using microwave irradiation technology. It was observed that about 24 min was the time required for chitin and chitosan extraction using microwave irradiation whereas the conventional extraction process took 6–7 h. The significant findings of the study suggested that microwave based extraction technology can be very well utilized to valorize shrimp shell waste for the production of chitin and chitosan. Another study by El-Knidri et al. (2019) have reported that microwave technology was very effective than the conventional extraction technique in terms of the time of degree of deacetylation of chitin. Zhao et al. (2009) have optimized the extraction conditions for obtaining astaxanthin from microalgae, *Haematococcus pluvailis* using microwave irradiation technology. The potential of MAE in extracting biopolymers from seaweeds has also been attempted by several researchers. Table 12.3 shows some of the biomolecules extracted using MAE from various sources.

12.5 OTHER EMERGING METHODS FOR EXTRACTION OF MARINE BIOMOLECULES

Apart from the most commonly employed methods like SFE, UAE, and MAE, several methods are currently being employed in isolation of marine bioactives. The most notable among them are the extraction using ionic liquids, DES, switchable solvents. DES are extraction mixtures comprising of a hydrogen bond acceptor (HBA) and hydrogen bond donor (HBD). Biodegradability, low toxicity, easiness in preparation, high thermal stability, low volatility, non-flammability, etc., makes DES one of the most efficient green extraction techniques (Xu et al., 2020). Das et al. (2020) have recently employed DES for extraction of kappa-carrageenan from red seaweed, *Kappaphycus alvarezii* and has reported that the hydrated DES were more effective in isolation.

TABLE 12.3 MAE in Extracting Marine Biomolecules

Biomolecules	Source of Extraction	References
Fish/Shell Fishes		
Astaxanthin	*Haematococcus pluvialis*	Zhao et al. (2009)
Nanohydroxyapatite	Mussel shells	Shavandi et al. (2015)
Chitin and chitosan	Shrimp shell waste	El Knidri et al. (2016)
Fatty acids	EeL fish (*Monopterus albus*)	Afolabi et al. (2018)
Nanohydroxyapatite	Fish bones and mussel shells	Irfan et al. (2020)
Seaweeds		
Fucoidan	Brown seaweed (*Fucus vesiculosus*)	Rodriguez-Jasso et al. (2011)
Fucoidan	Brown seaweed (*Ascophyllum nodosum*)	Yuan and Macquarrie (2015)
Polysaccharides	Brown seaweed (*Sargassum thunbergii*)	Ren et al. (2017)
Poly-phenolic compounds	Brown algae (*Ascophyllum nodosum, Laminaria japonica, Lessonia trabeculate* and *Lessonia nigrecens*)	Yuan et al. (2018)
Phenols and carrageenan	Red seaweed (*Mastocarpus stellatus*)	Ponthier et al. (2020)

Saravana et al. (2018) have reported the use of DES mixed with water as an extraction solvent for subcritical extraction of seaweed polysaccharides from brown seaweed, *Saccharina japonica*. The findings of the study suggest that the process can be an effective method for enhanced recovery of polysaccharides from seaweeds.

Switchable solvents (SS) are those solvents which can convert their polarity from hydrophilicity to hydrophobicity and vice-versa. The potentiality of SS in the extraction of marine bioactives is being explored presently. Al-Ameriand Al-Zuhair (2019) have reported the use of SS in extraction of lipids from microalgae for production of biodiesel. Liu et al. (2020) have explored the use of switchable hydrophilicity solvent (SHS) for extraction of astaxanthin from microalgae, *Haematococcus pluvialis*. It was reported that the SHSs were capable of extracting more than 80% of the total astaxanthin content. Unlike to DES, the SHS could also be removed from astaxanthin by simple addition of CO_2 and water.

12.6 CONCLUSION

Seafood and aquatic by-products can serve as excellent resources for isolation of many marine nutraceuticals. By envisaging a proper extraction protocol, such resources can be effectively valorized for obtaining nutraceuticals of high value such as chitosan, collagen, gelatin, hydroxylapatite, alginate, carrageenan, etc. The quality and quantity of the nutraceutical and depend to a greater extent on the type of extraction protocol followed. At present, there is an increasing trend among the researchers and industrialists to shift to greener extraction procedures as the conventional solvent-based extraction were reported to have serious setbacks. As a result, several innovative, greener, and sustainable extraction protocols are being developed for increasing the extraction efficiency. Extraction using supercritical carbon dioxide has been extremely successful in the isolation of bioactive compounds from several matrices. Apart from this, several other methods such as ultrasonic-assisted extraction, enzyme assisted extraction, extraction using DES, etc., is also being currently employed for isolation for marine bioactives. Though many of these methods were found to be successful on a lab-scale, further scaling up is needed. However, certain methods such as extraction using switchable solvents are still in its infancy. Hence, these methods have to be validated in terms of their safety, economical, and commercial aspects.

KEYWORDS

- **cavitation**
- **deep eutectic solvents**
- **free fatty acids**
- **marine bioactives**
- **polyphenoloxidase**
- **supercritical fluid extraction**
- **switchable solvents**

REFERENCES

Afolabi, H. K., Mudalip, S. K. A., & Alara, O. R., (2018). Microwave-assisted extraction and characterization of fatty acid from eel fish (*Monopterus albus*). *Beni-Suef Univ. J. Appl.*, 7465–7470.

Al-Ameri, M., & Al-Zuhair, S., (2019). Using switchable solvents for enhanced, simultaneous microalgae oil extraction-reaction for biodiesel production. *Biochem. Eng. J., 141*, 217–224.

Amiguet, V. T., Kramp, K. L., Mao, J., McRae, C., Goulah, A., Kimpe, L. E., Blais, J. M., & Arnason, J. T., (2012). Supercritical carbon dioxide extraction of polyunsaturated fatty acids from Northern shrimp (*Pandalus borealis* Kreyer) processing by-products. *Food Chem., 130*, 853–858.

Bruno, S. F., Ekorong, F. J. A. A., Karkal, S. S., Cathrine, M. S. B., & Kudre, T. G., (2019). Green and innovative techniques for recovery of valuable compounds from seafood by-products and discards: A review. *Trends Food SciTech., 85*, 10–22.

Conde, E., Moure, A., & Domínguez, H., (2015). Supercritical CO_2 extraction of fatty acids, phenolics and fucoxanthin from freeze-dried *Sargassum muticum. J Appl. Phycol., 27*, 957–964.

Couto, R. M., Simoes, P. C., Reis, A., Da Silva, T. L., Martins, V. H., & Sánchez-Vicente, Y., (2010). Supercritical fluid extraction of lipids from the heterotrophic microalga *Crypthecodinium cohnii. Eng. Life Sci., 10*, 158–164.

Das, A. K., Sharma, M., Mondal, D., & Prasad, K., (2016). Deep eutectic solvents as efficient solvent system for the extraction of κ-carrageenan from *Kappaphycus alvarezii. Carbohydr. Polym., 136*, 930–935.

El Knidri, H., Dahmani, J., Addaou, A., Laajeb, A., & Lahsini, A., (2019). Rapid and efficient extraction of chitin and chitosan for scale-up production: Effect of process parameters on deacetylation degree and molecular weight. *Int. J. Biol. Macromol., 139*, 1092–1102.

El Knidri, H., El Khalfaouy, R., Laajeb, A., Addaou, A., & Lahsini, A., (2016). Eco-friendly extraction and characterization of chitin and chitosan from the shrimp shell waste via microwave irradiation. *Process Saf. Environ. Prot., 104*, 395–405.

Fabrowska, J., Ibañez, E., Łęska, B., & Herrero, M., (2016). Supercritical fluid extraction as a tool to valorize underexploited freshwater green algae. *Algal Res., 19*, 237–245.

Feller, R., Matos, Â. P., Mazzutti, S., Moecke, E. H., Tres, M. V., Derner, R. B., Oliveira, J. V., & Junior, A. F., (2018). Polyunsaturated ω-3 and ω-6 fatty acids, total carotenoids and antioxidant activity of three marine microalgae extracts obtained by supercritical CO_2 and subcritical n-butane. *J. Supercrit. Fluids, 133*, 437–443.

Fiori, L., Solana, M., Tosi, P., Manfrini, M., Strim, C., & Guella, G., (2012). Lipid profiles of oil from trout (*Oncorhynchus mykiss*) heads, spines and viscera: Trout by-products as a possible source of *Omega*-3 lipids. *Food Chem., 134*, 1088–1095.

Flórez-Fernández, N., Domínguez, H., & Torres, M. D., (2019). A green approach for alginate extraction from *Sargassum muticum* brown seaweed using ultrasound-assisted technique. *Int. J. Biol. Macromol., 124*, 451–459.

Grosso, C., Valentão, P., Ferreres, F., & Andrade, P. B., (2015). Alternative and efficient extraction methods for marine-derived compounds. *Mar. Drugs., 13*, 3182–3230.

Haq, M., & Chun, B. S., (2018). Characterization of phospholipids extracted from Atlantic salmon by-product using supercritical CO_2 with ethanol as co-solvent. *J. Clean. Prod., 178*, 186–195.

Irfan, M., Suprajaa, P. S., Praveen, R., & Reddy, B. M., (2020). Microwave-assisted one-step synthesis of nanohydroxyapatite from fish bones and mussel shells. *Mater. Lett.*, 128685.

Kadam, S. U., Tiwari, B. K., Smyth, T. J., & O'Donnell, C. P., (2015). Optimization of ultrasound assisted extraction of bioactive components from brown seaweed *Ascophyllum nodosum* using response surface methodology. *Ultrason.Sonochem., 23*, 308–316.

Kuvendziev, S., Lisichkov, K., Zeković, Z., Marinkovski, M., & Musliu, Z. H., (2018). Supercritical fluid extraction of fish oil from common carp (*Cyprinus carpio* L.) tissues. *J. Supercrit. Fluids, 133*, 528–534.

Kwan, T. A., Kwan, S. E., Peccia, J., & Zimmerman, J. B., (2018). Selectively biorefining astaxanthin and triacylglycerol co-products from microalgae with supercritical carbon dioxide extraction. *Bioresour. Technol., 269*, 81–88.

Le Guillard, C., Dumay, J., Donnay-Moreno, C., Bruzac, S., Ragon, J. Y., Fleurence, J., & Bergé, J. P., (2015). Ultrasound-assisted extraction of R-phycoerythrin from *Grateloupia turuturu* with and without enzyme addition. *Algal Res., 12*, 522–528.

Leone, G. P., Balducchi, R., Mehariya, S., Martino, M., Larocca, V., Di Sanzo, G., Iovine, A., et al., (2019). Selective Extraction of ω-3 fatty acids from *Nannochloropsis* sp. using supercritical CO_2 extraction. *Molecules, 24*, 2406.

Létisse, M., Rozières, M., Hiol, A., Sergent, M., & Comeau, L., (2006). Enrichment of EPA and DHA from sardine by supercritical fluid extraction without organic modifier: I. Optimization of extraction conditions. *J. Supercrit. Fluids, 38*, 27–36.

Liu, H., Huang, W. C., Guo, N., & Mao, X., (2020). Application of secondary amine switchable hydrophilicity solvents for astaxanthin extraction from wet *Haematococcus pluvialis. Algal Res., 48*, 101892.

Liza, M. S., Rahman, R. A., Mandana, B., Jinap, S., Rahmat, A., Zaidul, I. S. M., & Hamid, A., (2010). Supercritical carbon dioxide extraction of bioactive flavonoid from *Strobilanthes crispus* (PecahKaca). *Food Bioprod. Process, 88*, 319–326.

López, M., Arce, L., Garrido, J., Rıos, A., & Valcárcel, M., (2004). Selective extraction of astaxanthin from crustaceans by use of supercritical carbon dioxide. *Talanta, 64*, 726–731.

Lordan, S., Ross, R. P., & Stanton, C., (2011). Marine bioactives as functional food ingredients: Potential to reduce the incidence of chronic diseases. *Mar. Drugs, 9*, 1056–1100.

Lorenzen, J., Igl, N., Tippelt, M., Stege, A., Qoura, F., Sohling, U., & Brück, T., (2017). Extraction of microalgae derived lipids with supercritical carbon dioxide in an industrial relevant pilot plant. *Bioproc. Biosyst. Eng., 40*, 911–918.

Medina-Torres, N., Ayora-Talavera, T., Espinosa-Andrews, H., Sánchez-Contreras, A., & Pacheco, N., (2017). Ultrasound assisted extraction for the recovery of phenolic compounds from vegetable sources. *Agron., 7*, 47.

Melgosa, R., Trigueros, E., Sanz, M. T., Cardeira, M., Rodrigues, L., Fernández, N., Matias, A. A., et al., (2020). Supercritical CO_2 and subcritical water technologies for the production of bioactive extracts from sardine (*Sardina pilchardus*) waste. *J. Supercrit. Fluids*, *164*, 104943.

Mezzomo, N., Martínez, J., Maraschin, M., & Ferreira, S. R., (2013). Pink shrimp (*P. brasiliensis* and *P. paulensis*) residue: Supercritical fluid extraction of carotenoid fraction. *J. Supercrit. Fluids, 74*, 22–33.

Michalak, I., Górka, B., Wieczorek, P. P., Rój, E., Lipok, J., Łęska, B., Messyasz, B., et al., (2016). Supercritical fluid extraction of algae enhances levels of biologically active compounds promoting plant growth. *Eur. J. Phycol., 51*, 243–252.

Mittal, R., Tavanandi, H. A., Mantri, V. A., & Raghavarao, K. S. M. S., (2017). Ultrasound assisted methods for enhanced extraction of phycobiliproteins from marine macro-algae, *Gelidium pusillum* (Rhodophyta). *Ultrason. Sonochem., 38*, 92–103.

Ojha, K. S., Aznar, R., O'Donnell, C., & Tiwari, B. K., (2020). Ultrasound technology for the extraction of biologically active molecules from plant, animal and marine sources. *Trends Analyt. Chem., 122*, 115663.

Ospina, M., Castro-Vargas, H. I., & Parada-Alfonso, F., (2017). Antioxidant capacity of Colombian seaweeds: 1. Extracts obtained from *Gracilaria mammillaris* by means of supercritical fluid extraction. *J. Supercrit. Fluids, 128*, 314–322.

Ponthier, E., Domínguez, H., & Torres, M. D., (2020). The microwave assisted extraction sway on the features of antioxidant compounds and gelling biopolymers from *Mastocarpus stellatus*. *Algal Res., 51*, 102081.

Raguraman, V., MubarakAli, D., Narendrakumar, G., Thirugnanasambandam, R., Kirubagaran, R., & Thajuddin, N., (2018). Unraveling rapid extraction of fucoxanthin from *Padina tetrastromatica*: Purification, characterization and biomedical application. *Process Biochem., 73*, 211–219.

Ren, B., Chen, C., Li, C., Fu, X., You, L., & Liu, R. H., (2017). Optimization of microwave-assisted extraction of *Sargassum thunbergii* polysaccharides and its antioxidant and hypoglycemic activities. *Carbohydr. Polym., 173*, 192–201.

Rodriguez-Jasso, R. M., Mussatto, S. I., Pastrana, L., Aguilar, C. N., & Teixeira, J. A., (2011). Microwave-assisted extraction of sulfated polysaccharides (fucoidan) from brown seaweed. *Carbohydr. Polym., 86*, 1137–1144.

Roh, M. K., Uddin, M. S., & Chun, B. S., (2008). Extraction of fucoxanthin and polyphenol from *Undaria pinnatifida* using supercritical carbon dioxide with co-solvent. *Biotechnol. Bioprocess. Eng., 13*, 724–729.

Roy, V. C., Getachew, A. T., Cho, Y. J., Park, J. S., & Chun, B. S., (2020). Recovery and bio-potentialities of astaxanthin-rich oil from shrimp (*Peneanus monodon*) waste and mackerel (*Scomberomorus niphonius*) skin using concurrent supercritical CO_2 extraction. *J. Supercrit. Fluids, 159*, 104773.

Rubio-Rodríguez, N., Sara, M., Beltrán, S., Jaime, I., Sanz, M. T., & Rovira, J., (2012). Supercritical fluid extraction of fish oil from fish by-products: A comparison with other extraction methods. *J. Food Eng., 109*, 238–248.

Sahena, F., Zaidul, I. S. M., Jinap, S., Jahurul, M. H. A., Khatib, A., & Norulaini, N. A. N., (2010). Extraction of fish oil from the skin of Indian mackerel using supercritical fluids. *J. Food Eng., 99*, 63–69.

Sánchez-Camargo, A. P., Martinez-Correa, H. A., Paviani, L. C., & Cabral, F. A., (2011). Supercritical CO_2 extraction of lipids and astaxanthin from Brazilian redspotted shrimp waste (*Farfantepenaeus paulensis*). *J. Supercrit. Fluids, 56*, 164–173.

Sánchez-Camargo, A. P., Meireles, M. Â. A., Ferreira, A. L., Saito, E., & Cabral, F. A., (2012). Extraction of ω-3 fatty acids and astaxanthin from Brazilian redspotted shrimp waste using supercritical CO_2+ ethanol mixtures. *J. Supercrit. Fluids, 61*, 71–77.

Sanzo, G. D., Mehariya, S., Martino, M., Larocca, V., Casella, P., Chianese, S., Musmarra, D., Balducchi, R., & Molino, A., (2018). Supercritical carbon dioxide extraction of astaxanthin, lutein, and fatty acids from *Haematococcus pluvialis* microalgae. *Mar. Drugs, 16*, 334.

Saravana, P. S., Cho, Y. N., Woo, H. C., & Chun, B. S., (2018). Green and efficient extraction of polysaccharides from brown seaweed by adding deep eutectic solvent in subcritical water hydrolysis. *J. Clean. Prod., 198*, 1474–1484.

Saravana, P. S., Getachew, A. T., Cho, Y. J., Choi, J. H., Park, Y. B., Woo, H. C., & Chun, B. S., (2017). Influence of co-solvents on fucoxanthin and phlorotannin recovery from brown seaweed using supercritical CO_2. *J. Supercrit. Fluids, 120*, 295–303.

Shavandi, A., Bekhit, A. E. D. A., Ali, A., & Sun, Z., (2015). Synthesis of nano-hydroxyapatite (nHA) from waste mussel shells using a rapid microwave method. *Mater.Chem. Phys., 149*, 607–616.

Siewe, F. B., Kudre, T. G., & Narayan, B., (2020). Optimization of ultrasound-assisted enzymatic extraction conditions of umami compounds from fish by-products using the combination of fractional factorial design and central composite design. *Food Chem., 334*, 127498.

Singh, R., Dhanani, T., & Kumar, S., (2017). Supercritical fluid extraction of bioactive compounds from fruits and vegetables. In: *Fruit and Vegetable Phytochemicals: Chemistry and Human Health* (2nd edn., pp. 749–762).

Sinthusamran, S., Benjakul, S., Kijroongrojana, K., Prodpran, T., & Agustini, T. W., (2018). Yield and chemical composition of lipids extracted from solid residues of protein hydrolysis of Pacific white shrimp cephalothorax using ultrasound-assisted extraction. *Food Biosci., 26*, 169–176.

Sivagnanam, S. P., Yin, S., Choi, J. H., Park, Y. B., Woo, H. C., & Chun, B. S., (2015). Biological properties of fucoxanthin in oil recovered from two brown seaweeds using supercritical CO_2 extraction. *Mar. Drugs, 13*, 3422–3442.

Solana, M., Rizza, C. S., & Bertucco, A., (2014). Exploiting microalgae as a source of essential fatty acids by supercritical fluid extraction of lipids: Comparison between *Scenedesmus obliquus, Chlorella protothecoides* and *Nannochloropsis salina. J. Supercrit. Fluids, 92*, 311–318.

Wang, L., Yang, B., Yan, B., & Yao, X., (2012). Supercritical fluid extraction of astaxanthin from *Haematococcus pluvialis* and its antioxidant potential in sunflower oil. *Innov. Food Sci. Emerg. Technol., 13*, 120–127.

Xu, H., Peng, J., Kong, Y., Liu, Y., Su, Z., Li, B., Song, X., et al., (2020). Key process parameters for deep eutectic solvents pretreatment of lignocellulosic biomass materials: A review. *Bioresour. Technol.*, 123416.

Youssouf, L., Lallemand, L., Giraud, P., Soulé, F., Bhaw-Luximon, A., Meilhac, O., D'hellencourt, C. L., et al., (2017). Ultrasound-assisted extraction and structural characterization by NMR of alginates and Carrageenans from seaweeds. *Carbohydr. Polym., 166*, 55–63.

Yuan, Y., & Macquarrie, D., (2015). Microwave assisted extraction of sulfated polysaccharides (fucoidan) from *Ascophyllum nodosum* and its antioxidant activity. *Carbohydr. Polym., 129*, 101–107.

Yuan, Y., Zhang, J., Fan, J., Clark, J., Shen, P., Li, Y., & Zhang, C., (2018). Microwave assisted extraction of phenolic compounds from four economic brown macroalgae species and evaluation of their antioxidant activities and inhibitory effects on α-amylase, α-glucosidase, pancreatic lipase, and tyrosinase. *Food Res. Int., 113*, 288–297.

Zhao, L., Chen, G., Zhao, G., & Hu, X., (2009). Optimization of microwave-assisted extraction of astaxanthin from *Haematococcus pluvialis* by response surface methodology and antioxidant activities of the extracts. *Sep. Sci. Technol., 44*, 243–262.

Zhu, Z., Wu, Q., Di, X., Li, S., Barba, F. J., Koubaa, M., Roohinejad, S., Xiong, X., & He, J., (2017). Multistage recovery process of seaweed pigments: Investigation of ultrasound-assisted extraction and ultra-filtration performances. *Food Bioprod. Process., 104*, 40–47.

Zou, Y., Wang, L., Cai, P., Li, P., Zhang, M., Sun, Z., Sun, C., et al., (2017). Effect of ultrasound assisted extraction on the physicochemical and functional properties of collagen from soft-shelled turtle calipash. *Int. J. Biol. Macromol., 105*, 1602–1610.

Zou, Y., Xu, P., Li, P., Cai, P., Zhang, M., Sun, Z., Sun, C., et al., (2017). Effect of ultrasound pre-treatment on the characterization and properties of collagen extracted from soft-shelled turtle (*Pelodiscus sinensis*). *LWT-Food Sci. Technol., 82*, 72–81.

Part III
Utilization of Fish Processing Wastes

CHAPTER 13

Fish Wastes: Present Status of Utilization and Scope of Valorization

RANENDRA K. MAJUMDER

College of Fisheries, Central Agricultural University (Imphal), Lembucherra – 799210, Tripura, India

ABSTRACT

Fish wastes are generated at different points in the value chain, viz. by-catch, onboard handling, landing centers, transportation, storage, processing, retailers, and consumers. Most of the wastes generated in the fish processing plants are discharged as untreated into the nearby coastal waters, which creates potentially hazardous environments in the receiving ends. There is a significant economic loss due to fish processing wastes, which otherwise could become a potential source in aquaculture practices such as in the form of fish meal and oil, which are recognized as an important source of protein and lipid in fish feed formulation. Solid fish waste includes head, tails, skin, gut, fins, and frames; these can be a great source for a variety of basic economical products such as proteins and amino acids, collagen, gelatin, oil, and enzymes. Besides these, an array of economically important products such as chitin, chitosan, taurine, creatine, antifreeze proteins, omega-3 PUFAs, chondroitin sulfate, etc., are presently obtained from fish wastes and commercialized. Development of different valorization processes for fish waste utilization would play a key role to decrease their indiscriminate disposal. Enzymatic hydrolysis of fish wastes and underutilized fish biomass is an ideal method for recovering proteins. Such recovered proteins after purification would have a wide variety of applications in the field of foods and pharmaceuticals. As a meaningful fish waste utilization and

Advances in Fish Processing Technologies: Preservation, Waste Utilization, and Safety Assurance.
Ranendra K. Majumder, PhD & Amjad K. Balange, PhD (Eds.)

management, proper implementation of eco-friendly recycling technologies to convert all the valuable components present in the waste into products useful for the mankind would not only reduce the amount of bio-waste going to disposal route, but would also substantially increase the economy of the fisheries sector.

13.1 INTRODUCTION

The global fish production reached to 177.8 million tons in 2019, a steady increase from 148.1 million tons in 2010 (FAO, 2020). In 2018, out of 179 million tons of total fish production about 88% and 12% were utilized for direct human consumption and non-food purposes, respectively. Due to growing recognition of fish as a key component for a balanced diet and a healthy lifestyle, consumption of fish products has been increasing steadily. For over 60 years, the global food fish consumption scenario has increased significantly and at a rate outpacing annual population growth. The per capita food fish consumption increased from 9.0 kg (live wt. equivalent) in 1961 to 20.5 kg in 2018, with an average annual growth rate of about 1.5% (FAO, 2020). Such growth in fish consumption rate is not only contributed by the substantial increases in production, but also there is a combination of other factors, these include technological advancements in processing, cold chain improvement, shipping, and distribution, etc. Moreover, rising household incomes worldwide strongly contribute to the increased demand for fish and fish products. Reduction in post-harvest loss and waste as well as increased consumers' awareness on the health benefits of fish are considered as other significant factors.

A serious global issue like food loss and waste is becoming a focus of Sustainable Development Goal (SDG), which aims a halving the wastage by 2030. Proper handling, adequate preservation infrastructure and expertise, hygiene, and maintenance of cold chain from harvest to consumption seem to be crucial at present in avoiding loss and waste as well as to preserve quality. Worldwide, total annual fish loss and waste range from 30 and 35% (FAO, 2020). Fish wastes are generated at different points in the value chain, viz. by-catch, onboard handling, landing centers, transportation, storage, processing, retailers, and consumers. In general practice, the waste generation begins with the discard of unintentional catches to the sea. During fish processing, except the muscle parts the rest is discarded. According to one estimate, the global fish waste generation is approximately in excess of 100 million tons, whereas, in India, the figure is more than 4 million tons. The fish fileting operation alone generates wastes of approximately 75% of the total fish weight. On an average, more than 50% of fish tissues consisting of fins, heads, skin, bones, and viscera are considered as wastes and are discarded.

Considering fish wastes a significant economic loss, its otherwise use as potential source of ingredients for feed in fish farming practices, in the form of fish meal and oil, could become another option for their judicious utilization/recycling. Economy of the aquaculture practices is significantly affected by the cost of fish feeds, which contributes a major share in the total costs of fish production. Annually about 60% of fish meal and 80% of fish oil of the world total production are utilized by the aquaculture farms, and this is expected to increase to cope with the growth of aquaculture in the coming future. Besides these, an extensive variety of products such as chitin, chitosan, collagen, gelatin, omega-3 PUFAs, taurine, creatine, antifreeze proteins, enzymes, chondroitin sulfate are obtained presently from fish wastes and commercialized. Development of different valorization processes would play a key role in decreasing the generation of fish processing wastes. Simultaneously, the possible adverse impacts of products developed from fish wastes on human health and the environment is equally essential to be evaluated. This would contribute to understanding whether the valorization processes are, in fact, an adequate solution for ensuring a sustainable reuse of fish wastes and byproducts instead of traditional waste management systems. Another important aspect is the way the fish wastes play a detrimental effect on aquatic ecosystems, because of rise of organic wastes could significantly alter the community structure, including biodiversity of the benthic assemblages. Therefore, it is obvious that the management of fish wastes is related to all other attempts intended to abate the pollution source. However, it is also essential to search for the efficient ways to solve this problem.

13.2 COMPOSITION OF FISH WASTES

The biochemical composition of fish is influenced by a number of factors, namely, type of species, nutritional status, sex, age, time of year and overall health condition. In general, proximate composition of fish is moisture-50–80%, fat-0–25% and protein-15–30%. Typically, head, tails, skin, gut, fins, and frames which are discarded in the processing plants and markets are considered as solid fish waste. These wastes can serve as potential sources of proteins and amino acids, collagen, gelatin, oil, and enzymes, as shown in Table 13.1. In general, fish waste contains proteins (58%), fat (19%), and essential minerals. Moreover, the fat content of fish waste is constituted mostly by palmitic acid, oleic acid and monounsaturated fatty acids (22%) (Esteban et al., 2007).

TABLE 13.1 Biochemical Composition of Fish Waste

Nutrient	Approx. Quantity
Crude protein (%)	57.92 ± 5.26
Fat (%)	19.10 ± 6.06
Crude fiber (%)	1.19 ± 1.21
Ash (%)	21.79 ± 3.52
Calcium (%)	5.80 ± 1.35
Phosphorous (%)	2.04 ± 0.64
Potassium (%)	0.68 ± 0.11
Sodium (%)	0.61 ± 0.08
Magnesium (%)	0.17 ± 0.04
Iron (ppm)	100.00 ± 42.00
Zinc (ppm)	62.00 ± 12.00
Manganese (ppm)	6.00 ± 7.00
Copper (ppm)	1.00 ± 1.00

*Values in % or mg/kg (ppm) on a dry matter basis.

Source: Esteban et al. (2007).

In fish fileting plants, the leftover frames contain significant amounts of proteins, which are highly nutritious and easily digestible. These proteins, instead of being discarded as waste, can be extracted by enzymatic hydrolysis. Compared to other animal protein sources, fish proteins consist of all dietary essential amino acids in well-balanced composition as required by human. Being a poikilothermic animal, fish muscle proteins are more heat-labile compared to mammalian muscle proteins. Moreover, proteins from the cold water fish are more susceptible to thermal denaturation when compared with their tropical counterpart.

Fish processing waste is also a rich source of omega-3 PUFAs, however, the quantity depends on the fish species and their habitat temperature. Wastes generated during fish processing is a great source for good quality fish oil which can be used for human consumption or even for production of biodiesel.

The visceral waste of fish serves as an important source for different digestive enzymes, viz., pepsin, trypsin, chymotrypsin, and collagenases. Pepsin, a proteolytic enzyme present in the stomach of fish, constitutes about 5% of the total fish weight, is wasted if not extracted and used for specific purposes. Digestive enzymes of fish are capable to exhibit high catalytic activities, even at relatively low concentrations. Additionally, fish skin is rich in collagen and gelatin, and recently there is a growing demand for these in food applications, cosmetic, and biomedical industries. Fishbones which are normally discarded after fileting operation and removal of adherent muscle, account for about 30% of the collagen in addition to fish skin. About 60–70% of fishbone is made up of minerals that include essential elements like calcium, phosphorous, and hydroxyapatite. One of the most significant aspects which attracted food researchers is the presence of a number of short peptides in fish proteins which have been reported to exhibit several bioactive properties like hypertensive, antithrombotic, antioxidative, and immunomodulatory.

13.3 IMPACTS OF FISH WASTES ON THE ENVIRONMENT

Environmental problems due to fisheries activity have become one of the important issues. In terms of management of environmental pollution also, fishery discards and byproducts may pose a significant problem. It is a fact that from the viewpoint of economic aspect, industry like fisheries not only fulfills the best performance, but simultaneously in the meantime, social, and environmental responsibilities too cannot be ignored. Fish processing plants generate heavy loads of both solid and liquid organic wastes, constituted by complex mixtures of many substances which includes remains of fish and shrimp after processing, soluble proteins, fats, and oils, partially decomposed organic matters, pathogenic, and other microbes, different chemical substances, inorganic nutrients particularly nitrogen and phosphorus along with many others. Usually, most of these wastes from the fish processing plants are discharged into the nearby barren lands or coastal waters without any pre-treatment, resulting potentially hazardous environment.

Fish processing wastes are primarily organic in nature and are therefore naturally degradable by microbial activity. The aerobic bacteria present in the water can quickly degrade the organic fish wastes in the presence of oxygen leading to an anoxic condition in water with simultaneous increase in BOD. The resultant anoxic aquatic environment initiates anaerobic decomposition of organic substances, especially proteins and other nitrogenous compounds, leading to release of obnoxious gases like hydrogen sulfide and ammonia, including greenhouse gases like carbon dioxide and methane and organic acids, all of which are potentially toxic and hazardous to the aquatic ecosystem even in their low concentrations. Moreover, the aquatic life will be deprived of the oxygen it needs. Nutrients released from the decaying organic matter promote plant growth and their rapid blooming together with depletion of oxygen results in modification in ecosystem structure, leading finally to a state known as eutrophication. Eutrophication has been shown to be responsible for major changes in the ecosystem which encompasses species composition, promotion of proliferation of macroalgae and filamentous algae, structure, and function of marine communities, etc., over large areas.

Effluents discharged from fish processing plants contain partially decomposed organic matters which harbor a variety of pathogens and harmful substances of both organic and inorganic in nature. The resultant short-term increase in nutrients load in water changes the food regime of fish with occasional creation of habitat complexity followed by population increase of fish species around the point of discharge. In addition to this, the presence of sufficient nitrogen, phosphorous, and ammonia in the fish wastes results pH variation, which causes increased turbidity of the water followed by decomposition of algae.

An increase of nutrient levels may cause the algal bloom development with subsequent toxin production leading to a decrease in dissolved oxygen. The algal toxins in the contaminated water will be absorbed by the fish species and may cause mass mortality. Eutrophication and its long-term effects will boost up phytoplankton biomass, and finally lead to extensive decrease in species diversity including benthic fish communities (Bonsdorff et al., 1997). Similar observation is noticed in zooplankton communities too, where herbivorous copepods are being replaced by small sized gelatinous zooplankton (Zaitsev, 1992). Eutrophication often affects benthic fauna, nursery, and feeding ground of fish, amenity, recreational uses and tourism (Riegman, 1995). Mass release of processing wastes and associated debris results in potential problems leading to great loss of amenities affecting the recreational use of water. Although, the intensity and scale of impacts of fish processing wastes may vary and depends on the quantity and nature of the waste output. Since wastes from fish processing industries are generally produced throughout the year, their local impacts are particularly obvious without giving any chance for the environment to recover. However, impacts are likely to be more detrimental if a particular ecosystem receives wastes from a cluster of processing plants.

13.4 PRESENT STATE OF UTILIZATION OF FISH WASTES

Presently, the most common utilization of fish wastes are production of fish meal, oil, fish silage and organic fertilizer. Additionally, chitin, and chitosan are also extracted from the shells of crustaceans.

13.4.1 FISH MEAL

Fish meal, being a highly concentrated fish protein is rich in essential amino acids, minerals, B group vitamins and other unknown growth factors. Fish meal is a highly nutritious feed supplement with wide application in animal feed industries. Recently, fishmeal production from the fisheries byproducts has increased drastically, and according to one estimate, in 2008 around 25% (1.23 million tons) of the total fishmeal was produced from fish byproducts (FOC, 2012).

Traditionally, the steps in fish meal production include cooking, pressing, drying, and grinding. The raw materials are usually, trash fish, by-catch, and miscellaneous fish, fileting waste including waste from canneries and various other processing operations. During cooking protein present in the fish waste is coagulated leading to rupture the fat deposits and release of oil and water. Pressing operation (through continuous screw press) removes large amounts of liquid from the cooked material, which is collected to separate oil from water. The liberated water, known as stick water, is further concentrated by evaporation to obtain thick syrup containing 30 to 40% solids. Finally press cake is fluffed and dried to a moisture level of ~8% followed by grinding to the desired particle size. In case of dry reduction process, especially for lean or non-oil fish, carcasses of shark, fish offal and fileting waste, raw material is dried to moisture content of 10% and pulverized. Presently, the major consumer of fish meal is aquaculture industry (63%) followed by pig feed (25%), poultry (8%) and feed of other animals (4%) (Chamberlain, 2011).

13.4.2 FISH OIL

Fish oils are enriched with polyunsaturated fatty acids, especially eicosapentaenoic acid (EPA) and docosahexaenoic acid (DHA), and can be easily extracted from the fish wastes or discarded whole fish. While producing fish meal by wet rendering process, the oil fraction of the liquid effluent obtained from the cooked mass is separated by centrifugation, which can be sold directly or after purification. Quantity of oil obtained generally depends on the fat content of the fish species. In general, fish contains oil in the range of 2 to 25%. In any case, fish processing waste serves as a potential source of good quality fish oil, which could be used even for human consumption in addition to other industrial applications.

13.4.3 FISH SILAGE

Fish silage is a liquid concentrate of fish protein having high biological properties to use in animal feed formulations by replacing fish meal. Any spoiled fish, by-catch fish, byproducts from fish markets and processing plants can be used to produce fish silage. Liquefaction is achieved by the activity of enzymes already present in the raw material, which break-down the fish proteins into smaller soluble units such as short peptides and amino acids. Enzymatic autolysis can be accelerated by an organic or inorganic acid. Acid maintains the proper pH for optimum enzymatic activity and also prevent bacterial spoilage. Presence of amino acids in fish silage makes it a most potential product for protein biosynthesis. In conventional practice, 98% formic acid is added at the rate of 3% by weight of the chopped or ground raw material, mixed well and stored for 48 days. The pH of the mixture should be kept less than 4 to prevent bacterial activity. Fish silage can also be prepared by fermentation using *Lactobacillus plantarum*, in which 5% (w/w) sugar beet molasses is used as substrate of the fermenting bacteria. After incubation for a specified period, complete liquefaction of the raw material is obtained. Fish silage can be co-dried with wheat bran or rice bran at 105°C. Fish silage co-dried with cereals improves its nutrients content and also prevent dried silage from spoilage. Moisture content of dried silage more than 12% can support bacterial, mold, and yeast growth. However, high water content which makes its transportation expensive, and additionally its disagreeable odor may be the reasons for limited use of fish ensilage in feed formulations. But co-drying with cereals could overcome many such problems including leaching of nutrients during storage and offensive odor. Fish ensilage has been encouraged to use as fish meal replacement for the production of animal feeds (Nair et al., 2004).

13.4.4 CHITIN AND CHITOSAN

Chitin is a polysaccharide and constituting about 15–20% of dried crustacean shell waste become one of the major components. It is non-toxic and biocompatible exhibiting a strong antimicrobial activity against a variety of microorganisms. Biodegradable properties of chitin find its wide applications as a food ingredient as well as in medical applications. Deacetylated form of chitin is chitosan. Chitosan is used as blends either as dietic product or as edible films (EFs) for the purposes of food preservation (Arvanitoyannis et al., 1998). Production of chitin and chitosan from the waste of food industries, specifically from the crustacean cannery made it environment friendly and economically feasible, especially when the recovery of carotenoids is included. A considerable amount of chitin is extracted from shellfish waste and is marketed as a fish food additive to provide dietary fiber.

13.5 RECENT ADVANCES ON THE UTILIZATION OF FISH WASTES

Presently, there is a social agreement that perceives solutions have to be implemented to avoid purposeless waste of this valuable biomass. Although some products are developed from the fish wastes, but research on waste management is gaining momentum to develop many more products of human use. Simultaneous to the development of different valorization processes for decreasing the fish wastes, it is also equally important to evaluate possible adverse impacts of the process on the environment but also on the human health, which will help to assess the perfectness of the present-day waste management systems over the traditional practices for ensuring a sustainable reuse of fish wastes/byproducts.

13.5.1 EXTRACTION OF ENZYMES

The digestive tract or the viscera of fish constitutes about 5 to 8% of the fish weight, is usually wasted and forms a bulk of fish wastes in the markets as well as in the processing plants. Fish viscera are a rich source of different digestive enzymes and many of them exhibit moderate to high catalytic activities at relatively low concentrations. Fish muscle and visceral enzymes include mostly pepsin, trypsin, chymotrypsin, and collagenases. It has been reported that digestive enzymes of fish exhibit better catalytic properties at lower temperatures, they have lower sensitivity to substrate concentrations and shows greater stability in a wide range of pH (Zhou et al., 2011). Acidic and alkaline proteases from fish viscera has been reported to exhibit their high activity between pH 2–4 and 8–10 respectively (Nasri et al., 2015). The alkaline proteases also perform with high activity and stability, and find suitable applications in the field of bioengineering and biotechnology. Fish trypsin and mammalian trypsin are similar in molecular weight and amino acid composition (Kishimura and Hayashi, 2002).

In detergent industry, the pH of the laundry detergents generally ranges from 9.0–12.0. In case of detergent formulations, the preferable proteolytic enzymes should exhibit satisfactory activity and stability at high pH and temperature, along with their compatibility with surfactants and oxidizing agents. Collagenolytic proteases or collagenases are present in epithelial, cartilaginous, bony tissues, and digestive tracts of fish and shellfish. Zinc-containing extracellular metallo-collagenases (MW 30 to 150 kDa) are involved in remodeling of the extracellular matrix, whereas serine collagenases (MW <60 kDa) (Daboor et al., 2010) are probably involved in many physiological functions which include protein digestion, blood-clotting, fibrinolysis, complement activation and fertilization in addition to production of hormones and pharmacologically-active peptides (Neurath, 1984). Collagenases show activities similar to both trypsin and chymotrypsin. Their pH and temperature requirement for optimum activity is 6.5–8.0 and 30°C respectively (Shahidi and Kamil, 2001). All these above mentioned enzymes can be commercially extracted from the fish visceral waste.

Some applications of fish enzymes have been found in caviar production, deskinning of fish and aquatic invertebrates, production of fish sauce, production of fish silage, production of 'maatjes herring,' production of FPH, chymosin in cheese manufacturing, enzymatic extraction of fish oil from raw material, meat tenderizing, etc. However, the economical feasibility for commercial-scale extraction of enzymes from fish byproducts depends on several factors: the amount and quality of the raw materials, the enzyme extraction technology and most importantly, the market for these enzymes. But, the largest potential application seems to be in the field of biotechnology and biochemistry.

13.5.2 NUTRITIONALLY VALUABLE PROTEINS/PEPTIDES WITH BIOFUNCTIONAL PROPERTIES

Proteins of fish wastes could be utilized as ingredients in food and feed industries as valuable nutritional supplements. Recently, number of biofunctional compounds have been identified from fish and shellfish wastes (Kim and Wijesekara, 2010). These bioactive compounds can be isolated and purified with different technologies available of varying degrees of complexity. With the development of new technologies for extraction of newer bioactive compounds from fish wastes and byproducts, the value of waste management would be drastically changed from what is understood today. In this way, the enzymatic processing of fish waste could be found as more beneficial to produce a broad spectrum of useful food ingredients and industrial products for a wide range of their applications (Kim and Park, 2006).

13.5.2.1 FISH PROTEIN HYDROLYSATES (FPH)

Enzymatic hydrolysis of fish wastes and underutilized fish biomass is an ideal method for recovering proteins for a number of products with multiple applications in the field of food and pharmaceuticals. Compared to autolysis, accelerated hydrolysis using commercial proteases allow better control over the end products' properties. FPH present a wide range of potential applications, which include the use in animal feeds, foods or microbial growth media. Fish protein hydrolysates (FPH) contribute considerably to different properties in food systems, which includes water-holding capacity, texture, gelling, foaming, and emulsification. FPH also acts as a potential antioxidant and in addition, provide several health beneficial molecules such as secretagogues, calciotropic hormones, and growth factors (Ravallec-Ple et al., 2001). In addition, protein hydrolysates also exhibit many bioactive properties, which are antioxidative, antihypertensive, antithrombotic, anti-proliferative, immuno-stimulant, and immunomodulatory activities (Kim and Mendis, 2006). Besides organoleptic and functional characteristics, the molecular weight of the peptides is very important as food related functional attributes like solubility, emulsifying capacity and bitterness depend on it. It has been reported that FPH can be utilized as a cryoprotectant to retard denaturation of protein of fish meat during frozen storage, because it suppresses the decrease of unfrozen water, maintain a high gel-forming ability and Ca-ATPase activity (Arvanitoyannis and Kassaveti, 2008).

In case of industrial process, there should be control over the raw material quality and composition including the degree of mincing/cutting, the activity and variety of endogenous enzymes, inactivation of endogenous enzymes, type, and specificity of enzymes including conditions for hydrolysis (amount of added water, time, temperature, pH). Enzymes which are commonly used to prepare hydrolysates from food proteins include trypsin, pepsin, α-chymotrypsin, papain, proteinase K, alcalase, protamex, flavourzyme, and neutrase. Factors influence the hydrolytic process such as enzyme to substrate ratio, temperature, pH, duration of hydrolysis, etc., are important for the yield and production of products with well-defined and desirable properties. However, some hydrolysates often taste bitter and this undesirable taste is one of the significant shortcomings on the way to successful performance of FPH in the market. The bitterness of the FPH may be due to the composition of raw material or the hydrolysis process used. Bitterness is related to the average hydrophobicity of the peptide (Mohr, 1979) as well as the degree of hydrolysis (DH) (Shahidi, 1994). Restricting the DH to values of 3–5% will usually result in a low bitterness. But high DH leading to complete hydrolysis resulting in more free amino acids decreases the bitterness, as hydrophobic peptides are considerably more bitter compared to the mixture of free amino acids. Bitterness may be removed by the application of specific peptidases, such as flavourzyme (Sugiyama et al., 1991), by isoelectric precipitation of bitter peptides (Adler-Nissen, 1984), and by active carbon (Suh et al., 2000). However, the debittering results in reduced yield and reduced nutritional quality of the hydrolysates (Dauksas et al., 2004).

13.5.2.2 BIOACTIVE PEPTIDES

The short peptides obtained by the enzymatic hydrolysis of the fish muscle protein exhibit several nutritional and functional properties and many of which are found as potent biologically active peptides. Short peptides with potent bioactivity remain buried deep inside the parent protein sequence. Generally bioactive peptides are smaller peptides containing 2–20 amino acids (Sarmadi and Ismail, 2010). Bioactive peptides are widely recognized as useful tools in the field of functional food and dietary remedies for health improvement and prevention of chronic health conditions. Expecting a much higher fish production in 2025 (~196 million tons), bioactive peptides from fish especially in the food industry would have great market potential (FAO, 2016). Bioactive peptides has been reported to perform various biological functions such as antioxidative (Sarmadi and Ismail, 2010), antihypertensive (Intarasirisawat et al., 2012), antimicrobial (Najafian and Babji, 2012), immunomodulatory (Chalamaiah et al., 2015), and anticancer (Ahn et al., 2012) effects. Hence, bioactive peptides play a significant role in the development of naturally derived nutraceuticals enriched functional foods which function by preventing and/or providing better management for treatment of human diseases to maintain optimal health. The most extensively researched bioactive peptides are the antihypertensive peptides, particularly which can inhibit the angiotensin-converting enzyme (ACE), involved in the regulation of blood pressure. ACE inhibitory peptides have been reported from several fish species, including shellfish, tuna, bonito, salmon, and sardine (Hai-Lun et al., 2006) and the majority of which were found in the solid fraction of the FPH (Theodore and Kristinsson, 2007).

It has been reported that bioactivity of peptides of low molecular weight are more potent than high molecular weight peptides because they are more easily absorbed by the gastrointestinal tract (Gu et al., 2011; Chi et al., 2015). Several studies suggested that peptides with molecular weight <20 kDa are more effective to perform biofunctional activities. More specifically, peptides with a molecular weight range of 3.5–10 kDa shows the strongest antihypertensive and antioxidative activity. The bioactive peptides that are easily absorbed can be used as functional ingredient in many food items to contribute to water holding capacity, oil absorption, gelling activity, foaming, and emulsification property. The controlled enzymatic hydrolysis of protein from under-utilized fish and fish byproducts is an efficient approach to obtain bioactive peptides. Bioactive peptides from FPH are separated through membrane separation techniques such as nanofiltration, ultrafiltration, and electro-membrane filtration, all of them provide a potentially suitable industrially relevant technology for the enrichment of fish peptides within specific molecular weight ranges (Picot et al., 2010).

13.5.2.3 GELATIN AND COLLAGEN

Bones and skins of the fish waste can be used for the production of gelatin and collagen. Gelatin is derived from collagen through partial hydrolysis, the principal constituent of connective tissue and bones. Although, fish gelatin has been produced since 1960, but till date it contributes only about 1% of the world gelatin production (Karim and Bhat, 2009). Due to unique functional and technological properties, gelatin is used in food, pharmaceuticals, cosmetics, etc. Gelatin especially from warm-water fish can replace pork gelatin for use in the coatings of food and drug, vitamins' microencapsulation and pharmaceutical additives as well as for the production of cosmetics, photographic films and low-set glues. The collagen and gelatin from the bovine sources are associated with the risk of mad cow disease, known as bovine spongiform encephalopathy (BSE), whereas fish skin's collagen and gelatin are devoid of the risks of BSE. Cold-water fish gelatin can be used in frozen or refrigerated foods. Fish waste, i.e., skin is treated with dilute acid or alkali solutions at 45–60°C to extract gelatin (Arnesen and Gildberg, 2007). But the extraction yield of fish gelatin (6–19% of dry gelatin per 100 g clean skin) is lower than mammalian gelatin. Fish gelatin's physical properties are dependent on the amino acid composition, which is species-specific, and also on the molecular weight distribution, which too depend on the extraction conditions. The enzymatically extracted gelatin from fish skin exhibits better biological functions as antioxidants and antihypertensive agents. Unlike peptides derived from fish muscle protein, the unique repeating sequence of gelatin, i.e., glycine-proline-alanine contributes mostly to their antioxidative property (Kim and Mendis, 2006). However, there are some problems associated with the functionalities of fish gelatin. To overcome these, several approaches are suggested, such as enzyme cross-linking using transglutaminase or chemicals (genipin), creating mixed gelling systems with other plant hydrocolloids (Haug et al., 2004).

13.5.3 OTHER VALUABLE PRODUCTS

13.5.3.1 MINERALS

Bones of fish wastes are a rich source of minerals (about 70% on dry wt. basis), where calcium make up around 60%. The bones in the fish processing waste are rich in calcium. To convert bones a fortified food, their hard texture is softened either by hot water treatment or by hot acetic acid solutions or by super-heated steam cooking to convert into edible form (Ishikawa et al., 1990). There is also an interest in the extraction of hydroxyapatite ($Ca_5(PO_4)_3OH$) from fish bones, which is also a source of phosphorous (Ferraro et al., 2010), that can be used in dental applications as bone graft material.

13.5.3.2 BIODIESEL

World is now facing the challenge of increasing demand for energy, finite reserves of fossil fuels and concern regarding pollution and GHG emissions. All of these led to intensive study of alternative fuels in recent years. Although raw fish oil has been found to be a suitable fuel for internal combustion engines, however, emissions of nitric oxides, carbon monoxide, and engine smoke were inferior to an ester fuel (biodiesel) manufactured from the same fish oil. Chemically biodiesel as a mono-alkyl ester of fatty acids, and in conventional method it is obtained by the reaction between triglycerides in the oil with alcohol. First of all oil is extracted from fish wastes through heating at 90°C for 2–3 hr. with continuous agitation followed by pressing. The extracted fish oil is converted to biodiesel through transesterification. This reaction is highly influenced by temperature, reaction time, molar ratio of methanol to oil, catalyst, and free fatty acids (FFAs) content of the oil. In the reaction, triglycerides combines with alcohol (methanol or ethanol @ 9% roughly) in the presence of catalyst (NaOH/KOH) to produce a mixture of biodiesel (mono-alkyl esters of fatty acids) and glycerol. There are several disadvantages to using chemical catalysts, which include high reaction temperature, formation of soap, waste generation and contamination of glycerol with alkali catalysts. Biodiesel can also be produced by enzymatic transesterification using microbial lipases from *Candida antarctica, Carica papaya, Rhizopus oryzae, Pseudomonas cepacia, Pseudomonas fluoresces, Rhizomucor miehei* and *Mucor miehei.* Enzymatic method is more advantageous due to its low temperature requirement, no soap formation, no waste generation and high quality of glycerol. However, some disadvantages of enzymatic transesterification are high reaction times (12–24 h) and high cost of enzymes. The advantage of biodiesel is that it can be directly utilized by vehicles fitted with diesel engines. Presently, biodiesel is given utmost consideration as next generation renewable fuel which can be produced from lipids of plant or animal origin. In future, biodiesel will be able to solve the energy related problems, because it is renewable, biodegradable, nontoxic, and most significantly could be able to reduce the emission of greenhouse gases which leads to global warming.

13.5.3.3 PHARMACEUTICAL AND NUTRACEUTICAL COMPOUNDS

Fish wastes are also sources of several pharmaceutical and nutraceutical compounds. Glucosamine can be extracted from the crustacean wastes. This is present in the exoskeleton of crustaceans as a part of the structural polysaccharides such as chitin and chitosan. Different derivatives of glucosamine such as glucosamine sulfate, glucosamine hydrochloride or N-acetyl-glucosamine has been proven as a promising biomolecule for the treatment of osteoarthritis, knee pain and back pain (Houpt et al., 1999). The anti-cancer, anti-inflammatory, and anti-bacterial effects of glucosamine have also been reported (Nagaoka et al., 2011).

Shark cartilage contains chondroitin sulfate, which is a repeated disaccharide units of glucuronic acid and N-acetyl galactosamine linked by β-(1→3) glycosidic bonds, sulfated in different carbon positions. In many vital biological processes, chondroitin sulfate plays major roles, which are function and elasticity of the articular cartilage, hemostasis, inflammation, cell development, cell adhesion, proliferation, and differentiation by acting as an essential element of the extracellular matrix of connective tissues (Schiraldi et al., 2010).

Hyaluronic acid is present in the fish eyeball and also in their cartilage matrix. It has several biomedical applications, especially, viscosupplementation in osteoarthritis treatment, eye surgery and wound regeneration. Besides these, hyaluronic acid finds its applications in drug delivery, tissue engineering applications, gene delivery applications, targeted drug delivery, tumor treatment, environmental applications and sensors (Mathew et al., 2017).

Squalene, present in shark liver oil, is a triterpenoid compound. It is also an intermediate in cholesterol synthesis. It has gained scientific attention due to diverse biofunctional activities which include antioxidant, anti-lipidemic, membrane stabilizing, cardioprotective, chemopreventive, anti-cancerous, antiaging properties, etc. (Ko et al., 2002). Further, it has the potential to protect human skin surface from oxidation.

13.6 FUTURE MARKET TRENDS

The market potential for products from fish wastes is yet to be fully realized. Fish meal and fish protein products are mainly used as feed and pet food. Even though the fish proteins are proven to have superior nutritional quality index in terms of amino acid composition and bioavailability, too little effort is put for preparation of protein isolate or hydrolysate. However, presently several FPH are commercially available for use in food, especially in American and Asian markets. Presently, there is increasing demand from the health and sports nutrition industry for high quality proteins and peptides, where proteins from fish processing wastes and by-catches could be easily placed. Several bioactive ingredients, namely, chitins, pigments, taurine, squalene, proteoglycans, polyphenols, probiotics, polysaccharides, enzymes, vitamins, and minerals obtained from fish wastes find applications as functional and medicinal supplements, and have already made suitable place in the beverage market. Besides these, enzymes from the fish visceral discards could find suitable applications in the detergent and pharmaceutical industries.

13.7 CHALLENGES AND WAY FORWARD

In near future, judicious utilization of fish wastes through proper implementation of eco-friendly recycling technologies with a view to convert all useful and valuable compounds present in the waste into products beneficial for the mankind would serve not only for reduction of bio-waste intended for proper disposal, but would also significantly move forward the overall economy of the fisheries sector. Nevertheless, there are many hurdles and challenges that must be overcome to achieve this objective (Binsi, 2018):

1. A major challenge is the consumer awareness and education. Without these, no sustainable eco-friendly fish waste utilization and management strategy would succeed. There should be adequate extension efforts from the research and extension organizations.
2. Due to the unorganized nature of this sector, highly scattered nature of fish processing operations which include domestic markets and processing facilities, poses problems in the collection and utilization of wastes.
3. High perishable nature of the fish as well as fish wastes, proper preservation of the raw materials is required for further utilization.
4. Since fish wastes is not categorized as "inactive/inert," therefore, stringent legal and environmental restrictions discourage the entrepreneurs to invest in this resource for valorization.
5. Processors seem to be least interested to invest on the cold chain management for the discards.
6. Due to lack of any baseline data on the availability and economics of valorization of fish wastes over the past years, there is uncertainty about economics and market demand of their recycled products.
7. Lack of clear legal classification of secondary products in the international market poses another major challenge to the investors.
8. Entrepreneurs for valorization of fish wastes are to be encouraged with technical and financial assistanceassistance.
9. Students researches should be initiated on the utilization of fish wastes for product development.
10. Industries like detergent, food, pharmaceuticals, etc., must be encouraged to utilize recycled products from fish wastes.

KEYWORDS

- **bioactive peptides**
- **biodiesel**
- **fish byproducts**
- **fish processing wastes**
- **fish protein hydrolysates**
- **supercritical fluid extraction**
- **switchable solvents**

REFERENCES

Adler-Nissen, J., (1984). Control of the proteolytic reaction and of the level of bitterness in protein hydrolysis processes. *J. Che. Technol. Biotechnol., 34B*, 215–222.

Ahn, C. B., Je, J. Y., & Cho, Y. S., (2012). Antioxidant and anti-inflammatory peptide fraction from salmon byproduct protein hydrolysates by peptic hydrolysis. *Food Res. Int., 49*(1), 92–98.

Arnesen, J. A., & Gildberg, A., (2007). Extraction and characterization of gelatin from Atlantic salmon (*Salmo salar*) skin. *Biores. Technol., 98*, 53–57.

Arvanitoyannis, I. S., & Kassaveti, A., (2008). Fish industry waste: Treatments, environmental impacts, current and potential uses. *Int. J. Food Sci. Technol., 43*, 726–745.

Arvanitoyannis, I., Nakayama, A., & Aiba, S., (1998). Chitosan and gelatin based edible films: State diagrams, mechanical and permeation properties. *Carbohy. Poly., 37*, 371–382.

Binsi, P. K., (2018). *Overview of Waste Generation in Fish and Shellfish Processing Industry*. ICAR-Central Institute of Fisheries Technology, Cochin.

Bonsdorff, E., Blomqvist, E. M., Mattila, J., & Norkko, A., (1997). Coastal eutrophication: Causes, consequences and perspectives in the Archipelago areas of the northern Baltic Sea. *Estuarine, Coastal Shelf Sci., 44*(Suppl A), 63–72.

Chalamaiah, M., Hemalatha, M. D., Jyothirmayi, T., Diwan, P. V., Bhaskarachary, K., Vajreswari, A., Ramesh, K. R., & Dinesh, K. B., (2015). Chemical composition and immunomodulatory effects of enzymatic protein hydrolysates from common carp (*Cyprinus carpio*) egg. *Nutrition, 31*, 388–398.

Chamberlain, A., (2011). *Fishmeal and Fish Oil: The Facts, Figures, Trends, and IFFO's Responsibility Supply Standard.* International Fishmeal and Fish Oil Organization.

Chi, C. F., Wang, B., Hu, F. Y., Wang, Y. M., Zhang, B., Deng, S. G., & Wu, C. W., (2015). Purification and identification of three novel antioxidant peptides from protein hydrolysate of bluefin leatherjacket (*Navodon septentrionalis*) skin. *Food Res. Int., 73*, 124–129.

Daboor, S. M., Budge, S. M., Ghaly, A. E., Brooks, S., & Dave, D., (2010). Extraction and purification of collagenase enzymes: A critical review. *Ame. J. Bioche. Biotechnol., 6*(4), 239–263.

Dauksas, E., Slizyte, R., Rustad, T., & Storro, I., (2004). Bitterness in fish protein hydrolysates and methods for removal. *J. Aqua. Food Prod. Technol., 13*, 101–114.

Erdmann, K., Cheung, B. W. Y., & Schroder, H., (2008). The possible roles of food-derived bioactive peptides in reducing the risk of cardiovascular disease. *J. Nutr. Biochem., 19*, 643–654.

Esteban, M. B., García, A. J., Ramos, P., & Márquez, M. C., (2007). Evaluation of fruit vegetable and fish wastes as alternative feedstuffs in pig diets. *Waste Manag., 27*, 193–200.

FAO, (2016). Global aquaculture production (fish stat). Protein hydrolysates prepared from alkaline-aided channel catfish protein isolate. *J. Sci. Food Agric., 87*(12), 2353–2357. http://www.fao.org/3/a-i5555e.pdf (accessed on 09 March 2022).

FAO, (2020). *The State of World Fisheries and Aquaculture 2020: Sustainability in Action.* Rome. https://doi.org/10.4060/ca9229en (accessed on 09 March 2022).

Ferraro, V., Cruz, I. B., Jorge, R. F., Malcata, F. X., Pintado, M. E., & Castro, P. M. L., (2010). Valorisation of natural extracts from marine source focused on marine byproducts: A review. *Food Res. Int., 43*, 2221–2233.

FOC, (2012). *The State of World Fisheries and Aquaculture.* Fisheries and Aquaculture Department. Food and Agriculture Organization of the United Nations, Rome, Italy.

Ghaly, A. E., Ramakrishnan, V. V., Brooks, M. S., Budge, S. M., & Dave, D., (2013). Fish processing wastes as a potential source of proteins, amino acids and oils: A critical review. *J. Microb. Biochem. Technol., 5*, 107–129. doi:10.4172/1948-5948.1000110.

Gu, R. Z., Li, C. Y., Liu, W. Y., Yi, W. X., & Cai, M. Y., (2011). Angiotensin I-converting enzyme inhibitory activity of low molecular–weight peptides from Atlantic salmon (*Salmo salar* L.) skin. *Food Res. Int., 44*(5), 1536–1540.

Hai-Lun, H., Xiu-Lan, C., Cai-Yun, S., Yu-Zhong, M., & Bai-Cheng, Z., (2006). Analysis of novel angiotensin-I-converting enzyme inhibitory peptides from protease-hydrolyzed marine shrimp *Acetes chinensis*. *J. Peptide Sci., 12*, 726–733.

Haug, I. J., Draget, K. I., & Smidsrød, O., (2004). Physical and rheological properties of fish gelatin compared to mammalian gelatin. *Food Hydrocol., 18*, 203–213.

Houpt, J. B., McMillan, R., Wein, C., & Paget-Dellio, S. D., (1999). Effect of glucosamine hydrochloride in the treatment of pain of osteoarthritis of the knee. *The J. Rheumatol., 26*(11), 2423–2430.

Intarasirisawat, R., Benjakul, S., Visessanguan, W., & Wu, J., (2012). Antioxidative and functional properties of protein hydrolysate from defatted skipjack (*Katsuwonous pelamis*) roe. *Food Che., 135*(4), 3039–3048.

Ishikawa, M., Kato, M., Mihori, T., Watanabe, H., & Sakai, Y., (1990). Effect of vapor pressure on the rate of softening of fishbone by super-heated steam cooking. *Nippon Suisan Gakkaishi, 56*, 1687–1691.

Karim, A. A., & Bhat, R., (2009). Fish gelatin: properties, challenges, and prospects as an alternative to mammalian gelatins. *Food Hydrocol., 23*, 563–576.

Kim, J. S., & Park, J. W., (2006). Mince from seafood processing byproducts and surimi as food ingredients. In: *Maximizing the value of Marine By-Products* (pp. 246–255). Woodhead Publishing Cambridge, London, UK.

Kim, S. K., & Mendis, E., (2006). Bioactive compounds from marine processing byproducts: A review. *Food Res. Int., 39*, 383–393.

Kim, S. K., & Wijesekara, I., (2010). Development and biological activities of marine-derived bioactive peptides: A review. *J. Func. Foods, 2*, 1–9.

Kishimura, H., & Hayashi, K., (2002). Isolation and characteristics of trypsin from pyloric ceca of the starfish *Asterina* pectinifera. *Comp. Biochem. Physiol. Part B: Biochem. Mol. Biol., 132*, 485–490.

Ko, T. F., Weng, Y. M., & Chiou, R. Y. Y., (2002). Squalene content and antioxidant activity of *Terminalia catappa* leaves and seeds. *J. Agricultl. Food Che., 50*(19), 5343–5348.

Mathew, S., Tejpal, C. S., Kumar, L. R., Zynudheen, A. A., & Ravishankar, C. N., (2017). Aquaceuticals for developing high value noble foods and dietary supplements. *Ind. J. Agricul. Biochem., 30*(1), 1–9.

Mohr, V., (1979). Enzymes technology in the meat and fisheries industries. In: *The International Microbiology and Food Industry Congress*. Paris.

Nagaoka, I., Igarashi, M., Hua, J., Ju, Y., Yomogida, S., & Sakamoto, K., (2011). Recent aspects of the anti-inflammatory actions of glucosamine. *Carbohy. Poly., 84*(2), 825–830.

Nair, K. G. R., Mathew, P. T., Joseph, A., Xavier, M., & Krishna, A., (2004). Livestock feed from cephalopod processing waste. In: *Report of National Agricultural Technology Project on Development of Value Added Products and Byproducts from Low Cost Fish and Processing Waste from Fish and Shell Fish* (pp. 1–11). Central Institute of Fisheries Technology, Cochin, India.

Najafian, L., & Babji, A. S., (2012). A review of fish-derived antioxidant and antimicrobial peptides: Their production, assessment, and applications. *Peptides, 33*(1), 178–185.

Nasri, R., Abed, H., Karra-Chaabouni, M., Nasri, M., & Bougatef, A., (2015). Digestive alkaline proteinases from *Serranus scriba* viscera: Characteristics, application in the extraction of carotenoproteins from shrimp waste, and evaluation in laundry commercial detergents. *Biocat. Agricult. Biotechnol., 4*(3), 355–361.

Neurath, H., (1984). Evolution of proteolytic enzymes. *Science, 224*(4647), 350–357.

Niemi, S., Vauhkonen, V., Hiltunen, et al., (2009). Results of an off-road diesel engine driven with different animal fat based biofuels. *ASME 2009 Internal Combustion Engine Division Fall Technical Conference* (pp. 27–30). Lucerne, Switzerland.

Picot, L., Ravallec, R., Fouchereau-Pe´ron, M., Vandanjon, L., Jaouen, P., Chaplain-Derouiniot, M., Gue´rard, F., et al., (2010). Impact of ultrafiltration and nanofiltration of an industrial fish protein hydrolysate on its bioactive properties. *J. Sci. Food Agric., 90*, 1819–1826.

Ravallec-Ple,´ R., Charlot, C., Pires, C., Braga, V., Batista, I., Van, W. A., et al., (2001). Presence of bioactive peptides in hydrolysates prepared from enzymatic processing waste of sardine (*Sardina pilchardus*). *J. Sci. Food Agric., 81*(11), 1120–1125.

Riegman, R., (1995). Nutrient-related selection mechanisms in marine plankton communities and the impact of eutrophication on the plankton food web. *Water Sci. Technol., 32*, 63–75.

Sarmadi, B. H., & Ismail, A., (2010). Antioxidative peptides from food proteins: A review. *Peptides, 31*(10), 1949–1956.

Schiraldi, C., Cimini, D., & De Rosa, M., (2010). Production of chondroitin sulfate and chondroitin. *Appl. Microbiol. Biotechnol., 87*(4), 1209–1220.

Shahidi, F., & Kamil, Y. V. A. J., (2001). Enzyme from fish and aquatic invertebrates and their applications in the food industry. *Trends Food Sci. Technol., 12*, 435–464.

Shahidi, F., (1994). Proteins from seafood processing discards. In: Sikorski, Z. E., Sun, P. B., & Shahidi, F., (eds.), *Seafood Proteins* (pp. 171–193). Chapman and Hall, New York.

Shahidi, F., (1994). Seafood processing byproducts. In: Shahidi, F., & Botta, J. R., (eds.), *Seafoods: Chemistry, Processing Technology and Quality* (pp. 320–330). Blackie Academic & Professional, London.

Sugiyama, K., Egawa, M., Onzuka, H., & Oba, K., (1991). Characteristics of sardine muscle hydrolysates prepared by various enzymatic treatments. *Nippon Suisan Gakkaishi, 57*, 475–479.

Suh, H. J., Bae, S. H., & Noh, D. O., (2000). Debittering of corn gluten hydrolysate with active carbon. *J. Sci. Food Agric., 80*, 614–618.

Theodore, A. E., & Kristinsson, H. G., (2007). Angiotensin-converting enzyme inhibition of fish protein hydrolysates prepared from alkaline-aided channel catfish protein isolate. J. Sci. Food Agric., *87*(12), 2353–2357.

Turid, R. T., Storrø, I., & Slizyte, R., (2011). Possibilities for the utilization of marine byproducts. *Int. J. Food Sci. Technol., 46*, 2001–2014.

Udenigwe, C. C., & Aluko, R. E., (2012). Food protein-derived bioactive peptides: Production, processing, and potential health benefits. *J. Food Sci., 77*(1), 11–24.

Zaitsev, Y. P., (1992). Recent changes in the tropic structure of the Black Sea. *Fish. Oceanograp.*, 180–189.

Zhou, L., Budge, S. M., Ghaly, A. E., Brooks, M. S., & Dave, D., (2011). Extraction, purification and characterization of fish chymotrypsin: A review. *Ame. J. Bioche. Biotechnol., 7*, 104–123.

CHAPTER 14

Production of Omega-3 Capsules from Fish Offal: Recycling of Resources for Sustainable Production and Consumption

BIMAL PRASANNA MOHANTY,[1,2] SATABDI GANGULY,[1] ARABINDA MAHANTY,[1,3] TANDRIMA MITRA,[1] SASMITA MOHANTY,[4] and BASANTA KUMAR DAS[1]

[1]*ICAR-Central Inland Fisheries Research Institute, Barrackpore, Kolkata – 700120, West Bengal, India*

[2]*Fisheries Science Division – ICAR, KAB II, Pusa, New Delhi – 110012, India*

[3]*Crop Protection Division, ICAR – National Rice Research Institute, Cuttack – 753006, Odisha, India*

[4]*Faculty of Science and Technology, Department of Biotechnology, Rama Devi Women's University, Bhubaneswar – 751022, Odisha, India*

ABSTRACT

Fish is a superfood rich in quality animal proteins, heart-, and brain-friendly omega (ω)-3 polyunsaturated fatty acids (PUFAs) and micronutrients. Fish is playing a big role and is going to play still a much bigger role in global food and nutritional security. A substantial amount of natural resources as inputs are spent towards fish production; however, a considerable amount of edible and inedible offal are lost in post-harvest processing. Sustainable development and economic growth require minimization of ecological

Advances in Fish Processing Technologies: Preservation, Waste Utilization, and Safety Assurance. Ranendra K. Majumder, PhD & Amjad K. Balange, PhD (Eds.)

footprint by modifying the manner in which goods and resources are produced and consumed in a responsible manner. Some of the offal thrown away as fish waste could be responsibly and intelligently utilized for health products thereby generating food supplements and health products at the same time keeping the environment clean and pollution-free. Thus, efficient management of the fish wastes, through recycling and reuse could reduce ecological footprints and have positive environmental impacts. Thus, the Sustainable Development Goals (SDGs) program of the United Nations, specifically Goal 12 – Responsible Production and Consumption, is well addressed through substantially reducing waste generation through prevention, reduction, recycling, and reusing the resources treated as waste. Keeping such issues in the backdrop, the present chapter highlights the utilization of fish wastes and offal from those species-rich in oils, especially PUFAs, for production of ω-3 capsules for human use as food supplement.

14.1 INTRODUCTION

Natural resources are limited; therefore, goods and resources need to be produced and consumed wisely and responsibly to keep a check on depletion of natural resources while ensuring sustainable development. Agriculture is the largest user of natural resources like water and land. At the same time, it also produces a huge amount of wastes. The efficient management of the wastes, through recycling and reuse could reduce ecological footprints and have positive environmental impacts. Thus, the Sustainable Development Goals (SDGs) program of the United Nations has put considerable emphasis on substantially reducing waste generation through prevention, reduction, recycling, and reusing by the year 2030 (Goal 12 – Responsible Production and Consumption) (World Bank, 2018) (Figure 14.1). This can be achieved through encouraging industries, businesses, and consumers to recycle and reduce the waste. A large share of the global population is consuming very little to meet their basic needs, whereas a major share of affluent population are wasting a lot. Drastically bringing down the per capita of global food waste is necessary for creating efficient production and supply chains that could lead towards food security, and a resource-efficient economy (https://www.undp.org/content/undp/en/home/sustainable-development-goals/goal-12-responsible-consumption-and-production.html).

FIGURE 14.1 Sustainable development goals (World Bank, 2018).

14.2 FOOD SECURITY

Food security, in simple terms, is a measure of the availability of food and individuals' ability to access it. Food security has been defined by the United Nations as a condition where "all people, at all times, have physical, social, and economic access to sufficient, safe, and nutritious food that meets their food preferences and dietary needs for an active and healthy life." Thus, food security has four dimensions: (a) food availability, (b) economic and physical access to food, (c) food utilization and (d) stability over time (FAO, IFAD, and WFP, 2015; https://www.ifpri.org/). Food insecurity is really a big concern in the developing and under developed world.

Food production needs to be increased significantly to meet the demands of an ever increasing world population. However, there is also a need for simultaneous reduction in food waste which would increase the efficiency of the whole food chain in an environmentally sustainable way. One-third of food produced, which roughly amounts to 1.3 billion tons of food is lost or wasted, and if managed properly, this could substantially contribute to global food security. Food loss in high- and middle-income countries is related mainly to consumer behavior, whereas food loss in low-income countries is often caused by logistical limitations in the food supply chain (FAO, 2011; World Bank, 2018).

There is less availability of resources like water, land, labor force, whereas the global population is increasing exponentially. Therefore, sustainable production and consumption in a responsible manner is need of the time. To feed the world sustainably, there is need for higher food production while keeping the environmental impacts such as soil, water, and nutrient loss, emission of greenhouse gases and ecosystems degradation under control. There is a need to shift to nutritious and safe diets while keeping the environmental footprints optimal. FAO has a greater role to play in coordinating global initiatives, activities, and projects on food losses and waste reduction in partnership with other organizations like UN agencies and other international organizations, private sector and civil society (http://www.fao.org/sustainable-development-goals/goals/goal-12/en/).

14.3 ω-3 POLYUNSATURATED FATTY ACIDS FROM FISH WASTES

Fish is a superfood, rich in proteins, heart-friendly ω-3 polyunsaturated fatty acids (PUFAs), and micronutrients (minerals and vitamins) (Mohanty, 2010, 2020; Mohanty et al., 2020). Fish, especially marine fishes are high in ω-3 fatty acids, which are heart- and brain-friendly and are conjectured to be a major contributing factor for the reduced risk of cardiovascular diseases (CVD) in Eskimos (Bang et al., 1976). Similarly, the longer lifespan of Japanese and Nordic populations are also attributed to their higher consumption of fish and seafood, besides other factors. Because of associated health benefits and advantages like easy digestibility, availability, and affordability, the demand for fish is ever increasing, so as its production. However, in between the demand-supply chain, what is often missed out is, the amount of fish wastes or inedible offal being generated. The fish waste constitutes the inedible parts of the fish such as intestine, scale, gills, etc. The offal could account for about 20–40% total biomass of fish depending upon the type of fish and the level of processing. The offal generated is either discarded as waste or is utilized as low value by-products such as fishmeal.

One of the feasible modes of utilizing the fish offal could be the production of ω-3 capsules or supplements. ω-3 PUFAs, especially, eicosapentaenoic acid (EPA, 20:5) and docosahexaenoic acid (DHA, 22:6) play important roles in keeping away several preventable human ailments coronary heart disease, atherosclerosis in adults, childhood asthma and attention-deficit hyperactivity disorder (ADHD) in pediatric population; and dementia, age-related macular degeneration (AMD) and Alzheimer's disease (AD) in the elderly population (Mohanty et al., 2016). The total oil (fat) content in fishes from different habitats in the Indian subcontinent showed that, the anadromous clupeid *Tenualosa ilisha* contains the highest amount of oil (10.5%)

followed by the oil sardines (*Sardinella longiceps* (9.2%). Depending upon the biology, habitat, and feeding behavior, the oil content of fish varies from species to species; accordingly these could be categorized as high fat fish, medium fat fish and low fat fish. The fishes categorized according to their oil content are listed in Table 14.1.

TABLE 14.1 Classification of Some Indian Food Fish Species based on Oil (fat) Content

Fat Content	Common Indian Food Fish Species
High fat fish (>8% fat)	*T. ilisha, S. longiceps*
Medium fat fish (4–8% fat)	*A. testudineus, G. chapra, N. japonicas, P. sophore, E. suratensis* *M. rosenbergii, A. mola, T. putitora, P. paradiseus, O. belangeri*
Low fat fish (2–4% fat)	*L. splendens, O. mykiss, C. batrachus, T. lepturus, M. cephalus, N. hexagonolepsis, C. carpio, C. catla, C. mrigala, C. madrasensis, H. fossilis*
Lean meat fish (<2% fat)	*E. affinis, Ailia coila, P. virdis, R. kanagurta, R. rita, Epinephelus* Spp., *S. seenghala, F. indicus, P. monodon, T. albacores, L. gonius, J. coitor, L. parsia*

The fatty acid fingerprints by GC/MS showed that *Tenualosa ilisha*, *Sardinella longiceps*, *Harpadon nehereus*, *Nemipterus japonicas*, *Stolephorus waitei*, *Rastralliger kanagurta* (marine fishes) and *Oncorhynchus mykiss*, *Neolissochilus hexagonolepsis* (coldwater fishes) are rich in ω-3 PUFAs, EPA and DHA (Mohanty et al., 2016, 2019) (Table 14.2). Koi (*Anabas testudenius*), pengwa (*Osteobroma belangiri*) are some of the freshwater fishes that are rich in the ω-3 PUFAs. Moreover, deficiency of ω-3 PUFAs has been a major nutritional challenge of the developing and underdeveloped countries. Supply of ω-3 fatty acids decrease significantly with decreasing gross domestic product (GDP) and in some of countries with the lower GDP, total ω-3 PUFAs are below or very close to the lower end of recommended intake range. There is huge demand for omega-3 products and the monetary value for such product has been estimated to be USD 2.29 billion in 2018, globally, and is expected to grow at a Compound annual growth rate (CAGR) of 7.4% from 2019 to 2025 (Grandviewresearch, 2019). ω-3 PUFAs are associated with many health benefits including brain health, memory, and cognition; heart health, prevention of CVD; vision, prevention of AMD and thus could greatly help in achieving SDG Goal 3 – Good health and well-being (World Bank, 2018).

The chief sources of ω-3 fatty acids are oil extracted from the liver and muscle of oily fishes. However, if the waste being generated from fishes could be utilized for extracting ω-3 fatty acids, it could have multifarious advantages. For instance, we carried out to judge the feasibility of the approach taking oil rich fish *Tenualosa ilisha* (hilsa) fish waste (gill and liver) (Mohanty et al., 2012, 2016). Total fat content and fatty acid composition analysis was carried out by GC/MS. Oil was extracted using solvent extraction method.

TABLE 14.2 DHA, EPA, PUFA, and Crude Fat Content of Commercially Important Food Fishes (in %)

Fish Species	Habitat	DHA	EPA	PUFA	Crude Fat
Rastralliger kanagurta	Marine	28.52 ± 5.6	5.2 ± 1.2	40.1	1.7 ± 0.0
Nemipterus japonicus	Marine	26.5 ± 1.6	6.6 ± 1.3	39.8	5.1 ± 0.0
Stolephorus waiti	Marine	23.2 ± 8.7	5.6 ± 1.3	39.5	1.1 ± 0.0
Harpodon nehereus	Marine	20.26 ± 3.3	7.97 ± 2.3	38.07	2.2 ± 0.2
Trichiurus lepturus	Marine	12.24 ± 2.9	4.38 ± 1.0	23.26	3.4 ± 4.1
Fenneropenaeus indicus	Shellfish	10.05 ± 1.0	10.65 ± 3.0	35.6	0.7 ± 0.4
Perna virdis	Shellfish	9.5 ± 2.1	10.21 ± 4.5	28.57	1.7 ± 0.0
Ailia coila	Freshwater	9.3 ± 0.1	–	36.6	1.8 ± 0.0
Tenualosa ilisha	Freshwater	8.9 ± 2.5	2.9 ± 0.9	22.0	10.5 ± 4.6
Thunnus albacares	Marine	8.3 ± 2.3	3.0 ± 0.2	28.4	0.6 ± 0.0
Epinephelus spp.	Marine	8.17 ± 2.1	3.64 ± 1.1	19.77	0.9 ± 0.5
Crassostrea madrasensis	Shellfish	7.43 ± 2.6	7.35 ± 2.1	28.86	2.7 ± 0.2
Leignathus splendens	Marine	7.16 ± 1.3	7.13 ± 1.2	23.13	3.8 ± 3.7
Sardinella longiceps	Marine	6.99 ± 1.5	12.35 ± 1.3	26.78	9.2 ± 5.8
Oncorhynchus mykiss	Coldwater	6.4 ± 1.6	2.3 ± 0.6	31.4	3.8 ± 0.1
Penaeus monodon	Shellfish	6.39 ± 1.4	12.88 ± 1.5	38.4	0.7 ± 0.2
Sperata seenghala	Freshwater	6.2 ± 1.3	4.4 ± 0.9	35.5	0.8 ± 0.4
Mugil cephalus	Brackish water	6.15 ± 0.1	5.79 ± 0.3	17.9	3.3 ± 0.7
Gudusia chapra	Freshwater	6.0 ± 0.5	–	31.8	5.7 ± 0.0
Stolephorus commersoni	Marine	5.8 ± 1.9	1.6 ± 0.3	13.1	1.2 ± 0.0
Neolissochilus hexagonolepsis	Coldwater	5.2 ± 1.1	7.4 ± 2.3	31.2	3.3 ± 0.1

TABLE 14.2 *(Continued)*

Fish Species	Habitat	DHA	EPA	PUFA	Crude Fat
Lates calcarifer	Brackishwater	5.14 ± 0.5	6.28 ± 0.5	25.5	2.6 ± 0.5
Cyprinus carpio	Coldwater	5.1 ± 1.9	–	23.7	3.0 ± 0.0
Ethynnus affinis	Marine	5.0 ± 1.6	3.0 ± 0.5	12.7	1.9 ± 0.0
Rita rita	Freshwater	5.0 ± 0.6	3.8 ± 0.6	11.4	–
Catla catla	Freshwater	4.7 ± 0.9	6.8 ± 1.2	31.9	2.8 ± 0.3
Schizothorax richardsonii	Coldwater	3.8 ± 1.2	9.6 ± 2.3	19.4	2.5 ± 0.0
Puntius sophore	Freshwater	3.3 ± 1.1	6.2 ± 1.2	42.9	4.9 ± 0.5
Amblypharyngodon mola	Freshwater	3.1 ± 1.1	2.2 ± 0.9	24.8	4.3 ± 0.0
Anabas testudineus	Freshwater	2.7 ± 0.6	0.4 ± 0.0	23.7	6.9 ± 0.6
Tor putitora	Coldwater	2.7 ± 0.9	4.7 ± 1.2	18.3	4.3 ± 0.1
Heteropneustes fossilis	Freshwater	2.2 ± 0.6	1.5 ± 0.2	14.2	2.7 ± 0.5
Clarius batrachus	Freshwater	0.47 ± 0.8	–	25.49	3.7 ± 0.4
Labeo rohita	Freshwater	0.4 ± 0.0	0.9 ± 0.1	22.5	2.7 ± 0.2
Cirrhinus mrigala	Freshwater	–	1.50 ± 0.3	43.55	2.8 ± 0.3
Katsuwonus pelamis	Marine	–	5.11 ± 1.9	12.3	1.2 ± 1.1
Macrobrachium rosenbergii	Shellfish	–	7.42 ± 2.1	35.21	4.4 ± 0.2

Source: Mohanty et al. (2016).

Fractionation of EPA and DHA from crude fish oil extracted from fish waste involved refinement of crude fish oil, ethyl esterification and separation through silica gel chromatography followed by molecular distillation (Figure 14.2). Employing these methods, 1 kg of liver and 1 kg of gill tissue yielded 80 g and 70 g fat, respectively. If the results are extrapolated, an estimated annual landing of hilsa of 21,901 metric tons in India (2011 report) can yield 21024.96 Kg and 42,049 kg oil from liver and gill tissues, respectively (Ganguly et al., 2018; Mohanty et al., 2019).

The advantages of producing ω-3 fatty acids from fish could be summarized as follows:

- Cost of raw material used would be low which in turn would lower the cost of production making it more affordable especially in the developing countries;
- Greater financial benefits to fish traders, fish vendors;
- Reduce the volume of waste and the environmental pollution associated with it;

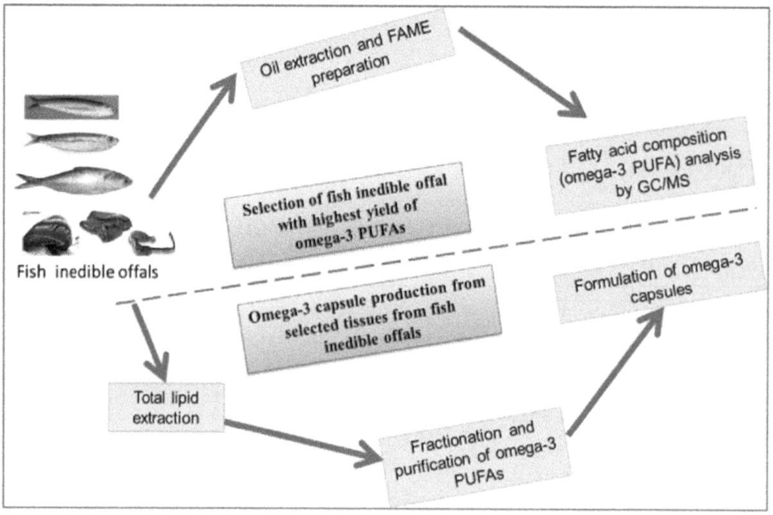

FIGURE 14.2 Schematic representation of production of omega-3 capsules from fish offal.

- Could possibly contribute to biodiversity conservation; no additional fish catch being involved in the process of ω-3 fatty acids and would also serve as a classic example of 'Waste to Wealth' concept.

There has been a number of reports on describing the process of extraction of ω-3 fatty acids from fish offal (Fiori et al., 2017; Simat et al., 2019); however, examples of large scale production of these fatty acid from offal is scanty which could be attributed to two important reasons; the first one is lack of a system for large scale collection of the offal from fish markets or any other such sources on a regular basis and second one being the infrastructure for oil extraction and product development. Addressing these two issues would certainly help in the development of the industry.

ACKNOWLEDGMENTS

This work was presented for Biotechnology Ignition Grant (BIG) of Biotechnology Industry Research Assistance Council (BIRAC), Department of Biotechnology, Ministry of Science and Technology, Government of India under BIRAC/KIIT0175/BIG-10/17 and BIRAC/cCAMP0658/BIG-12/18. The encouragement and support from ICAR-Central Inland Fisheries Research Institute, Barrackpore is thankfully acknowledged.

KEYWORDS

- **docosahexaenoic acid**
- **fish offal**
- **fish waste**
- **omega-3 capsules**
- **recycling of resources**
- **sustainable development goal**
- **sustainable production and consumption**

REFERENCES

Bang, H. O., Dyerberg, J., & Hjoorne, N., (1976). The composition of food consumed by Greenland Eskimos. *Acta medica Scandinavica, 200*, 69–73.

FAO, (2011). *Global Food Losses and Food Waste-Extent, Causes and Prevention*. Rome.

FAO, IFAD and WFP, (2015). The state of food insecurity in the world 2015. *Meeting the 2015 International Hunger Targets: Taking Stock of Uneven Progress*. Rome, FAO. https://www.ifpri.org. Food Security | IFPRI: International Food Policy Research Institute (accessed on 09 March 2022).

Fiori, L., Volpe, M., Lucian, M., Anesi, A., Manfrini, M., & Guella, G., (2017). From fish waste to omega-3 concentrates in a biorefinery concept. *Waste Biomass Valor*. doi: 10.1007/s12649-017-9893-1.

Ganguly, S., Mahanty, A., Mitra, T., Mohanty, S., Das, B. K., & Mohanty, B. P., (2018). Nutrigenomic studies on hilsa to evaluate flesh quality attribute and genes associated with fatty acid metabolism from rivers Hooghly and Padma. *Food Res. Int., 103*, 21–29.

Grandview Research, (2019). https://www.grandviewresearch.com/industry-analysis/omega-3-market/request/rs1 (accessed on 09 March 2022).

Mohanty, B. P., (2010). Fish as health food. In: Ayyappan, S., Moza, U., Meenakumari, B., Jena, J. K., & Pandey, A. K., (eds.), *Handbook of Fisheries and Aquaculture* (Vol. 2, pp. 843–861). ICAR-DKMA: New Delhi.

Mohanty, B. P., (2020). Pradhan Mantri Matsya Sampada Yojana-a boon for the Indian fisheries and aquaculture sector, editorial. *J. Inland Fish. Soc. India, 52*(2), 117–120.
Mohanty, B. P., Ganguly, S., Karunakaran, D., Chakraborty, K., Sharma, A. P., Mohapatra, P. K. R., & Nayak, N. R., (2012). Maternal fish consumption and prevention of low birth weight in the developing world. *Nat. Acad. Sci. Let., 35*(5), 433–438.
Mohanty, B. P., Mahanty, A., Ganguly, S., Mitra, T., Karunakaran, D., & Anandan, R., (2019). Nutritional composition of food fishes and their importance in providing food and nutritional security. *Food Chem., 293*, 561–570.
Mohanty, B. P., Mohanty, S., & Mahanty, A., (2020). Opinion: Malnutrition-free India. *RDWU Bull. Sci., 2*, 7–11.
Mohanty, B. P., Sankar, T. V., Ganguly, S., Mahanty, A., Anandan, R., Chakrabarty, K., et al., (2016). DHA, EPA content and fatty acid profile of 39 food fishes from India. *Biomed. Res. Int.* doi: 10.1155/2016/4027437.
Simat, V., Vlahovic, J., Soldo, B., Skroza, D., Ljubenkov, I., & Mekinic, I. G., (2019). Production and refinement of omega-3 rich oils from processing by-products of farmed fish species. *Foods, 8*(125), 14.
World Bank, (2018). *Atlas of Sustainable Development Goals 2018: World /Development Indicators.* Washington, DC: World Bank. doi: 10.1596/978-1-4648-1250-7.

CHAPTER 15

Bioactive and Functional Nutraceuticals from Shrimp Processing Wastes

K. A. MARTIN XAVIER and KASTURI CHATTOPADHYAY

Department of Post-Harvest Technology, Fishery Resource Harvest and Postharvest Management Division, ICAR-Central Institute of Fisheries Education, Versova, Mumbai – 400061, Maharashtra, India

ABSTRACT

Exploration of processing waste as a source of secondary raw material for the development of functional and nutraceutical products are increasing over the years. Wastes from the processing operations of crustaceans dumped by the processing industries are one of the alarming environmental concerns in terms of its massive disposal rate. However, these wastes possess a huge reserve of unique bioactive compounds especially lipids (shrimp oils), polysaccharides (chitin, chitosan, glucosamine), carotenoproteins, minerals, various enzymes (protease, lipase), pigments (astaxanthin) which have immense marketable efficacy in the export market as new functional foods, nutraceuticals, and dietary supplements in promoting consumer health and nutrition. Hence, effective extraction and recovery of these shellfish processing wastes is required in order to minimize the impact of environmental hazards and to explore their potency as an effective nutraceutical. The nutraceuticals discussed in this chapter are characterized as polymers (chitin, chitosan, glucosamine hydrochloride), protein (enzymes, protein hydrolysate, shrimp paste and sauces) and lipid-based (Shrimp oils and their components) on the material used from the shrimp processing wastes.

Advances in Fish Processing Technologies: Preservation, Waste Utilization, and Safety Assurance.
Ranendra K. Majumder, PhD & Amjad K. Balange, PhD (Eds.)

15.1 INTRODUCTION

Modern-day era revolves around new health paradigms. The present lifestyles have deprived everyone of healthy food intakes like consumption of junk foods, etc. This severely hampers the human nutrition leading to increased health issues like osteoporosis, arthritis, cancer, obesity, diabetes globally. Treatments of these health disorders with expensive medications and high tech technologies have annoyed the present population. So intake of supplements from harmless sources can be a preventive measure against any disease or physiological deviations. Nutraceuticals from natural bioderivates are emerging as effective substitutes that can potentially bridge the gap between recent foods and drugs. In the year 1989, Dr. Stephen De Felice coined the word "nutraceutical." Nutraceuticals are defined as any component which can be considered as a whole food or a part of food that imparts medical or health benefits, including treatment and prevention of diseases.

Arthropod is the largest phylum in the biome and crustaceans constitutes a major group among its which thrives around a diverse habitat. Among them, prawns, shrimp, lobsters, and crabs form a large economy in the seafood sectors where these are processed and exported globally. Currently, shrimp is a high value exported commodity worldwide due to its delicacy wide consumer preferences. They are marketed both in processed and raw forms. This has directed shrimp-based aquaculture to acquire popularity in the international export trade. As per FAO, global production of shrimp is estimated at around 5 million metric tons (MMT) from which *Penaeus vannamei* (white shrimp) itself contributes for around 53% of the total crustacean trade (FAO, 2020). Vannamei culture has attracted the farmers to choose it as a profitable business venture due to its fast-growing nature and outstanding disease resistivity. It is exported in different processed forms like headless (HL), peeled tail on (PTO), peeled un-deveined (PUD) and peeled deveined (PD) forms. But prior to exporting as a commodity, shrimps are thoroughly processed by eliminating its non-edible parts like carapace, tail, cephalothoraxes, etc. Carapaces solely account for about 40–50% of total weight that has negligible market value and is thrown off.

Recently, wastes produced from shrimp processing industries is one of the largest industrial wastes in India which is a also an alarming environmental concern. Byproducts from seafoods are usually either thrown away directly into the open sea, or landfilled or are burned together causing deleterious environmental effects. Similarly, wastes produced from aquaculture farms are also dumped back as feeds for the raised species. In general, approximately 45–75% of the gross weight of all major seafood varieties like shrimp, lobster, krill, cray fish and crabs are disposed off as wastes or byproducts which are a source of huge environmental hazard.

However, these wastes possess a huge reserve of unique bioactive compounds, especially lipids (shrimp oils), polysaccharides (chitin, chitosan, glucosamine), carotenoproteins, minerals, various enzymes (protease, lipase), pigments (astaxanthin) which have immense marketable efficacy in the export market. Studies also reported the presence of 30–40% proteins, 30–50% minerals (mostly calcium carbonate), and 20–30% chitin from different shellfish wastes. These have drawn a huge global interest for intensifying the market for new dietetic supplements, functional foods and nutraceuticals that can promote consumer health and nutrition. Hence, sustainable utilization of these valuable components from shell processing wastes can be utilized as promising bio-derived nutraceuticals as well as minimize the impact of its associated environmental hazards from their direct disposals.

15.2 POLYMER-BASED

15.2.1 CHITIN

Among the diverse array of polymers found naturally, chitin stands in the second place just after cellulose as natural and most abundantly found polymer on earth. It is the chief organic component of the exoskeletal in many marine invertebrates (crab, squilla, lobster, crayfish, clams, cuttlefish beaks and bones and squid pens) where it's tough and resilient property is utilized as skeletal support and body armor for their self-defense. The yield of chitin present in dried shell waste is roughly estimated to be around 14–27% and 13–15% for shrimps and crabs, respectively. It was first discovered by Professor Henri Braconnot in 1811 from mushrooms and was named as fungine. Chitin subsists in three polymorphic forms, namely, α chitin; β-chitin; and ɤ-chitin. The most stable form amid the three crystalline variations is α chitin whose occurrence resembles as antiparallel strands. β-chitin is less stable and which is arranged in parallel chains. They are dominant in the extracellular fibers of diatoms, squid pens and in certain annelids. The least common form is ɤ-chitin is a blend of both α and β structures and is found in squid intestinal linings.

Chitin is N-acetyl 2-amino 2-deoxy D-glucose, which is a macromolecular and linear polymer of anhydro-N-acetyl-D-glucosamine units linked by β (1–4) glycosidic bonds. It is unsolvable in most organic solvents as well as water. Chitin serves as a basic material for the industrialized manufacture of its derivatives like glucosamine (65%), chitosan (25%), chitooligosaccharides (9%) and *N*-acetyl glucosamine (1%). Shrimp processing industry serve as the most economical source of raw material for the industrial production of chitin, from which almost 75% of its share is used by the nutraceutical industries (Figure 15.1).

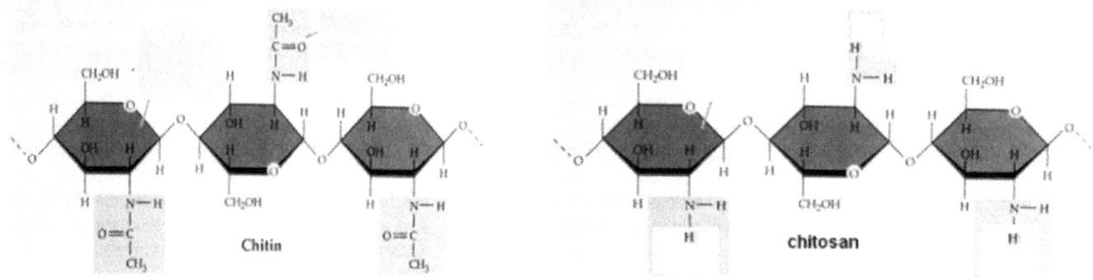

FIGURE 15.1 Structure of chitin and chitosan.

15.2.1.1 EXTRACTION OF CHITIN FROM SHRIMP SHELL WASTE

15.2.1.1.1 Chemical Method

Preparation of chitin from shell waste involves two methods such as demineralization and deproteinization, which employs the use of chemicals. In case, protein extraction is envisaged solely for getting shrimp extracts, then fresh and hygienically collected prawn shells must be processed by deproteinization followed by demineralization.

Deproteinization is the process by which proteins are eliminated from the complex chitin. Shells after demineralization and washing free of acid are shifted to the deproteinizing tank where it is treated with 3% aqueous caustic soda. It is continuously stirred using a propeller-type stirrer at about 100 rpm for nearly 60–120 minutes at 80–90°C by an open steam jacketed cylindrical vessel. By this time, the protein from the shell gets dissolved in the alkali which can be discarded. The residue is repeatedly washed free of alkali by potable grade water.

Demineralization is the process followed after deproteinization in which the minerals are eliminated from shells by treating with commercial mineral acids (hydrochloric acid). The shells are demineralized for about one hour by adding 1.25 N hydrochloric acid at a ratio of 1:2 (w/v) accompanied with occasional stirring in PVC tanks. If wet shell is used as the starting material, the volume of water used for dilution of the acid should be adjusted considering that the water in the wet shell remains around 80%. The excess acid-containing minerals was drained off, and the residues were washed repeatedly with good quality ground/potable water along with stirring to make it free from any adhering acid. At least three washings are essential to make the final mass completely acid free, and this is further pressed under screw press to remove the water. The end mass obtained is free from minerals and protein. The product obtained is the wet chitin.

15.2.1.1.2 Biotechnological Methods

Currently, isolation and production of chitin and its derivatives from shellfish wastes encompass tough chemical treatments with the use of strong alkalis for deproteinization and strong mineral acids for demineralization which generates undesired by-products like irregularly deacetylated polymers. Studies on alternative disposal methods are being extensively conducted to overcome the drawbacks associated with harsh chemical usages. Biological conversion of wastes is possibly the most lucrative and eco-friendly procedure for waste utilization. Chitin has been well evaluated to be prepared through microbiological and fungal actions. Fermentation of shrimp biowastes using lactic acid bacteria with added carbohydrate sources (cassava or molasses), organic acids and salt supply was also studied.

Deproteinizations using several proteases like papain, pancreatin, devolvase, alcalase, pepsin, and trypsin are experimented for chitin production. They are either added before or after the demineralization step. These enzymes minimize the deacetylation and depolymerization during isolation of chitin. Deproteinization by means of protease-generating microorganisms for example *Pseudomonas aeruginosa* K-187, *Pseudomonas maltophilia* LC-102, *Candida parapsilosis*, and *Bacillus subtilis* are also reported from shellfish wastes. However, further treatment is still needed for demineralization.

15.2.2 CHITOSAN

Recently, chitosan has gained a huge attention due to its increasing demand as a useful biopolymer with a broad application base. Deacetylation of chitin offers chitosan, a linear polymer of amino-D-glucose of high molecular weight.

15.2.2.1 PREPARATION OF CHITOSAN

The wet chitin obtained from the deproteinizing tank was transferred to the steam jacketed vessel after dewatering by centrifuging. The dewatered mass of chitin is shifted to the deacetylation tank where it is treated with 50% (w/w) sodium hydroxide solution at 95–100°C for 1.5 to 2 hours or more until the degree of deacetylation reaches above 70%. A minimum concentration of 40% by weight of caustic soda in the reaction mass is required for conversion of chitin to chitosan. Since the wet chitin contains a good amount of water, care is to be taken to add sufficiently high concentration alkali in sufficient quantity so as to give a final concentration of 40% (w/w) in the reaction mass. The ratio of wet chitin to sodium hydroxide (w/v) should be around 1:2. After checking the complete deacetylation of the chitin mass by its solubility in 1% acetic acid, the alkaline mass is washed thoroughly till made alkali-free and properly centrifuged to remove any adhering water. The final alkali-free wet chitosan with moisture content around 50% are spread in clean trays and hot air-dried at 50–60°C temperature. Alternative sun drying methods can also be employed for drying by hygienically spreading in open cemented floor, without any external contaminations. The final moisture content must be always less than 10%.

15.2.2.2 NUTRACEUTICAL APPLICATIONS

Chitosan is approved as dietary additives and functional food supplements in many developed countries like Japan, United States, European countries like England, Italy, Portugal, and Finland. It is reported to impart desirable health attributes like prevention of illness, elevation of longevity, immunity enhancement, and biorhythm control. It has been sold for more than 20 years in Japan and European nations as a non-prescription weight-loss supplement and is recently radiating towards the U.S. market over the past few years. It hence reduces the incidence of high cholesterol, obesity, and hypertension by interfering with assimilation of neutral lipids present in body via hydrophobic interactions.

Though chitosan is soluble in the acidic conditions of the stomach, it starts precipitating once it passes to the small intestine where the pH is

6.3–6.8. As the polysaccharide chains of chitosan combine together, they entrap all the bound acids, cholesterols, and lipids making a non-digestible mass which is excreted out as feces. This biological mechanism proposed for chitosan may promote weight loss in individuals by reducing dietary metabolism of body fats and sterols. Similarly, chitosan may also aid to decrease the level of total serum cholesterol by entrapping the bile acids. Cholic acid and other bile acids are normally produced by the oxidation of liver cholesterol and are stored in the gallbladder and afterwards secreted into the intestine. Typically, around 90% of the bile acids are reabsorbed and recycled by entero-hepatic circulation in the small intestine. However, entrapment of bile acids by chitosan causes their subsequent removal from this recycling system, compelling the liver to manufacture extra bile acids from cholesterol. Similar mechanism is also reported for other soluble fibers, like the fiber supplement *Psyllium mucilloid* and oat brans which reduces the total and LDL cholesterol levels through enhanced excretion of bile acids and decreased release of liver cholesterols and fatty acids.

Chitosan gel in batter formulations have also proved excellent in reducing oil uptake in coated fish sticks. In extruded snacks, the flavors and spices are added to the product through an oil medium. As a result, extruded snacks was having 20–30% oil in its composition. This issue can be managed with chitosan incorporation in extrusion base. Chitosan in the form of acid hydrogel at minimal concentrations has helped in developing a functional fish mince emulsion sausage with desirable functional and sensory attributes.

15.2.3 GLUCOSAMINE HYDROCHLORIDE

Acid hydrolysis of chitin with concentrated mineral acids like hydrochloric acid breaks down the polymer releasing a monomer glucosamine as hydrochloride. Glucosamine hydrochloride ($C_6H_{13}NO_5$=HCl) is a small molecule comprising of sugar (glucose), amine, and hydrochloride groups. Glucosamine hydrochloride is white crystalline in its purest form, and at elevated temperatures, it undergoes degradation causing a color change. It has a melting point of 190–194°C and possesses a molecular weight of 215.63 Da. It is having a CAS Registry Numbers of (66-84-2) and is sold by many trade names like 2-Amino-2-deoxy-D-glucopyranose; chitosamine hydrochloride; D-glucosamine hydrochloride; D-(+)-glucosamine hydrochloride. The absolute configuration of α-D-glucosamine hydrochloride by X-ray diffraction method was validated and the results revealed that the molecule exists in a chain form. There are three major forms of glucosamine supplements in the market, i.e., glucosamine hydrochloride, glucosamine sulfate and N-acetyl glucosamine (Figure 15.2).

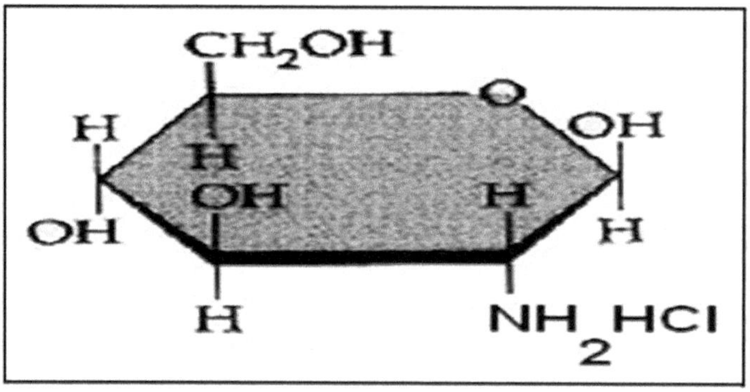

FIGURE 15.2 Structure of glucosamine hydrochloride.

15.2.3.1 PREPARATION OF GLUCOSAMINE HYDROCHLORIDE

According to work done by Xavier (2006) hydroxylation of dry chitin powder with hydrochloric acid of 10 N concentration in a glass reactor armed with reflux condenser in a thermostat digital water bath forms glucosamine hydrochloride. The temperature of the reaction mix increases slowly to 90–95°C with occasional stirring. During the hydrolysis process, HCl gas fumes may liberate, and it can be recovered for other purposes. The excess residual acid can be vacuum distilled after the termination of the reaction. Any undissolved residue still remaining in the reaction mixture is strained after adding an equal volume of water. Activated charcoal (10% weight of chitin) is added to this mixture followed by warming at 60°C for 30 min and filtered. The ultimate pale yellow colored solution was vaporized in a flash evaporator under reduced pressure. The mixture was washed with alcohol at the end of complete crystallization and dried. Yield of glucosamine was calculated on the basis of raw chitin using the following equation:

$$\text{Yield of GHC (\%)} = (\text{Weight of dry glucosamine hydrochloride} / \text{Weight of dry chitin}) \times 100$$

15.2.3.2 NUTRACEUTICAL APPLICATIONS

Researches on glucosamine dates back to several decades, and its applications are well established. Glucosamine in the human body stimulates the cartilage to generate fundamental constructive material of collagen and cartilage proteoglycans proving it as an anti-inflammatory compound used as a drug in the treatment of rheumatic diseases. It prevents production of cartilage damaging enzymes. Clinical applications of glucosamine include its growth inhibition capacity of cancerous cells and its enhanced rate of antibiotic absorption when administered along with other antibiotics like tetracycline or similar. It plays vital physiological functions in the human body like detoxification of important body organs (liver and kidney); serve in anti-coagulation and immuno-modulatory activities and act as anti-pepsin or anti-ulcer agents. Heart disorders are also reported to be improved by this rare sugar derivative. Glucosamine is also being suggested as a valuable ingredient in the preparation of diet meals for infants by producing Bifidobiogen in the infant intestines. Glucosamine derivatives serve as intermediates in the amalgamation of oligosaccharides and glycoconjugates as well as in the preparation of emulsifiers.

15.3 PROTEIN BASED

15.3.1 ENZYMES

Enzymes are biocatalysts and are an important biological component having huge potential for biotechnological applications. Shellfish processing wastes usually contain several vital digestive enzymes that include proteinases and lipases. Adaption of marine aquatic animals to varying environmental atmospheres makes their body enzymes unique in several attributes when compared with other terrestrial enzymes. Several enzymes from processing wastes have been isolated and characterized, which has paved the way for their new industrial applications. Extraction and recovery of these unique enzymes may significantly help to reduce the hurdle of massive environment pollution ending up with discovery of new valuable products.

Proteinases are well known in degrading proteins through hydrolysis of peptide bonds. Hepatopancreas is the major digestive organ in shellfish, which serve as an excellent source of the proteinases. Among all, trypsin is the major proteinase which apart from digestive functions, also activates other pancreatic enzymes by cleaving a short activation peptide from the amino-terminus of inactivate zymogens. It can catalyze three essential amino acid residues such as serine (Ser), histidine (His), and aspartate (Asp). It is highly specific in cleaving at the carboxyl side of the peptide bond in amino acid (lysine or arginine) residues.

Lipase is another rich enzyme also found in shellfish hepatopancreas. This enzyme from the waste derivatives encourages the production of free fatty acids (FFAs) and glycerol by hydrolysis of the triacylglycerols. It works at the oil-water interface.

15.3.2 PROTEIN HYDROLYSATES

Byproducts from shellfish processing like head, shell, and undersized shellfish harbor significant quantities of essential amino acids that make them a rich protein source that can be utilized as feed supplements in human and animal nutrition. Bioactive peptides from shell wastes discarded have been proven to inherit several properties, including anti-hypersensitivity by angiotensin-converting enzyme (ACE) inhibitions, antioxidant as well as antimicrobial properties. Antioxidant property is attributed by preventing oxidative stress by scavenging the free radicals and reactive oxygen species (ROS) generated in lipid oxidation chain reaction. These special properties spread their applications in producing various food and nutraceutical for health promotions. Food industries have widely applied its diverse use as protein supplements, stabilizers in food beverages, and flavor enhancers in confectionery products.

15.3.2.1 PREPARATION OF PROTEIN HYDROLYSATE

These proteins are mostly isolated by the process of isoelectric pH solubilization precipitation and further processed into liquid, paste, and powder that can be consumed as food supplement. Extraction efficiency of these waste derivative proteins varies and is highly dependable on the extraction methods. Biological and chemical methods are commonly employed for the hydrolysate preparations. However, the chemical extraction methods form products having lowered nutritional attributes, poor functional properties and restricted use as flavor enhancers. Additionally, the use of sodium hydroxide during isolation restricts its utilization. Alternative methods like biocatalysts or enzymatic methods and fermentation have been popularized for the recovery of shellfish wastes in order to combat the negative aspects associated with the practice of chemical treatments.

Enzymatic recovery of shrimp waste protein has several advantages associated with them as the hydrolysis process involved can be controlled in a desirable manner giving rise to minimal undesirable reactions. This can be executed by both digestive enzymes derived from the crustacean itself as well as by incorporating external enzymes sources. Additionally, recovery of potent bioactive peptides is reported as highly efficient in enzymatic hydrolysis of proteins and yields better reclamation of modified fish proteins, functional food and ingredients. Fermentation is one of the eco-friendly, cost-effective, and interesting technology employed for waste protein hydrolysis, which involves the use of microorganisms. Shrimp shell waste fermented by proteolytic enzymes from a few microorganisms such as *Pediococcus acidolactici, Bacillus licheniformis* forms peptides and amino acids by hydrolysis.

15.3.3 SHRIMP PASTE AND SAUCES

Fresh crustacean shells have to be first treated with dilute caustic soda (0.5% NaOH) and the excess protein water is drained off. The alkaline drained mass is neutralized with acetic acid and concentrated as semi-solid paste. Hence alkaline deproteinization yields shrimp peptone. In some countries this solubilized protein from high quality shrimp shell paste are concentrated and canned for edible purpose. It can be used for flavoring agent. Fermented sauce is also possible to be made with salt fermentation of tiny shrimps. The shrimps are mixed with salt and allowed to ferment for 4–6 months. The hydrolyzed proteins are filtered and centrifuged to make clear sauce which is rich in peptones and used in food applications.

15.4 LIPID-BASED

15.4.1 SHRIMP OIL

Shrimp is one of the most nutritious balanced complete seafood items in human nutrition with a rich protein, essential fatty acids, and mineral composition. Compared to other fish oils, shrimp derived oils possesses numerous additional beneficial bioactive compounds such as omega-3 essential fatty acids (EPA and DHA), lipid-soluble pigments or carotenoids (β carotene, astaxanthin), fat-soluble vitamins (Vitamin A, D, E), phospholipids, and cholesterol. Shrimp oil is derived mainly from the head (cephalothorax region) wastes of shrimps and is an evolving nutraceutical product. Generally, the total content of lipid in shrimps is estimated to be between 1.0% and

8.0%, which varies according to species. Apart from species variation, lipid-based components extracted from the shrimp oil also vary greatly between gender, season, area of availability, diet composition, etc., which can alter the amount and composition of oil extracted. Here are some important components from shrimp oils having excellent nutraceutical values.

15.4.2 CAROTENOIDS

Carotenoids constitute a set of fat-soluble pigments present in many living organisms, including plants, yeast, algae, photosynthetic bacteria, fungi, and crustaceans. Carotenoids were classified as carotenes (e.g., β-carotene and lycopene) and xanthophylls (e.g., zeaxanthin, and canthaxanthin). In nature, carotenoids exist as protein-pigment complex in shellfish where their stability enhances by protein-pigment interaction. Conferring to a definition, when any molecule encloses a molecule of carotenoid with a protein in stoichiometric proportion, it is termed as carotenoprotein. These were further classified into two categories: True carotenoproteins (the complexes having no lipids and depicting a carotenoid-protein stoichiometry) and carotenolipoproteins (the complexes comprising of lipids and not necessarily possess any stoichiometric combination).

Astaxanthin is a red-colored carotenoid found in crustacean shells which is having an exceptional potency as an antioxidant. It had singlet oxygen scavenging activity 1,000 times higher than coenzyme Q10 and around 500 times better than Vitamin C. Comparative studies done on various shrimp species shells to estimate the content of astaxanthin revealed that 88% of astaxanthin as carotenoprotein is present in the black tiger shrimp shells along with 20% other lipid components. Shrimp derived carotenoids have gained high commercial importance as they are of biological origin with no toxicity. Astaxanthin is now used as an immunomodulatory and anti-inflammatory agent; as a medicine in pharmaceutical industries for treating cardiovascular diseases, UV induced skin problems and antiaging creams; in food industries as antioxidant, etc. Recently cancer-treating drugs are also supplemented by such derivatives.

Several methods are reported by many researchers in their studies to recover the carotenoid from wastes of shellfish processing that includes the process of fermentation, supercritical fluid extraction, extraction using organic solvents and vegetable oils. As the carotenoid pigment is soluble in fats and oils, various kinds of vegetable oils were being used for its extraction like sunflower oil, palm oil, rice bran oil, groundnut oil, sesame oil, mustard oil, soybean oil, and coconut oil. The yield of carotenoid obtained from the experiments was at highest quantity when sunflower oil was used as extracting solvent in contrast to other vegetable oils used in the studies.

15.4.3 ESSENTIAL FATTY ACIDS (PUFA)

Shrimp oils are one of the promising sources of essential fatty acids, i.e., polyunsaturated fatty acids (PUFA) like eicosapentaenoic acid (EPA) and docosahexaenoic acid (DHA). They are followed by saturated fatty acids (palmitic acid) and monounsaturated fatty acids (oleic acid) in their compositional reserves. As the total fatty acid composition in shrimp oil are mainly dominated by PUFAs, its high susceptibility to oxidation makes its application a challenging one as it can develop a typical fishy odor and off-flavors.

Encapsulation of shrimp oil in nanoliposomes is a recent new technology, which can be utilized to protect the oil against oxidation and mask any off-odors. It also preserves the bioactive components which are highly sensible. Another method of encapsulation most widely applied in nutraceutical industries is the liposome technology. They are utilized in fortifying a diverse range of products such as n-3 fatty acids (PUFAs), vitamins (vitamin-D), oils from both fish and crustaceans. Nanoliposome powder from shrimp oils could be used in the fortification of several foods, particularly beverages. Spray-drying and freeze-drying operations are employed and when added with anticaking agents offer numerous advantages. Shrimp oils encapsulated by nanoliposomes offers enhanced oxidative stability, better solubility, flowability higher retentions of essential fatty acids when stored for weeks. This technology thus proves as an efficient preservation of bioactive and functional nutraceutical elements.

15.4.4 FAT-SOLUBLE VITAMINS

Crustaceans constitute an outstanding source of fat-soluble vitamins, including vitamin A (Retinol), D, and E (α-Tocopherol) that is very important in many human physiological functions. Hepatopancreas is not only the organ that aids in the storage of shrimp lipids, but also stores higher levels of these fat-soluble vitamins than other organs. Vitamin A (Retinol) helps in improving the vision and boosts the human immune system. Vitamin E functions as a natural antioxidant in the human body and is an important vitamin that helps in reproductive and muscular development. Additionally, this vitamin plays an exceptional role in the proper regulation of body calcium as it is the principal steroid of human body.

15.4.5 PHOSPHOLIPIDS AND CHOLESTEROL

Phospholipids are the polar lipids found in shrimp having phosphorus as a nutrient source. Phospholipids from shrimps mainly comprises of phosphatidylcholine and phosphatidylethanolamine followed by sphingomyelin, phosphonolipids, cardiolipin, lysophosphatidylcholine, phosphatidylserine, and phosphatidylinositol. Shrimps also contain cholesterol as the most important sterol group which is associated with several illnesses like hypercholesterolemia. Recently, beta-cyclodextrin is used as adsorption component in the shrimp cholesterol extraction from the cephalothorax region of Pacific white shrimp. There was a 95% reduction in the level of shrimp cholesterol as well as in lipid oxidation substances along with an increase of 86% in the level of astaxanthin when compared with shrimp oils in crude form.

15.5 SHRIMP SHELL MEAL

Generally, the shrimp shell wastes after drying and crushing are made into fine powder consistency which is used as a feed ingredient in different aquaculture sectors. These shells are laid on open beaches for drying which has maximum chances for contamination due to the rapid growth of microorganisms which produces ammoniacal nitrogen. This can produce a foul smell and gives a black color appearance on the dried meal making it non-palatable for feeding purposes. So now modified practices are followed where the shrimp wastes are cooked at 80–90°C and then mechanically dried and pulverized into fine shell meal powder.

KEYWORDS

- **chitin**
- **chitosan**
- **glucosamine hydrochloride**
- **nutraceuticals**
- **protein hydrolysates and enzymes**
- **shrimp oils**

REFERENCES

Andlauer, W., & Fürst, P., (2002). Nutraceuticals: A piece of history, present status and outlook. *Food Res. Int., 35*(2, 3), 171–176.

Bernal, J., Mendiola, J. A., Ibáñez, E., & Cifuentes, A., (2011). Advanced analysis of nutraceuticals. *J. Pharmaceu. Bbiomed. Analy., 55*(4), 758–774.

Chattopadhyay, K., Xavier, K. M., Balange, A., Layana, P., & Nayak, B. B., (2019a). Chitosan gel addition in pre-emulsified fish mince-effect on quality parameters of sausages under refrigerated storage. *LWT-J. Food Sci. Technol., 110*, 283–291.

Chattopadhyay, K., Xavier, K. M., Layana, P., Balange, A. K., & Nayak, B. B., (2019b). Chitosan hydrogel inclusion in fish mince based emulsion sausages: Effect of gel interaction on functional and physicochemical qualities. *Int. J. Biol. Macromol., 134*, 1063–1069.

Chattopadhyay, K., Xavier, K. M., Layana, P., Balange, A. K., & Nayak, B. B., (2020). Effect of chitosan hydrogel inclusion on centesimal composition of pangasius (*Pangasianodon hypophthalmus*) (Sauvage, 1878) emulsion sausages under refrigerated storage. *Fish. Technol., 57*, 22–27.

Dhanabalan, V., Xavier, M., Kannuchamy, N., Asha, K. K., Singh, C., & Balange, A., (2017). Effect of processing conditions on degree of hydrolysis, ACE inhibition, and antioxidant activities of protein hydrolysate from Acetesindicus. *Env. Sci. Pollu. Res., 24*(26), 21222–21232.

FAO, (2020). *The State of World Fisheries and Aquaculture 2020: Sustainability in Action* (pp. 224). Rome. https://doi.org/10.4060/ca9229en (accessed on 09 March 2022).

Klomklao, S., Benjakul, S., Visessanguan, W., Kishimura, H., & Simpson, B. K., (2009). Extraction of carotenoprotein from black tiger shrimp shells with the aid of bluefish trypsin. *J. Food Bioche., 33*(2), 201–217.

Kuepethkaew, S., Sangkharak, K., Benjakul, S., & Klomklao, S., (2017). Laundry detergent-stable lipase from Pacific white shrimp (*Litopenaeus vannamei*) hepatopancreas: Effect of extraction media and biochemical characterization. *Int. J. Food Prop., 20*(4), 769–781.

Mao, X., Guo, N., Sun, J., & Xue, C., (2017). Comprehensive utilization of shrimp waste based on biotechnological methods: A review. *J. Clean. Prod., 143*, 814–823.

Pattanaik, S. S., Sawant, P. B., Xavier, K. M., Dube, K., Srivastava, P. P., Dhanabalan, V., & Chadha, N. K., (2020). Characterization of carotenoprotein from different shrimp shell waste for possible use as supplementary nutritive feed ingredient in animal diets. *Aquacul., 515*, 734594.

Poonsin, T., Simpson, B. K., Benjakul, S., Visessanguan, W., Yoshida, A., & Klomklao, S., (2019). Albacore tuna spleen trypsin: Potential application as laundry detergent additive and in carotenoprotein extraction from Pacific white shrimp shells. *Biocat. Agricul. Biotechnol., 17*, 638–646.

Raushan, K., Martin, X., Manjusha, L., Balange, K. A., & Gudipati, V., (2018). Fortification of extruded snacks with chitosan: Effects on techno-functional and sensory quality. *Carbohy. Polym., 194*, 267–273.

Wanichpongpan, P., & Attasat, S., (2016). Optimum conditions for preparation of glucosamine hydrochloride and glucosamine sulfate from shrimp-shell chitin. *KMUTNB Int. J. Appl. Sci. Tech., 6*, 24–29.

Xavier, K. A. M., (2006). *Standardization of optimum conditions for the production of Glucosamine Hydrochloride from Chitin.* PhD thesis. Cochin University of Science and Technology, Kochi, India.

Xavier, K. M., Kannuchamy, N., Balange, A. K., Chouksey, M. K., & Gudipati, V., (2017). Functionality of chitosan in batter formulations for coating of fish sticks: Effect on physicochemical quality. *Carbohyd. Polym., 169*, 433–440.

Xavier, M., Kannuchamy, N., Balange, K. A., & Gudipati, V., (2019). Development of enrobed fish products: Improvement of functionality of coated materials by added aquatic polymers. *J. Food Proc. Engg., 42*(3), e12999.

CHAPTER 16

Edible Packaging from Fishery Waste: A Future Prospective

SATHISH KUMAR KANNAIYAN and J. BINDU

ICAR-Central Institute of Fisheries Technology (CIFT), Cochin, Ernakulam, Kerala, India

ABSTRACT

Current trends in food packaging research are aimed towards the improvement of food quality and safety by using environmental friendly bio-based packaging materials, having biodegradable characteristics. In fish processing industries, a huge amount of processing waste is being generated in the form of head, viscera, fins, scale, bones, etc., which have high levels of nutrients such as protein, lipids, polysaccharides, etc. Majority of these waste materials are being dumped as waste on land or sea, which cause environmental pollution. In order to avoid this and to recover these nutritious materials, modern seafood industries started focusing on waste utilization and converting the waste into wealth in the form high value products. Edible packaging is one of the grave areas that has not been focused so far. It can address the three major problems namely, converting fishery waste into edible packaging, which can provide additional nutrition to the consumer and revenue to the industries, addressing environmental pollution by converting the waste into wealth and as an alternative for synthetic polymers due to their environmental friendliness and finally ensuring food safety and improving quality by retarding microbial growth, prevent the oxidation of lipids and loss of moisture, and functions as a carrier of additives such as antimicrobial and antioxidant agents, flavors, aroma components, etc. This chapter reviews the importance of fishery

Advances in Fish Processing Technologies: Preservation, Waste Utilization, and Safety Assurance.
Ranendra K. Majumder, PhD & Amjad K. Balange, PhD (Eds.)

waste and their utilization for edible packaging development and applications in the different sectors.

16.1 INTRODUCTION

In modern-day life, waste management is a major problem in all the processing sectors. All industries contribute a huge amount of biodegradable and non-biodegradable waste including seafood industries. In this decade, consumer demand for fish and fishery products have increased significantly. On the other hand, quality and safety are major concerns for the health-conscious and the perception of fish as healthy food with high amount of digestible protein, PUFA, and other nutrients. During fish harvesting and processing a huge amount of waste is generated in terms of head, viscera, fins, scale, bones, etc. Generally, seafood waste finds its way directly into the sewers or nearby water bodies. Modern seafood industries have scaled up their production capacities and generates enormous amounts of waste. Hence, we must look for better technologies of waste management and convert these fishery waste into wealth.

Food processing industry wastes are one of the major causes of environmental contamination, and several studies have been carried out to convert these wastes into valuable products. Overall, more than 50% of the raw material (total fish capture) is left over as waste during harvesting and processing. In India, more than 2 million metric tons of waste is generated every year and its categorized as certified waste which is even very expensive to dispose of due to their nutritional content. Disposal of wastes generated from seafood processing plants have always been a problem for the industries which directly contributes to the CO_2 production in all the stages, starting from culturing/capturing until it reaches the consumers. Direct disposal of these fishery solid waste leads to environmental pollution and wasting its high nutritive content.

Solid waste from the seafood industry, including head parts, shells, stomach parts, and bycatches is now in huge demand by many industries, which could be a source of raw material for developing a high value fishery products and nutraceutical products. An important strategy of waste reduction is to find marketable byproducts from fish wastes, and it could address the environmental pollution problems and provide additional revenue to the industries. During processing, nearly 50–80% of the raw material generates solid wastes, which contain a high amount of nutrients such as protein, lipids, carbohydrates, etc. An estimate of waste generation profile during fish processing is given in Table 16.1.

TABLE 16.1 Waste Generation Profile During Fish Processing in India

Products	Percentage of Waste Generated (w/w)
Shrimp products	50
Fish fillets	65
Fish steaks	30
Whole and gutted fish	10
Surimi	70
Cuttlefish rings	50
Cuttlefish whole	30
Cuttlefish fillets	50
Squids whole	20
Squid tubes	50
Squid rings	55

In application, 80–90% of the fish meal industries wholly depend on seafood processing waste as a primary raw material for protein source. Similarly, it could be used for the poultry and pet foods industries. Skin and bone waste are excellent sources of raw material for the gelatin and collagen industries. Fish protein hydrolysates (FPH) from fishery wastes can be used for fish or pig meal and fertilizer components, etc. Prawn and crab shell waste can be utilized to prepare chitin, chitosan, and their derivative products which are having huge role in nutraceutical applications. In Europe, Fish Silage (liquefied Fish) production offers the most promising aspect for fish processing waste utilization, which is being fed in its liquid form to pigs and cattle. Fish oils recovered from seafood wastes contain an abundance of long-chain, polyunsaturated fatty acids (PUFA) having several nutraceutical applications. Overall, the utilization of fishery waste offers additional benefits to the processing industries by converting waste into wealth.

Though fishery waste has several applications, industries, and researchers have been recently showing interest to develop a novel biodegradable and edible packaging or coating from seafood processing waste. Edible films (EFs) and edible coatings (ECs) are one of the most promising and recent approaches to enhance the quality and shelf-life of food products. Edible film/coating can be defined as a thin, continuous layers of edible coating or film material placed between food components. EFs can be used as wraps, covers, packages, and also to separate layers for food products. ECs are in liquid form applied on the surface of food materials to form a thin layer on the food surface or between the different layers of food components to prevent the moisture loss, oxygen transmission, and solute into the food materials. EFs and ECs constitute an environment friendly technology that may enhance food safety and quality, stability, and their mechanical properties by providing a semi-permeable barrier to oxygen, CO_2 and water vapor between the food and the surrounding. EFs and coatings can act as a carrier of antimicrobials and antioxidants, flavoring, coloring agents, etc., that will improve the quality and safety of food products.

EFs and ECs provide a feasible alternative to synthetic packaging and coatings that cause environmental pollutions. Rising public demands for healthier packaging without synthetic additives, together with awareness on environmental pollution have promoted a significant interest in EFs and ECs to work on advanced developments. This is an emerging area that utilizes the EFs and ECs as a food matrix for the delivery of active agents (antimicrobials and antioxidant, nutraceuticals, flavors, and aroma compounds, etc.), which may effectively extend shelf life, improve its quality and food safety.

In this chapter, a brief description of the content, properties, and applications of EFs and ECs, and utilization of fishery waste for making EFs and ECs and their application in fish processing is provided.

16.2 MATERIALS FOR EDIBLE FILMS (EFS) AND EDIBLE COATINGS (ECS) MATRICES

The major features of EFs' and ECs' are determined by a main component formulation's (matrix). EFs and ECs matrix-forming materials can be classified into three groups: (1) proteins, (2) lipids, and (3) polysaccharides. In addition, films or coating matrix can be combined with several materials to form a composite film or coating.

16.2.1 IMPORTANCE OF PROTEIN-BASED EDIBLE PACKAGING

There are three major materials used for development of biopolymers such as polysaccharides, proteins, or lipids. Polysaccharides are abundant in nature and has several advantages such as availability, low cost, non-toxicity, and thermo-process ability. However, hydrophilicity nature of polysaccharides, display poor barrier properties to water vapor. On the other hand, lipids have hydrophobicity properties, which can improve the film barrier properties. However, lipid films are commonly used for coating purposes because of the inelastic properties of lipid-based films. Whereas, proteins have several advantages, such as relatively abundance in nature, excellent film-forming ability and high nutritional value for preparing biodegradable films. Similarly, protein-based films show an excellent gas barrier properties compared to polysaccharides and lipids.

Several proteins sources such as gelatin, casein, corn zein, wheat gluten, whey, soy, mung bean and peanut protein, and more can be used as EFs and ECs matrix. Proteins can easily dissolve in aqueous solutions and interact to form ionic and hydrogen bonds, or hydrophobic interactions, depending on their functional groups. If particular proteins tend to create ionic or hydrogen bonds, the EFs and ECs will usually exhibit better mechanical properties. Proteins-based EFs and ECs are generally less permeable to gases and vapors due to their poor water solubility and also offers a variety of properties such as film-forming ability, better physical and mechanical properties, which can fit to any kind of food products.

Besides, due to the intrinsic properties, proteins are used as an excellent material for edible film preparation and can effectively cover the food materials with EFs to prevent the moisture and flavors loss, control the gas (O_2, CO_2 and ethylene) exchange, transport, and deliver the active substances such as antimicrobials, antioxidants, or nutraceuticals. Moreover, the major advantage of the protein-based film is that it is an important source of nitrogen that can act as fertilizer, which is provided during the degradation of protein-based films. Though protein-based films are the most promising materials for food packaging as they are having an excellent gas barrier and good mechanical properties, it possesses poor barrier characteristics to moisture. The addition of some other biomaterials into protein-based films can reduce moisture sensitivity.

During film formation, protein-based biopolymers exhibits fragility and brittleness, because of strong cohesive energy density, which cause poor mechanical properties regarding their processability and end-use application. However, the addition of natural and/or biodegradable plasticizers into films which possess excellent compatibility with protein-based biopolymers and also could improve their viscoelasticity and extensibility, to solve the above problem.

16.2.1.1 COLLAGEN

Among the protein sources, collagen is one of the most commercially successful natural polymers extracted from animal sources, mainly used as a packaging material in the food industries, especially meat product or sausage casing, and drug delivery carriers or wound dressings in the medical fields. Collagen fibers and collagen powder are most suitable for biocomposite films development in which the fibers act as filler. With the outbreak of diseases like bovine spongiform encephalopathy (BSE) and foot/mouth diseases, collagen from fish processing wastes has started receiving much more attention as a safe and an alternate to mammal collagen. A number of studies have focused on the extraction and characterization of collagen or gelatin from different species of fish, such as cod and hake, plaice, perch, megrim, cuttlefish, tilapia, squid, surf smelt, deep-sea redfish, black carp, eel-fish, Amur sturgeon cartilage, and seabass, seabream, mackerel, and yellowtail. However, the research on the development of biodegradable films from collagen is very limited. Fishes like tilapia, grass, and silver carp, Jumbo squid (*Dosidicus gigas*), Shark catfish (*Pangasius pangasius*) skin, Irish cod (*Gadus morhua*) and ling (*Molva molva*) have already been characterized and utilized for biodegradable films.

16.2.1.2 GELATIN

Gelatin is a protein, that can be extracted from collagen by alkali or acid hydrolysis at higher temperatures. Most commonly produced gelatin is from animal sources such as bovine or fish, either in granulated or powder form. The main attributes of gelatin are its physicochemical properties like solubility, color, transparency, odorless, and tastelessness, which define its commercial applications. The most important physical properties like gel strength and viscosity are major properties that influence its quality. Gelatin can be divided into two categories based on its functional properties: (i) surface characteristics such as emulsion and foam formation, stabilization, adhesion, and cohesion, and film-forming property; and (ii) gelation behavior such as texturizing thickening and water holding capacity. Other applications of gelatin include packaging, pharmaceutical, food, cosmetics, and photography, especially as a gelling, stabilization, texturization, and emulsification agent for bakery, dairy products and beverages in the food industry. However, lack of thermal stability and mechanical properties, are limitations in its application.

Film-forming properties have been extensively utilized to protect food from dryness, exposure to light and oxygen during storage. Due to the highly hygroscopic nature, gelatin has a tendency to swell or to be dissolved when it contacts with the food surface. However, the addition of other substances such as surfactants, crosslinkers, strengthening agents, plasticizers can improve the mechanical properties of films. Research has been conducted to study the effect of incorporating other substances, in gelatin-based films to improve its functional properties and the shelf-life of food products. Gelatin with other biopolymer combinations such as whey proteins, starch, chitosan or pectin, natural anti-oxidants and/or antimicrobial substances and surfactant addition have varied characteristics and could be a good strategy to improve its water resistance and mechanical properties. Compounds that are used as gelatin matrices and active additives in gelatin-based films and coatings are shown in Table 16.2.

TABLE 16.2 Active Additive Compounds Used in Gelatin-based Films and Coatings

Gelatin Type	Additive	Mode of Application	Major Benefits
Fish	*Origanumvulgare L.* essential oil	Film	Solubility, barrier to UV light enhanced antimicrobial properties
Fish	Nano encapsulated *Origanumvulgare L.* essential oil	Film	Thermal stability, more flexible films, decreased water vapor permeability, antimicrobial properties
Skate skin	Thyme essential oil	Chicken tenderloin (wrap)	Antimicrobial properties, extension of shelf-life, increase in elongation at break
Grouper bone	Chitosan, clove, and pepper essential oils	Fish steaks (coating)	Antimicrobial properties, extension of the shelf-life
Fish	Cinnamon essential oil, nanoliposomes	Film	Lesser tensile strength, water-solubility, antimicrobial property
Fish skin	Peppermint and citronella	Film	Antimicrobial properties
Fish	Green tea, ginkgo leaf grape seed, ginger	Film	Antioxidant properties
Tilapia and triggerfish skin and bone	Glycerol and sorbitol	Film	Enhancement in mechanical properties
Fish	Gellan and k-carrageenan	Film	Enhancement in mechanical properties

Recent studies have focused on the development of active packaging (AP) films and coatings, which includes antimicrobial, antioxidant, and other agents to enhance or improve the biological features of food. These components are most commonly essential oils or extract and come under the category of GRAS compounds (generally recognized as safe). These natural components minimize the food spoilage that occurs due to oxidative and microbiological changes thereby extending the shelf-life. Table 16.3 gives an account of the gelatin and additives used to preserve the fishery products.

16.2.1.3 FISH PROTEINS (MYOFIBRILLAR PROTEIN/FISH PROTEIN ISOLATES (FPI))

Among biopolymers, proteins are the most attractive material commonly used for film development because of their film-forming ability with better mechanical and gas barrier properties. Myofibrillar proteins from fish muscle have been widely used for film formation. The insoluble nature of protein in water can be made soluble by adjusting the pH. During processing of fish, particularly production of fish filets, a substantial amount of byproducts is generated which are in protein form. Thus, the recovery of these proteins and their utilization to develop the biodegradable films represents as an alternative for upgrading of byproducts from fish processing industries. Several researches has been carried out for developing protein-based films such as transparent and flexible EFs from blue marlin myofibrillar proteins, films from surimi of threadfin bream, and bigeye snapper.

Fish protein isolate (FPI) is another promising material with lower haem-protein and lipid, that enhances the physio mechanical properties of film. However, FPI based film is very rigid, due to the strong covalent bonding, particularly disulfide bond. In addition to this, it shows poor water vapor barrier property, owing to hydrophilicity of amino acids in protein molecules and hence significant amounts of hydrophilic plasticizers need to be added for adequate film flexibility. Various studies have been carried out to improve the properties of FPI films including chemical treatment, enzymatic treatment and thermal treatment. Blending of FPI with other miscible biopolymers is another way to improve the strength and flexibility and requires less amount of plasticizer, and this would be an inexpensive and effective alternative approach to improve those properties of FPI films.

TABLE 16.3 Common Gelatin Matrices for Edible Films and Coatings for Fishery Products

Food Applicability	Product Matrix	Final Product
Rainbow trout	Coldwater fish skin and chitosan	Film and coating
Cod filets	Bovine gelatin, chitosan, and sorbitol with clove essential oil	Film
Minced trout filets	Fish skin gelatin, chitosan, red grape seed extract and *Ziziphora linopodioides* essential oil	Film
Rainbow trout fillets	Food grade gelatin, sorbitol, and laurel leaf essential oil	Film
Tuna meat	Gelatin, red pepper seed meal protein and several plasticizers (glycerol, sorbitol, fructose, and sucrose)	Film
Fish sausages	fish gelatin, chitosan, shrimp concentrate	Film
Atlantic salmon	Warm-water fish gelatin, lignin, and sorbitol	Film
Salmon	Porcine skin gelatin, barley bran protein, sorbitol, and grapefruit seed extract	Film
Cold smoked salmon	Pork gelatin, chicken feather protein, sorbitol, and clove oil	Film
Shrimps	Gelatin, glycerol, sorbitol, and orange leaf extract	Coating

16.2.2 IMPORTANCE OF LIPID-BASED EDIBLE PACKAGING

Lipids were the first materials used for the development of EFs and coatings. Waxes were initially applied on fruits to prevent the moisture loss and later on to cover the fresh meat surface, which is the process called "larding." Among the lipid materials, paraffin, waxes, organic fatty acids, resins, and acetoglycerides, are capable of forming EFs or ECs. They are hydrophobic in nature and exhibit high moisture barrier properties, which prevents or retards the deterioration of food products. However, lipid-based films have very poor mechanical and physical properties and are thick, brittle, and nontransparent Another limitation of lipid-based coatings is less adhesion to hydrophilic foods (Table 16.4).

The efficiency of lipid-based EFs and coatings depend on the nature of the lipid, and particularly on its structure, hydrophobicity, physical state, and lipid interactions with the other components. Lipids are generally combined with other film-forming materials, like proteins or polysaccharides and act as emulsion particles or multilayer coatings to decrease water penetration.

The incorporation of lipids into films and coatings can improve cohesiveness, hydrophobicity, and film flexibility with better moisture barriers. This can also help to extend the freshness, aroma, color, tenderness, and microbiological quality of fresh and processed seafood. However, only a few researchers have studied the benefit of lipids as a protective coating for fish-based products.

TABLE 16.4 Fish Protein-based Films

Fish Species	Characterized
Silver carp (*Hypophthalmichthys molitrix*)	Characterization and their application in Silver carp filets.
Red tilapia	Characterization and yellow discoloration
Hake proteins and thyme oil	Characterization of films
Surimi based film	Acid and alkaline solubilization on properties
bigeye snapper (*Priacanthus tayenus*) surimi	Effect of pH on the film properties
Round scad (*Decapterus maru-adsi*)	Film properties
Fish myofibrillar protein-based film	Film-forming properties

16.2.3 IMPORTANCE OF POLYSACCHARIDES BASED EDIBLE PACKAGING

Polysaccharides such as cellulose, chitosan, starches, and gums have been widely used as raw materials for developing EFs and coatings for food preservation. Polysaccharides are having several advantages such as most abundant material in the world, low cost, easy availability with some functional or specific properties. It is also one of the most potential materials for EC and EF formation, due to their excellent mechanical and structural properties, particularly as an effective barrier against gas transmissions, but they are hydrophilic in nature resulting in poor barrier property against moisture.

Alginate and carrageenan are polysaccharides having poor water vapor barrier properties but can be applied as relatively thick films on the food surface to aid water absorption and enhance protection to further moisture loss from the food until the coating is dehydrated. Moreover, the gas barrier properties of the polysaccharides-based films are important characteristics to retard the loss of organic vapors (aroma compounds) or to prevent solvent penetration in foods during storage, which could result in toxicity or quality loss. To overcome the vapor properties of the polysaccharide EFs materials composite films are developed by adding protein or lipid to enhance the storage stability of the foods.

16.2.3.1 CHITOSAN

Chitosan $(C_8H_{13}O_5)_n$ is a natural polysaccharide that is derived from the exoskeleton of many crustaceans and cell wall of fungi. For this reason, it is commercially available from renewable sources, primarily from the shellfish processing industry which is generated as waste material. It has several attractive properties such as nontoxic nature, biofunctional, biodegradable, and biocompatible properties have a vast potential in the food industry for several applications. It has been extensively used as a films and coatings material due to their mechanical properties, antioxidant, and antimicrobial properties. These transparent films cover the food products and act as a barrier against oxygen, CO_2 and microorganisms, resulting in increased shelf life and product quality improvement.

Though the exact method of antimicrobial activity of chitosan is not cleared yet, the most accepted hypothesis is the interactions between the positively charged chitosan and the negatively charged microbial cell membrane causing the changes in cell permeability. The antimicrobial activity due to the interaction of diffused hydrolysis products with microbial DNA, leads to inhibition of the mRNA and synthesis of protein. Generally, chitosan films are adhesive and cohesive in nature with a smooth surface. The major drawback of chitosan film is their sensitivity towards environmental humidity and leads to poor moisture barrier, which limits their use in food applications. Several strategies have emerged to improve the functional properties of chitosan-based films such as modifications of the deacetylation degree, pH, solvent type or plasticizers. However, chitosan has the ability to mix with additives and plasticizers (propylene glycol, glycerol, and polyethylene glycol) which can improve the mechanical strength of chitosan-based edible film. There are a few factors that reduce the mechanical and functional properties and affect the antimicrobial activity of chitosan-based film such as molecular weight, concentration, and properties of chitosan, pH, concentration of antimicrobials, and component of food products. Cross-linking of chitosan with several reagents such as genipin, glutaraldehyde or formaldehyde to prevent dissolving and/or swelling of chitosan-based films has been evaluated.

16.2.3.2 ALGINATE

Alginates ($C_{12}H_{20}O_{12}P_2$) are the salts of alginic acid obtained from brown seaweeds and are hydrophilic in nature. It has a unique colloidal property includes thickening property, suspending, stabilizing, film-forming, emulsion stabilizing gel producing properties. The most important property of alginates is its capability to react with di- and trivalent cations to form films and coatings with or without gelation and electrolyte cross-linking to enhance its physio-mechanical and barrier properties, cohesiveness, and stiffness and to delay the release of active material. Due to fast crosslinking process with calcium ions, localized gelling areas are produced and form a uniform film.

The alginate-based edible film has attractive properties to improve the quality (moisture retention, texture, color, and odor) and food products shelf life by reducing dehydration, lipid oxidation and controlling the respiration rate. Sodium alginate ($C_6H_9NaO_7$) is an ecofriendly polymer, it was the first byproduct produced from brown seaweeds and widely used in food industries, particularly in packaging industries and other applications in pharmaceutical industries.

16.2.3.3 CARRAGEENAN

Carrageenan is water-soluble polysaccharides extracted from the red seaweeds. It has high potential activities for film-forming material due to the linear chain with partially sulfated galactans. Carrageenan has countless applications in food and other industries; it is used as a high value functional ingredient in foods products. Carrageenan can be used as a stabilizer, as oxygen barrier, pet food, infant food and nutritional supplement, suspending agent, antimicrobial substance for printing, cosmetics, pharmaceuticals, and textile industries. Carrageenan-based film has poor transparency as comparable to starch-based film. Carrageenan is generally used as a surface coating material. There are few studies on the development of improved structural and mechanical properties of carrageenan-based films even with some active components.

16.2.3.4 AGAR

Agar ($C_{14}H_{24}O_9$) is an unbranched hydrocolloid consisting of polysaccharide derivative of agarose and agaro-pectin which is extracted from red algae and widely used as gelation agent in food processing industries. It forms a strong gel by melting point far above the initial temperature, called gelation formation. The agar-based edible film/coating have good mechanical strength, physical, and functional properties, transparent, homogenous, and flexible. It is widely used as an edible packaging in food and food products due to their many functional properties. The concentration of plasticizer is an important factor to determine the characteristics of agar-based edible film, which help to improve film characteristics such as yield, thickness, tensile strength, elongation, and color.

16.3 EDIBLE PACKAGING PARAMETERS THAT EFFECT FOOD QUALITY

16.3.1 MOISTURE LOSS

Loss of moisture from food products causes poor texture and appearance, which may lead to economic losses for the processor. Application of coatings and films prevent the loss in moisture, particularly lipids-based EFs and ECs retard the moisture loss because of their hydrophobic properties. The moisture barrier properties of hydrocolloid-based films and coatings are relatively less, even though, they delay surface moisture loss. There is a gravimetric method used to measure water vapor transmission of EFs. However, it is difficult to measure the water vapor transmission of the coatings due to the irregular shapes of food products, which complicated the accurate measurement. Therefore, it is best to measure directly the weight loss from coated foods.

16.3.2 MECHANICAL PROPERTIES

Mechanical properties of food products are referred to as firmness, crispiness, juiciness, and toughness. Generally, food texture serves as a direct indicator of the product quality and it can be evaluated directly by the consumer. Accurate measurements can be obtained using instruments such as a texture profile analyzer, a universal tensile strength tester, or a dynamical mechanical analyzer, that provide a series of data on textural properties such as tensile strength, elongation, deformability, and elastic properties of the food product. Inferior mechanical properties result in loss of moisture and microbial deterioration. Application of ECs and EFs can reduce or prevent these damages and maintain the texture of foods products.

16.3.3 GAS PERMEABILITY

Gas permeability is an important attribute to study the properties of EFs and ECs and it is mainly used for application on fresh fruit, vegetables, and meat products in some extent. EFs or ECs can control the gas exchange to avoid the creation of anaerobic conditions, that may lead to alcoholic fermentation and formations of off-flavors. The major components that affect the gas permeability of EFs' and ECs' are the formulations used and the thickness of the films or coatings.

16.3.4 ADHESION

Application of EFs and ECs are to form a protective layer on the food surface and facilitate better adhesion to the food surface. Matrix components is an important parameter for good adhesion, that dictate the density, viscosity, surface tension and roughness. Hydrophilic ECs such as protein and polysaccharide-based coating, show good adhesion to fish and fishery products.

16.3.5 APPEARANCE

The term "appearance" refers to color, glossiness, and absence of any defects in food products. Appearance of the food products is the most important feature that will eventually make the consumer to decide, whether to buy or not. If any negative effects during their storage, which may hinder their appearance such as oxidative reactions and enzymatic browning, that leads to undesired color changes. Loss of moisture might lead to product shrinkage, irregular shape and unattractive product. Microbial deterioration and mechanical damages may cause spots on the food products. Applications of EFs or ECs can prevent these harmful effects by serving as a protective layer over the food surface and also carry active agents like antimicrobial or antioxidants compounds to extend the shelf life. Some coating materials like gelatin, chitosan, and waxes can enhance the product's glossy appearance and increase the visual attractiveness of the consumers.

16.4 ACTIVE COMPOUNDS INCORPORATION INTO THE EDIBLE FILMS (EFS) AND COATINGS

The most important functions of EFs and ECs is to serve as a matrix to the food products and deliver the active agents that have beneficial properties. For this reason, it's termed as "active EFs or coatings," and the most common active agents for the incorporations, are antimicrobials and antioxidants compounds. However, several other active compounds like aroma and flavor compounds, texture enhancers, and nutraceuticals are being used in the food system. These active agents should be edible and GRAS.

16.4.1 ANTIMICROBIALS

Currently, antimicrobials are widely used in the food system and incorporated as an active compound into EFs and ECs to increase the shelf-life of food products. The antimicrobial agents include organic acids (acetic, benzoic, citric, sorbic, fumaric, propionic, tartaric, and succinic acid), polypeptides (nisin, peroxidase, lysozyme, lactoferrin) and essential oils (lemongrass, oregano, cinnamon, etc.). Organic acids are more active and can penetrate the microorganism cytoplasm and destroy the cell metabolism. Essential oils originated from plants, has an excellent antimicrobial activity against all foodborne pathogen, including human pathogens, and may also affect organoleptic characteristics of foods. The main objective for the incorporation of antimicrobials into EFs and ECs is to carrying matrices effectively, improve the quality and safety, and to extend the food shelf life.

16.4.2 ANTIOXIDANTS

Antioxidants are active compounds extracted from plants, that can absorb or bind free radicals and protect the product from damaging oxidative reactions and ultimately increasing food safety and quality. Enzymatic oxidation is one of the major problems in the food industries, which cause an undesired color change, alter the flavor and odors, softening, and loss of nutrients. Several types of antioxidants were studied as an active compound in EFs and coatings to protect food products from oxidation. A growing demand for natural components in foods and food safety has induced researchers to identify the new natural antioxidants from plant extracts and essential oils.

16.4.3 NUTRACEUTICALS

Nutraceuticals are derived from food sources that can provide health benefits to the consumers. Incorporation of nutraceuticals into the EFs and ECs can compensate the loss of nutrients during their processing and storage of food. However, direct incorporation of nutraceuticals may lead to oxidation and undesired reactions with food component, which is problematic to the consumers. Therefore, EFs, and ECs may serve as a protective and delivery matrix for these compounds. Currently, very few studies have reported the incorporation of nutraceuticals in EFs and ECs to enhance the nutritional quality of the food products.

16.4.4 TEXTURE ENHANCERS

Decline of mechanical properties, particularly textural collapse is a common phenomenon that happens during processing and storage of food products due to moisture loss and enzymatic reactions. Incorporation of textural enhancers such as calcium salts into EFs and ECs can effectively prevent the textural softening with good moisture barrier features. Generally, ECs or EFs have very good moisture barrier properties which can prevent water loss of the products during the storage. While applying the textural enhancers like calcium ions (or other divalent cations) into the ECs or EFs, which can interact with carboxylate containing polymers to create a cross-linked network and improve the firmness of the food products.

16.4.5 FLAVOR AND AROMA COMPOUNDS

Flavor is an important key contributor for food acceptance and maintaining the same flavor thought out the storage period is the biggest challenge for the food processors which is essential for later consumption. Direct applications of flavor components onto the food surface may alter or escape from the product due to their volatile nature or degradation during the prolonged storage period. Essential oils are one of the classic examples of natural flavoring compound and has wide diversity of aroma components. Usually, natural aroma compounds are very sensitive, highly volatile and very difficult to apply evenly on the surface of the food. EFs and ECs are one of the best innovation/methods for the effective addition of natural flavor compounds into food products.

16.5 CONCLUSIONS AND FUTURE APPLICATIONS

Utilizing fish processing waste by converting into EFs and coating and their application on food products to increase the shelf life is an ideal option for modern-day fish processing industries to convert their own processing waste into edible materials. It provides additional benefits to the industries by solving the environment pollution, additional nutrition to the consumers and revenue generation. Scientifically, ECs and films exhibit various functions, such as control the migration of moisture, gases like O_2 and CO_2, flavors, aromas, and lipids; ability to carry food additives such as antioxidants, antimicrobial agents, nutraceuticals, texture enhancer; and the ability to improve the functional and physico-mechanical properties of the food products. There are many functions or properties of EFs and coatings similar to the synthetic packaging films; however, food packaging must be chosen based on the specific applications, the types of food material, and the mechanisms of quality deterioration. Biodegradable EFs and coatings have the vast potential to minimize the use of synthetic biopolymers. However, in order to do so, bio-based edible packaging must perform functions like conventional packaging or synthetic packaging materials and be cost-effective.

KEYWORDS

- **biopolymer**
- **edible coating and films**
- **edible packaging**
- **fishery waste**
- **polysaccharide**
- **protein**

REFERENCES

Anon, (2005). *Assessment of Harvest and Post-Harvest Losses in Marine Fisheries* (p. 122). Central Institute of Fisheries Technology, Cochin.

Bourtoom, T., (2008). Edible films and coatings: Characteristics and properties. *Int. Food Res. J., 15*, 237–248.

Bourtoom, T., (2011). Edible protein films: Properties enhancement. *Int. Food Res. J., 16*, 1–9.

Dhall, R. K., (2013). Advances in edible coatings for fresh fruits and vegetables: A review. *Crit. Rev. Food Sci. Nutr., 53*(5), 435–450.

Embuscado, M. E., & Huber, K. C., (2009). *Edible Films and Coatings for Food Applications*. Springer.

Falguera, V., Quinterob, J. P., Jimenez, A., Munoz, J. A., & Ibarz, A., (2011). Edible films and coatings: Structures, active functions and trends in their use. *Trends Food Sci. Technol., 22*(6), 292–303.

Marina, R., Arantzazu, V., Ana, & BMaría, C. G., (2016). Gelatin-based films and coatings for food packaging applications. *Coatings, 6*, 41. doi: 10.3390/coatings6040041.

Mathew, S., Raman, M., Parameswaran, K. M., & Rajan, D. P., (2019). Waste management in seafood industry. In: *Fish and Fishery Products Analysis*. Springer, Singapore. https://doi.org/10.1007/978-981-32-9574-2_6.

Sahu, B. B., Barik, N. K., Agnibesh, A., Paikaray, A., Mohapatra, S., Senapati, S., Das, K. C., & Sundaray, J. K., (2017). Recycled fish processing waste for welfare of aquaculture industry: A review. *Int. J. Fish. Aqua. Res., 2*(2), 37–42.

Umaraw, P., & Verma, A. K., (2015). Comprehensive review on application of edible film on meat and meat products: An eco-friendly approach. *Crit. Rev. Food Sci. Nutr*. https://doi.org/10.1080/10408398.2014.986563.

Part IV

Emerging Seafood Safety Concern and Quality Management

CHAPTER 17

Antimicrobial Resistant Bacteria: An Emerging Seafood Safety Concern

JERUSHA STEPHEN,[1] MANJUSHA LEKSHMI,[1] PARVATHI AMMINI,[2] BINAYA BHUSAN NAYAK,[1] and SANATH H. KUMAR[1]

[1]*QC Laboratory, Post-Harvest Technology Department, ICAR-Central Institute of Fisheries Education, Mumbai – 400061, Maharashtra, India, E-mail: sanathkumar@cife.edu.in (S. H. Kumar)*

[2]*Department of Biotechnology, Cochin University of Science and Technology, Kochi-682022, Kerala, India*

ABSTRACT

Seafood is one of the widely consumed foods of animal origin. Due to the protein-rich biochemical composition and the nature of the habitat, management of quality and safety of seafood from a microbial perspective is a challenge. Fish and shellfish harvested from coastal-marine environment are known to harbor numerous pathogens, acquired either naturally from the seawater or due to the contamination of the aquatic environment. The problem of human pathogens in seafood is confounded by them being multiple drug-resistant (MDR). Infections by MDR bacteria can be complicated by the fact that these bacteria are recalcitrant to many clinically relevant antibiotics. Many known pathogens have acquired abilities to tolerate multiple antibiotics. Such bacteria can contaminate seafood *via* different pathways and end up in the human food chain. Thus, seafood as vehicles of multiple drug-resistant human pathogenic bacteria is an emerging problem that can potentially compromise the safety of seafood for consumption as well as its acceptance in the global market. The current trend in seafood safety assurance is limited to screening for pathogenic bacteria while grossly overlooking

Advances in Fish Processing Technologies: Preservation, Waste Utilization, and Safety Assurance.
Ranendra K. Majumder, PhD & Amjad K. Balange, PhD (Eds.)

the possibility of the occurrence of antibiotic-resistant bacteria. Hence, it is imperative from the food safety point of view to determine the extent to which antibiotic-resistant bacteria are prevalent in seafood, the molecular basis of antimicrobial resistance, and the sources of such bacteria. This will ultimately help in understanding the dynamics of microbial interaction in the aquatic environment, identifying the critical routes of contamination, developing methods to detect them, and devising strategies to prevent the contamination of seafood.

17.1 INTRODUCTION

Seafood is a rich source of protein, essential fatty acids, vitamins, and minerals. The global consumption of seafood is increasing annually at 2.5%. Seafood has profound significance in promoting the health and food security of the global population. Nearly 20% of the animal protein consumed by a quarter of 7.5 billion population of the world population is met by seafood (FAO, 2014). The per capita consumption of seafood has increased from 10 kg in 1960 to more than 20 kg in 2014 (FAO, 2016). By 2050, the projected global population of 9 billion is expected to steeply increase the demand for food fish to about 270 million tons. In developing nations, seafood is not only an indispensable part of the diet but also a cheap source of protein, essential fats, and vitamins. Consumption of seafood improves general health and wellbeing, particularly in the younger population. Notwithstanding the importance of seafood in the human diet and health, seafood has also been involved in human illness. Seafood-borne illnesses are responsible for a considerable number of morbidity and mortality worldwide. In the United States, seafood was responsible for 857 outbreaks, 4,815 illnesses, 359 hospitalizations, and four deaths between 1998 and 2015 (Barrett et al., 2017). Data from developing countries are mostly unavailable due to the lack of scientific monitoring systems. Also, since most illnesses arising out of seafood consumption are self-limiting, they are largely unreported.

17.2 EMERGENCE OF ANTIBIOTIC-RESISTANT BACTERIA

The discovery of antibiotic penicillin in the 1920s heralded an era of antimicrobial chemotherapy. Several infectious diseases that haunted humans for generations were conquered and cured with antibiotics. The next couple of decades were the golden period of antibiotic discovery, with several more antibiotics being discovered. However, along with antibiotic use came the problem of antibiotic resistance. Bacteria resistant to antibiotics were isolated, and with the increased use of antibiotics, more such bacteria started appearing in clinical settings. During the same time, the observation that the use of antimicrobials in animal feed enhanced the growth of food animals profoundly promoted the use of antibiotics in animal feed. As a result, antibiotic-resistant bacteria started appearing in the animal food production sector (Lekshmi et al., 2017). Such bacteria soon entered the food chain and then into the community. Methicillin-resistant *Staphylococcus aureus* (MRSA) is the best example. The bacterium was initially associated with livestock (Livestock-associated or LA-MRSA). Through the food chain and livestock workers, the bacterium gained entry into the community and became CA-MRSA (community-associated MRSA).

The emergence of antibiotic-resistant bacteria in the clinical setting has been rapid. Through various sources of contamination, such bacteria enter into the environment and the human food chain. In a study conducted in the retail fish markets of Kerala, India, out of the 133 *S. aureus* isolated from seafood, 29.3% were multidrug-resistant (MDR) with resistance to 3 to 8 classes of antibiotics (Murugadas et al., 2016). The emergence of antibiotic-resistant human pathogenic bacteria such as the NDM-1 (New Delhi Metallo-β-lactamase) producing *Klebsiella pneumoniae* and *Escherichia* coli (Yong et al., 2019; Walsh et al., 2011), has raised concerns on resistant bacteria entering into the human food chain and transferring these resistance mechanisms to other related bacteria, both pathogenic and non-pathogenic (Cuzon et al., 2013; Das et al., 2019; Singh et al., 2016). Several human pathogenic bacteria have become extremely resistant to clinically relevant antibiotics. The most important group of such MDR bacteria is termed as ESKAPE pathogens, representing *Enterococcus faecium*, *Staphylococcus aureus*, *Klebsiella pneumoniae*, *Acinetobacter baumannii*, *Pseudomonas aeruginosa,* and *Enterobacter* spp. (Varela and Kumar, 2019).

17.3 MECHANISMS OF ANTIBIOTIC RESISTANCE

17.3.1 ANTIBIOTICS AND THEIR MECHANISMS OF ACTION

Antibiotics are classified into different classes such as β-lactams (penicillin, cephalosporins, carbapenems), quinolones, fluoroquinolones, aminoglycosides, phenicols, peptide antibiotics, etc. Antibiotics can be bactericidal or bacteriostatic. They are also generally classified as broad-spectrum (acting on a range of Gram-positive and -negative bacteria) and narrow spectrum (active against a specific group of bacteria). Different classes of antibiotics have different mechanisms of bacterial inhibition, which again decide whether an antibiotic is bactericidal or bacteriostatic. β-lactams such as penicillins and cephalosporins inhibit bacterial cell wall synthesis. The glycopeptide antibiotic vancomycin employs a similar mechanism. Polymyxins and daptomycin disrupt bacterial membrane structure and destabilize them. Antibiotics such as macrolides and tetracyclines inhibit protein synthesis in bacteria. Fluoroquinolones and rifampin inhibit bacteria by interfering with their nucleic acid replication, while trimethoprim-sulfamethoxazole inhibits specific metabolic pathways (Kumar and Varela, 2013). Antibiotics belonging to a single class generally have a similar mechanism of bacterial inhibition.

17.3.2 ANTIMICROBIAL RESISTANCE

Although the use of antibiotics concomitantly results in the development of antibiotic resistance, a mechanism termed as acquired resistance, and bacteria may also be intrinsically resistant to certain antibiotics. Intrinsic resistance is developed in natural bacteria as a strategy to ward off the inhibitory effects of antibiotic-producing bacteria in the same environment. Alternatively, depending on the mechanism of antibiotics, certain bacteria may naturally be impermeable to antibiotics or lack the antibiotic target (Andersen et al., 2015).

17.3.2.1 MECHANISMS OF ANTIBIOTIC RESISTANCE

Bacterial mechanisms of antibiotic resistance include enzymatic hydrolysis of antibiotics, alteration of antibiotic targets, reduced permeability, and antibiotic extrusion by efflux pumps (Zaman et al., 2017). Enzymatic hydrolysis of the antibiotics is used against β-lactams antibiotics. The group of hydrolyzing enzymes, commonly referred to as β-lactamases, cut open the β-lactam ring, thus hindering the binding with the target molecule (Reygaert, 2018). Esterases hydrolyze the lactone ring present in the macrolide antibiotics such as erythromycin (Nakajima, 1999). Modification of the antibiotic target is seen against most of the antibiotic classes. For example, modification of PBPs (penicillin-binding proteins) is a resistance mechanism against β-lactam drugs. A similar mechanism of antibiotic resistance includes methylation of the ribosomal subunit and ribosomal mutation against aminoglycosides; and modification of DNA gyrase or topoisomerase IV against fluoroquinolones. The resistance due to reduced permeability is achieved by the reduction and or the modification of the porin channels. Efflux pumps remove the harmful compounds from the bacterial cell. Specific efflux pumps are capable of transporting a large variety of compounds that are termed as a multidrug (MDR) efflux pump (Figure 17.1) (Andersen et al., 2015; Kumar et al., 2020).

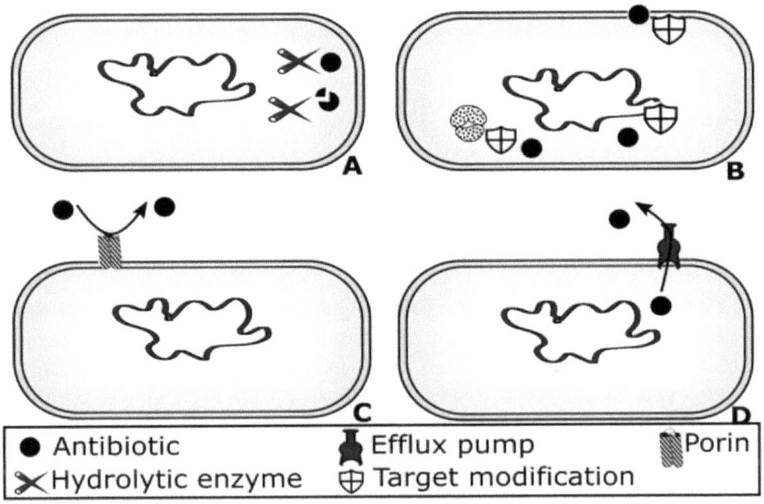

FIGURE 17.1 Mechanism of antibiotic resistance.
Note: (A) enzymatic hydrolysis of antibiotics; (B) alteration of antibiotic targets; (C) reduced permeability; and (D) antibiotic extrusion by the efflux pumps.

17.3.2.2 DISSEMINATION OF THE ANTIBIOTIC RESISTANCE

Exposure to sublethal levels of antibiotics promotes resistance development. The genetic factors responsible for antibiotic resistance usually reside on mobile genetic elements such as plasmids and transposons. Related or unrelated bacteria can acquire the genes through horizontal gene transfer (HGT), a mechanism responsible for the rapid dissemination of resistance genotypes. HGT results in the rapid spread of the MDR phenotype in hospital environments. However, a similar phenomenon can occur in the aquatic environment when MDR bacteria are introduced *via* sewage contamination.

Resistance genes can be transferred to co-occurring pathogenic and non-pathogenic bacteria in the aquatic environment. Contamination of coastal water, rivers, and lakes with antibiotic-resistant human pathogenic bacteria results in such bacteria being found in fish and shellfish harvested from these waters (Figure 17.2).

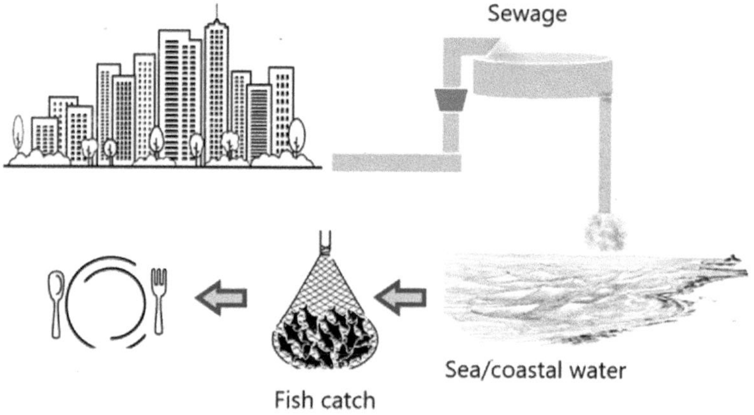

FIGURE 17.2 In thickly populated cities, a significant portion of urban sewage is let into coastal waters untreated. Land runoff during monsoon, human activities in landing centers, and beaches are also crucial sources of MDR bacteria into these environments. Fish and shellfish harvested from contaminated water can harbor MDR bacteria, which ultimately enter the human food chain and can pose a significant human health risk (Singh et al., 2017, 2020).

17.4 SEAFOOD-BORNE ILLNESSES

The consumption of seafood is generally safe, but it depends on its source, the way seafood is handled and prepared before consumption. Contamination of seafood with chemical and biological agents harmful to human health can occur in the sea or after it is harvested. The biological agents of concern in seafood are pathogenic bacteria, viruses, and algal toxins. In the United States, bacterial pathogens are the predominant causes of seafood-associated outbreaks being responsible for 76% of the outbreaks, followed by viruses (21.3%) and parasites (2.6%) (Iwamoto et al., 2010). Certain groups of bacteria are naturally associated with seafood, such as the vibrios, some of which are potentially pathogenic. Other human pathogenic bacteria such as *E. coli*, *Salmonella*, *Campylobacter*, *Listeria monocytogenes,* etc., are introduced into seafood as secondary contaminants. Fecal contamination of coastal waters is the primary source of pathogenic *E. coli* and *Salmonella* in seafood (Prakasan et al., 2018; Prabhakar et al., 2020). Contamination can occur during handling in landing centers and retail markets, as well as during transportation and processing. The majority of seafood-borne infections are attributed to the consumption of raw or improperly cooked seafood. Pasteurization is sufficient to kill pathogenic bacteria and viruses in seafood and make them safe for consumption. However, when contamination occurs in ready-to-eat (RTE) products, the human infection can occur when such products are consumed without further processing. Studies from India have reported the incidence of human pathogenic bacteria in seafood. Hygienic handling, low-temperature preservation, and proper handling and processing practices guided by HACCP principles have been advocated to reduce the risk of pathogenic bacteria in seafood.

17.4.1 ANTIBIOTIC-RESISTANT BACTERIA IN SEAFOOD

While the pathogens associated with seafood have always been a concern, the recent surge in antimicrobial resistance of pathogens has stressed on the need to monitor the aquatic environment and the fish and shellfish for the occurrence of drug-resistant bacteria. With the emergence of antibiotic-resistant pathogenic bacteria becoming a global problem, there is an immediate need to address this issue of development and spread of antibiotic resistance. The contribution of aquaculture towards the global burden of antibiotic resistance is evident from various works (Pham et al., 2018). Antibiotic residues found in the environment also play a vital role in the development of antibiotic resistance in bacteria. In Portugal, aquaculture systems and surrounding areas were loaded with fluoroquinolones and tetracycline antibiotics, with norfloxacin occurring in high frequency and concentration (Pereira et al., 2015). Antibiotic-resistant bacteria are estimated to cause around 300 million premature deaths by 2050, with a loss of up to $100 trillion to the global economy (Munita and Arias, 2018). Seafood is known to be vehicles for numerous bacterial pathogens, acquired either naturally (e.g., *Vibrio* spp.) from the seawater or via the human contamination of the aquatic environment (e.g., *E. coli*, *Salmonella,* etc.). The presence of multiple drug-resistant (MDR) pathogens in seafood is an intriguing problem. Treatment of infections caused by such bacteria could be unmanageable owing to them being resistant to the antibiotics used. Resistance to multiple antibiotics has been reported in many of the enteric human pathogens. Contamination of seafood can occur with MDR pathogens through various routes. Hence, seafood acts as a potential vehicle for transmission of such bacteria, thereby posing a serious threat to its safety.

17.4.2 QUINOLONE RESISTANCE

Quinolones consist of antibiotics such as norfloxacin, ciprofloxacin, nalidixic acid etc. The resistance is mainly due to the modification of DNA gyrase enzyme (encoded by *gyrA* and *gyrB* genes) and topoisomerase IV enzyme (encoded by the *parC* and *parE* genes) involved in bacterial DNA replication (Miranda et al., 2013). The modifications of enzymes are affected by the chromosomal mutation of the respective genes. DNA gyrase enzyme has a tetrameric structure, which is composed of two A and B subunits, respectively, encoded by *gyrA* and *gyrB*. The Quinolone-Resistance Determining Region (QRDR) is present between positions 67 to 106 in *gyrA* subunit-containing some specific codons. Even a single mutation in this region interferes with the attachment of quinolone antibiotics with its target. In *gyrB* subunit, some mutations have been observed in positions such as 426 and 447, which result in the inability of quinolone antibiotics to bind with their targets. Topoisomerase IV, which is another important target, is also a tetrameric enzyme similar to DNA gyrase. It also consists of two types of subunits, such as ParC and ParE, each of which is present in two numbers. Some specific positions in both the subunits have been reported, such as 78, 80, and 84 in ParC, and 445 in ParE, where mutations have been known to occur, resulting in the alteration of binding sites so that the quinolone antibiotics cannot bind with their targets. The genes are chromosomally encoded. Later, quinolone resistance encoded on transmissible plasmid were reported. The plasmid carrying *qnr* gene is responsible for quinolone resistance. The genes encode for pentapeptide repeat proteins that block the target of quinolone antibiotic, i.e., DNA gyrase and Topoisomerase IV. As a result, the antibiotic cannot bind with its target. The plasmid-mediated quinolone resistance (PMQR) genes like *qnr, qeqA, oqxAB*, or *qnrVC* are also responsible for the quinolone antibiotics resistance (Po et al., 2018).

17.4.2.1 OCCURRENCE OF FLUOROQUINOLONE RESISTANT BACTERIA IN SEAFOOD

Fluoroquinolone resistance in *Vibrio* spp., especially *V. parahaemolyticus,* is a serious concern for the seafood consumers since fluoroquinolone is a major choice of drug in combating *Vibrio* infections. A recent study has shed light on 10 new genes that could possibly play a major role along with point mutation in the *gyrA* gene and *parC* genes, inducing ciprofloxacin resistance in *V.* parahaemolyticus (Zhou et al., 2019). In Hong Kong, the *V. parahaemolyticus* clinical isolates exhibiting point mutation in the *gyrA* gene was most prevalent, followed by *parC* gene (Chen et al., 2020). In China, 784 food samples collected from 12 cities showed the occurrence of fluoroquinolone-resistant *V. parahaemolyticus* harboring point mutation in *gyrA* and *parC* genes along with PMOR genes *qnrD* and *qnrS* (Lei et al., 2020). The wide use of biofertilizers (biologically treated pig manure and poultry dung) for shrimp aquaculture in Thailand and certain parts of China create conducive ambience for HGT. The higher abundance of PMQR *oqxB* in the biofertilizer samples and its likelihood for HGT is high. This gives a different perspective of using the so-called natural or organic manure in aquaculture systems (Zhao et al., 2020).

Nalidixic acid-resistant *Salmonella* was isolated from tilapia or catfish imported from Taiwan into the USA (Zhao et al., 2003). *Salmonella* Weltevreden and *Salmonella* Newport isolated from seafood showed the point mutation in the *gyrA* gene, which resulted in the nalidixic acid resistance (Deekshit et al., 2015). Two *qnrS1*-positive isolates of *Salmonella* Corvallis and *S. Typhimurium* were reported from imported seafood in the USA. The plasmid carrying *qnrS1* was similar to those of the clinical isolates reported in Taiwan and the United Kingdom (Akiyama et al., 2012). This shows the extent of antibiotic resistance gene dissemination occurring in various ways. In South Korea, 95% of the *Aeromonas* spp. isolated from the frozen white-leg shrimp (*Litopenaeus vannamei*) showed resistance to nalidixic acid (De Silva et al., 2018).

17.4.3 COLISTIN RESISTANCE IN SEAFOOD-BORNE PATHOGENS

Colistin, a polycationic peptide antibiotic, is used as the last resort antibiotic for treating MDR Gram-negative pathogenic infections. The dissemination of mobile colistin-resistance (*mcr*) genes globally through travel as well as through imported food commodities is raising concern worldwide (Hassan and Kassem, 2020). To date, six slightly different variants of *mcr-1*, and a second gene *mcr-2*, have been reported or annotated in the GenBank database. MDR *V. parahaemolyticus* carrying plasmid-mediated *mcr*-1 gene along with the β-lactamase gene and PMOR was isolated from the food sample in China

(Lei et al., 2019). Colistin resistance was recorded in 83% of the *vibrios* isolated from shellfishes in Cochin (Sudha et al., 2014). Farm animals and aquaculture systems act as a reservoir of *mcr* genes (Shen et al., 2020). The occurrence of colistin-resistant *Enterobacteriaceae* members from various sources is a serious concern since colistin is used for treating carbapenem-resistant *Enterobacteriaceae* (CRE) (Almeida et al., 2018; Ellis-Iversen et al., 2020; Ovejero et al., 2017; Slettemeas et al., 2017).

17.4.4 EXTENDED SPECTRUM-Β-LACTAMASE (ESBL)-PRODUCING BACTERIA

Enzymatic hydrolysis of antibiotics, a strategy employed by β-lactamase producing Gram-negative bacteria, is an essential mechanism of antibiotic resistance. β-lactam antibiotics, particularly cephalosporins of different generations, are the most commonly used antibiotics against infectious diseases. However, rapidly growing resistance to cephalosporins in the form of ever-expanding ESBL group of lactamases is threatening the future of antimicrobial therapy involving cephalosporins. ESBLs are plasmid-mediated enzymes that hydrolyze several β-lactam antibiotics, including third-generation cephalosporins, penicillins, and aztreonam. ESBLs have evolved from TEM (first discovered in an *Escherichia coli* isolate from a patient named Temoneirain Athens, Greece), and SHV (sulphydryl variable active site) type β-lactamases by point mutation. Presently, the total number of ESBLs exceeds 200. ESBLs are generally susceptible to β-lactamase inhibitors like clavulanates and sulbactam. Among these, CTX-M is a relatively new ESBL with more than 172 variants and is widely distributed throughout the world. CTX-M enzymes hydrolyze cefotaxime (CTX) better than ceftazidime (CAZ). All ESBL-producing bacteria are resistant to penicillins and cephalosporins and susceptible to cephamycins and aztreonam (Ranjana et al., 2017).

Carbapenems are another important group of antibiotics that remain as last resort antibiotics against ESBL producing bacteria. However, CRE that produce one or more carbapenemases have evolved to hydrolyze carbapenem antibiotics. Carbapenemases are Metallo-β-Lactamases (MBLs) that have a metal cofactor in their active sites. These belong to the Ambler class A (KPC) β-lactamases such as New Delhi Metallo-β-lactamase (NDM), imipenemase (IMP), Verona Integron-encoded Metallo-β-lactamase (VIM) and the Class D oxacillinases (OXA).

17.4.5 OCCURRENCE OF ESBL-PRODUCING BACTERIA IN SEAFOOD

It is estimated that nearly 50% of the healthy Southeast Asian population carry the ESBL producing *E. coli* (Harada, 2018). The ESBL producing *Enterobacteriaceae* members, especially *E. coli,* have become a significant concern in the seafood industry across the globe (Singh et al., 2020; Ir et al., 2019; Vu et al., 2018; Sapugahawatte et al., 2020; Harada et al., 2018; Baothman et al., 2020; Sivakumar et al., 2020; Yamaguchi et al., 2018; Boss et al., 2016; Boonyasiri et al., 2014). The occurrence of ESBL producing bacteria in various stages of food processing from the culture site (Hoa et al., 2020), capture site (Hassen et al., 2020; Sellera et al., 2018; Brahmi et al., 2017), processing facilities, retail marketing (Harada et al., 2018; Janecko et al., 2016) and processing effluent (Divyashree et al., 2019; Sivaraman et al., 2018) reflects the critical outcome of antibiotics overuse. A meta-analysis of published reports on ESBL bacteria in food animals determined the overall prevalence of ESBL producing isolates in animal samples in India to be 10% (Kuralayanapalya et al., 2019). Zone-wise analysis revealed the highest prevalence in the North zone (26%), 11% in the East, 6% in the South, and 8% in central zones. According to this study, the highest prevalence of ESBL is in *K. pneumoniae* (10%), followed by *E. coli* (9%) and *Pseudomonas* spp. (5%) (Kuralayanapalya et al., 2019). A few studies on seafood isolates of *Enterobacteriaceae* suggest the presence of ESBL-producing *Enterobacteriaceae*. A study reported that 78.60% of *Enterobacteriaceae* isolated from seafood in Mumbai were ESBL-positive, with *E. coli* as the predominant species (Singh et al., 2017). The ESBL-producing isolates also harbored the metallo-β-lactamase-encoding genes bla_{OXA}, bla_{NDM}, and bla_{VIM} (Singh et al., 2020). At Bizerte lagoon, among the 126 wild fish sampled, *E. coli* was the predominant bacteria among the isolated ESBL producing *Enterobacteriaceae* (Hassen et al., 2020). Four ESBL positive *E. coli* were isolated from retail fish outlets at Veraval, India (Sivaraman et al., 2015). Elhadi (2014) analyzed the frozen freshwater fish imported into Saudi Arabia and found that 27.2% of the samples harbored ESBL-producing *E. coli.* Harada et al. (2018) reported that about 36% of *E. coli* isolated from fish gut contents in the Mekong Delta, Vietnam was ESBL-positive. Nguyen et al. (2016) found that 29.3% of fish/shrimp samples in the Ho Chi Minh City of Vietnam harbored ESBL-producing *E. coli.* In Brazil, Two strains of *Enterobacter cloacae* isolated from cultured shrimp tested positive for ESBL production (Almeida et al., 2017). In Phnom Penh, Cambodia, 17% of the fish samples harbored ESBL positive *Salmonella enterica* with predominant serovar being

Saintpaul and Newport (Nadimpalli et al., 2019). ESBL positive *Vibrio spp.* in hard-shelled mussels (*Mytilus couscous*) were reported from Korea, and the bla_{CTX} gene was the most prevalent in the isolated *Vibrios* (Hossain et al., 2019). In Vietnam, the ESBL producing *E. coli* was prevalent in 40.6% of the retail meat, and shrimp samples and B1 was the most commonly isolated ESBL phylogroup (Le et al., 2015). More studies are necessary to understand the prevalence of ESBL-producing pathogens in seafood and how much do they contribute to the disease burden involving MDR bacteria in the community. The cycle involving anthropogenic contamination of harvesting or growing water, contamination of fish and shellfish with MDR bacteria and their dissemination in the community through seafood chain is a critical aspect of antibiotic resistance spread that has received little prominence.

17.5 DETECTION OF ESBLS AND MBLS

17.5.1 DETECTION OF ESBLS USING CHROMOGENIC AGAR

Chromogenic agars such as the HiCrome ESBL agar (HiMedia, Mumbai) are useful for preliminary identification of the ESBL-positive phenotype. Bacteria that exhibit resistance to one or more of indicator cephalosporins viz., CTX, CAZ, cefoxitin (CX), cefpodoxime (CPD) are streaked or spotted on chromogenic agar. A blue or bluish-green colony color usually indicates ESBL production.

17.5.2 DOUBLE DISC SYNERGY TEST (DDST) FOR ESBL PRODUCTION

ESBLs are generally inhibited by the β-lactamase inhibitor clavulanic acid. In the double-disk synergy test, clavulanic acid-mediated inhibition of ESBL, leading to potentiation of cephalosporin against ESBL-positive bacterium, is tested (Chaudhary and Aggarwal, 2004). If the bacterium is an ESBL producer, the zone of inhibition around cephalosporin will get extended towards the amoxicillin-clavulanate disc. This extension of the inhibition is described as a "keyhole effect" (Wayne, 2011). A second method, called the combination disc method, employs two cephalosporin discs, one of which contains clavulanic acid [e.g., CAZ (30 μg) alone and ceftazidime + clavulanic acid (30 μg/10 μg)] placed 25 mm apart. An increase in zone diameter by 5 mm or more around the clavulanic acid-containing disc is considered indicative of ESBL production (Figure 17.3) (Feraro et al., 2002).

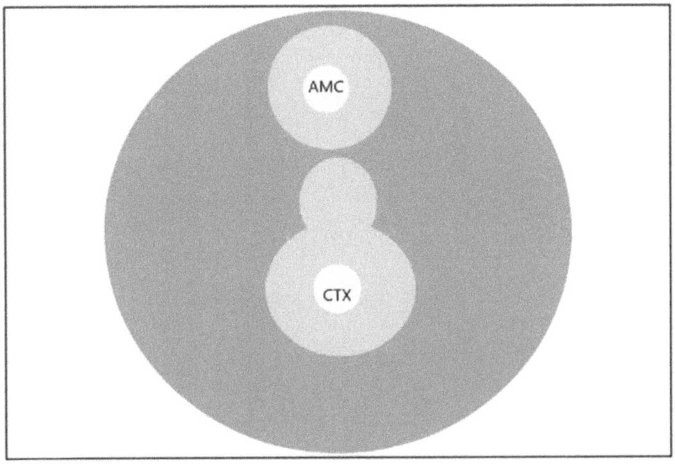

FIGURE 17.3 Double disc synergy test. The inhibitory effect of clavulanic acid on ESBLs results in the extension of the inhibition zone around a cephalosporin (CTX) in the form of a "keyhole."

17.5.3 ESBL STRIPS

Gram-negative bacteria that produce both ESBLs and AmpC β-lactamase are being increasingly reported worldwide. AmpC β-lactamases are cephalosporinases that differ from ESBL by being resistant to β-lactamase inhibitors. They hydrolyze cephamycins but not the 4th generation cephalosporins such as cefepime. When a bacterium produces both AmpC and ESBLs, conventional methods such as DDST can show erroneous results. Ezy MIC™ strip–Mix+/Mix (Hi-Media, India) can help detect the production of AmpC and ESBLs simultaneously. The upper half of the strip denoted as "mixture+" is coated with CAZ, CTX, cefepime, plus clavulanic acid. In contrast, the lower half of the strip marked as "mixture" is coated with the same antibiotics in a reverse concentration gradient, but without an inhibitor. The strip is placed on the lawn culture of the test strain on MH agar, and MICs are read from mixture and mixture+ ends of the strip. The bacterium is considered ESBL-positive if the mix/mix+ value is ≥8. Alternatively, AmpC β-lactamase can be inhibited by adding 200 μg/mL cloxacillin in the agar medium (Figure 17.4).

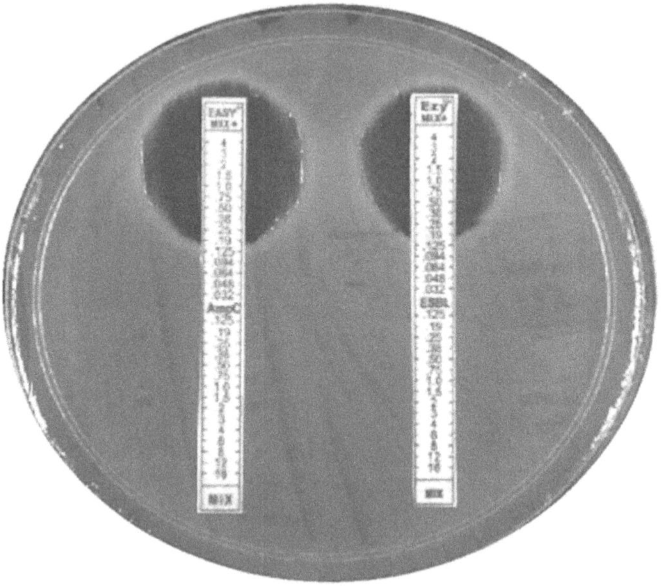

FIGURE 17.4 EZY MIC™ strips detect production of both AmpC and ESBL enzymes.

17.5.4 COMBINED DISK METHOD FOR DETECTION OF MBL PRODUCTION

This method employs the susceptibility of metallo-β-lactamases to metal sequestering agents such as the EDTA. Two imipenem discs, one with imipenem alone and the second with imipenem+ 292 μg EDTA are placed 25 mm apart on the MH plate. An optimum distance of the disc is critical for a successful result. A strain producing ≥4 mm zone of inhibition around imipenem-EDTA compared to imipenem alone is considered MBL-positive (Nordmann et al., 2011) (Figure 17.5).

17.5.5 DETECTION OF ANTIBIOTIC RESISTANCE GENES

Detection of antibiotic resistance genes, particularly those encoding ESBLs and MBLs, ensures rapid detection of resistant bacteria. PCR protocols for ESBL- and MBL-encoding genes such as bla_{TEM}, bla_{SHV}, bla_{CTX}, bla_{NDM}, bla_{OXA}, etc., have been well established. Sequencing of PCR products can potentially help to identify the type of ESBL genes most prevalent in a geographical region, as well as the emergence of a new variant gene. Considering the rapid evolution of ESBL genes, this approach is necessary to monitor changes in antibiotic resistance genotypes that result in changes in MDR phenotypes of pathogenic bacteria.

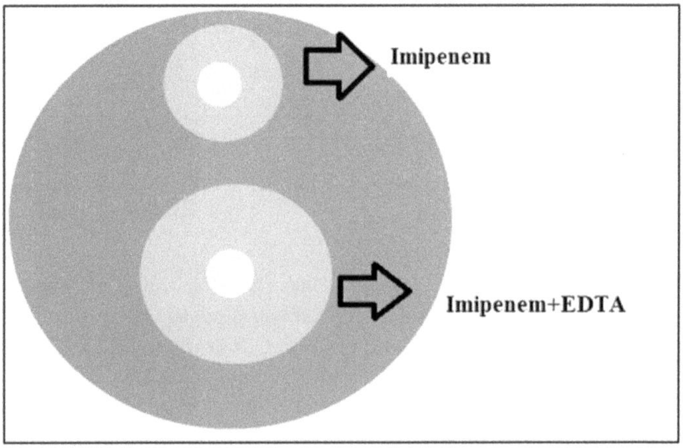

FIGURE 17.5 Combined disc method for MBL detection. EDTA enhances the effect of a carbapenem such as imipenem, resulting in a larger zone of inhibition.

17.6 CONCLUSIONS

Anthropogenic activities are exerting extreme pressure on the fragile ecological balance within the natural environments. The urban population in coastal cities is increasing at a phase faster than the infrastructural growth, particularly in terms of sanitation. Coastal pollution with sewage is a significant source of pathogenic as well as MDR bacteria. The consequence of coastal pollution is evident from the frequent incidence of pathogenic and MDR bacteria in seafood. The presence of multidrug-resistant, human-adapted pathogenic bacteria, especially *E. coli* carrying multiple antibiotic resistance genes, pose a considerable health risk to the seafood consumer community. The seafood harboring MDR bacteria can also act as vehicles of transmission for resistant bacteria into the community. This can potentially threaten the quality and safety credentials of seafood industries and can also affect the trade. This situation requires urgent attention of all concerned, so that suitable policies to protect the coastal ecosystem as a critical source of human food can be formulated and implemented.

KEYWORDS

- **antibiotic resistance**
- **extended-spectrum β-lactamase (ESBL)**
- **multiple drug resistance (MDR)**
- **quinolone-resistance determining region**
- **ready-to-eat**
- **seafood**

REFERENCES

Akiyama, T., & Khan, A. A., (2012). Isolation and characterization of small *qnrS1*-carrying plasmids from imported seafood isolates of *Salmonella enterica* that are highly similar to plasmids of clinical isolates. *FEMS Immunol. Med. Microbiol., 64*(3), 429–432.

Almeida De M. V. A., Cangussú, Í. M., Carvalho, A. L. S. D., Brito, I. L. P., & Costa, R. A., (2017). Drug resistance, AmpC-β-lactamase and extended-spectrum β-lactamase-producing Enterobacteriaceae isolated from fish and shrimp. Rev. Inst. Med. Trop. São Paulo, 59, doi.org/10.1590/s1678-9946201759070.

Almeida De, M. V. A., Brito, I. L. P., Carvalho De, A. L. S., & Costa, R. A., (2018). *In vitro* resistance of *Enterobacter cloacae* isolated from fresh seafood to colistin. *Rev. Soc. Bras Med. Trop.*, 674, 675.

Andersen, J., He, G. X., Kakarla, P., Kc, R., Kumar, S., Lakra, W., Mukherjee, M., et al., (2015). Multidrug efflux pumps from *Enterobacteriaceae*, *Vibrio cholerae* and *Staphylococcus aureus* bacterial food pathogens. *Int. J. Environ. Res. Public. Health, 12*(2), 1487–1547.

Baothman, O. A., Alshamrani, Y. A., & Al-Talhi, H. A., (2020). Prevalence of extended-spectrum β-lactamases in *Enterobacteriaceae* isolated from polluted wild fish. *Open Biochem. J., 14*(1), doi: 10.2174/1874091X02014010019.

Barrett, K. A., Nakao, J. H., Taylor, E. V., Eggers, C., & Gould, L. H., (2017). Fish-associated foodborne disease outbreaks: The United States, 1998–2015. *Foodborne Pathog. Dis., 14*(9), 537–543.

Boonyasiri, A., Tangkoskul, T., Seenama, C., Saiyarin, J., Tiengrim, S., & Thamlikitkul, V., (2014). Prevalence of antibiotic resistant bacteria in healthy adults, foods, food animals, and the environment in selected areas in Thailand. *Pathog. Glob. Health, 108*(5), 235–245.

Boss, R., Overesch, G., & Baumgartner, A., (2016). Antimicrobial resistance of *Escherichia coli, Enterococci, Pseudomonas aeruginosa*, and *Staphylococcus aureus* from raw fish and seafood imported into Switzerland. *J. Food Prot., 79*(7), 1240–1246.

Brahmi, S., Touati, A., Dunyach-Remy, C., Sotto, A., Pantel, A., & Lavigne, J. P., (2017). High prevalence of extended-spectrum β-lactamase-producing *Enterobacteriaceae* in wild fish from the Mediterranean Sea in Algeria. *Microb. Drug Resist., 24*(3), 290–298.

Chaudhary, U., & Aggarwal, R., (2004). Extended-spectrum–lactamases (ESBL): An emerging threat to clinical therapeutics. *Indian J. Med. Microbiol., 22*(2), 75.

Chen, S., Chan, E., Po, K., Ye, L., & Li, R., (2020). Molecular mechanisms of fluoroquinolone and expanded-spectrum cephalosporin resistance in *Vibrio parahaemolyticus*: Abridged secondary publication. *Hong Kong Med. J., 26*(3), 5.

Cuzon, G., Bonnin, R. A., & Nordmann, P., (2013). First identification of novel NDM carbapenemase, ndm-7, in *Escherichia coli* in France. *PLoS One, 8*(4), e61322.
Das, U. N., Singh, A. S., Lekshmi, M., Nayak, B., & Kumar, S., (2019). Characterization of bla NDM-harboring, multidrug-resistant *Enterobacteriaceae* isolated from seafood. *Environ. Sci. Pollut. Res., 26*(3), 2455–2463.
De Silva, B. C. J., Hossain, S., Dahanayake, P. S., & Heo, G. J., (2018). Frozen white-leg shrimp (*Litopenaeus vannamei*) in Korean markets as a source of *Aeromonas* spp. harboring antibiotic and heavy metal resistance genes. *Microb. Drug Resist., 24*(10), 1587–1598.
Deekshit, V. K., Kumar, B. K., Rai, P., Karunasagar, I., & Karunasagar, I., (2015). Differential expression of virulence genes and role of *gyrA* mutations in quinolone-resistant and susceptible strains of *Salmonella* Weltevreden and Newport isolated from seafood. *J. Appl. Microbiol., 119*(4), 970–980.
Divyashree, M., Vijaya, K. D., Ballamoole, K. K., Shetty, A. V., Chakraborty, A., & Karunasagar, I., (2019). Occurrence of antibiotic resistance among gram-negative bacteria isolated from effluents of fish processing plants in and around Mangalore. *Int. J. Environ. Health Res.*, 1–8.
Elhadi, N., (2014). Prevalence and antimicrobial resistance of *Salmonella* spp. in raw retail frozen imported freshwater fish to eastern province of Saudi Arabia. *Asian Pac. J. Trop. Biomed., 4*(3), 234–238.
Ellis-Iversen, J., Seyfarth, A. M., Korsgaard, H., Bortolaia, V., Munck, N., & Dalsgaard, A., (2020). Antimicrobial-resistant *E. coli* and *Enterococci* in *Pangasius* fillets and prawns in Danish retail imported from Asia. *Food Control, 114*, 106958.
FAO, (2014). *The State of World Fisheries and Aquaculture 2014: Opportunities and Challenges*. Food & Agriculture Org.
FAO, (2016). *The State of World Fisheries and Aquaculture: Contributing to Food Security and Nutrition for All*. Food Agric. Organ Rome.
Feraro, M. J., Craig, W. A., Dudley, M. N., Eliopoulos, G., Hecht, D. W., Hindler, J. F., Reller, L. B., et al., (2002). Performance standards for antimicrobial disk susceptibility tests, twelfth informational supplement. *Natl. Comm. Clin. Lab. Stand. NCCLS, 22*(1), 33–117.
Harada, K., (2018). Antibiotic residue in environmental water in Vietnam. *Yakugaku Zasshi, 138*(3), 271–275.
Harada, T., Taguchi, M., Kawahara, R., Kanki, M., & Kawatsu, K., (2018). Prevalence and antimicrobial susceptibility of bacterial pathogens in ready-to-eat foods retailed in Osaka prefecture, Japan. *J. Food Prot., 81*(9), 1450–1458.
Hassan, J., & Kassem, I. I., (2020). Audacious Hitchhikers: The role of travel and the international food trade in the global dissemination of mobile colistin-resistance (*mcr*) genes. *Antibiotics, 9*(7), 370.
Hassen, B., Jouini, A., Elbour, M., Hamrouni, S., & Maaroufi, A., (2020). Detection of extended-spectrum β-lactamases (ESBL) producing *Enterobacteriaceae* from fish trapped in the lagoon area of Bizerte, Tunisia. *BioMed Res. Int.* doi.org/10.1155/2020/7132812.
Hoa, T. T. T., Nakayama, T., Huyen, H. M., Harada, K., Hinenoya, A., Phuong, N. T., & Yamamoto, Y., (2020). Extended-spectrum beta-lactamase-producing *Escherichia coli* harboring *sul* and mcr-1 genes isolates from fish gut contents in the Mekong delta, Vietnam. *Lett. Appl. Microbiol., 71*(1), 78–85.
Hossain, S., Wickramanayake, M. V. K. S., Dahanayake, P. S., & Heo, G. J., (2019). Occurrence of virulence and extended-spectrum β-lactamase determinants in *Vibrio* spp. isolated from marketed hard-shelled mussel (*Mytilus coruscus*). *Microb. Drug Resist., 26*(4), 391–401.

Ir, I., Hn, O., Ib, M., Ae, N., En, U., & Ude, U. I., (2019). Isolation of *Pseudomonas* species and extended-spectrum beta-lactamase-producing *Escherichia coli* from retail imported mackerel frozen fishes sold in Abakaliki metropolis. *Arch. Clin. Microbiol., 10*(3).

Iwamoto, M., Ayers, T., Mahon, B. E., & Swerdlow, D. L., (2010). Epidemiology of seafood-associated infections in the United States. *Clin. Microbiol. Rev., 23*(2), 399–411.

Janecko, N., Martz, S. L., Avery, B. P., Daignault, D., Desruisseau, A., Boyd, D., Irwin, R. J., et al., (2016). Carbapenem-resistant *Enterobacter* spp. in retail seafood imported from Southeast Asia to Canada. *Emerg. Infect. Dis., 22*(9), 1675–1677.

Kumar, S., & Varela, M. F., (2013). Molecular mechanisms of bacterial resistance to antimicrobial agents. *Chemotherapy, 14*, 18.

Kumar, S., Mukherjee, M. M., & Varela, M. F., (2013). *Modulation of Bacterial Multidrug Resistance Efflux Pumps of the Major Facilitator Superfamily*. https://www.hindawi.com/journals/ijb/2013/204141/; https://doi.org/10.1155/2013/204141 (accessed on 09 March 2022).

Kuralayanapalya, S. P., Patil, S. S., Hamsapriya, S., Shinduja, R., Roy, P., & Amachawadi, R. G., (2019). Prevalence of extended-spectrum beta-lactamase-producing bacteria from animal origin: A systematic review and meta-analysis report from India. *PloS One, 14*(9), e0221771.

Le, Q. P., Ueda, S., Nguyen, T. N. H., Dao, T. V. K., Van, H. T. A., Tran, T. T. N., Hirai, I., et al., (2015). Characteristics of extended-spectrum β-lactamase–producing *Escherichia coli* in retail meats and shrimp at a local market in Vietnam. *Foodborne Pathog. Dis., 12*(8), 719–725.

Lei, T., Jiang, F., He, M., Zhang, J., Zeng, H., Chen, M., Pang, R., et al., (2020). Prevalence, virulence, antimicrobial resistance, and molecular characterization of fluoroquinolone resistance of *Vibrio parahaemolyticus* from different types of food samples in China. *Int. J. Food Microbiol., 317*, 108461.

Lei, T., Zhang, J., Jiang, F., He, M., Zeng, H., Chen, M., Wu, S., Wang, J., Ding, Y., & Wu, Q., (2019). First detection of the plasmid-mediated colistin resistance gene *mcr*-1 in virulent *Vibrio parahaemolyticus*. *Int. J. Food Microbiol., 308*, 108290.

Lekshmi, M., Ammini, P., Kumar, S., & Varela, M. F., (2017). The food production environment and the development of antimicrobial resistance in human pathogens of animal origin. *Microorganisms, 5*(1), 11.

Miranda, C. D., Tello, A., & Keen, P. L., (2013). Mechanisms of antimicrobial resistance in finfish aquaculture environments. *Front. Microbiol.*, 4.

Munita, J. M., & Arias, C. A., (2016). Mechanisms of antibiotic resistance. *Virulence Mech. Bact. Pathog.*, 481–511.

Murugadas, V., Joseph, T. C., & Lalitha, K. V., (2016). Antibiotic resistance pattern of *Staphylococcus aureus* isolated from seafood. *Fish. Technol., 53*, 232–237.

Nadimpalli, M., Fabre, L., Yith, V., Sem, N., Gouali, M., Delarocque-Astagneau, E., Sreng, N., & Le Hello, S., (2019). The BIRDY study group. CTX-M-55-type ESBL-producing *Salmonella enterica* are emerging among retail meats in Phnom Penh, Cambodia. *J. Antimicrob. Chemother., 74*(2), 342–348.

Nakajima, Y., (1999). Mechanisms of bacterial resistance to macrolide antibiotics. *J. Infect. Chemother., 5*(2), 61–74.

Nguyen, D. T. A., Kanki, M., Nguyen, P. D., Le, H. T., Ngo, P. T., Tran, D. N. M., Le, N. H., et al., (2016). Prevalence, antibiotic resistance, and extended-spectrum and AmpC β-lactamase productivity of *Salmonella* isolates from raw meat and seafood samples in Ho Chi Minh City, Vietnam. *Int. J. Food Microbiol., 236*, 115–122.

Nordmann, P., Poirel, L., Carrër, A., Toleman, M. A., & Walsh, T. R., (2011). How to detect NDM-1 producers. *J. Clin. Microbiol., 49*(2), 718–721.
Ovejero, C. M., Delgado-Blas, J. F., Calero-Caceres, W., Muniesa, M., & Gonzalez-Zorn, B., (2017). Spread of mcr-1-carrying *Enterobacteriaceae* in sewage water from Spain. *J. Antimicrob. Chemother., 72*(4), 1050–1053.
Pereira, A. M. P. T., Silva, L. J. G., Meisel, L. M., & Pena, A., (2015). Fluoroquinolones and tetracycline antibiotics in a Portuguese aquaculture system and aquatic surroundings: Occurrence and environmental impact. *J. Toxicol. Environ. Health A, 78*(15), 959–975.
Pham, T. T. H., Rossi, P., Dinh, H. D. K., Pham, N. T. A., Tran, P. A., Ho, T. T. K. M., Dinh, Q. T., & De Alencastro, L. F., (2018). Analysis of antibiotic multi-resistant bacteria and resistance genes in the effluent of an intensive shrimp farm (Long An, Vietnam). *J. Environ. Manage., 214*, 149–156.
Po, K. H. L., Wong, M. H. Y., & Chen, S., (2015). Identification and characterization of a novel plasmid-mediated quinolone resistance gene, *qnrVC7*, in *Vibrio cholerae* of seafood origin. *Int. J. Antimicrob. Agents, 45*(6), 667, 668.
Prabhakar, P., Lekshmi, M., Ammini, P., Nayak, B. B., & Kumar, S., (2020). *Salmonella* contamination of seafood in landing centers and retail markets of Mumbai, India. *J. AOAC Int., 103*, 1661–1665.
Prakasan, S., Prabhakar, P., Lekshmi, M., Nayak, B. B., & Kumar, S., (2018). Isolation of Shiga toxin-producing *Escherichia coli* harboring variant Shiga toxin genes from seafood. *Vet. World, 11*(3), 379–385.
Ranjana, K., Shrestha, U., Kumar, S., Ranaweera, I., Kakarla, P., Mukherjee, M. M., Barr, S. R., et al., (2017). Molecular biology of multidrug resistance efflux pumps of the major facilitator superfamily from bacterial food pathogens. *Food Borne Pathog. Antibiot. Resist.*
Reygaert, W. C., (2018). An overview of the antimicrobial resistance mechanisms of bacteria. *AIMS Microbiol., 4*(3), 482–501.
Sapugahawatte, D. N., Li, C., Zhu, C., Dharmaratne, P., Wong, K. T., Lo, N., & Ip, M., (2020). Prevalence and characteristics of extended-spectrum-β-lactamase-producing and carbapenemase-producing *Enterobacteriaceae* from freshwater fish and pork in wet markets of Hong Kong. *mSphere., 5*(2), e00107–20. https://doi.org/10.1128/mSphere.00107-20.
Sellera, F. P., Fernandes, M. R., Moura, Q., Carvalho, M. P., & Lincopan, N., (2018). Extended-spectrum-β-lactamase (CTX-M)-producing *Escherichia coli* in wild fishes from a polluted area in the Atlantic coast of South America. *Mar. Pollut. Bull., 135*, 183–186.
Shen, Y., Zhang, R., Schwarz, S., Wu, C., Shen, J., Walsh, T. R., & Wang, Y., (2020). Farm animals and aquaculture: Significant reservoirs of mobile colistin resistance genes. *Environ. Microbiol., 22*(7), 2469–2484.
Singh, A. S., Lekshmi, M., Prakasan, S., Nayak, B. B., & Kumar, S., (2017). Multiple antibiotic-resistant, extended-spectrum-β-lactamase (ESBL)-producing enterobacteria in fresh seafood. *Microorganisms, 5*(3), 53.
Singh, A. S., Nayak, B. B., & Kumar, S. H., (2020). High prevalence of multiple antibiotic-resistant, extended-spectrum β-lactamase (ESBL)-producing *Escherichia coli* in fresh seafood sold in retail markets of Mumbai, India. *Vet. Sci., 7*(2), 46.
Singh, P., Pfeifer, Y., & Mustapha, A., (2016). Multiplex real-time PCR assay for the detection of extended-spectrum β-lactamase and carbapenemase genes using melting curve analysis. *J. Microbiol. Methods, 124*, 72–78.
Sivakumar, M., Abass, G., Vivekanandhan, R., Singh, D. K., Bhilegaonkar, K., Kumar, S., Grace, M. R., & Dubal, Z., (2020). Extended-spectrum beta-lactamase (ESBL) producing

and multidrug-resistant *Escherichia coli* in street foods: A public health concern. *J. Food Sci. Technol.*, 1–15.

Sivaraman, G. K., Vanik, D., Prasad, M. M., Visnuvinayagam, S., Basha, K. A., Nadella, R. K., Jha, A. K., & Vaja, C., (2018). Detection of chromosomal AmpC, *bla CTX-M* in extended-spectrum beta-lactamase-producing *Escherichia coli* in seafood processing effluent. *Proc. Natl. Acad. Sci. India Sect. B Biol. Sci., 88*(3), 1265–1273.

Sivaraman, G., Lalitha, K., Jha, A., Remya, S., Visnuvinayagam, S., & Renuka, V., (2015). *Incidence of Extended-Spectrum Beta-Lactamase (ESBL) Producing Multidrug-Resistant Escherichia coli in Seafood* (pp. 13, 14). Fishtech Reporter.

Slettemeås, J. S., Urdahl, A. M., Mo, S. S., Johannessen, G. S., Grave, K., Norström, M., Steinbakk, M., & Sunde, M., (2017). Imported food and feed as contributors to the introduction of plasmid-mediated colistin-resistant *Enterobacteriaceae* to a 'low prevalence' country. *J. Antimicrob. Chemother., 72*(9), 2675–2677.

Sudha, S., Mridula, C., Silvester, R., & Hatha, A. A. M., (2014). Prevalence and antibiotic resistance of pathogenic *Vibrios* in shellfishes from Cochin market. *Ind. J. Mar. Sci., 43*(5), 10.

Varela, M. F., & Kumar, S., (2019). Strategies for discovery of new molecular targets for anti-infective drugs. *Curr. Opin. Pharmacol., 48*, 57–68.

Vu, T. T. T., Alter, T., Roesler, U., Roschanski, N., & Huehn, S., (2018). Investigation of extended-spectrum and AmpC β-lactamase–producing *Enterobacteriaceae* from retail seafood in Berlin, Germany. *J. Food Prot., 81*(7), 1079–1086.

Walsh, T. R., Weeks, J., Livermore, D. M., & Toleman, M. A., (2011). Dissemination of NDM-1 positive bacteria in the New Delhi environment and its implications for human health: An environmental point prevalence study. *Lancet Infect. Dis., 11*(5), 355–362.

Wayne, P., (2009). Clinical and laboratory standards institute (CLSI) performance standards for antimicrobial disk diffusion susceptibility tests approved standard (19th edn). *CLSI Doc. M100-S19, 29*(2011), M100–S21.

Yamaguchi, T., Kawahara, R., Harada, K., Teruya, S., Nakayama, T., Motooka, D., Nakamura, S., et al., (2018). The presence of colistin resistance gene *mcr-1* and *-3* in ESBL producing *Escherichia coli* isolated from food in Ho Chi Minh City, Vietnam. *FEMS Microbiol. Lett., 365*(11).

Yong, D., Toleman, M. A., Giske, C. G., Cho, H. S., Sundman, K., Lee, K., & Walsh, T. R., (2009). Characterization of a new metallo-β-lactamase gene, *blaNDM*-1, and a novel erythromycin esterase gene carried on a unique genetic structure in *Klebsiella pneumoniae* sequence type 14 from India. *Antimicrob. Agents Chemother., 53*(12), 5046–5054.

Zaman, S. B., Hussain, M. A., Nye, R., Mehta, V., Mamun, K. T., & Hossain, N., (2017). A review on antibiotic resistance: Alarm bells are ringing. *Cureus, 9*(6).

Zhao, S., Datta, A. R., Ayers, S., Friedman, S., Walker, R. D., & White, D. G., (2003). Antimicrobial-resistant *Salmonella* serovars isolated from imported foods. *Int. J. Food Microbiol., 84*(1), 87–92.

Zhao, S., Wei, W., Fu, G., Zhou, J., Wang, Y., Li, X., Ma, L., & Fang, W., (2020). Application of biofertilizers increases fluoroquinolone resistance in *Vibrio parahaemolyticus* isolated from aquaculture environments. *Mar. Pollut. Bull., 150*, 110592.

Zhou, H., Liang, Y., Gao, L., Ren, J., Xue, F., Guo, D., Jiang, Y., Yang, Z., Lian, L., & Dai, J., (2019). Identification and expression analyses of new genes associated with ciprofloxacin resistance in *Vibrio parahaemolyticus*. *Food Res. Int., 125*, 108629.

CHAPTER 18

Emerging Pathogens of Public Health Significance Associated with Seafood

L. MANJUSHA, H. SANATH KUMAR, and BINAYA BHUSAN NAYAK

Department of Post-Harvest Technology, ICAR-Central Institute of Fisheries Education, Mumbai – 400061, Maharashtra, India

ABSTRACT

Seafood acts as a potential vehicle for the transmission of foodborne pathogens among humans. Contamination of seafood through various portals, mainly by unhygienic handling and use of unclean equipment and contact surfaces is a persistent problem. *Vibrio* spp., *E. coli*, *Salmonella enterica*, *Listeria monocytogenes, Clostridium botulinum*, and *Staphylococcus aureus* are among the important pathogens that are reported to cause foodborne illnesses as a result of consumption of contaminated seafood. Emerging pathogens such as *Arcobacter* spp., *Cronobacter* spp., enteric viruses, etc., that have not previously been thought to be foodborne have become considerable threats to food safety and human health later. Microaerophilic *Arcobacter* spp. are highly virulent pathogens of seafood origin that cause gastroenteritis, peritonitis, and bacteremia in infected individuals. *Cronobacter sakazakii* is an important bacterium that contaminates a variety of ready-to-eat and minimally processed foods, including seafood. Its exceptional ability to thrive in dried and dehydrated foods adds to its virulence potential. Enteric viruses are natural inhabitants of the human and animal gastrointestinal tract, which contaminate seafood through the fecal-oral route. Norovirus, adenovirus, enterovirus, and hepatitis A (HA) virus are the significant groups of enteric viruses that cause gastroenteritis disorders in humans post-consumption of contaminated shellfish, especially the bivalve mollusks. Cautious handling

Advances in Fish Processing Technologies: Preservation, Waste Utilization, and Safety Assurance.
Ranendra K. Majumder, PhD & Amjad K. Balange, PhD (Eds.)

of seafood post-harvest and meticulous management practices could be probable strategies for the prevention of contamination of fish and shellfish by enteric pathogens. This is a pre-requisite for ensuring the supply of safe seafood.

18.1 INTRODUCTION

Foodborne illnesses associated with consumption of seafood are of potential public health concern around the world. Seafood-associated infections are caused by bacteria, viruses or parasites, which are acquired by fish and shellfish either naturally from the seawater or *via* the human contamination of the aquatic environment. Human infections associated with fish and fishery products occur when they are consumed raw or undercooked. Fresh and processed seafood are responsible for a significant proportion of outbreaks in the developed countries such as the United States, Europe, and Canada, mainly from molluscan shellfish, followed by finfish and crustaceans (Iwamoto et al., 2010). Gram-negative bacteria such as *Escherichia coli, Salmonella enterica* and vibrios are the important bacterial agents of seafood-borne illness, while norovirus, hepatitis A (HA) virus, adenovirus are significant among the viral pathogens (Amagliani et al., 2012). Vibrios such as *Vibrio cholerae*, *Vibrio parahaemolyticus* and *Vibrio vulnificus* are naturally present in marine environments and are always associated with fish and shellfish harvested from the sea. However, the presence of enteric pathogens such as *Escherichia coli, Salmonella enterica* and enteric viruses is solely due to secondary contamination. Fecal contamination of coastal water by sewage, anthropogenic activities, agricultural, and land run off introduce enteric pathogens of human and animal origin which enter into seafood chain (Kumar et al., 2017). Microbial contamination of seafood can occur through different portals along the food chain from its harvest to consumption. It can occur in the fishing deck, landing centers, processing, and packing centers, distribution chain, retail markets and even during the preparation of food by the consumer. The problem of seafood contamination with pathogenic microorganisms is becoming increasingly important in developing and highly populous countries, where anthropogenic contamination of aquatic environments is quite common. Pathogen contaminated seafood is not only a health risk for domestic consumers but also can threaten the health of consumers in geographical regions far from where the seafood is harvested. Seafood being a globally traded food commodity, the presence of pathogenic microorganisms can have far reaching economic implications in the form of seafood rejections and trade bans.

The important pathogens of human health significance in seafood are *Vibrio* spp., different pathogroups of *E. coli*, *Salmonella enterica*, *Listeria monocytogenes* and *Shigella* spp. Others like *Clostridium botulinum* and *Staphylococcus aureus* are toxin-producing bacteria that cause gastrointestinal illnesses when foods containing their pre-formed toxins are consumed. In addition to these known pathogenic or toxigenic bacteria, many new species of bacteria such as *Aeromonas hydrophila*, *Yersinia enterocolitica*, *Arcobacter* spp., *Cronobacter* spp. and *Acinetobacter* spp. are emerging as important seafood-borne pathogens. Enteric viruses are increasingly becoming important in seafood, since these are found in seafood as a consequence of fecal contamination or contamination during handling. Although some of the above-listed bacteria are not known to have been involved in seafood-borne illnesses thus far, considering their potential to cause human infections and frequent occurrences in seafood, studies are necessary to understand their distribution, sources, virulence genes, and antimicrobial resistance.

18.2 EMERGING PATHOGENS

While the pathogens associated with seafood have always been a concern, the recent emergence and re-emergence of novel pathogens pose new threats to public health. The epidemiology of emerging pathogens has changed in recent years as a consequence of the globalization of the food chain. This has led to microorganisms that have not previously been thought to be foodborne becoming considerable threats to food safety and human health in the future. Emerging foodborne microbial risks refer to those microorganisms that have recently appeared in a population, pathogens extended to new hosts or new vehicles of transmission, or rapid increase in incidences of a particular microbial infection or in a new geographic range. Such emerging pathogens are considered to be crucial microbiological public health hazards (Vouga and Greub, 2016). These also include the microbial strains that are currently prevalent, yet striving to survive under exposure to stress factors in novel environments. Besides, emerging foodborne pathogens are reported to be more virulent and associated with new food products, and also these might have caused infections previously but were wrongly assigned to other microorganisms as causative agents. Some of the emerging pathogens are considered to have newly appeared, which means they are associated with foodborne transmission recently, although they were recognized as pathogens for several years (Mor-Mur and Yuste, 2010).

18.2.1 FACTORS CONTRIBUTING TO EMERGENCE OF PATHOGENS

There has been a surge in the global trade and movement of fishery products as a result of increased demand and consumption of seafood across the world. The year-round availability of fresh and processed seafood caters to this demand of increasing global population (Kumar et al., 2017). However, changes in consumer preferences and the introduction of new and exotic foods can lead to the introduction of novel microbial risks into the food chain as well. The transmission of pathogenic microorganisms in this manner can also favor cross-species infections that are otherwise not apparent, along with the increased ability of these pathogens to adapt to human hosts (Koopmans, 2012). The emergence of new foodborne risks are driven by the acquisition of new virulence or resistance characteristics by demographic changes and associated increases in population susceptibility, or by the emergence of pathogens in a new host species or geographic region. Increase in population density, commercial food trade across the countries, population migration and cross-border travel are some of the important factors responsible for the emergence of microbial pathogens. In addition, alterations in the behavior of microbial pathogens, changing dietary habits of consumers, modified fishing practices and locations also enable rapid global spread of pathogens. Antigenic drifts in the microbial population due to random mutations also result in the emergence and spread of new pathogens. Furthermore, developments in molecular techniques and pathogen detection methods allow increased recognition of the presence of new pathogens. Use of advanced and sensitive molecular amplification techniques help in identification of novel, unknown pathogens that previously remained unidentified (Vouga and Greub, 2016).

The increased human exposure to microbes that were once restricted to animal hosts is considered to be one of the foremost reasons for the emergence of microbial diseases. The intricate food web that interlinks humans, animals, plants, and microbial communities across the globe serves as the major source of human foodborne infections (Behravesh et al., 2012). Disruption of the vital balance among these components of the food chain due to socio-demographic and environmental changes has resulted in enhanced exposure of humans to pathogenic microbes and their transmission among individuals.

Majority of the emerging pathogens are zoonotic in nature, which means healthy food animals act as their reservoirs. They do not cause diseases in the infected host animals. Often, most of them remain sustained in the host animal populations, and incidentally, they spread to different kinds of foods such as milk, meat or eggs, or cause fecal contamination of food and water, thus infecting humans. Once adapted in human hosts, they can rapidly spread around the globe. The identification and determination of pathogenicity and epidemiological characteristics of these emerging pathogens are carried out as a part of foodborne outbreak investigations (Behravesh et al., 2012). Climate change associated with global increase in temperature leading to ecological modifications also result in changes in behavior of animals and pathogens, and thereby cause an increase in the vulnerability of humans to pathogenic infections.

Most of these emerging zoonotic pathogens are increasingly becoming resistant to antibiotics, owing to the extensive use of antibiotics in their food animal settings (Tauxe, 1997). Some of them, which are multi-drug resistant such as the methicillin-resistant *Staphylococcus aureus* (MRSA) and extended-spectrum β-lactamase (ESBL)-producing *E. coli* are reported to be rapidly disseminating among human populations and are considered emerging threats to public health (Vouga and Greub, 2016). MRSA and ESBL-producing enterobacteria have been reported from seafood in India (Kumar et al., 2016; Singh et al., 2017, 2020). Gram-negative and Gram-positive bacteria, viruses, and parasites, mostly of zoonotic origin constitute the emerging pathogens associated with seafood-borne infections (James and Pina, 2018).

18.3 EMERGING BACTERIAL PATHOGENS

18.3.1 ARCOBACTER SPP.

Many species of *Arcobacter* are now known to be human pathogens and have zoonotic potential. They are microaerophilic members of the family *Campylobacteriaceae*, and are Gram-negative, motile, non-spore-forming and spiral-shaped. The genus *Arcobacter* comprises of at least 29 recognized species isolated from different environments and hosts (Chieffi et al., 2020). Of these, *Arcobacter butzleri, Arcobacter cryaerophilus, Arcobacter cibarus* and *Arcobacter skirrowii* are considered as pathogens with potential public health significance (Levican et al., 2015). The infections with *Arcobacter* spp. are manifested with symptoms such as gastroenteritis, peritonitis, bacteremia, etc. They are aerotolerant and can survive at temperatures as low as 15°C (Figueras et al., 2011).

Virulence-associated genes reported from *A. butzleri* include *ciaB, cj1349, cadF, hecA, hecB, mviN, pldA, tlyA, and irgA* (Collado et al., 2014; Rathlavath et al., 2017). Among these, *ciaB* codes for *Campylobacter* invasive antigen, *cadF*, and *cj1349* code for fibronectin-binding proteins. The filamentous hemagglutinin, HecA protein is coded by *hecA* and the hemolysin activation protein is coded by *hecB*. The protein MViN essential for biosynthesis of peptidoglycan is coded by *mviN*, while *pldA* codes for outer membrane phospholipase PldA associated with lysis of erythrocytes. Their on-regulated outer membrane protein IrgA is encoded by *irgA* and *tlyA* codes for hemolysin. The incidence of virulence-associated genes is reported to be higher in *A. butzleri* when compared to *A. cryaerophilus* (Zacharow et al., 2015). This variation in the virulence genes composition may be attributed to their high genome heterogeneity or the variation in their pathogenic behavior. There are reports of large genetic heterogeneity of *Arcobacter* spp. population and different virulence genes are detected in even closely related isolates (Douidah et al., 2012).

The intestinal tract and feces of healthy and diseased animals are common sources of *Arcobacters* and therefore, food products of animal origin are potential sources of transmission of these emerging pathogens. *Arcobacter* was first isolated from aborted bovine and later from porcine fetuses. Livestock animals and birds harbor *Arcobacter* spp., and their isolation has been reported from both healthy and debilitated animals with mastitis, septicemia, reproductive abnormalities and abortion (Patyal et al., 2011). Several studies have isolated *Arcobacter* spp. from varieties of foodstuffs of animal origin, shellfish, and water. Seafood, especially molluscan shellfish such as clams and oysters, are natural reservoirs of *Arcobacter* spp. and hence are considered as potential sources of human infections, especially when they are eaten raw or partially cooked (Collado and Figueras, 2011). *Arcobacter* spp. was isolated from 25% of the fresh fish samples (Rathlavath et al., 2015), 14.7% of shellfish and 20.8% of seawater samples in Mumbai, India (Laishram et al., 2016). Several new species of *Arcobacter* of unknown pathogenic potential have also been reported from shellfish.

18.3.1.1 ISOLATION AND IDENTIFICATION OF ARCOBACTER SPP.

A two-step enrichment and selective plating method is used for the isolation of *Arcobacter* from samples (Rathlavath et al., 2017). Around 48 hour enrichment in the Arcobacter enrichment broth supplemented with antibiotics cefoperazone, amphotericin B and teicoplanin (CAT) and 5% laked horse blood is performed for the selective enrichment. For isolation, *Arcobacter* selective agar supplemented with CAT supplement and 5% laked horse blood is used. Typical small gray–white and round colonies from the Arcobacter blood agar plates are purified by streaking on brain heart infusion (BHI) agar and subjected to biochemical tests such as Gram staining, motility, catalase, oxidase, urease, indoxyl acetate hydrolysis, salt tolerance tests, and growth on MacConkey agar. Genus-specific 16S ribosomal ribonucleic acid (16SrRNA)-based polymerase chain reaction (PCR) is available for the identification of *Arcobacter* spp. (Harmon and Wesley, 1996). A multiplex-PCR assay can be used to differentiate among three pathogenic *Arcobacter* species (Houf et al., 2000).

18.3.2 CRONOBACTER SPP.

Cronobacter spp. (formerly *Enterobacter* spp.), belonging to the *Enterobacteriaceae* family, are ubiquitous, motile, non-spore-forming, Gram-negative facultative anaerobes. They are catalase positive, oxidase negative, and generally positive for α-D-glucosidase. Although all species excluding *Cronobacter condimenti* have been related to clinical diseases, *C. sakazakii, C. malonaticus* and *C. turicensis* are frequently associated with human illnesses (Iversen and Forsythe, 2004). *C. sakazakii* causes necrotizing colitis, meningitis, and bloodstream infections in neonates and infants, and pneumonia, septicemia, osteomyelitis, splenic abscesses, and wound infections in adults (Healy et al., 2010). These bacteria contaminate and survive in a wide range of foodstuffs, including fruits, meat, water, dairy, fresh and dried seafood, eggs, ready-to-eat foods and minimally processed vegetables. The International Commission for Microbiological Specifications for Foods (ICMSF) ranked *Cronobacter* as a "Severe hazard for restricted populations, life-threatening or substantial chronic sequelae or long duration" (ICMSF, 2002). It is considered more resistant to dry and osmotic stresses than other members of the Enterobacteriaceae, which may be attributed to the accumulation of trehalose, an osmoprotectant, inside the cells (Breeuwer et al., 2003). *C. sakazakii* can grow at pH 4.5 and temperature range of 6–45°C, with an optimal range of 37–43°C (Iversen et al., 2006). This wide range of tolerance to physical and chemical conditions makes this bacterium physiologically competent to survive in a wide range of foods and storage conditions. *Cronobacter* infections are often underreported, especially in developing and less-developed countries (Mashoufi et al., 2017).

Different virulence factors such as attachment to intestinal cells, capsule production, biofilm formation and production of enterotoxin are displayed by *C. sakazakii* strains (Iversen and Forsythe 2004). Characteristics like adherence and biofilm formation enable *Cronobacter* spp. to survive as free-living bacteria in different environments, while invasion and the presence of some putative virulence genes (*hly* and *cpa*) allow them to be successful opportunistic pathogens (Cruz et al., 2011). One of the best-characterized virulence factors is the outer membrane protein A encoded by the *ompA* gene which helps in binding and invasion of brain endothelial cells during pathogenesis.

18.3.2.1 ISOLATION AND IDENTIFICATION OF CRONOBACTER SPP.

A two-step enrichment and selective plating method is used for the isolation of *Cronobacter* spp. from seafood samples (U.S. Food and Drug Administration (USFDA), 2004). Pre-enrichment of samples in buffered peptone water at 37°C for 18 to 24 hours is followed by selective enrichment in Cronobacter screening broth supplemented with 0.9 ml of vancomycin and incubation for 24 hours at 37°C. Isolation of the bacterium is undertaken on Hi-Crome *Enterobacter sakazakii* agar. Presumptive isolates of *Cronobacter* exhibiting small, bluish-green, circular colonies are identified by a series of biochemical tests such as Gram's staining, motility test, indole, Voges-Proskauer, catalase, oxidase, urease, and malonate utilization tests. For molecular identification of *Cronobacter* spp., PCR using *Cronobacter* species-specific oligonucleotide primers targeting the 16SrRNA, *dnaG*, and *gluA* genes can be used (Hassan et al., 2007; Iversen et al., 2007).

18.3.3 EMERGING PATHOGENIC VIBRIO SPP.

Vibrios are comma-shaped, Gram-negative, fermentative bacteria that are common inhabitants of estuarine, marine, and freshwater environments. There are over 100 species in the genus *Vibrio*. Species such as *V. cholerae*, *V. parahaemolyticus* and *V. vulnificus* are pathogenic to humans and cause infections via consumption of both farmed and wild-caught aquatic produce (Iwamoto et al., 2010). *V. cholerae* is prevalent in freshwater and more than 200 serotypes are identified. Among these, the serotypes O1 and

O139 are reported to be pathogenic to humans, causing the disease cholera. However, some of the non-O1 and non-O139 serotypes may occasionally cause gastroenteritis or extra-intestinal infections in humans. *V. parahaemolyticus* causes gastroenteritis and *V. vulnificus* causes wound infections, cellulitis, and septicemia, especially in immuno-compromised individuals (Karunasagar and Karunasagar, 2018). Another emerging pathogenic species of *Vibrio* is *V. mimicus*, which is found in fresh and brackish water environments. Gastrointestinal symptoms such as profuse diarrhea and dehydration are characteristic of *V. mimicus* infections in humans (Ali et al., 2020). Being a native inhabitant of marine waters, sediment, fish, and shellfish, and other aquatic animals, *V. parahaemolyticus* is naturally present in seafood, especially shellfish such as shrimps and bivalves. Consumption of raw or partially cooked seafood, particularly bivalves like oysters and clams are often associated with human infections by *V. parahaemolyticus* (Scharer et al., 2011).

The genes responsible for virulence traits of human pathogenic *Vibrio* species are several. Seven pathogenicity islands (PAI) have been reported in *V. parahaemolyticus*, of which the Vp-PAI representing the *tdh* gene and the genes encoding the type III secretion system is the most important one contributing to its virulence (Okada et al., 2009). The hemolysins, thermostable direct hemolysin (TDH), and TDH-related hemolysin (TRH) play an important role in imparting the virulence potential of *V. parahaemolyticus* (Khouadja et al., 2014). *V. parahaemolyticus* infections are usually associated with the consumption of raw or inadequately cooked molluscan shellfish. The serotypes involved in human infections were not given much emphasis until 1996, when a new serotype O3:K6 producing only TDH but not TRH and possessing point mutations in *toxRS* sequences was reported from India as responsible for sudden increase in the cases of gastroenteritis (Okuda et al., 1997). Subsequently, several other serovars such as O4:K68, O1:K25 and O1:KUT were found to be genetically similar to O3:K6 and these were collectively called pandemic clones of *V. parahaemolyticus*. Since then, the pandemic clones of *V. parahaemolyticus* have been responsible for cases of seafood-borne infections all over the world. Based on the variations in the *toxRS* sequences of new clones of *V. parahaemolyticus* compared to the pre-1996 (non-pandemic) clones, a new PCR was developed to specifically identify newly emerged pandemic clones irrespective of their serotypes (Okuda et al., 1997; Matsumoto et al., 2000). Studies from India have reported the occurrence of pandemic clones of *V. parahaemolyticus* in fish and shellfish (Karunasagar et al., 2016; Prabhakaran et al., 2020).

Other *Vibrio* spp. implicated in sporadic cases of human illness includes *Vibrio alginolyticus*, *V. mimicus*, *V. hollissae*, *V. fluvialis* and *V. metschnikovii.* Among these, *V. alginolyticus* causes ear and wound infections, and rarely gastroenteritis, while other species are involved in gastroenteritis, wound, ear, and eye infections (Baker-Austin et al., 2018). Gastroenteritis is caused by consumption of seafood, whereas wound, ear or eye infections can occur while handling seafood or upon exposure to seawater. Epidemiological data on *Vibrio* infections related to consumption of seafood are rare from developing countries. In the United States, systematic data collection was initiated in 1988 with the establishment of cholera and other vibrio information service (COVIS). In temperate countries, *Vibrio* infections are generally observed during warmer summer months, when the water temperature is above 20°C.

Among emerging potentially pathogenic *Vibrio* spp., *V. fluvialis* is considered important. Cases of human infections by *V. fluvialis* are rampant in Asian countries as well as in the United States, causing cholera-like gastroenteritis (Chowdhury et al., 2012; Lesmana et al., 2002; Levine et al., 1993). The bacterium exhibits yellow (sucrose-fermenting) colony morphology on Thiosulfate Citrate Bile salt Sucrose Agar (TCBS agar), but detailed biochemical tests are necessary to identify *V. fluvialis* due to its biochemical resemblance with *V. cholerae* and *Aeromonas hydrophila.* Although not much is known about the virulence of *V. fluvialis*, some of the putative virulence factors include a cytolysin, hemolysin, cell adherence factors, etc., (Ramamurthy et al., 2014).

18.3.3.1 ISOLATION AND IDENTIFICATION OF VIBRIO SPP.

Owing to the large diversity among different species of *Vibrio*, elaborate biochemical and physiological tests are required for their identification. However, some of the common pathogenic vibrios can be identified to the species level with a few biochemical tests. Alkaline peptone water (APW) is used to selectively enrich vibrios in seafood samples, employing their distinctive ability to withstand high pH. The APW enrichment are performed at 37°C for 6–8 hours for isolation of *V. cholera* and for 18 hours for *V. parahaemolyticus* and *V. vulnificus*, respectively. Isolation of vibrios from APW is undertaken on the selective TCBS agar which contains sucrose and the pH indicator dye bromothymol blue. The sucrose fermenting *V. cholerae* and *V. alginolyticus* appear as yellow colonies and non-sucrose fermenting *V.*

parahaemolyticus, *V. harveyi* exhibit green colony morphologies. Commonly employed biochemical tests to distinguish different *Vibrio* spp. include oxidase test, salt tolerance tests, carbohydrate fermentation tests, amino acid decarboxylase tests, gelatinase, urease tests, etc. (FDA, 2001). For molecular identification of vibrios, PCR assays specific to *toxRS*, *16SrDNA*, *ctxA* or *gyrB* genes are used (Kim et al., 1999; Lago et al., 2009; Kumar et al., 2006).

18.4 EMERGING ENTERIC VIRAL PATHOGENS

Enteric viruses are a group of pathogenic viruses that are naturally present in the mammalian gastrointestinal tract and pathogenic to humans and other mammals. They enter the environment through anthropogenic activities such as inappropriate waste discharge, sewage contamination and land runoff. Enteric viruses are transmitted from infected to healthy but susceptible individuals, by the fecal-oral route (Greening and Cannon, 2016). Generally, they are shed in high numbers and display longer survivability in the environment (Fong and Lipp, 2005). Important groups of these enteric viruses belong to the families Caliciviridae (noroviruses, caliciviruses, and astroviruses), Picornaviridae (polioviruses, enteroviruses, coxsackieviruses, HA virus and echoviruses), Adenoviridae (adenoviruses), and Reoviridae (reoviruses and rotaviruses). Most of these viruses have been associated with foodborne disease outbreaks. Except the human adenovirus which has double stranded DNA as genome, all other enteric viruses have either single stranded (enterovirus, hepatovirus, norovirus, etc.), or double stranded RNA (Reovirus and Rotavirus) as their genetic material. The genome sizes vary from being very small, 5 kb in parvovirus, to a relatively larger genome size of 28–45 kb in adenovirus (Fong and Lipp, 2005). Among these, the emerging viral pathogens, noroviruses (NoV) and hepatitis A viruses (HAV) are frequently recognized as causes of foodborne gastroenteritis and acute hepatitis, respectively.

18.4.1 NOROVIRUS

Human noroviruses (HuNoVs) are among the most important enteric viral pathogens that pose a risk to public health. HuNoVs are RNA viruses of the family Caliciviridae, which are small (25–35 nm in diameter), non-enveloped with icosahedral symmetry. The human norovirus genome consists of a linear, positive-sense RNA of 7.6 kb in length. Based on the differences in the phylogenetic analysis and amino acid sequence of viral capsid protein 1, norovirus is classified into five genogroups from genogroup I to V, with more than 40 genotypes within these genogroups (Zheng et al., 2006). Among these, genogroup II (GII) is reported to cause infections in humans frequently, followed by GI and GIV (Lindesmith et al., 2008). GIII and GV infect bovine and murine species, respectively (Ramani et al., 2014). Under favorable conditions, genetic mixing can occur, making the host barrier flexible (Souza et al., 2007). In humans, NoVs cause gastroenteritis, while the animal viruses can cause a range of different clinical syndromes, including oral lesions, systemic disease with hemorrhagic syndromes, upper respiratory tract infections, and others (Koopmans, 2012).

Noroviral gastroenteritis has been reported to be the second most common cause of enteric infections caused by viruses worldwide. Normally, the average incubation period of these viruses is 24–48 h and the symptoms include vomiting, diarrhea, nausea, abdominal cramps, malaise, and low-grade fever. Despite the geographic variability and annual fluctuations, the majority of norovirus infections have been reported in the winter months. Some of the major factors contributing to the high communicability of noroviruses include their low infectious dose (as low as 10–100 particles), high levels of virus shedding (> 10^9 particles/ml of feces in the initial phase of infection) and stability of these viruses in the ambient temperature conditions (Melhem et al., 2016). Moreover, the high rate of mutations and recombinations among the enteric viruses leading to antigenic diversity is very alarming.

18.4.2 HEPATITIS A VIRUS

Among the enteric viruses causing human infections, hepatitis A virus is of utmost importance after norovirus. Humans and primates are the only natural reservoirs of hepatitis A virus, which belongs to the family Picornaviridae. There are six genotypes of hepatitis A virus classified based on their genetic diversity. Of these, genotypes I, II, and III affect humans (Koopmans, 2012). The primary site of infection of hepatitis A virus is hepatocytes causing disruption of liver function, leading to inflammation of the liver and consequently resulting in jaundice (Franco et al., 2012). In addition, symptoms like mild gastrointestinal disorders, fever, nausea, loss of appetite and abdominal pain also occur (Cuthbert, 2001). The transmissibility and infection rate of HAV is less than that of HuNoV and hence in regions with appropriate hygienic and sanitary conditions, there is reduced incidence of HAV infection. As a result, depending on the socioeconomic status, great variation exists between the occurrences of HAV infections across the world. However, in highly populated developing and underdeveloped countries, infections are quite prevalent owing to their poor sanitary standards. In HAV endemic regions, the majority of persons are infected in early childhood, generally asymptomatic, and virtually all adults are immune (Mohd Hanafiah et al., 2011). Hepatitis A virus can survive for prolonged periods on contaminated food, water, and other environments and remains stable outside living host. The infections are transmitted *via* contaminated food, water, as well as from person to person. Filter feeding bivalve mollusks such as oysters, mussels, and clams, especially those harvested from contaminated waters are considered to be important sources of HAV infections in humans (Pinto et al., 2009).

18.4.3 ENTERIC VIRUSES IN SEAFOOD

Viral contamination of foods can occur during all stages of the food supply chain, and transmission occurs by consumption of food contaminated during the production process (primary production, or during further processing), or contaminated by infected food handlers. Viruses do not multiply in foods, but may persist for extended periods of time as infectious particles in the environment, or in foods (EFSA, 2011). Fresh seafood, especially bivalve mollusks can act as vehicles for transmission and spread of norovirus in the community. The high level of enteric viral contamination of seafood not only constitutes a health risk to the domestic consumers, but also compromises the microbiological qualities of seafood meant for international trade (Das et al., 2020). Enteric viruses enter shellfish through contaminated seawater or by contamination during handling and processing, resulting in outbreaks ranging from isolated to epidemic. A large number of shellfish-associated outbreaks have been attributed to enteric viruses, particularly norovirus, responsible for an estimated 60–80% of human gastroenteritis outbreaks worldwide (Suffredini et al., 2008). Molluscan shellfish, being filter feeders, can filter large volumes of water as part of their feeding activities and are able to accumulate and concentrate different types of pathogens in fecally polluted waters including viruses (Le Guyader et al., 2009). Crustaceans such as crabs and shrimps acquire enteric viruses when they feed on contaminated oysters or other organisms and thus act as sources of transmission to humans

(Umesha et al., 2008). High prevalence of enteric adenovirus in filter feeding bivalves, such as clams and oysters, as well as in fish suggests persistent fecal contamination of coastal waters (Ghosh et al., 2019). The prevalence of enteric viruses in seafood samples also shows a strong positive correlation with the bacteriological indicators of fecal contamination, suggesting that fecal indicator bacteria may be used to monitor the presence of adenoviruses (ADV) in seafood (Lekshmi et al., 2018).

Norovirus and hepatitis A virus are predominant causes of illness linked to shellfish consumption in many countries (Koopmans, 2012). Suffredini et al. (2020) studied the viral contamination in oysters and clams marketed in Vietnam and detected NoV in 81.8% and HAV in 2% of the samples. In Mexico, 57.3% of raw oysters from retail seafood markets were contaminated with norovirus (Villicana et al., 2019). A report from France revealed the presence of norovirus in 9% of the 242 oysters sampled over a period of 16 months (Schaeffer et al., 2013). Noroviral RNA was detected in 12.1% of the samples from open-air markets and fish shops in Italy (Terio et al., 2010). The presence of NoV in 37% and HAV in 33% of the shellfish sampled from harvesting areas of Portugal was reported by Mesquita et al. (2011).

Anbazhagi and Kamatchiammal (2010) reported 24% of shellfish samples from waterbodies and fish markets in Chennai to be contaminated with norovirus by RT-PCR and Southern hybridization. ADV were detected in 17% of oyster and 27% of clam samples collected off the south west coast of India (Umesha et al., 2008). A few recent studies from Mumbai report the occurrence of these emerging viruses from samples of fish and shellfish. Enterovirus was detected in 32 (35.95%) of seafood samples, all of which were from bivalve molluscan group (Lekshmi et al., 2018), of which clams were the most contaminated ones. Ghosh et al. in 2019 reported the incidence of adenovirus in 21.3% of the analyzed samples, and Norovirus Group II (GII) was detected in 41.34% of seafood samples contaminated with fecal indicator bacteria (Das et al., 2020).

18.4.4 METHOD FOR DETECTION OF ENTERIC VIRUSES

The ability to detect traces of viral contamination in foods and related samples is essential for the foodborne disease investigations and possible prevention of viral disease outbreaks. Traditional methods for the detection of enteric viruses in foods involve their extraction from the sample matrix and concentration, followed by viral multiplication in sensitive cell cultures in which characteristic cytopathic effects can be visualized. However, these methods are time consuming, laborious, less sensitive while also requiring expertise in handling cell cultures and cytopathic studies. Molecular biology techniques have been proposed for the sensitive and specific detection of many enteric viruses in foods which include polymerase chain reaction (PCR)-based methods and nucleic acid hybridization techniques. For viruses with RNA as their genetic material, the reverse transcription-PCR (RT-PCR) technique offers a powerful method for direct detection of enteric viruses in foods. In RT-PCR, the viral RNA is first converted to its complementary double-stranded DNA (cDNA) in a reverse transcription step, followed by PCR amplification of the target cDNA sequences to a level detectable by a DNA probe or agarose gel electrophoresis (Jean et al., 2001).

Nevertheless, the detection of viruses in food matrix is constrained by their low numbers in samples which requires laborious concentration methods to concentrate these viruses into smaller sample volumes to enhance the usefulness of detection assays. The development and application of methods for concentrating viruses have contributed significantly to the detection of diverse viruses using culture- or molecular-based assays (Cashdollar and Wymer, 2013). Most methods are designed to concentrate and purify viruses and remove inhibitors before performing the RT-PCR. For these purposes, multi-step elution and extraction procedures are employed using a combination of several reagents. Virus concentration followed by RT-PCR could be successfully used to detect the presence of enteric viruses in seafood (Lekshmi et al., 2018). According to one protocol suggested by Sdiri et al. (2006), the fish and shellfish samples are first homogenized followed by concentration of virus using PEG precipitation method. Then RNA is extracted from the virus concentrates followed by conversion to cDNA by reverse transcription. Finally, enteric viruses are detected using specific primers and hybridization.

A typical method for detection of viruses in food involves three different steps: (1) elution and clarification of the virus from the foodstuff; (2) concentration of the virus particles; and (3) nucleic acid extraction and detection. Molecular techniques such as PCR are the preferred techniques for detection of viral pathogens that are noncytopathic or nonculturable. Successful PCR detection relies on effective removal of natural inhibitors and efficient recovery of viruses from oysters during processing. The initial step commonly used to elute viruses from oyster tissue includes direct alkaline elution and acid adsorption-elution. Different eluants containing glycine, beef extract, and other compounds have been compared to determine their virus recovery and PCR compatibility characteristics. When the levels of viral contamination in shellfish are low, it is necessary to include a concentration step, such as polyethylene glycol (PEG) precipitation, organic flocculation, or ultracentrifugation. After concentration, additional steps to remove PCR inhibitors from the processed concentrates are critical in order to have a sensitive PCR test. RNA extraction and nested PCR have been found to improve significantly the sensitivity of virus detection with environmental and clinical samples (Croci et al., 2000).

18.5 CONCLUSION

In order to reduce the risk of human infections transmitted *via* seafood, prolonged investigations on the mechanism of microbial pathogenesis, physiology of survival in food, resistance mechanisms, prevention, and control strategies are necessary. Thorough identification of the sources of microbial infection, the critical points (CPs) of contamination, and the development of effective methods for their elimination are important steps towards controlling microbial infections. It is imperative to understand the persistence of these pathogens in their animal reservoirs as well. The recent surge in emergence and re-emergence of pathogens has emphasized the need to monitor the aquatic environment and the fish and shellfish for the occurrence of such microbes. Adoption of suitable measures to prevent contamination of seafood, such as maintaining proper sanitation and hygiene in landing centers and retail markets, is critical to ensure the safety of seafood for human consumption.

KEYWORDS

- **arcobacter**
- **cronobacter**
- **emerging seafood**
- **enteric pathogens**
- **enteric virus**
- **polymerase chain reaction**

REFERENCES

Anbazhagi, S., & Kamatchiammal, S., (2010). Contamination of seafood by norovirus in India. *Int. J. Virol., 6*(3), 138–149.

Das, O., Lekshmi, M., Kumar, S., & Nayak, B. B., (2020). Incidence of norovirus in tropical seafood harboring fecal indicator bacteria. *Mar. Pollut. Bull., 150*, 110777.

Fong, T. T., & Lipp, E. K., (2005). Enteric viruses of humans and animals in aquatic environments: Health risks, detection, and potential water quality assessment tools. *Microbiol. Mol. Biol. Rev*., 69 (2), 357–371.

Ghosh, S. K., Lekshmi, M., Das, O., Kumar, S., & Nayak, B. B., (2019). Occurrence of human enteric adenoviruses in fresh tropical seafood from retail markets and landing centers. *J. Food Sci., 84*, 2256–2260.

Iversen, C., Lehner, A., Mullane, N., Marugg, J., Fanning, S., Stephan, R., & Joosten, H., (2007). Identification of "*Cronobacter*" spp. (*Enterobacter sakazakii*). *J. Clin. Microbiol., 45*(11), 3814–3816.

Iwamoto, M., Ayers, T., Mahon, B. E., & Swerdlow, D. L., (2010). Epidemiology of seafood-associated infections in the United States. *Clin. Microbiol. Rev., 23*, 399–411.

Karunasagar, I., & Karunasagar, I., (2018). Ecology, virulence factors and global spread of *Vibrio parahaemolyticus*. *Asian Fish. Sci., 31S*, 15–28.

Kumar, H. S., Parvathi, A., Karunasagar, I., & Karunasagar, I., (2006). A *gyrB*-based PCR for the detection of *Vibrio vulnificus* and its application for direct detection of this pathogen in oyster enrichment broths. *Int. J. Food Microbiol., 111*(3), 216–220.

Kumar, L. R. G., Kasim, A. K., Lekshmi, M., Nayak, B. B., & Kumar, S., (2016). Incidence of methicillin-resistant staphylococci in fresh seafood. *Adv. Microbiol., 6*(6), 399–406.

Kumar, S., Lekshmi, M., Parvathi, A., Nayak, B. B., & Varela, M. F., (2017). Antibiotic resistance in seafood-borne pathogens. In: Om, V. S., (ed.), *Foodborne Pathogens and Antibiotic Resistance* (1st edn., pp. 397–415). John Wiley & Sons, Inc.

Laishram, M., Rathlavath, S., Lekshmi, M., Kumar, S., & Nayak, B. B., (2016). Isolation and characterization of *Arcobacter* spp. from fresh seafood and the aquatic environment. *Int. J. Food Microbiol., 232*(2), 87–89.

Lekshmi, M., Das, O., Kumar, S., & Nayak, B. B., (2018). Occurrence of human enterovirus in tropical fish and shellfish and their relationship with fecal indicator bacteria. *Vet. World, 11*(9), 1285–1290.

Pintó, R. M., Costafreda, M. I., & Bosch, A., (2009). Risk assessment in shellfish-borne outbreaks of hepatitis A. *Appl. Environ. Microbiol., 75*(23), 7350–7355.

Prabhakaran, M. D., Ramamurthy, T., & Thomas, S., (2020). Genetic and virulence characterization of *Vibrio parahaemolyticus* isolated from Indian coast. *BMC Microbiol.*, 20.

Rathlavath, S., Kohli, V., Singh, A. S., Lekshmi, M., Tripathi, G., Kumar, S., & Nayak, B. B., (2017). Virulence genotypes and antimicrobial susceptibility patterns of *Arcobacter butzleri* isolated from seafood and its environment. *Int. J. Food Microbiol., 263*, 32–37.

Rathlavath, S., Kumar, S., & Nayak, B. B., (2017). Comparative isolation and genetic diversity of *Arcobacter* spp. from fish and the coastal environment. *Lett. Appl. Microbiol., 65*(1), 42–49.

Rathlavath, S., Mishra, S., Kumar, S., & Nayak, B. B., (2015). Incidence of *Arcobacter* spp. in fresh seafood from retail markets in Mumbai, India. *Ann. Microbiol., 66*(1), 65–170.

Sanjit, S. A., Lekshmi, M., Prakasan, S., Nayak, B. B., & Kumar, S., (2017). Multiple antibiotic-resistant, extended-spectrum-β-lactamase (ESBL)-producing enterobacteria in fresh seafood. *Microorganisms, 5*(3).

Sdiri, K., Khelifi, H., Belghith, K., & Aouni, M., (2006). Comparison of cell culture and RT-PCR for the detection of enterovirus in sewage and shellfish. *Pathologie-Biologie, 54*(5), 280–284.

Singh, A. S., Nayak, B. B., & Kumar, S. H., (2020). High prevalence of multiple antibiotic-resistant, extended-spectrum β-lactamase (ESBL)-producing *Escherichia coli* in fresh seafood sold in Retail Markets of Mumbai, India. *Vet. Sci., 7*(2), 46.

Suffredini, E., Le, Q. H., Di Pasquale, S., Pham, T. D., Vicenza, T., Losardo, M., To, K. A., & De Medici, D., (2020). Occurrence and molecular characterization of enteric viruses in bivalve shellfish marketed in Vietnam. *Food Cont., 108*, 106828.

Terio, V., Martella, V., Moschidou, P., Di Pinto, P., Tantillo, G., & Buonavoglia, C., (2010). Norovirus in retail shellfish. *Food Microbiol., 27*, 29–32.

Vouga, M., & Greub, G., (2016). Emerging bacterial pathogens: The past and beyond. *Clin. Microbiol. Infect., 22*(1), 12–21.

CHAPTER 19

Understanding Food Safety in Fish and Fishery Products

T. V. SANKAR

ICAR-Central Institute of Fisheries Technology, Cochin – 682029, Ernakulam, Kerala, India

ABSTRACT

Food safety is for providing safe food, food without any chemical and microbiological hazards, so that the consumers are not put into difficulties. Food provides nutrition for the body to grow and energy for metabolism and sustaining life, and any contaminant present would have negative impact in one form or other. The chapter discusses food safety with special reference to fish and fish products. The hazards normally encountered, both biological and chemical, the types of biological hazards, the different classes of chemicals hazards including additives, some of which become hazards at levels beyond maximum residual level recommended by the regulatory agencies, the incidence of foodborne illness and the related information including management measures. This will be useful for the students at undergraduate and Postgraduate levels for academic purposes, as a source of ready reckoner for the professions in the discipline and any other reader to know about food safety as a concept.

19.1 INTRODUCTION

Food safety is defined by Codex Alimentarius as assurance that food will not cause harm to the consumer when it is prepared and/or eaten according to

Advances in Fish Processing Technologies: Preservation, Waste Utilization, and Safety Assurance.
Ranendra K. Majumder, PhD & Amjad K. Balange, PhD (Eds.)

its intended use (FAO, 2014). According to the United States Department of Agriculture, food safety refers to the conditions and practices that preserve the quality of food to prevent contamination and foodborne illnesses (https://ask.usda.gov/s/article/What-does-food-safety-mean).

Food is defined as any substance which consists of protein, carbohydrate, and fat required for growth, repair, and essential body functions and to furnish energy together with minerals and vitamins. Food in fact maintains body and sustains life. With the population growing disproportionately, the availability of the food for the populace is an important issue. The food leads to illness in consumer, if not clean, hygiene, and free of any hazard. Therefore, when we talk of food, the related issue also needs to be kept in mind, and if food is not there in good quality and quantity, the existence of entire life is in question. FAO reiterates that food safety is integral to Sustainable Development Goals and sustainable human development cannot take place with safe food. Also, unsafe food impacts the food security as well (FAO, 2019).

When food becomes an easy commodity, then the questions food security, availability, access, utilization, and stability become a concern. Food availability refers to the accessibility of quality food in sufficient quantity to the consumers. Food utilization is the supply of food including water with sufficient sanitation and hygiene and such a condition where food is available to the population at all times is considered as food secure state.

19.2 FOODBORNE ILLNESS

Food availability is an issue in the globe today, but at the same time, hygiene, and safe food with minimum hazards which is the requirement for every citizen. In the absence of safe food, the health of the citizen gets affected which in turn impact the economy of a country. In the absence of any control, the food safety hazards take the toll. Foodborne diseases result from the consumption of unhealthy and contaminated food and break in the hygiene practices necessitating the need for consumer education about the relationship between food preparation, hygiene, and foodborne illnesses (Altekruse et al., 1996). Foodborne disease impacts the consumer due to the consumption of food contaminated with hazards, particularly biological hazards. In general, the impact of foodborne illness may be severe to people or the producers or the country where the food is produced (Adley and Ryan, 2016).

Food hazards are disease causing agents, whether biological or chemical, reducing the safety and hence quality of the food being consumed. Typically, foodborne illness is caused by microorganisms such as bacteria, fungi, viruses, and parasites and the toxins produced by them also often leads to illness, the severity of which varies with the type of agent (Pigott, 2008). Besides, toxic environmental contaminants such as heavy metal residues, pesticides, etc., also contribute significantly to foodborne illness. Foodborne diseases, particularly from biological hazards, are usually associated with improper handling, preparation, and unhygienic food storage conditions along the food value chain.

It is estimated that in industrialized counties, more than 30% of the population suffer from foodborne illness (WHO, 2003), and a significant number of deaths do occur in severe cases and often lead to medical expenditure of over billions of dollars (Duff et al., 2003). The Foodborne disease burden Epidemiology Reference Groups (FERD) of WHO has estimated over 600 million illnesses leading to 0.42 million deaths worldwide in 2010 as a result of over 31 types of foodborne diseases (Havelaar et al., 2010). A quantitative understanding on the hazards is critical for developing effective approaches to their prevention (WHO, 2006) and often such information provides the data required to developing a risk-based food safety system supported by a strong surveillance data (Hoffmann et al., 2017). Generally, the risk of getting foodborne illness depends upon many factors including the type of food consumed, cross-contamination, consumption of partially cooked or uncooked raw food, health conditions of the consumer viz., ill, elderly, immunocompromised, pregnant, or infant (Bender et al., 1999). Poverty, lack of awareness, poor hygiene practices, contaminated food, water, and eating environment and hygiene conditions, etc., are the common factors which expose the consumer to foodborne disease in developing countries (WHO, 1999; King et al., 2000).

There are two types of foodborne illness reported namely foodborne illness and foodborne intoxication. Disease caused due to the presence or consumption of a microbial vegetative cell (contagion) is referred to as foodborne illness while the toxins elaborated in the food by the microorganism causes intoxication (CDCP, 2011; WHO, 2015). *Clostridium botulinum,* for example, is gram-positive rods, spore-forming anaerobic bacteria which cannot be killed by boiling. The organism produces an enterotoxin in the gastro-intestinal tract of the consumer once ingested and causes a food infection (Morris and Fernández-Miyakawa, 2009). The vegetative cells of bacteria form spores in the intestine. Food infection starts when enterotoxin attach epithelial cells in the digestive tract, altering the membrane permeability (Ramachandran, 2014). The practical method of forbidding foodborne intoxication is by killing the vegetative cells (Robinson et al., 2000).

19.3 HAZARDS IN FOOD

Hazards are an agent or condition which are likely to harm the consumer and need to be identified and controlled. The agents responsible for food related issues can be chemical, microbiological, or physical in nature. The food is sensitive to incidence of hazards which can happen at any point along the value chain and the impact on the consumer depends on the frequency of occurrence and the severity of the hazard. In the present-day environment, food (fish and fish based products) travels across borders to cater to the need of consumers (free trade) and in the process carry the pathogens specific for a particular location or region. The pathogen, if associated with the consignment, is either detected and mitigated or missed and carried forward to the consumers posing food safety threat. Besides the increasing number of emerging pathogens, antibiotic-resistant pathogens, etc., have potentially made the situation difficult demanding more serious approaches for their extenuation (Wang et al., 2016).

19.3.1 BIOLOGICAL HAZARDS

Biological hazards are basically pathogenic microorganisms (e.g., salmonella), viruses (e.g., hepatitis A) or parasites (trichinella), whose presence causes illness in humans, either by their presence (infection) or due to the effect of toxins produced by them (intoxication). There could be many undesirable conditions encountered in food. Among the pathogens, *Listeria monocytogenes,* Campylobacter spp.*,* *Escherichia coli*, Staphylococcus spp. deserve urgent surveillance and attention by the food industry owing to their ubiquitous nature. *E. coli*, a facultative anaerobic microorganism is common to the human intestine, are of great concern as several strains with virulence capable of causing illness in humans have been reported in the recent past. *Bacillus cereus,* a common pathogen and *Cronobacter sakazakii, another* opportunistic foodborne pathogen are potential threats to infants (Song et al., 2016).

Viruses are obligate intracellular parasites that require susceptible host cells for propagation and host infection. Norovirus (NoV) is the leading agent responsible for acute gastroenteritis globally (Hall et al., 2014). Foodborne virus infections are predominantly transmitted through ingested contaminated food and/or water, or through person-to-person (secondary route of infection). Norovirus, rotavirus, sapovirus, herpesvirus, and enterovirus are some of the examples of viruses responsible for causing illness among the human beings (Bosch et al., 2018).

Parasites, organisms that can live in a host and derive benefit at the expense of its host. Protozoa and helminths (worms) are the two common parasites found in food such as meat and fish, including the fresh produce. Protozoa are single-cell microscopic organisms that can multiply only in host, say humans, and can cause infection even at low doses and in some cases with only a few cysts. *Toxoplasma gondii is a common* protozoa reported in raw or undercooked meat and *Giardia lamblia* in contaminated water. Helminths are larger organisms and flukes (*Clonorchis sinensis, Paragonimus westermani*), tapeworms (*Taenia solium*) and roundworms (*Angiostrongylus cantonensis*) are common food associated helminths. The parasite infections are generally associated with the consumption of raw and partially cooked meat and fish and thorough washing and cooking before eating can effectively prevent illness due to parasite infection.

19.3.1.1 EMERGING PATHOGENS AND FOODBORNE DISEASES

With globalization, international trade makes food available away from its location of production and the vulnerability of population to infections, in the absence of food safety measures. Pathogens may become emerging when they are linked to the first report of disease or a new strain emerges from the existing strain and causing foodborne illness (Mor-Mur and Yuste, 2010; Sofos, 2008). The emergence of pathogens is a very complex phenomena, depending on the interrelationship between different factors including microbial adaptation to natural selection and evolution, technological changes in the production systems, demographic changes, environmental changes related to trade and climate change, availability of new vectors create a condition for emerges of pathogens (Koutsoumanis et al., 2014; Broglia and Kapel, 2011). Antigenic drift due to random mutation also leads to the reemergence of resistance microorganisms and hence to newer diseases (Vouga and Greub, 2016). Such appearance of pathogens in food web is a challenge to the community and need a calculated approach to protect the consumers from foodborne illness.

There are reports indicating incidence of emerging pathogen across food groups with specificity for certain food, for example, *Aeromonas* spp. in fish and salt-cured fish (Vivekandan et al., 2005; Udgata et al., 2000), *Clostridium difficile* in seafood (Norman et al., 2014) and Cronobactor spp. in marine products (Chon et al., 2012). Similarly, *E. coli* (O157:H7 and non-O157) have the potential to infect fish-eating population even though fish is not the identified source of habitat but mainly through contamination or cross-contamination. Similarly, the emergence of specific strain of Salmonella, particularly S. Enteritidis and S. Typhimurium contributing to the incidence of non-typhoidal salmonellosis among the humans is a concern (CDC, 2011a; Scallan et al., 2011). The use or abuse of antibacterial substances in production systems for prophylactic purposes growth promotion facilitated to the development of resistant strains. Therefore, the understanding of the genetic elements responsible for antimicrobial resistance and epidemiology would help to learn about pathogens or emerging pathogen (Louden et al., 2012) besides addressing the antibiotic management and food safety measure (Hur et al., 2012; Oteo et al., 2010). Besides gastroenteritis-causing noroviruses (NoVs) and HA virus and some parasites are identified as major concerns due to their infective epidemiology (Koopmans and Duizer, 2004; Koustoumanis et al., 2014). Unlike bacterial contamination, viral infection is not generally expected through the food value chain and maintaining good food safety management system (FSMS) would be the best solution to contain the incidence of pathogens and emerging pathogens under check. The challenge due to food pathogens could be addressed by continuous surveillance in addition to adopting mitigation measures to prevent the entry of pathogens in to food system.

19.3.2 CHEMICAL HAZARDS

Chemical hazards are chemicals or compounds used in food production systems leading to health concerns in the absence of their control. There are different types of hazards, namely hazards or toxins naturally occurring in food, namely aflatoxins (peanuts), poisonous neurotoxins (mushrooms) and scrombotoxins (histamine in fish). The second category of chemicals are unintentionally added or accidentally contaminating due to sanitation or maintenance, pest repellent or even from the environment. There are also certain chemicals formed in food during processing of certain raw materials namely formation of trans fatty acids following chemical hydrogenation of unsaturated fats, or 3-monochloropropanediol from the chemical hydrolysis of proteins or formation of acrylamide during the baking of bread and frying of potato (South et al., 2008). Intentionally added chemicals are those added to food for specific technological reasons, namely additives permitted by regulatory measures, namely food additives like sodium nitrate and sodium benzoate in certain foods. The last but important category of chemical hazards are food allergens, which cause allergic reactions in susceptible people and examples include peanuts, fish, dairy products. However, there are also hazards becoming part of the food as a result of the food processing.

19.4 THE NATURALLY OCCURRING TOXIN

Lections in bean, glycoalkaloids solanine and chaconine in potatoes, oxalic acid and oxalate, phytic acid phytates and cyanogenic glycosides present in certain vegetable foods are good examples of such naturally occurring toxins. Mycotoxins and marine toxins are other natural toxins from algae sources and other marine lives (South et al., 2008).

19.4.1 MYCOTOXINS

Mycotoxins, belong to a group of chemical compounds with a diverse range of physiological effects on human being, are elaborated by a number of fungal species, including Aspergillus, Fusarium Penicillium, etc. Mycotoxins are produced by fungus on selected products stored under inappropriate storage conditions. These toxins are of great significance as they are elaborated by certain fungi on cattle feed and ultimately find their way onto the ilk, a good source of nutrients for children. The formation of fungi in specific fish products namely dried fish/shellfish products due to low water activity, is a concern among the fish-eating population.

19.4.2 MARINE TOXINS

Marine toxins, otherwise called shellfish toxins are toxins produced by phyto-plankton that accumulate in fish and shellfish, find their way into the food leading to food safety issues in consumers. Some of these toxins responsible for known intoxications are amnesic shellfish poisoning (ASP), paralytic shellfish poisoning (PSP), neurologic shellfish poisoning (NSP), diarrhetic shellfish poisoning (DSP) and azaspiracid shellfish poisoning (AZP) and also pectenotoxins and yessotoxins (Krock et al., 2018). These toxins are reported to be low molecular weight (between 500 to 200 Da) compounds and are either lipophilic or hydrophilic in nature. ASP due to domoic acid produced by certain diatoms namely Pseudo-nitzschia and the subspecies *pungens* is a neurotoxin and leads to nausea, vomiting, diarrhea with abdominal pain when the level in the food exceeds the maximum residual limit (MRL) 20 mg/kg. Saxitoxin, produced by the dinoflagellate belonging to the genus Alexandrium, is a classic example for PSP. The poisoning causes similar symptoms besides causing respiratory difficulties and muscular paralytic effects eventually leading to death in severe cases. The permitted level in fish/shellfish is 800 μg saxitoxin 2-HCl equivalents/kg shellfish. NSP caused by brevotoxins is produced by algae belonging to the species *Karenia*. The ingestion leads to diarrhea, vomiting, cramps, respiratory, and cardiac problems which can be life-threatening. The USFDA regulatory limit is 800 ug brevotoxin-2 equivalents/kg shellfish. The DSP is caused by okadaic acid, dinophysistoxin-1 (DTX1) and dinophysistoxin-2 (DTX2), as well as the esterified forms of the compounds and are produced by algae belonging to Dinophysis genus. Intoxication leads to gastrointestinal disorder, diarrhea, nausea, and vomiting. European Food Safety Authority (EFSA) has given a regulatory limit of 45 μg OA-equivalents/kg shellfish for human consumption (EFSA, 2008). Azaspiracids are produced by a minute dinoflagellate called Azadinium spinosum. The toxicity is reported to be more common in American and European waters and EFSA has suggested a safe level below 30 μg AZA-1 equivalents/kg shellfish. Structure of important marine toxins are given in Figure 19.1.

19.4.2.1 CIGUATERA POISONING

Ciguatera foodborne illness is caused due to the consumption of fish that is contaminated with ciguatera toxin formed in the fish by biotransformation of less polar Gambier toxins produced by a marine dinoflagellate *Gambierdiscus toxicus*. The heat-stable lipid-soluble toxin is reported from fishes such as moray eel, barracuda, grouper, kingfish, jacks, snapper, surgeonfish, parrotfish, wrasses, hogfish, Spanish mackerel, coral trout, flowery cod and red emperor and ingestion leads to cardiac and neurological symptoms in humans at very low concentrations (0.08 to 0.1 μg/kg) (Friedman et al., 2008).

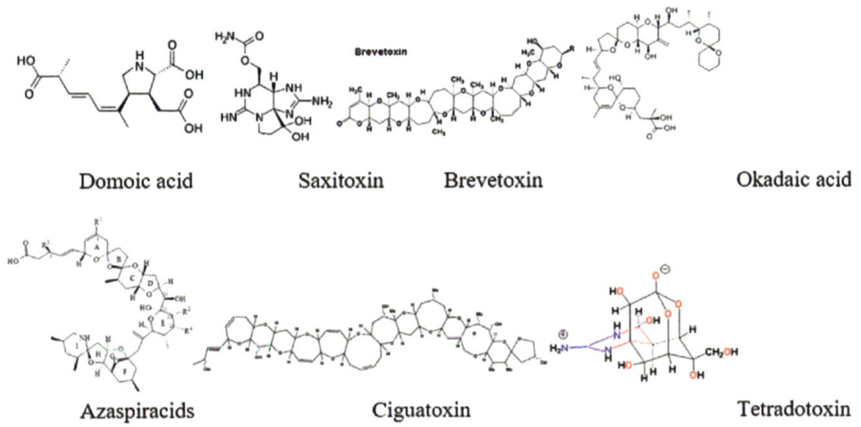

FIGURE 19.1 Structure of marine toxins.

19.4.2.2 TETRODOTOXIN POISONING

Tetrodotoxin is a low molecular weight neurotoxin isolated from toxic pufferfish belonging to the family Tetraodontidae. Liver, gonad, and skin of pufferfish are reported to contain the toxin and leads to death in 50–60% of the ingested persons through its neurotoxin effect. The lethal potency is 5,000 to 6,000 MU/mg and 2 MU approximately equivalent to 2 mg is the estimated minimum lethal dose for humans (Noguchi and Ebesu, 2001).

19.4.2.3 SCOMBROID POISONING

Scombroid (Histamine) poisoning is a foodborne illness with and the symptoms more or less similar to allergic reactions. Some of the symptoms of histamine poisoning are hypotension, flushing, and headache, conditions such as urticaria, hemoconcentration, and eyelids edema either due to dilation of blood vessels or increased capillary permeability; abdominal discomforts due to the contraction of smooth muscles. Increased palpitations and tachycardia due to increased cardiac stimulation by histamine. Histamine is also a potent stimulant of sensory and motor neurons leading to pain and itching associated with the rash (FAO/WHO, 2012).

Histamine is a naturally occurring biogenic amine, found in fish, fish products, meat, cheese, and fermented foods. Biogenic amines are low molecular weight organic bases formed in foods by either microbial decarboxylation of the corresponding amino acids or amino acid transaminases action. Biogenic amines such as histamine, tyramine, tryptamine, putrescine, and cadaverine are formed from free amino acids namely histidine, tyrosine, tryptophan, ornithine, and lysine, respectively (Zhai et al., 2012; Zarei, 2011). The other biogenic amines noticed in fish and other meat-based products namely Spermidine and spermine are other two biogenic amines formed during the spoilage of fish and fish-based products and are formed from putrescine (Zarei, 2011). The biogenic amines are structurally unrelated viz., putrescine, cadaverine, spermidine, and spermine have an aliphatic structures, histamine, and tryptamine have heterocyclic structures and tyramine and phenylethylamine have aromatic structures but exhibit allergic reactions (Mahmoudi and Mardani, 2015).

Production of histamine in susceptible species is a time-temperature related process. The production of biogenic amine is due to exposure to conditions that support microbial proliferation and in high moisture products like fish exposed to temperature abuse favor microbial growth and biogenic amine production. The food safety limit is suggested as 200 mg/kg histamine for susceptible fish belonging to the family Scombridae, Clupeidae, Engraulidae, Coryphaenidae, Pomatomidae, and Scomberesocidae (CAC, 2018), even though there are differences in the standards from different countries.

As per the Hazard Analysis Critical Control Point guidelines, a core temperature of 4.4°C or less is to be achieved and maintained throughout the value chain to prevent histamine production. Histamine is not found in fresh fish but formed during spoilage (Figure 19.2), particularly at high abusive temperatures, 21.1°C or higher near 32.2°C (FDA, 2011). However, high temperature treatment or cooking can inactivate histamine decarboxylase, the enzyme responsible as well as the microorganisms responsible. At the same time, histamine, already formed in the meat due to favorable conditions cannot be eliminated because of its thermostability (FDA, 2011).

During the post-harvest deterioration of fish, the pH of the meat changes consequent to the anaerobic glycolysis and the ensuing condition activates the proteolytic enzymes in the system leading to proteolysis forming free amino acids. Fish is a food commodity with high water content and deserves cold chain to keep it safe and in edible condition. In the absence of a cold chain, the formation of minor nutrients due to enzymatic hydrolysis, favor the multiplication of bacteria leading to the formation of biogenic amines from the corresponding amino acids, by the bacterial amino acid decarboxylases enzymes. The presence of biogenic amines in foods could be an indices for bacterial contamination in food. Fish, belonging to the Scombridae family are rich in the amino acid histidine, which is decarboxylated under unfavorable conditions to histamine. Fish being susceptible to spoilage in the absence of cold chain, formation of histamine in fish indicate unfavorable preservation conditions and therefore is an indicator of hygiene as well.

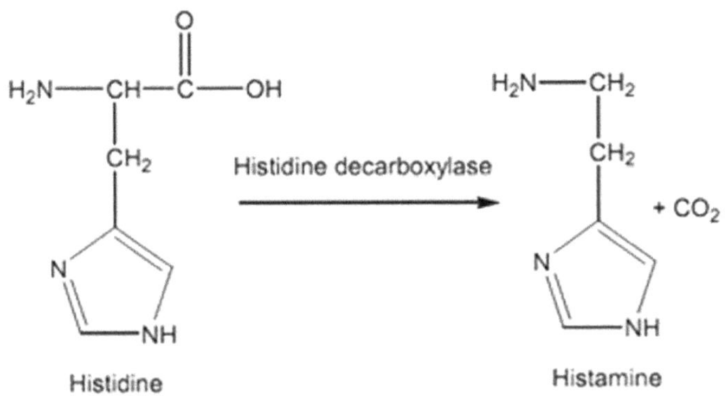

FIGURE 19.2 Formation of histamine from histidine.

19.5 ENVIRONMENTAL CONTAMINANTS

There are many potentially toxic substances in the environment, entering either accidently or through anthropogenic activities, contaminating foods consumed by people. This global problem with many industrial chemicals is their stability and their characteristic to bioaccumulate, bioconcentrate, and biomagnify leading to dangerous manifestations in animals and human beings. Some of these chemicals include persistent organic chemicals like pesticides, polychlorobiphenyls, heavy metals, veterinary drug, antibiotic residues, etc.

19.5.1 PESTICIDE RESIDUES

Pesticides, the chemicals used to control or kill pests, are extensively used chemicals in the globe, particularly in the food production systems. These dangerous chemicals are available in large variety and the contamination and potency in food depends on its nature of exposure, concentration, clinical, and pathological effect. The health risk associated with these chemicals include simple allergic reactions to life threatening danger necessitating the need to establish maximum residue limit (MRL), the amount of contaminant over the safe limit for human consumption and their control.

Water bodies are the ultimate dumping yards for the persistent chemicals, which settles down to the bottom sediment. The plankton, aquatic invertebrates and other lives including fish consume them from the water column or sediments along with their nutrients creating a scenario for accumulation along the food chain. The adverse effects of DDT demonstrated in experimental animals include infertility, cancer, neurologic developmental disorders and mortality (Ogunfowokan, 2012). This has led to the development of maximum residue level (MRL) and no observable adverse effect level (NOAEL) for various pesticides in food and water by the Codex Alimentarius Commission (CAC) of the FAO. Residue surveillance studies demonstrated the incidence of pesticides in sediments and fish species across the globe (Kalyoncu et al., 2009; Shalaby et al., 2018; Teklit, 2016; Salah El-Dien and Mahmoud, 2011; Darko et al., 2008).

Today's world uses different composition of pesticides namely organochlorine, organophosphorus, and carbamate pesticides depending on the need. Though the application of organochlorine pesticides has been restricted for their health impacts, the other pesticides find extensive use in agricultural activities as well as public health programs. Pesticides, in general, with low polarity, low to medium aqueous solubility and high lipophilicity, tend to bioaccumulate in the ecosystem and in the human body and tend to have deleterious impacts on health including reproduction, immune system dysfunction, endocrine disruptions and cancer. Different action level of 300 mg/kg for most of the organochlorines and 5,000 mg/kg for DDT and related compounds have been notified by international organizations (FDA, 2001; FAO, 1983) while much lower limits of 1,000 mg/kg for DDT and related compounds and 100 mg/kg for others, was recommended by NAS/NES. These levels are, however, still under debate in different countries.

19.5.2 POLYCHLOROBIPHENYLS

Polychlorinated biphenyls (PCBs) are basically man-made organic chemicals containing chlorine and the number of chlorine atoms and their location in a PCB molecule determine its physical and chemical properties. Generally, PCBs are tasteless and odorless with oily to waxy consistency and are identified as congeners (Table 19.1) and a PCB congener is any single, unique well-defined chemical compound in the PCB category (https://www.epa.

gov/pcbs/learn-about-polychlorinated-biphenyls-pcbs). PCBs are derived from biphenyl, a substituted benzene, by substitution of hydrogen atoms by chlorine atoms. The nomenclature depends on the total number of chlorine substituents and the position of each chlorine, for example, 4,4'-Dichlorobiphenyl is a congener having a biphenyl with two chlorine substituents – one on each of the #4 carbons of the two rings. As of today, there are 209 PCB congeners.

PCBS have wide industrial applications as heat transfer fluids, additives in plastics, dyes, and carbonless copying paper, etc., and used worldwide in large quantities, but the production ceased around the late 1970s because of their adverse effects to the environment due to their persistency, bioaccumulation, and toxicity. Certain congeners of PCB show toxicity similar to much toxic polychlorinated dibenzo-p-dioxines (PCDDs) and dibenzofurans (PCDFs) and have created concern on its presence in the food from environmental contamination. PCBs have been reported to cause promote cancer as well as a number of serious non-cancer health effects, including effects on the immune system, reproductive system, nervous system, endocrine system in animals (EPA, 2020). PCBs are highly lipophilic and rapidly accumulate in aquatic organisms with bioaccumulation factor increasing with chlorine content up to hexachlorobiphenyls and then decreases (EPA, 1999). The tolerance limit for PCB notified by FDA and Canada is 2,000 mg/kg while EPA has established two limits: (i) based on the potential to cause cancer in the life span of an individual (0.05 ug/kg body weight per day); and (ii) tolerance limit (0.02 ug/kg body weight per day) though it is understood, that the concentration of PCBs in fish is much less than the limits (EPA, 1999).

TABLE 19.1 List of Common PCB Congeners

CASRN	Congener Number	IUPAC Name	Type
2051-60-7	1	2-Chlorobiphenyl	Congener
33146-45-1	10	2,6-Dichlorobiphenyl	Congener
55712-37-3	25	2,3,'4-Trichlorobiphenyl	Congener
62796-65-0	50	2,2,'4,6-Tetrachlorobiphenyl	Congener
39485-83-1	100	2,2,'4,4,'6-Pentachlorobiphenyl	Congener
68194-08-1	150	2,2,'3,4,'6,6'-Hexachlorobiphenyl	Congener
52663-73-7	200	2,2,'3,3,'4,5,6,6'-Octachlorobiphenyl	Congener
52663-77-1	208	2,2,'3,3,'4,5,5,'6,6'-Nonachlorobiphenyl	Congener
2051-24-3	209	Decachlorobiphenyl	Congener

19.5.3 POLYCYCLIC AROMATIC HYDROCARBONS

Polycyclic aromatic hydrocarbons (PAHs) are a ubiquitous group of chemically related organic compounds with environmental persistence with varied toxicity and enter the environment through different applications (Figure 19.3). Biologically PCBs intervene with the functions of cellular membranes as well as with enzyme systems associated (Abdel-Shafy and Mansour, 2016). Available evidences indicate the carcinogenicity and mutagenicity associated with PAHs and are potent immunosuppressants as well (Armstrong, 2004). PAHs are formed in the environment consequent to incomplete natural or manmade combustion and are commonly detected almost everywhere (Baklanov et al., 2007). PAHs are moderately persistent and can be bio-accumulated. The concentrations of PAHs found in fish and shellfish are expected to be much higher than in their environment and the metabolism of these compounds, generally, prevent biomagnifications (Inomata et al., 2012). But at the same time, their solubility in both polar and nonpolar solvents facilitate its transportation and distribution as well as their uptake in to living organisms. The chemical reactivity of PAHs influences its adsorption to organic material or degradation in the environment as well as its persistence and capacity to bio-accumulate in the food chain (Kipopoulou et al., 1999; Jiang et al., 2011).

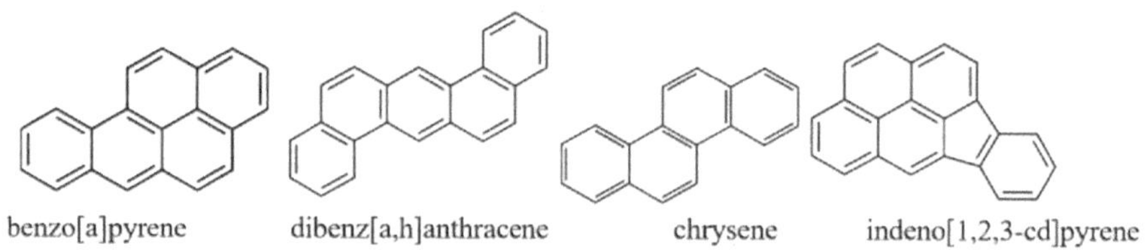

FIGURE 19.3 Structure few hazardous PAHs.

It is reported that dietary exposures are the major source of human exposure to PAHs and some of the processing techniques like curing, drying, smoking, roasting, grilling, barbecuing, and refining contribute to the incidence of PAHs in fish and other commodities and smoking and smoked products are one of the important sources of PAH (Palm et al., 2011; Tongo et al., 2017).

There are regulatory standards (Table 19.2) for eating, drinking, and breathing of PAH contaminated food, water as well as working environment. The United States Environmental Protection Agency prescribes a limit for

PAH in drinking water (0.2 ug per liter) while National Institute for Occupational Safety and Health (NIOSH) and the Occupational Safety and Health Administration (OSHA) made exposure limits for human in different industrial environments. There are limits prescribed for products from seafood as well (EFSA, 2008). *In vitro* and *in vivo studies* have demonstrated the genotoxic nature of 15 PAHs namely benz[a]anthracene, benzo[a]pyrene, benzo[b] fluoranthene, benzo[ghi] perylene, benzo[j]fluoranthene, benzo[k] fluoranthene, chrysene, cyclopenta[cd]pyrene, dibenz[a,h]anthracene, dibenzo[a,e]pyrene, dibenzo[a,h]pyrene, dibenzo[a,i]-pyrene, dibenzo[a,l] pyrene, indeno[1,2,3-cd]pyrene and 5-methylchrysene (FAO/WHO, 2006), though they are metabolized in mammals and do not tend to bioaccumulate (EFSA, 2008).

TABLE 19.2 MRL for Benzo[a]pyrene and the Sum of Benzo[a]pyrene, Benz[a]anthracene, Benzo[b]fluoranthene, and Chrysene

Food Commodity	**Maximum Levels [µg/kg]**	
	Benzo[a] pyrene	**∑ benzo[a]pyrene, Benz[a] Anthracene, Benzo [b] Fluoranth-ene and Chrysene**
Smoked meat and smoked meat products	2.0	2.0
Muscle meat of smoked fish and smoked fishery products	2.0	12.0
Smoked sprats and canned smoked sprats; bivalve mollusks (fresh, chilled or frozen); heat-treated meat and heat-treated meat products sold to the final consumer	5.0	30.0
Bivalve mollusks (smoked)	6.0	35.0

Source: Regulation (EU) No 835/2011.

19.5.4 HEAVY METAL RESIDUES

Heavy metals, released in to the ecosystem naturally and by anthropogenic activities connected to the agricultural and industrial activities, impact the lives due to their persistence, bioaccumulation, biomagnification, and the resultant toxicity (Papagiannis et al., 2004; Sankar et al., 2006). Fish being at the highest level of the food chain bioaccumulate significantly which is important considering the role of fish as an important food commodity (Sankar et al., 2006; Rajeshkumar and Li, 2018). The accumulation in fish is reported to be related to sex, age, size, reproductive cycle, swimming pattern, feeding behavior, and geographical location (Mustafaand Guluzar, 2003; Zhao et al., 2012; Voegborlo et al., 2012; Al-Busaidi et al., 2011; Rahman et al., 2012) (Altekruse et al., 1996; Broglia and Kapel, 2011) but the commodity reaches the consumer where it significantly impacts the health (Romeoa et al., 1999; Canli and Atli, 2003; Chan 1995; Sankar et al., 2006; Bawuro et al., 2018).

Heavy metals accumulated in aquatic environments are transported into the human body through the food chain and increased levels of metals like lead (Pb), cadmium (Cd), and Chromium (Cr) leads to food safety risk (Table 19.3). The incidence of renal failure and liver damage in lead toxicity, increased incidence of cadmium leading to impaired kidney function, infertility, hypertension, tumors, toxicity of cadmium leading to hepatic dysfunction and disruption of cellular integrity and functions due to Chromium are global concerns due to heavy metal toxicity. Even though certain metals are required at certain concentrations for normal body functions, their presence higher levels pose potential toxic conditions. The Agency for Toxic Substances and Disease Registry (ATSDR) notified toxicity by metals as arsenic followed by lead, cadmium, nickel, zinc, chromium, copper, and manganese, in that order. The organic form of the element is less harmful and exhibit a wide distribution, while the inorganic form, released into the environment through various industrial and natural processes, are more harmful to the consumers causing detrimental effects. Cadmium is a rare element and found in nature along with other elements. Cd has wide industrial applications and reach the environment by one way or other. Vegetables and plants are important sources of Cd and gain entry into the human body through food. Cd in fish is extensively reported and ingestion of Cd and its compounds is reported to be highly toxic as is a suspected carcinogen.

Mercury is an important metal that gain entry into the body from the consumption of fish. The metal is naturally present in the ecosystem besides industrial anthropogenic activities contributing to its incidence in the aquatic ecosystem. It is from the ecosystems that the metal finds its way into fish and to higher forms. It is important to note that majority of mercury in fish is in the form of methylmercury which gets into the human being causing gastrointestinal disturbances (NAS-NRC, 2000). Mercury is also a neurotoxicant and is considered as a developmental toxicant as well. An action level of 0.2 ug/kg/body weight per day has been suggested for human beings, and high-end fish eaters have been reported to have symptoms of mercury poisoning.

TABLE 19.3 Maximum Permissible Limits (mg/kg) for Heavy Metals in Fish/Shellfish by Different Regulatory Bodies

	Pb	Cd	Hg	As	Cr	Ni	Cu	Zn	References
Fish	0.2–0.4	0.05–0.1	0.5–1.0	–	–	–	–	–	EC (2001)
	1.5	–	–	–	–	–	–	–	FDA (2001)
	0.5	–	–	–	–	–	30	40	FAO (1983)
	0.3	–	–	76	–	–	–	–	Codex (2009)
Crustacean	0.5	0.5	–	–	–	–	–	–	EC (2001)
	1.5	3.0	–	76	12	70	–	–	FDA (2001)
	0.5	0.5	–	–	–	–	–	–	FAO (1983)
Bivalves/ clam/oysters/ mussels	1.0	1.0	–	86	13	80	–	–	EC (2001)
	–	2.0	–	–	–	–	–	–	Codex (2009)
	1.7	4.0	–	–	–	–	–	–	FDA (2001)
Cephalopods	1.0	1.0	–	–	–	–	–	–	EC (2001)
	–	2.0	–	–	–	–	–	–	Codex (2009)

19.5.5 ANTIMICROBIAL DRUGS

The antimicrobial drugs (Figure 19.4) are extensively used for the health management in animal husbandry and aquaculture but indirectly lead to a global health concern. It is reported that about 63,151 tons of antimicrobial drugs are used in food animals, including aquaculture during 2010 and may rise by 67% in 2030 (Okocha et al., 2018). The unregulated use of antibiotics in aquaculture production of fish and shrimps is a cause of concern in consumers and could lead to the development of antimicrobial resistance besides causing toxicity related symptoms of morbidity and in extreme cases causes mortality. The presence of chloramphenicol, a banned antibiotic, in the fish meat could lead to cancer in susceptible populations and at lower concentrations could trigger the development of aplastic anemia and further complications in extreme cases. Similarly, presence of sulphathiazole, oxytetracycline, and furazolidone could lead to immunopathological effects and carcinogenicity, gentamicin could lead to mutagenicity and nephropathy and allergy by penicillin (EC, 2001).

The antibacterial residues get accumulated in an animal or human system, irrespective of the purpose of usage, before getting metabolized by the body as noticed in aquaculture products harvested immediately after treatment or before the lapse of the withdrawal period (Tollefson and Miller, 2000). Development of antibacterial resistance along the food chain is the consequence of such unscientific practices. The extra-label usage or usage beyond the prescribed limits and duration and prescription by unapproved persons, etc., causes such drastic conditions (Woodward, 1991). Chloramphenicol, dimetridazole, ipronidazole, nitroimidazoles, furazolidone, nitrofurazone, and fluoroquinolones are some of the antibiotics prohibited for use in aquaculture industry as per the recommendations of the CAC of the (FAO/WHO. Maximum residue limits (MRLs) are established for approved veterinary drugs by National Council regulation EEC/2377/90 (EEC, 1990). The MRL values for antibiotics in fish and across of forms of food-producing animals has been established by European Union (EU) legislation and it includes the list of prohibited pharmacologically active compounds namely Chloramphenicol, Chloroform, Chlorpromazine, Colchicine, Dapsone, Dimetridazole, Metronidazole, Nitrofurans (including furazolidone) and Ronidazole (EC, 2010).

Among other industries, antimicrobial agents or antibiotics are also used extensively to manage diseases and enhance growth as growth promotors in aquaculture, particularly through feed. The data available are quite alarming, and among other countries, Brazil, Russia, India, and South Africa have the highest estimated global antimicrobial consumption (Van Boeckel et al., 2015). Antibiotics are often used irresponsibly without any caution and this unregulated use could pose human health and food safety concerns that remain largely unaddressed in most parts of the world. Non-adherence to recommended directions or dosage, non-observance of recommended withdrawal periods, administration of too large a volume at a single injection site, chemical interactions between drugs and environmental contamination are found to be the reasons for their presence in food animals (CFIA, 2015), their presence in products could result in the development of bacterial resistance and toxicity to consumers leading to morbidity and/or death.

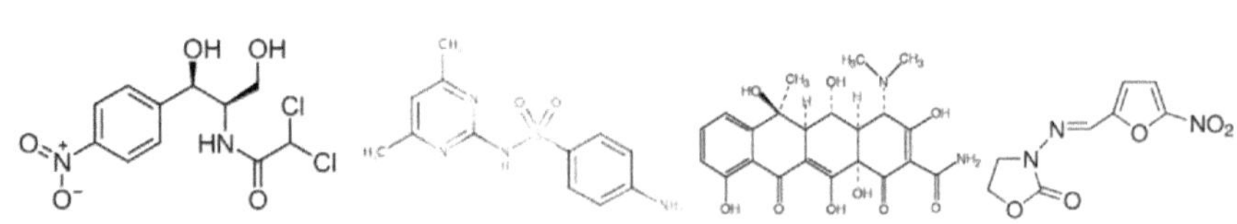

Chloramphenicol Sulphamethazine Oxytetracycline Furazolidone

FIGURE 19.4 Structure of important antibiotics.

19.5.6 FOOD ADDITIVES

FOOD additives are defined, by the Joint WHO/FAO Committee, as substances which are added intentionally to food, generally in small quantities, to improve its appearance, flavor, texture, or storage properties (FAO, 1956) but excludes substances which find their way into foods as contaminants (Spencer, 1974). One of the most common food additives are used in food from time immemorial, is common salt. The main objective of using food additive is to ensure receipt of safe food by the consumer. Substances added to a food for a specific purpose are direct additives and are to be identified on the ingredient label of the food to which the ingredient is added (Pressman et al., 2017). For example, addition of sweetener to soft drinks, etc., implies direct addition while indirect addition include the addition of certain additive during processing, packaging, or storage to maintain the nutritional quality of the food, to enhance keeping quality or stability, to make food attractive to consumers; and to provide essential aids during processing. According to USFDA, the definition includes any substance used in the food production value chain but with the approval of the competent authority (CA). However, the substances which are approved as safe (GRAS chemicals) by international bodies (FAO) are exempted from taking prior approval, if the concentration of the chemicals fall within the prescribed maximum residual limit.

The WHO and FAO identifies additive into three broad categories namely flavoring agents, enzyme preparations and others (WHO, 2018). The flavoring agents, either natural or artificial, are used in food preparations to impart or improve the aroma and to make food more acceptable. Enzyme preparations are proteins from plant, animal or microbial origin, used in the processing of products namely in baking, fruit, and dairy industry for a particular purpose to improve the products quality. Other additives are used for a variety of reasons like extending shelf life, impart color and other forms of quality improvements, protect during storage and transportation and eventually become part of the food. According to FDA, a color additive, can be a dye, pigment, or other substance that can impart color to a food, drink, pharmaceutical, or cosmetic or to the human body, certified or not, but subject to rigorous safety standards prior to their approval and listing for use in foods (Pressman et al., 2017). Some examples of natural colors include annatto extract (yellow) often used in butter, dehydrated beets (bluish-red to brown), caramel (yellow to tan) often used in confections and soft drinks, beta-carotene (yellow to orange) and grape skin extract (red, green) (Pressman et al., 2017).

According to USFDA website there are nearly 4,000 additives under different categories, namely pH control agents, anticaking agents, emulsifiers, humectants, stabilizers, thickeners, leavening agents, gelling agents, etc., each on performing a technological role to keep the food tasty and acceptable for the consumers (Pressman et al., 2017).

The permitted additives in different categories, the purpose of their addition and the food commodities where they are generally used, as approved by USFDA is given in Table 19.4.

TABLE 19.4 Food Ingredients, Their Functions and Uses as per USFDA*

Type	Purpose	Examples of Uses
Preservatives	Prevent food spoilage; slow or prevent changes in color, flavor, or texture and delay rancidity; maintain freshness	Fruit sauces and jellies, beverages, baked goods, cured meats, oils, and margarines, cereals, dressings, snack foods, fruits, and vegetables
Sweeteners	Add sweetness with or without extra calories	Beverages, baked goods, confections, table-top sugar substitutes, many processed foods
Color additives	Offset color loss; correct natural variations in color; enhance colors that occur naturally; provide color to colorless and "fun" foods	Many processed foods
Flavors and spices	Add specific flavors	Pudding and pie fillings, gelatins, dessert mixes, cake mixes, salad dressings, candies, soft drinks, ice cream, etc.
Flavor enhancers	Enhance flavors already present in foods	Many processed foods
Fat replacers	Provide expected texture and a creamy "mouth-feel" in reduced fat foods	Baked goods, dressings, frozen desserts, confections, cake, and dessert mixes, and dairy products
Nutrients	Replace vitamins and minerals lost in processing, add nutrients that may be lacking in the diet (enrichment and fortification)	Flour, breads, cereals, rice, macaroni, margarine, salt, milk, fruit beverages, energy bars, and instant
Emulsifiers	Allow smooth mixing of ingredients, keep emulsified products stable, reduce stickiness, control crystallization, keep ingredients dispersed	Salad dressings, peanut butter, chocolate, margarine, and frozen desserts

TABLE 19.4 *(Continued)*

Type	Purpose	Examples of Uses
Stabilizers and thickeners, binders, and texturizers	Produce uniform texture and improve "mouth-feel"	Frozen desserts, dairy products, cakes, pudding, and gelatin mixes, dressings, jams, and jellies, and sauces
pH Control agents and acidulants	Control acidity and alkalinity and prevent spoilage	Beverages, frozen desserts, chocolate, low acid canned foods and baking powder
Leavening agents	Promote rising of baked goods	Breads and other baked goods
Anti-caking agents	keep powdered foods free-flowing and prevent moisture absorption	Salt, baking powder, and confectioner's sugar
Humectants	Retain moisture	Shredded coconut, marshmallows, soft candies, and confections
Yeast nutrients	Promote growth of yeast	Breads and other baked goods
Dough straighteners and conditioners	Produce more stable dough	Breads and other baked goods
Firming agents	Maintain crispness and firmness	Processed fruits and vegetables
Enzyme preparations	Modify proteins, polysaccharides, and fats	Cheese, dairy products, and meat enzymes
Gases	Serve as propellant, aerate, or create carbonation	Oil cooking spray, whipped cream, and carbonated beverages

*Adopted from: Overview of Food Ingredients, Additives, and Colors, USFDA (2010).

19.6 PHYSICAL HAZARDS

Physical hazards include foreign objects, glass or metal fragments. Hard or sharp foreign objects in food may cause injury of the mouth and other organs and are considered as hazards. As per FDA Guidelines foods are considered adulterated if the product contains a hard or sharp foreign object that measures 7 mm to 25 mm, in length, and the product is ready-to-eat, or according to instructions or other guidance or requirements, it requires only minimal preparation steps, e.g., heating, that would not eliminate, invalidate, or neutralize the hazard prior to consumption (FDA, 1999).

19.7 FOOD PROTECTION SYSTEMS

Food safety is a primary requirement to protect the consumers from the impact of foodborne illnesses by removing the incidence of chemical, biological, and physical hazards mentioned above and to protect the human health. The FSMS emphasizes the importance of controlling food safety aspects and help controlling them. There are many theories to establish the FSMS, namely SQF, FSSC 22000, BRC, IFS, etc., and the Hazard Analysis and Critical Control Point (HACCP) is the fundamental requirement, irrespective of the system.

The FSMS demonstrates a systematic approach to controlling food safety hazards and all food entrepreneurship are expected to execute FSMS based on the principles of HACCP. The International Standard ISO 22003, defines FSMS as a set of interrelated elements of the establishment and establish policy and objectives to direct and control the activities to achieve the objectives of the FSMS system. The International Organization for Standardization created ISO 22000 standard to regulate food safety and security, and the essentials for a FSMS involves interactive communication, system management, prerequisite programs and HACCP principles. The prerequisite program and the HACCP are the basic elements that govern the requirements of food safety through the hazard analysis and control from raw material production, procurement, and handling, to manufacturing, distribution, and consumption of the finished product.

The HACCP is a science-based system that identifies and prevents hazards in a food system which otherwise becomes unsafe. The advantage of HACCP is that it makes the produced responsible for the safety of the consumer and at the same time facilitate the food entrepreneurs to compete more effectively in the open world market. Therefore, HACCP is not concerned with standardization of production process but a monitoring process for the production of a safe food by minimizing the risk of food safety hazards. There are seven basic principles that need to be followed for compliance of the system namely:

- Principle 1: Conduct a hazard analysis.
- Principle 2: Determine the critical control points (CCPs).
- Principle 3: Establish critical limits (CL).
- Principle 4: Establish monitoring procedures.
- Principle 5: Establish corrective actions.
- Principle 6: Establish verification procedures.
- Principle 7: Establish record-keeping and documentation procedures.

In order to have the system established the mandatory steps are assembly of team with appropriate understanding on the product with reference to the hazards to be taken care during the subsequent stages of production. The step which need to be followed include assembling of HACCP team, describing the product (composition, production process, packaging, shelf-life, etc.), identifying the intended use (end-user, nature of preparation/consumption, etc.), construction of a flow chart of production (steps in production) and onsite verification (finalizing the process) of production process to ascertain the possibilities of hazard introduction.

Hazard analysis is an important principle in the system and all the possible and expected hazards with reference to the product in question are identified, from the raw material collection to distribution to consumer. It is also important to identify the specific hazard in the product, its incidence, survival, stage of elimination, possibilities of production of other hazards during the processing stages, etc., and the basic for the HACCP need to be dealt with at this point. Risk analysis is an important part of the hazard analysis as the hazard may be minor or deadly depending on its concentration and frequency of incidence.

The *critical control point* is the place where control measure is applied so that no hazard appears in the product. There could be multiple CCPs for hazards in a product and the control step is to be identified and controlled at that step or any later stage to include the control measure.

The *critical limit* needs to be specified and validated for each critical control point to mitigate the hazard which may be control of temperature, pH, water content, water activity, etc. Appropriate critical limit is important for eliminating the hazard.

Monitoring is an important step where observations of CCPs in relation to the critical limit is measured to identify the effectiveness of control. Deviations, if any, need to be addressed through corrective actions in order to ensure safety of the product.

Once the system is set, then *verification procedure* needs to establish to confirm that the system is under control and the HACCP system is working perfectly and the hazards are taken care of.

Once the system is established, proper *documentation and record-keeping* is absolutely required to sustain the system. Hazard analysis, CCP determination, and critical limit determination are processes that require documentation and activities concerned with CCP monitoring, deviations, and corrective actions and modifications of the system are the steps that need to be recorded for future reference and alterations.

Prerequisite programs (PRPs) are foundations for food safety or HACCP system. The PRPs include all the steps or procedures for controlling the operational conditions environmental conditions within a food establishment favorable for the production of safe food. Appropriate gap analysis is to be carried out to confirm the PRP requirement and should be made in place for ensuring the safety of the product. Some of the important PRPs are GMP, GHP, SSOP, raw material control, production control, pest control, etc., and PRPs vary from plant to plant and between product to product.

19.8 REGULATORY BODIES

All most all countries in the world have an organization dedicated to food safety and consumer health. The CAC is concerned with the safety of food meant for consumers and its trade across the globe. European Commission of EU, Australian Food Safety Commission, Canadian Commission on Food Safety are few to name and India has Food Safety Standards Authority of India (FSSAI) to address the issues.

19.8.1 CODEX ALIMENTARIUS

The CAC (FAO website) was created in 1963 by FAO and WHO to develop food standards, guidelines, codes of practice, etc., to protect the health of consumers, ensure fair trade practices in the food trade, and promote coordination of all food standards work undertaken by international governmental and non-governmental organizations. Currently, the CAC has 189 Codex Members made up of 188 Member Countries and 1 Member Organization (The European Union). Besides regular members, Codex also has the support of237 observers, expert technical bodies, industry, and consumer associations, contributing to the standard-setting process in a spirit of openness, collaboration, and transparency which includes 58 IGOs, 163 NGOs, 16 UN bodies. The Joint FAO/WHO Expert Committee on Food Additives (JECFA) is an international expert scientific committee administered jointly by the Food and Agriculture Organization of the United Nations (FAO) and the World Health Organization (WHO). The Joint Expert Committee for Food Additives (JECFA) serves as an independent scientific committee that performs risk assessments and provides advice to FAO, WHO, and the member countries of both organizations. The requests for scientific advice are channeled through the CAC in their work to develop international food standards and guidelines under the Joint FAO/WHO Food Standards Program.

Besides every country has got their own mechanism for addressing food safety. EU has got European Commission, Canada has the Canadian food Inspection Agency and so on and so forth. FSSAI, based at New Delhi is the apex organization In India with Food Safety Commissionerate in all the states to implement the regulations. The Food Safety and Standards Act, 2006 (Act No. 34 of 2006) came into existence during August 2006. The Food Safety and Standards Authority of India (FSSAI) has been established under Food Safety and Standards (2006) and has been created for laying down science-based standards for articles of food and to regulate their manufacture, storage, distribution, sale, and import to ensure availability of safe and wholesome food for human consumption. The Food Safety and Standards Rules (2011) came into force on 1st August 2011. The FSSAI is mandated to perform seven important activities for developing food safety for citizens of India namely, Setting globally benchmarked regulations, standards, and guidelines, facilitating compliance through optimized procedures and improved laboratory network, building capacity of regulatory staff as well as food business operators (FBOs), driving public health initiative in time spirit of convergence, leveraging frontier communication techniques to build food safety culture, embracing technology to streamline processes and forging strategic partnerships to generate exchange knowledge and best practices.

KEYWORDS

- **additives**
- **environmental pollutants**
- **food pathogens**
- **food safety**
- **hazard management**
- **hazards**
- **marine toxins**

REFERENCES

Abdel-Shafy, H. I., & Mansour, M. S. M., (2016). A review on polycyclic aromatic hydrocarbons: Source, environmental impact, effect on human health and remediation. *Egyptian Journal of Petroleum, 25,* 107–123.

Adley, C. C., & Ryan, M. P., (2016). The nature and extent of foodborne disease. Chapter 1. In: *Antimicrobial Food Packaging* (January, pp. 1–10). University of Limerick, Limerick, Ireland. http://doi.org/10.1016/B978-0-12-800723-5.00001-2.

Al-Busaidi, M., Yesudhason, P., Al-Mughairi, S., et al., (2011). Toxic metals in commercial marine fish in Oman with reference to national and international standards. *Chemosphere, 85*(1), 67–73.

Altekruse, S. F., Street, D. A., Fein, S. B., & Levy, A. S., (1996). Consumer knowledge of foodborne microbial hazards and food-handling practices. *J. Food Prot., 59*, 287–294.

Armstrong, B. G., Hutchinson, E., Unwin, J., & Fletcher, T., (2004). Lung cancer risk after exposure to polycyclic aromatic hydrocarbons: A review and meta-analysis. *Environ. Health Perspect., 112*(9), 970–978.

Baklanov, A., Ha¨nninen, O., Slørdal, L. H., Kukkonen, J., Bjergene, N., Fay, B., et al., (2007). Integrated systems for forecasting urban meteorology, air pollution and population exposure. *Atmos. Chem. Phys., 7*, 855–874.

Bawuro, A. A., Voegborlo, R. B., & Adimado, A. A., (2018). Bioaccumulation of heavy metals in some tissues of fish in Lake Geriyo, Adamawa State, Nigeria. *Hindawi Journal of Environmental and Public Health*, 7. Article ID 1854892. https://doi.org/10.1155/2018/1854892

Bender, J., Smith, K., Hedberg, C., & Osterholm, M., (1999). Foodborne disease in the 21st century. What challenges await us? *Postgrad. Med., 106*, 109–112, 115–116, 119.

Bosch, A., Gkogkab, E., Le Guyaderc, F. S., Loisy-Hamond, F., Lee, A., Lieshoutf, L., Marthi, B., et al., (2018). Foodborne viruses: Detection, risk assessment, and control options in food processing. *International Journal of Food Microbiology, 285*(20), 110–128.

Broglia, A., & Kapel, C., (2011). Changing dietary habits in a changing world: Emerging drivers for the transmission of foodborne parasitic zoonoses. *Veterinary Parasitology, 182*, 2–13.

CAC, (2018). *Codex Alimentarius Commission, Joint FAO/WHO Food Standards Program Codex Committee on Food Hygiene*. Fiftieth Session Panama.

Canli, M., & Atli, G. C., (2003). The relationship between heavy metal (Cd, Cr, Cu, Fe, Pb, Zn) levels and the size of six Mediterranean fish species. *Environ. Pollut., 121*(1), 129–136.

CDC, (2011). Vital signs: Incidence and trends of infection with pathogens transmitted commonly through food—foodborne diseases active surveillance network, 10 U.S. Sites, 1996–2010. *Morbidity and Mortality Weekly Report, 60*(22), 749–755. U.S. Department of Health and Human Services Centers for Disease Control and Prevention.

CDCP, (2011). *Estimates of Foodborne Illness in the United States, Center for Disease Control and Prevention*. https://www.cdc.gov/foodborneburden/index.html (accessed on 09 March 2022).

CFIA, (2014). Canadian food inspection agency: Chemical residues. In: *Meat Hygiene Manual of Procedures: Chapter 5-Sampling and Testing Procedures*. Meat Hyg. Dir. Canada: Russell A. Farrow Limited. http://www.inspection.gc.ca/food/meat-andpoultry-products/manual-of-procedures/chapter-5/eng/1395150894222/1395150895519?chap=2 (accessed on 09 March 2022).

Chan, K. M., (1995). Concentration of Cu, Zn, Cd and Pb in rabbitfish (*Signus oramin*) collected in Victoria Harbour, Hong Kong. *Mar. Pollut. Bull., 31*, 277–280.

Chon, J. W., Song, K. Y., Kim, S. Y., Hyeon, J. Y., & Seo, K. H., (2012). Isolation and characterization of cronobacter from desiccated foods in Korea. *Journal of Food Science, 77*(7), M354–M358.

Darko, G., Akoto, O., & Oppong, C., (2008). Persistent organochlorine pesticide residues in fish, sediments and water from Lake Bosomtwi, Ghana. *Chemosphere, 72*(1), 21, 22.

Duff, S. B., Scott, E. A., Mafilios, M. S., Todd, E. C., Krilov, L. R., Geddes, A. M., & Ackerman, S. J., (2003). Cost-effectiveness of a targeted disinfection program in household kitchens to prevent foodborne illness in the United States, Canada and the United Kingdom. *J Food Protect., 66*(11), 2103–2115.

EC, (2001). European Commission: Establishment of maximum residue limits (MRLs) for residues of veterinary medicinal products in foodstuffs of animal origin. *Volume 8, Notice to Applicants and Note for Guidance.* https://ec.europa.eu/health/documents/eudralex/vol-8_en (accessed on 30 March 2022).

EC, (2010). European Commission. Commission Regulation (EU) No 37/2010. *Off. J. Eur. Union.* 2009 L 15/1, https://ec.europa.eu/health/system/files/2016-11/reg_2010_37_en_0.pdf (accessed on 09 March 2022).

EEC, (1990). EEC European union council regulation (EEC). *Journal of Eur. Comm., 16*, 2377–2390.

EFSA, (2008). Marine biotoxins in shellfish – okadaic acid and analogues1 Scientific Opinion of the Panel on Contaminants in the Food chain. *The EFSA Journal, 589*, 1–62.

EFSA, (2008). *Polycyclic Aromatic Hydrocarbons in Food, Scientific Opinion of the Panel on Contaminants in the Food Chain.* European Food Safety Authority.

El-Dien, W. M. S., & Mahmoud, H. A., (2011). Survey on some chemical pollutant residues in *Clarias gariepinus* at Sharkia Governorate, Egypt. *Journal of American Science, 7*, 286–293.

EPA, (1999). EPA fact sheet. *Polychlorinated Biphenyls (PCBs) Update: Impact on Fish Advisories.* Office of Water 4305, EPA-823-F-99-019, United States Environmental Protection Agency EPA website. https://www.epa.gov/pcbs/learn-about-polychlorinated-biphenyls-pcbs#what (accessed on 09 March 2022).

EPA, (2020). https://www.epa.gov/pcbs/learn-about-polychlorinated-biphenyls-pcbs (accessed on 09 March 2022).

FAO, (2006). *Food Security, Policy Brief.* Food Agriculture Organization of the United Nations.

FAO, (2014). Food safety and quality program: Horizon scanning and foresight an overview of approaches and possible applications in food safety, background paper 2. *FAO Early Warning/Rapid Alert and Horizon Scanning, Food Safety Technical Workshop.* Food Agriculture Organization of the United Nations.

FAO, (2019). *The Future of Food Safety.* Brochure, FAO.

FAO, (2022). Website http://www.fao.org/fao-who-codexalimentarius/about-codex/observers/observers/ (accessed on 09 March 2022).

FAO/WHO, (2012). Food and agriculture organization/world health organization (FAO/WHO). *Joint FAO/WHO Expert Meeting on the Public Health Risks of Histamine and Other Biogenic Amines from Fish and Fishery Products* (pp. 1–111). Joint FAO/WHO expert meeting report. Rome: FAO Headquarters.

FDA, (1999). *Foods–Adulteration Involving Hard or Sharp Foreign Objects.* FDA/ORA Compliance Policy Guide, Chapter 5, Sub Chapter, 555, Section 555.425 (Issued: 3/23/1999). Department of Health and Human Services, Public Health Service, Food and Drug Administration, Washington, DC. https://www.fda.gov/media/71953/download (accessed on 30 March 2022).

FDA, (2011). *Fish and Fishery Products Hazards and Controls Guidance* (4th edn.). Washington, DC: Department of Health and Human Services, Food and Drug Administration, Center for Food Safety and Applied Nutrition.

Hall, A. J., Wikswo, M. E., Pringle, K., Gould, L. H., & Parashar, U. D., (2014). Vital signs: Foodborne norovirus outbreaks—United States, 2009–2012. *MMWR Morb. Mortal. Wkly Rep., 63*, 491–495.

Havelaar, A. H., Kirk, M. D., Torgerson, P. R., Gibb, H. J., Hald, T., Lake, R. J., et al., (2015). World Health Organization global estimates and regional comparisons of the burden of foodborne disease in 2010. *PLoS Med., 12*(12), e1001923. https://doi.org/10.1371/journal.pmed.1001923.

Hoffmann, S., Devleesschauwer, B., Aspinall, W., Cooke, R., Corrigan, T., Havelaar, A., et al., (2017). Attribution of global foodborne disease to specific foods: Findings from a world health organization structured expert elicitation. *PLoS One 12*(9), e0183641. https://doi.org/10.1371/journal. pone.0183641.

Hur, J., Jawale, C., & Lee, J. H., (2012). Antimicrobial resistance of *Salmonella* isolated from food animals: A review. *Food Research International, 45*, 819–830.

Inomata, Y., Kajino, M., Sato, K., Ohara, T., Kurokawa., J. I., et al., (2012). Emission and atmospheric transport of particulate PAHs in Northeast Asia. *Environ Sci. Technol., 46*(9), 4941–4949.

Jiang, Y. F., Wang, X. T., Wu, M. H., Sheng, G. Y., & Fu, J. M., (2011). Contamination, source identification, and risk assessment of polycyclic aromatic hydrocarbons in agricultural soil of Shanghai, China. *Environ. Moni.t Assess, 183*, 139–150.

Kalyoncu, L., Agca, I., & Aktumse, A., (2009). Some organochlorine pesticide residues in fish species in Konya, Turkey. *Chemosphere, 74*(7), 885–889.

King, J. C., Black, R. E., Doyle, M. P., Fritsche, K. L., Halbrook, B. H., Levander, O. A., Meydani, S. N., et al., (2000). Foodborne illnesses and nutritional status: A statement from an American society for nutritional sciences working group. *J. Nutr., 130*, 2613–2617.

Kipopoulou, A. M., Manoli, E., & Samara, C., (1999). Bioconcentration of polycyclic aromatic hydrocarbons in vegetables grown in an industrial area. *Environ Pollut., 106*(3), 369–380.

Koopman, M., & Duizer, E., (2004). Foodborne viruses. An emerging problem. *International Journal of Food Microbiology, 90*, 23–41.

Koutsoumanis, K. P., Lianou, A., & Sofos, J. N., (2014). Food safety: Emerging pathogens. *Encyclopedia of Agriculture and Food Systems, 3*, 250–272. doi: 10.1016/B978-0-444-52512-3.00049-8.

Krock, B., Ferrario, M. E., Akselman, R., & Montoya, N. G., (2018). Occurrence of marine biotoxins and shellfish poisoning events and their causative organisms in Argentine marine waters. *Oceanography, 31*(4), 132–144. https://doi.org/10.5670/oceanog.2018.403.

Louden, B. C., Haarman, D., Han, J., Foley, S., & Lynne, A. M., (2012). Characterization of antibacterial resistance in *Salmonella enterica* serovar *Typhimurium* isolate from food animals. A review. *Food Research International, 45*, 968–972.

Mahmoudi, R., & Mardani, K., (2015). Histamines and foods: A review on importance, detection and controlling in foods. *Malaysian Journal of Science, 34*(1), 103–107.

Mohamed, R., Livia, S. S., Hassan, S., Soher, E. S., & Ahmed-Adel, E. B., (2009). Changes in free amino acids and biogenic amines of Egyptian salted fermented fish (Feseekh) during ripening and storage. *Food Chemistry, 115*, 635–638.

Mor-Mur, M., & Yuste, J., (2010). Emerging bacterial pathogens in meat and poultry: An overview. *Food Bioprocess Technol., 3*, 24–35 doi: 10.1007/s11947-009-0189-8.

Morris, W. E., & Fernández-Miyakawa, M. E., (2009). Toxinas de *Clostridium perfringens*. *Revista Argentina de Microbiologia, 91*, 251–260 (cross referred).

Mustafa, C., & Guluzar, A., (2003). The relationships between heavy metal (Cd, Cr, Cu, Fe, Pb, Zn) levels and the size of six Mediterranean fish species. *Environmental Pollution, 121*(1), 129–136.

NAS-NRC (National Academy of Sciences–National Research Council), (2000). *Toxicological Effects of Methylmercury* (p. 22). Accessed at https://nap.nationalacademies.org/catalog/9899/toxicological-effects-of-methylmercury (accessed on 30 March 2022).

Noguchi, T., & Ebesu, J. S. M., (2001). Puffer poisoning: Epidemiology and treatment. *J. Toxicol.-Toxin Rev., 20*, 1–10.

Norman, K. N., Harvey, R. B., Andrews, K., Hume, M. E., Callaway, T. R., Anderson, R. C., & Nisbet, D. J., (2014). Survey of *Clostridium difficile* in retail seafood in college station, Texas. *Food Additives & Contaminants: Part A, 31*(6), 1127–1129. doi: 10.1080/19440049.2014.888785.

Ogunfowokan, A. O., Oyekunle, J. A. O., Torto, N., & Akann, M. S., (2012). A study on persistent organochlorine pesticide residues in fish tissues and water from an agricultural fish pond. *Emir. J. Food Agric., 24*(2), 165–184.

Okocha, R. C., Olatoye, I. O., & Adedeji, O. B., (2018). Food safety impacts of antimicrobial use and their residues in aquaculture. *Public Health Reviews, 39*(21), 1–22. https://doi.org/10.1186/s40985-018-0099-2.

Oteo, J., Perez, V. M., & Campos, J., (2010). Extended-spectrum B–lactamase producing *Escherichia coli*: changing epidemiology and clinical impact. *Current Opinion in Infectious Diseases, 23*(4), 320–326. (Abstract only).

Palm, L. M. N., Carboo, D., Yeboah, P. O., Quasie, W. J., Gorleku, M. A., & Darko, A., (2011). Characterization of polycyclic aromatic hydrocarbons (PAHs) present in smoked fish from Ghana. *Adv. J. Food Sci. Technol., 3*(5), 332–338.

Papagiannis, I., Kagalou, I., Leonardos, J., Petridis, D., & Kalfakakou, V., (2004). Copper and zinc in four freshwater fish species from Lake Pamvotis (Greece). *Environment International, 30*(3), 357–362.

Pigott, D. C., (2008). Foodborne illness. *Emergency Medicine Clinics of North America, 26*, 475–497.

Pressman, P., Clemens, R., Hayes, W., & Reddy, C., (2017). Food additive safety: A review of toxicologic and regulatory issues. *Toxicology Research and Application, 1*, 1–22. doi: 10.1177/2397847317723572 journals.sagepub.com/home/tor.

Rahman, M. S., Molla, A. H., Saha, N., & Rahman, A., (2012). Study on heavy metals levels and its risk assessment in some edible fishes from Bangshi River, Savar, Dhaka, Bangladesh. *Food Chemistry, 134*(4), 1847–1854.

Rajeshkumar, S., & Li, X., (2018). Bioaccumulation of heavy metals in fish species from the Meiliang Bay, Taihu Lake, China. *Toxicology Reports, 5*, 288–295.

Ramachandran, G., (2014). Gram-positive and gram-negative bacterial toxins in sepsis, *Virulence, 5*, 213–218.

Robinson, R., Batt, C., & Patel, P. D., (2000). *Encyclopedia of Food Microbiology* (5th edn.). Academic Press. San Diego.

Romeoa, M., Siaub, Y., Sidoumou, Z., & Gnassia, B. M., (1999). Heavy metal distribution in different fish species from the Mauritanian coast. *Sci. Total Environ., 232*, 169–175.

Sankar, T. V., Zynudheen, A. A., Anandan, R., & Nair, P. G. V., (2006). Distribution of organochlorine pesticides and heavy metal residues in fish and shellfish from Calicut region, Kerala, India. *Chemosphere, 65*, 583–590.

Scallan, E., Hoekstra, R. M., Angulo, F. J., Tauxe, R. V., Widdowson, M., Roy, S. L., Jones, J. L., & Griffin, P. M., (2011). Foodborne illness acquired in the United States—major pathogens. *Emerging Infectious Diseases, 17*(1), 7–15.

Shalaby, S. E. M., El-Saadany, S. S., Abo-Eyta, A. M., Abdel-Satar, A. M., Al-Afify, A. D. G., & El-Gleel, W. M. M. A., (2018). Levels of pesticide residues in water, sediment, and fish samples collected from Nile River in Cairo, Egypt. *Environmental Forensics, 19*(4), 228–238. doi: 10.1080/15275922.2018.1519735.

Sofos, J. N., (2008). Challenges to meat safety in the 21st century. *Meat Science, 78*, 3–13.

Song, X., Shukla, S., Lee, G., Park, S., & Kim, M., (2016). Detection of cronobacter genus in powdered infant formula by enzyme-linked immunosorbent assay using anti-cronobacter antibody. *Front. Microbiol.* https://doi.org/10.3389/fmicb.2016.01124.

South, J., Edwards, M., & Hutton, T., (2008). Chemical hazards in food. *Campden BRI IUFoST Scientific Information Bulletin.* http://iufost.org/iufost-scientific-information-bulletins-sib (accessed on 30 March 2022).

Spencer, M., (1974). Food additives. *Postgraduate Medical Journal, 50*, 620–624.

Teklit, G. A., (2016). Residues analysis of organochlorine pesticides in fish, sediment and water samples from Tekeze Dam, Tigray, Ethiopia. *Journal of Environmental & Analytical Toxicology, 6*, 342.

Tollefson, L., & Miller, M. A., (2000). Antibiotic use in food animals: Controlling the human health impact. *J. AOAC Int., 83*, 245–256.

Tongo, I., Ogbeide, O., & Ezemonye, L., (2017). Human health risk assessment of polycyclic aromatic hydrocarbons (PAHs) in smoked fish species from markets in Southern Nigeria. *Toxicology Reports, 4*, 55–61.

Udgata, S. K., Dash, G., Parida, S. K., & Mishra, R., (2009). Isolation and identification of human pathogenic *Aeromonas hydrophila* in seafood of Gopalpur coast, Orissa. *Indian J. Anim. Res., 43*(1), 45–48.

USFDA, (2010). *Food Ingredients and Colors*. International food information council (IFIC), United States Food and Drugs Administration, USA.

Van, B. T. P., Brower, C., Gilbert, M., Grenfell, B. T., Levin, S. A., Robinson, T. P., Teillant, A., & Laxminarayan, R., (2015). Global trends in antimicrobial use in food animals. *Proc Natl. Acad. Sci. U S A.*, 5649–5654.

Vivekanandhan, G., Hatha, A. A. M., & Lakshmanaperumalsamy, P., (2005). Prevalence of *Aeromonas hydrophila* in fish and prawns from the seafood market of Coimbatore, South India. *Food Microbiology, 22*, 133–137.

Voegborlo, R. B., Atta, A., & Agorku, E. S., (2012). Total mercury distribution in different tissues of six species of freshwater fish from the Kpong hydroelectric reservoir in Ghana. *Environmental Modeling & Assessment, 184*(5), 3259–3265.

Vouga, M., & Greub, G., (2016). Emerging bacterial pathogens: The past and beyond. *Clin. Microbiol. Infect., 22*, 12–21.

Wang, S., Weller, D., Falardeau, J., Strawn, L. K., Mardones, F. O., Adell, A. D., et al., (2016). Food safety trends: From globalization of whole-genome sequencing to application of new tools to prevent foodborne diseases. *Trends Food Sci. Technol., 57*, 188–198. doi: 10.1016/j.tifs.2016.09.016.

WHO, (2006). *Food Safe Risk Analysis: A Guide for National Food Safety Authorities.* FAO Food and Nutrition Paper 87. World Health Organization.

WHO, (2015). *Estimates of the global Burden of Foodborne Diseases.* World Health Organization.

WHO, (2018). *Food Additives.* https://www.who.int/news-room/fact-sheets/detail/food-additives (accessed on 09 March 2022).

Woodward, K. N., (1991). Hypersensitivity in humans and exposure to veterinary drugs. *Vet. and Hum. Toxicol., 33,* 168–172.

World Health Organization, (1999). *Removing Obstacles to Healthy Development—World Health Organization Report on Infectious Diseases* (p. 68). WHO, Geneva, Switzerland, World Health Organization. Fact Sheet No. 237: Food safety and foodborne illness.

Zarei, M., Najafzadeh, H., Enayati, A., & Pashmforoush, M., (2011). Biogenic amines content of canned tuna fish marketed in Iran. *American Eurasian Journal Toxicology Science, 3,* 190–193.

Zhai, H., Yang, X., Li, L., Xia, G., Cen, J., Huang, H., & Hao, S., (2012). Biogenic amines in commercial fish and fish products sold in southern China. *Food Control, 25,* 303–308.

Zhao, S., Feng, C., Quan, W., Chen, X., Niu, J., & Shen, Z., (2012). Role of living environments in the accumulation characteristics of heavy metals in fishes and crabs in the Yangtze River Estuary, China. *Marine Pollution Bulletin, 64*(6), 1163–1171.

CHAPTER 20

Developments in Safety and Quality Management of Fish and Fishery Products

G. JEYASEKARAN and R. JEYA SHAKILA

Department of Fish Quality Assurance and Management, Tamil Nadu Dr. J. Jayalalithaa Fisheries University, Tuticorin – 628008, Tamil Nadu, India

ABSTRACT

Fishery products are important for a balanced diet and contribute positively to the nutritional status of humans. Fish is known to be an excellent source of proteins, long-unsaturated chains of omega-3 fatty acids (EPA and DHA), vitamin D, vitamin B12, and many useful trace elements such as selenium and iodine. Fatty fish are the most important sources of omega-3 fatty acids, and vitamin D. Many epidemiological studies have shown that the frequent consumption of fish results in a lower risk of heart disease, improved neurological development in fetuses, and reduced levels of disease including obesity, diabetes type 2, multiple sclerosis and osteoporosis. Though fishery products are highly traded food commodities at the international level with perceived health benefits, they can pose numerous health risks. Fish products may harbor various contaminants, including biological, chemical, and physical hazards, with the most prevalent being the biogenic amines, biotoxins, heavy metals, pathogenic bacteria, and viruses. High concentrations of these contaminants when present in fish may constitute serious hazards.

Advances in Fish Processing Technologies: Preservation, Waste Utilization, and Safety Assurance.
Ranendra K. Majumder, PhD & Amjad K. Balange, PhD (Eds.)

20.1 INTRODUCTION

The overexploitation of marine fish stock around the world has reduced the production of fish products significantly, which in turn has led to a substantial increase in the aquaculture production of fish. Although the development of aquaculture can provide an enhanced and regular supply of seafood, it also introduces new risks. Without appropriate farming methods, it may become necessary to use the veterinary drugs for disease control in aquaculture. In some situations, their use can also have an economic impact as they can increase the rate at which the fish grow. However, the use of these chemicals also has the potential to harm consumers, if the level in the consumed fish is above a safe level. An additional concern is that their use leads to the development of antimicrobial resistance in bacterial pathogens.

20.2 FISH PRODUCTION AND CONSUMPTION

The world's per capita consumption of fish products has grown steadily to 19.7 kg in 2013 and 20 kg in 2016 from 9.9 kg in the 1960s as a result of the high aquaculture supply. For developing countries, the annual per capita consumption has grown to 18.8 kg in 2013 from 5.2 kg in 1961, whereas in the low-income food-deficit countries, it increased to 7.6 kg in 2013 from 4.9 kg in 1961. World fish production destined for human consumption has been marketed as 46% in live, fresh, and chilled state (67 MMT), 30% as frozen (44 MMT), 12% as cured (17 MMT), and 13% as prepared (19 MMT). Frozen fish and fishery products, however, have been the foremost processing method used for marketing fish intended for human consumption and accounted for 26% of total fish production in 2014. Fish contributed around 17% of the global intake of protein of animal source and 6.7% of the entirely protein consumed Worldwide.

20.3 FISH TRADE

The international trade in fishery products has been expanding rapidly and has continued to be the most traded food commodity in recent decades valuing US $130 billion in 2013. The rising demand for fishery products combined with enhanced systems, technological advances, and trade liberalization have increased fish trade, making it a significant source of foreign exchange and employment for many countries. In 2014, fishery products accounted for around 9% of total agriculture exports and 1% of World commodities in value. The world trade of exported fish and fishery products as feed and food grew considerably from 25% of total production in 1976 to around 38% in volume (58.1 MMT) in 2012, with values within the same periods showing an increase from just US $8 billion to reach a level of US $129 billion. In 2014, the value of exports increased further to US $148.147 billion. The developing countries supplied the world market with 37% of the fishery products in value terms in 1976 but this had increased by 2014 to 54% in value term and 60% of the quantity (live weight). In 2014 alone, the fishery exports from the developing countries were valued at US $80 billion. The largest fish producer and exporter in the global market is China with around US $20.98 billion in value terms in 2014. The EU, USA, and Japan are the major importers of fishery products and their combined imports consist of 63% by value and 59% by quantity of world imports, with the EU being the largest market for fish products in 2014.

20.4 DEVELOPMENT IN FISH SAFETY AND QUALITY

There are many challenges specifically in issues concerning food safety and quality for the globalization and liberalization of the world fish trade. The international food safety regulators have adopted risk-based analysis by means of implementing the HACCP system, so as to maintain fish safety that reflect recent scientific knowledge. The safety levels of the different environmental pollutants and contaminants have been set by the regulators for achieving a high level of protection in the health of the consumers. Furthermore, consumer educational programs are vital to enhance awareness and improve transparency among the different stakeholders in the fish supply chain. Similarly, the international standards in food safety and quality have to be met for the fish products traded internationally. Otherwise, they could be detained, rejected, and even destroyed if in non-conformance to international import and export safety and quality requirements.

The fishery products shall be cautiously handled, as they are highly perishable. Postharvest losses can be minimized by implementing fish safety management systems. Effective hygiene practices are essential throughout the supply chain so as to produce safe fish products. Most food poisoning outbreaks in relation to microbial contaminants are caused by consumption of mishandled raw fish, insufficiently cooked or harvested in contaminated areas. Aquatic food-associated infections are mainly caused by bacteria from the aquatic environment (*Vibrio parahaemolyticus, V. vulnificus, Clostridium botulinum*) or terrestrial sources (*Salmonella* spp., *Shigella* spp., *Staphylococcus* spp., *Vibrio cholerae, Escherichia coli, Clostridium perfringens*), or due to recontamination during storage and distribution after processing. Viruses such as Norovirus, HA virus, and parasites like protozoa, nematodes, trematodes, and cestodes can also be involved resulting in diverse clinical symptoms. When compared to chemical hazards, the microbial hazards are more dynamic as the microbial pathogens get multiplied in foods. The "scombrotoxin fish poisoning" (SFP) or "histamine poisoning" is due to the ingestion of certain marine fish species with high levels of free histidine in their tissue. Under temperature abuse, a heat-stable scombrotoxin is produced by specific bacteria known to possess the histidine decarboxylase enzyme, which leads to the formation of histamine. SFP accounts for a large portion of seafood-borne illnesses worldwide. A variety of symptoms have been observed on different patients but these are rarely fatal.

The marine environment encompasses potentially hazardous chemicals released from industrial and domestic sources. These contaminants have been classified as inorganic contaminants such as mercury, lead, cadmium, zinc, arsenic, copper, and sulfites; organic contaminants like dioxins, polychlorinated biphenyls (PCBs), chlorinated hydrocarbons and insecticides; naturally occurring compounds such as polycyclic aromatic hydrocarbons (PAHs) and marine biotoxins; processing-related compounds like sulfite compounds used in the shrimp processing, nitrosamines; and residues of veterinary drugs such as antibiotics or hormones used in aquaculture.

Marine biotoxins particularly ciguatera (ciguatoxin), paralytic shellfish poisoning (saxitoxin), neurotoxic shellfish poisoning (brevetoxin), pufferfish poisoning (tetradotoxin), diarrhetic shellfish poisoning (DSP) (dinophysistoxin), and amnesic shellfish poisoning (ASP) (domoic acid) are also important causative agents in many foods poisoning cases worldwide. The increased level of these in the aquatic environment has raised the levels that are found concentrated in fish tissues. More contaminants get accumulated in predator fish at the higher levels of the food chain due to the bio-magnification process. Therefore, the level of contamination depends on the species sizes, feeding patterns, geographic location, and the chemical's solubility in the aquatic environment.

20.5 INTERNATIONAL REGULATIONS ON FISH AND FISHERY PRODUCTS

The regulations of the European Union (EU), USA, Japan, FAO/WHO are most important to be followed by the fish exporters around the world for the trading of fishery products at the international level.

20.5.1 EUROPEAN UNION (EU) REGULATIONS

The biggest importer of fishery products from wild and aquaculture sources in the world is the EU, accounting for 40% of total world imports. The responsibility of maintaining food safety in the EU rests with Directorate-General for Health and Food Safety (DG-SANTE), who has to ensure that the quality and safety of farmed and captured seafood supplied to the EU markets meet the standards set in EU legislation. The control of seafood safety to a competent authority (CA) in a country shall be delegated by the EU so as to confirm that the farms, vessels, processors, and other necessary facilities in the country are under a system equivalent to that specified in the EU. The establishments from third countries satisfying the EU requirements have to obtain the agreement from their CA so as to export fish products to the EU. Once the establishments from these countries have been registered with their CA, the details of those establishments need to be passed to the European Commission, which in turn publishes it via its website. Moreover, the CA of third countries with approval is regularly audited by officials from DG-SANTE (formerly known as the Food and Veterinary Office (FVO). Alternatively, the third countries are classified, based on the Commission Decision 2006/766/EC, either in Annex I or in Annex II. Annex I gives the list of third countries that are permitted to export bivalve molluscans, echinoderms, tunicates, and marine gastropods in any form to the EU for human consumption. Annex II contains the list of third countries and territories that are permitted to export fishery products in any form to the EU for human consumption.

The food safety legislation consisting of food regulations, directives, decisions, codes of good practices and official food standards governs food safety in the EU. Over many years, the legislation has undergone change and amendment based on emerging issues and the availability of scientific evidence of food hazards. In particular, the EU has produced controls that generally regulate food safety including fish and aquaculture production and trade. In early 2000, a general food control law at a high level was proposed in the EU, which was then made in 2002 as Regulation 178/2002. The EU in July 2000 published proposals for a package of five new hygiene regulations to provide consistent control in hygiene from farm to fork. The resulting four regulations adopted in 2004 are as follow: 852/2004, 853/2004, 854/2004 and 822/2004. In January 2006, the hygiene package legislation in the EU came into force featuring the main food safety requirements for primary producers, food business operators (FBOs), and traders within the EU and other countries. Other issues of food safety have been addressed in the non-hygiene food regulations. The provision of these EU regulations cover issues concerned with general food products, food from animal origin and their health, fisheries product with emphasis on the aquaculture products, and food contaminants. They also tackle pesticides, labeling, organic agriculture, and genetically modified organisms (GMO) amongst other things.

Previously Council Directives 91/67/EEC, 93/53/EEC and 95/70/EC were the main legal documents governing the animal health conditions of aquaculture fish placed on the market and the control of certain diseases affecting fish and bivalve molluscans. However, these directives have been updated by Council as Directive 2006/88/EC containing the animal health requirements for aquaculture animals and their products, including the control of certain diseases in aquatic animals. The EU has issued two controls on food contaminants: EC Council Directives 96/22/EC and 96/23/EC determine the unauthorized usage of prohibited chemical substances in the animal production operations, which apply to aquaculture products. The Maximum Residual Limits (MRL) for veterinary medicinal products in foodstuffs of animal origin has been laid down in Regulation 2377/90 lays down, while the MRL for pesticides in or on food and feed of plant and animal origin is given in Regulation 396/2006. The earlier Regulation 466/2001 has been replaced by Regulation 1881/2006 containing maximum levels for certain contaminants like metals (lead, cadmium, mercury, and inorganic tin), nitrates, dioxins, and dioxin-like PCBs, mycotoxins, and polycyclic aromatic hydrocarbons in food. However, further amendment has been made for some foodstuffs and feed on the maximum levels of some of the contaminants. Such Regulations are the Regulation 2015/1006 for inorganic arsenic in foodstuffs; Regulation 1259/2011 for dioxins, dioxin-like PCBs and non-dioxin-like PCBs in foodstuffs; Regulation 853/2004 for natural toxins such as marine bio-toxins. The Annex I of Regulation 2073/2005 on microbiological criteria for foodstuffs including aquatic organisms was amended as Regulation 1019/2013 by including the maximum levels of histamine in fishery products so as to meet the recommendation set by the Codex Alimentarius Commission (CAC) on histamine.

Article 8 of Directive 92/59/EEC on general product safety (later amended by Directive 2001/95) has initially established the Rapid Alert System for Food and Feed (RASFF). But, now Regulation 178/2002 is the legal basis for RASFF. The scope and procedures of the RASFF are defined in Articles 50, 51, and 52 of the Regulation. The scope of the RASFF to include animal feeds and the border inspection posts (BIPs) network, and aims to provide the control authorities with an effective tool to exchange information in response to serious food and feed risks detected in food has been extended as per the new legal basis. The identification of the members of the system and setting out the requirement for a RASFF notification as per Article 50 of the Regulation. The RASFF members have to notify any information on serious health risks deriving from food or feed and, in particular, any measures taken to withdraw or recall food or feed products from the market to the European Commission so as to protect the health of the consumers. The Commission also notifies third countries involved in the RASFF system. The scope of RASFF on feed related to food only was extended to include all types of feed and resultant risks on animal health and environment, as a result of the Feed Hygiene Regulation entering into force in 2006. In 2008, the system introduced the 'Consumer complaint' and 'Food poisoning' provisions. The provision of 'Official control' was introduced in the system in non-member countries in 2010. Over the years, the RASFF network has grown with 10 more new members joined in 2004 and integrated smoothly into the network.

The RASFF reported a total of 2,924 original notifications received from the different RASFF members in 2016, with a total of 519 notifications reported on fish products and subdivided into the different categories. The 519 notifications on fish and fishery products accounted for 18% of the total notified products in 2016. The fishery products represented the highest notification received of 327 notifications (63% of the total notifications), which was the highest over the last three years. Bivalve molluscan products represented 84 notifications accounting for about 16%; 69 (13%) for crustacean products; 38 (7.0%) for cephalopods and products and finally just one notification for gastropods products with a 0.2% of the whole notified products.

Most of the 2016 notifications extracted from the RASFF system which related to the fish products were due to higher concentrations of heavy metals, in particular total mercury, followed by histamine, the most implicated biogenic amine. The latter was mostly detected on tuna, anchovies, and sardine, with the highest concentration (1,774 ppm) detected in tuna (*Thunnus albacares*) loins imported from Spain. Microbial contamination was mostly caused by *Listeria monocytogenes* on smoked fish products and *Vibrio* spp. on the other fish products. Norovirus and HA were the mostly detected viruses on bivalve molluscans and their products alongside the bacteria *Escherichia coli* and *Salmonella*. DSP toxin was the most common biotoxin detected on bivalve molluscans products among other biotoxins. As for fish products, the ciguatera poisoning was predominant. Food poisoning notifications were firstly introduced in the RASFF database in 2008 and have been valuable for risk assessment and management by providing a link between occurrence data and causative agents of food poisonings in notified food and adverse health effects of consumers. The most frequently mentioned countries reported in the notifications are China (433), India (257), and Thailand (88), followed by the EU and then Latin America.

The DG SANTE's Audit Department (formerly the FVO) is responsible for the proper implementation and enforcement of Community legislation for ensuring food and feed safety, food quality, animal health, animal welfare and plant health. The Department's main tasks are to assure the effectiveness of the control systems and evaluation of their compliance with the standards of EU both within the EU and in non-EU countries through inspecting the Member States and in third countries exporting to the EU. An inspection program is developed yearly by the Department for the identification of the countries and priority areas for necessary inspection and information to the stakeholders of the outcome of its audits and inspection programs. Captured and aquacultured products are among the most important food imports in the EU countries. So, the EC has issued legislation specifically targeting these products. The animal health conditions and certification requirements for imports of live fish, their eggs and gametes intended for farming, and live fish of aquacultured origin and products thereof intended for human consumption has been laid down in the Commission Decision 2003/858/EC. Later, two Commission Decisions 2003/804/EC and 2003/858/EC were amended, and a new Commission Decision 2006/767/EC was issued on the certificate requirements for live molluscans and live fish of aquacultured origin as products intended for human consumption. Imports of fishery products from the non-EU countries into the EU must be recognized by their internal CA and through an approved export certification.

The Commission Decision 2009/821/EC listed the BIPs and veterinary units in EU Member States. It was later amended as Commission Implementing Decision 2014/187/EU, which state that there are 300 approved BIPs in the EU. These posts carry out official checks at the Community border on products imported into the member states from the third countries in order to verify their compliance with EU legislation. These veterinary checks are categorized into three types: documentary, identity, and physical checking. The BIPs have implemented a computerized system specifically for imports ("TRACE") to facilitate the procedures of imports and transit and to ease the communication between these posts and the different EU Member States. Regular auditing is also carried out by the Commission's Audit Department on the functioning of the BIPs.

20.5.2 USA REGULATIONS

The USA is the single largest importing country of seafood to the value of USD 20,317 million imported in 2014. In the USA, food safety and quality is governed under the Acts of Congress, federal laws, regulations, administrative procedures and guidelines of 15 federal agencies. The US Food and Drug Administration (FDA), and the US Department of Agriculture (USDA) share the primary responsibility of regulating the safety of the food supply chain and some functions overlap between the two organizations, particularly in the inspection, enforcement, research, and execution of rules for both domestic and imported foods. The Federal Center for Disease Control (CDC) is, however, the main agency for collecting data on food-borne diseases, conducting investigations on these diseases and illnesses and researching effective ways to prevent and reduce them. The Federal CDC also plays a major role in the State and Local Health Departments' laboratories, epidemiological, and environmental work by providing them funding and support to strengthen their building and technical capacity in food-borne disease surveillance and outbreak response. The Emerging Infections Program of CDC established the Foodborne Diseases Active Surveillance Network (FoodNet) in 1995. The organizations involved in the network are CDC, USFDA, 10 State Health Departments, and USDA. The program serves as the primary food-borne disease surveillance system collecting information and data from sites within the 10 States on seven food-borne bacteria and two parasites.

The US Department of Health and Human Services established the USFDA with the responsibility of assuring the safety, quality, effectiveness, and security of human and veterinary drugs, vaccines, medical devices, and other biological products. It has the Office of the Commissioner and four

Directorates discharging the core functions of USFDA. They are Medical Products and Tobacco, Foods, and Veterinary Medicine, Global Regulatory Operations and Policy, and Operations. The US Federal Food, Drug, and Cosmetic Act (FFDCA) has provisions for USFDA inspections of food, including fisheries products, imported through the USA entry ports. All the foods, including imported and domestically-produced have to meet the similar legal requirements, and the USFDA shall detain shipments of imported food products, which are not in compliance with the said legal requirements. The FFDCA mandates the USFDA to verify and approve the veterinary drug residues used in the aquaculture industry. The Section 123 in 21 of the US Code of Federal Regulations (CFR) has provisions on the control of import of fish and specifies regulations for safety and sanitary processing for imported fishery products. The USFDA established the HACCP system in 1994, and made it mandatory in the seafood industry by 1997. In order to prevent and control any issues jeopardizing food safety, the aquaculture industry in 1995 was subjected to HACCP. With current mandatory HACCP in place, any fishery product (including aquaculture) processed in the absence of this system is considered adulterated, and processors (local or foreign) are subjected to penalties in accordance with FDA regulations.

The passage of the new Food Safety Modernization Act (FSMA) of 2010 by the Congress in December 2010 has made the USFDA into proactive rather than reactive. The new Act focuses on prevention by stressing the requirement that imported food shall meet the same standards as that of domestically produced foods. Under the new FSMA law, the USFDA has been appointed as the responsible agency for applying comprehensive and prevention-based controls to the food supply. It also entitles USFDA to establish science-based standards and makes the food industries responsible for the safe production of foods. It can be noted that an estimated 15% of food consumed in the US is imported, which includes 80% of fish products and 60% of fresh fruits and vegetables. The new legislation has enhanced the USFDA's ability to manage domestic and imported products. With respect to fish and fishery products inspection, the Memorandum of Understanding (MOU) signed between the USFDA and the US National Marine Fisheries Service (NMFS) allowed the NMFS to carry out a Seafood Inspection Program of the different establishments internally and globally. The USFDA has also set up a number of permanent overseas posts in Latin America, China, the Middle East, India, and Europe to effectively control and manage the safety of the imports. These include posts.

The Bioterrorism Preparedness and Response Act (BTA) under the Public Health Security of USA passed in 2002 strengthened the regulatory and administrative control of USFDA. The BTA was intended to improve the control and traceability of food flow within the US and to further protect local consumers from emerging threats such as biological agents and toxins. The Act, however, required certain conditions to be complied with by local and foreign food facilities including: registration with US FDA and the allocation of a US agent/sponsor for the foreign firms intended to export food to US; recordkeeping for traceability; detention of suspected food with food hazards and having a food safety system in place such as HACCP for the seafood facilities. Another Act was passed in 2002, the US Farm Security and Rural Investment Act, requiring labeling of all the perishable food (meat and fish) and peanuts with "country of origin labeling" (COOL) with labeling of the wild and farmed fish becoming compulsory in 2008.

The CDC and Prevention (CDC, 2016) of the USA has reported that in 2014 the outbreaks due to aquatic animals were 64, which was 31% of the total outbreaks. In the aquatic animals, ciguatoxin in fish (19 outbreaks) and histamine in fish (16) being the most usual seafood vehicles responsible for the outbreaks and 4 outbreaks resulted in product recall for raw oysters and 3 outbreaks for tuna. Within 2014, there were 141 illness and 16 hospitalizations with one death reported due to *Clostridium botulinum* in fish products. In general, the outbreaks rate is in increasing trend, which is mostly linked to imported food of Asia and Latin America. In fact, 16% of the consumed food in the USA is imported with seafood accounting for more than 80% of this consumed food.

20.5.3 JAPANESE REGULATIONS

Japan depends heavily on imports of fish to satisfy high local consumption. Total fish import is USD 14,844 million and is the second-largest importers. The Ministry of Health, Labor, and Welfare (MHLW) has authorized the Department of Food Safety functioning under the Pharmaceutical and Food Safety Bureau to have a control on the administration of food safety in Japan. The Japan has issued two laws for the regulation of quality and safety of food. The Food Sanitation Law (No. 233) promulgated in 1947 covered safety of foods, packages, and containers as well as human health, which was amended by Law (No. 87) in July 2005. Japan adopted a Law on Standardization and Proper Labeling of Agricultural and Forestry Products

(JAS Law) in 1950 with the aim of preventing the distribution of poor quality food in the markets by means of common food standards, including labeling requirements. However, the main objective of the JAS Law focused on the consumer protection. The Ministry of Agriculture, Forestry, and Fisheries (MAFF) in Japan established the JAS Standards in 1970 for the overall control of safety of food. The JAS system consisted of the Standards System, and the Quality Labeling Standards System. The Food Sanitation Law and the JAS Law governed the safety and quality of food in Japan for many years until problems caused by bovine spongiform encephalopathy (BSE) reached Japan. This incident triggered a complete restructuring of the existing judicial and administrative systems governing the safety and quality aspects of food in Japan. The fundamental of the Food Safety Basic Law issued (No. 48) issued in May 2003 and adopted is the Risk analysis and traceability. It was amended in June 2006 with Law (No. 50). The approach adopted the principle of 'farm to fork' to cover the whole food supply chain. Similar integration was done with the laws that covered materials related to agricultural and livestock industries, pesticides, veterinary drugs, and feed. The principle of the Food Safety Basic Law is very similar as to the EU Regulation 178/2002. This law is based on fundamental principles such as: consumers' protection on the basis of scientific principles and farm to fork concept. Japan has established a Food Safety Commission for independently assessing the food-related risks in the hope to regain back the confidence of Japanese consumers that was severely damaged by the BSE incident.

Japan has introduced the HACCP-based food control regulations for maintaining the sanitary and hygienic requirements in fish handling, processing, transportation, and storage. The principle of risk analysis is incorporated, making the spot checks compulsory at the border and sending the quality control staff of major importing companies to work on food hygiene, and grading in the fish exporting companies in order to control the imports at source itself. The Lawson Food Sanitation, Plant Protection, Customs, Quarantine, Domestic Animal Infectious Diseases Control, and Standardization and Proper Labeling of Agricultural and Forest Products are the major Laws of Japan that control the entry of foods into Japan. Recently, Japan updated the Food Sanitation Law in order to meet the challenges of non-satisfaction of consumers due to the BSE crisis as well as false labeling and pesticide residue issues. Japan made the Inspection and Safety Division (ISD) belonging to the Department of Food Safety as the Central Competent Authority (CCA) for seafood and its by-products exported to the EU, which is functioning under the Pharmaceutical and Food Safety Bureau of the

Ministry of Health, Labor, and Welfare. The ISD is in collaboration with the Food Safety and Consumer Affairs Bureau functioning under the Ministry of Agriculture, Forestry, and Fisheries along with other Fisheries Agencies to ensure proper controls of seafood safety in different establishments, fishing, and freezer vessels, and farming sites. With the risk-based approach in place, the Japanese government monitors very closely food imports with well-qualified inspectors of veterinarians, pharmacists, medical doctors, and agricultural and fisheries degrees that must be approved by the MHLW. In the case of exports to the EU, Japan permitted only the Designated Food Sanitation Inspectors for conducting the inspection. However, the Regional Bureaus of Health and Welfare formed in 2001 has been involved in the enforcement and approval of processing establishments, and laboratories. The MHLW collaborates with a network of regional health offices functioning under the Local Governments as well as with the National Institute of Infectious Diseases (NIID) for controlling the outbreaks of food-borne illness. The NIID is responsible for the collection, analysis, and monitoring of the incidences of infectious diseases in Japan.

The Japanese Ministry of Health and Welfare reported 1,048 food poisoning outbreaks in 2009. The implicated foods were considered in 11 categories: meat and meat products, fish, and fish products, milk, and milk products, vegetable, and vegetable products, egg, and egg products, grain, and grain products, compound dishes, confectioneries, and others. Fish and fish products have been implicated in 94 incidences of food poisoning outbreaks. Of which, natural poison caused 39 incidents consisting of 6 incidents due to shellfish, 24 incidents with fugu (pufferfish) and 9 with other fish or shellfish and followed by 33 incidents of norovirus. Considering fish products as a proportion of the total number of notifications, Japan appears to have the highest (25%) due to fish and fish products, followed by the USA (17%), and the EU (16%).

20.5.4 FAO/WHO REGULATIONS

The increase in the global consumption of fish and fish products and their growth in trade at international level led to an increased interest in the requirements for their safety. National regulatory authorities impose restrictions to protect their citizens but these can have an impact on trade. The Food and Agriculture Organization (FAO) was founded by the United Nations as an Intergovernmental Organization in October 1945 and acts as an international forum on food, agriculture, and forestry issues. A Committee on Fisheries (COFI) was established in 1965 under the FAO Council. The Committee is an Intergovernmental Forum meeting periodically for discussion on the major global problems and issues in fisheries and aquaculture, and then communicating the outcomes to the National Governments, Non-Governmental Organizations (NGOs), Regional Fisheries Organizations, FAO, fisherfolk, fish farmers, and the international community, who are the stakeholders in the field of fisheries and aquaculture. The COFI in 2014 outlined the major global fisheries and marine conservation issues. Food safety and quality have been given a priority with the Committee's Members, with most having an effective fish and seafood safety and quality assurance system in place. Post-harvest losses, by-catch, and illegal processing and trading issues were also highlighted and mitigation measures were suggested to overcome the issues. Many technical and scientific reports on Assessment and Management of Seafood Safety and Quality, and Causes of Detentions and Rejections in International Fish Trade were published with a clear description on the international regulatory framework, and also covering a wide range of factors affecting the safety and quality of fish and fish products such as chemical and microbial contaminants.

The FAO established the CAC in 1963, which is a Joint FAO/WHO Intergovernmental Body. The principle objectives of the Commission are protecting the health of the food consumers, as well as ensuring fair food trade practices by setting international standards on different food commodities. Codex had a major influence on National Regulatory Agencies, food producers and manufacturers, international food trade and end-users, by providing the Countries with the opportunities in the formulation and harmonization of food standards in compliance with Codex Standards besides ensuring their global implementation. The Codex Alimentarius is the Global Reference Point for Countries, which use the CAC standards in order to provide the protection for their consumers. With the increased global market, the need to have harmonized and uniform food standards has become much more prominent. Therefore, the World Trade Organization (WTO) was created with its related Agreements on the SPS, and TBT, which are essential for the harmonization of international food standards. In particular, the CAC develops standards containing maximum limits for additives and contaminants, general as well as specific guidelines, and Codes of Practices. Much of the work of the CAC is handled within committees. The FAO formed a Codex Committee for Fish and Fishery Products (CCFFP), which is a Commodity Committee responsible for the global standards for fresh, frozen or processed crustaceans, fish, and molluscans. A Code of Practice for fish and fishery products was developed by the Committee combining individual codes that deal with fish and fishery products inclusive of wild and aquaculture products. These Codes consist mainly of fish product standards that deal with defects of a commercial nature. The general principles of good hygiene practices (GHP)/HACCP have also been adopted by the CAC and have been incorporated in the Code of Practice for wild and aquaculture products for integrating the principles into the fish industry. The Codes mainly offer general advice on the different processes in the supply chain such as handling of fish products on-shore and on-board the fishing vessels, distribution, and retail display of the products. Different horizontal committees dealing with different topics are also involved in issues relevant to fish safety. The codex committee on food hygiene (CCFH) has dealt with the work on biological risk management, including developing risk profiles for a number of aquatic foods such as the presence of parasites in fish products, and measures to control pathogenic *Vibrio* spp. in aquatic food.

A meeting was conducted under the Codex Alimentarius by FAO and WHO in 2012 for addressing the issues of histamine and other biogenic amines in fish and fish products, which affect the public health, and impact the international trade. The experts concluded that SFP can be controlled and alleviated by applying basic GHPs and preferably a HACCP system. The Committee recommended establishing an appropriate histamine safety limit in the Codex Standard as well as suggesting a suitable sampling plan in relation to the histamine safety limit in standards and to closely communicate on this issue with the CCFFP delegations.

The Codex Alimentarius is generally recognized as the key body developing international food standards. But, the Codex on its own does not have any power or authority to persuade countries to incorporate the standards into national controls. However, the situation changed after the trade negotiations in Uruguay that led to the establishment of the World Trade Organization (WTO) in 1995. The WTO is an international organization that deals with the global rules and policies of trade between the nations. The key component of the Uruguay Round of negotiations is the trade liberalization of agricultural products. The negotiations also led to two specific binding agreements such as Agreement on Sanitary and Phytosanitary (SPS) Measures, and Agreement on Technical Barriers to Trade (TBT), which have on impact food safety and quality.

The right of members of WTO for implementing the measures in the protection of humans, animals, and plants from diseases, pests, and risks from contaminants, toxins, and additives in foods and feed is confirmed by the SPS agreement. The framework is provided on the rules for guiding the WTO member countries in the development, adoption, and the application of sanitary (relating to humans, animals, and fish) and phytosanitary (relating to plants) measures that are consistent with the general rules of the WTO. The key change introduced by the SPS Agreement is that, with respect to food safety and quality measures, the member countries should base their SPS measures on guidelines, and standards, and the other recommendations of the CAC. This recognition raised the status of the CAC, although the SPS Agreement does include the right for member countries to adopt stricter measures if they can provide a valid scientific justification. In a similar manner, the SPS Agreement encourages WTO Member countries to take suitable measures in harmonization with that of the International Office of Epizootics (OIE) in the case of animal (including fish) health. The SPS Agreement also requires that national SPS measures are based on risk assessment with clauses relating to transparency, protectionism, dispute settlement and harmonization of international standards.

The prevention of the use of national or regional technical requirements or standards as barriers to trade is the main purpose of the TBT Agreement. Standards related to different commodities, including food, are covered under this Agreement except measures that come under the SPS Agreement. The Agreement has provision for the protection of its members against deception and economic fraud. The nutritional requirements, quality provisions, packaging, analytical methods, and labeling and product content regulations pertaining to food standards are covered under the TBT Agreement. The TBT Agreement does not indicate any specific international organizations as benchmarks to assess their members' compliance to the requirements of the Agreement. The TBT Agreement ensures that the regulations, standards, testing, and certification procedures are not unnecessarily blocking the international trade. The right to implement the measures for achieving the objectives of the legitimate policy for the protection of the environment, and human safety and health is provided in the TBT Agreement.

20.5.5 INDIAN REGULATIONS

The Government of India has set up the Export Inspection Council (EIC) for ensuring sound development of export trade from India through the system of Quality Control and Inspection. The EIC, as the advisory body to the Government of India, has been empowered for notifying the commodities that are subject to inspection and/or quality control before export, for establishing the standards of quality for such notified commodities, and for specifying the type of inspection and/or quality control to be applied to. The national standards of importing countries/international standards have been recognized by the EIC. However, such specifications are not lower than the minimum standard of specifications prescribed in the order/notification of EIC issued by the Government of India for the said products.

The EIC establishes the Export Inspection Agencies (EIAs) located at the major cities of India viz. Mumbai, Kolkata, Chennai, Cochin, and Delhi. The EIA is the field organization for EIC. The list of approved inspection agencies and laboratories along with their scope/the indicative products for which approval is granted by the EIC is provided in its website. The EIC through EIAs is rendering the following services:

- Installation of quality assurance systems in the exporting units as well as consignment-wise inspection system for the certification of quality of export commodities;
- Installation of FSMS in the food processing units as per international standards for the certification of quality of food items for export;
- Issuing various types of Certificates such as Health, Authenticity, etc., under various product schemes to the exporters for their export products;
- Issuing Certificates of Origin under various preferential tariff schemes to the exporters for their export products;
- Recognizing the Inspection Agencies for the purpose of export inspection and testing as per ISO 17020 and Laboratories as per ISO 17025.

20.6 FISH SAFETY MANAGEMENT SYSTEM (HACCP)

The FAO defines Food hygiene as the conditions and measures necessary for ensuring the suitability and safety of food at different stages of the food chain. Upon implementation of the measures such as personnel hygiene, cleaning, sanitation, and pest control, the fish contamination and microbial growth could be significantly reduced. Increased levels of food-borne disease globally have prompted action led by international organizations such as FAO, WHO, and the CAC. This has included the recognition of the HACCP system as an important tool in protecting consumers from food-related hazards. Many guidelines have been developed to ensure the basis of implementing these practices in the seafood industry, including publications by the CAC of FAO. Its adoption in 1993 by the CAC stimulated its use as a preventive tool to assess the biological, chemical, and physical hazards throughout the different phases of food production – from the primary producer to the consumer. The implementation of HACCP utilizes the scientific evidence and a risk analysis approach to protect human health, instead of relying on the testing of end products. The International Code of Practice on food hygiene general principles developed by the Codex Alimentarius has been adopted into the food legislations of different countries.

The importance of HACCP has been underlined by the International Organization for Standardization (ISO). The ISO also developed an international standard (ISO 22000), which incorporates HACCP as a core element. The Codex Committee on Fish and Fishery Products has developed a Code of practice for fish and fishery products (CAC/RCP 52-2003). The Code provides assistance in compliance with the Codex Standards for those involved in the process of production, handling, distribution, storage, import, and export of fish and fishery products on national and international markets. The said Code of Practice was amended in 2016 by merging other Codes dealing with these products. The new Code incorporates the different practices from growing fish (in freshwater and marine), through different stages of harvesting, handling, processing to transportation and distribution and display of the fishery products at retail meant for human consumption. The Section 3 of this Code emphasizes the importance of implementing the prerequisite programs (PRPs) before the adoption of HACCP system. The Code highlights the significance of establishing PRPs in the food business, as they enable the HACCP team to concentrate more on the food safety hazards emerging from the intended product and its production process. The PRPs mainly consist of GMPs (Good Manufacturing Practices) and GHPs. They are required to ensure food products are produced and controlled during processing to minimize any risks so as to produce food safely. The GMPs deal mainly with the infrastructure and layout of the food production units, condition of the food production surrounding environment, sufficient maintenance and calibration of equipment and machinery, etc. The GHPs mostly deal with the personal hygiene of the workers, cleanliness of the food production units, equipment, and machinery, utensils, food contact surfaces, walls, and ceilings, floors, management of waste, and pest control, storage, and transportation, etc. These programs generally cover the same concepts and in most cases often used as GHPs. The GHPs can be tailored to fit into the individual food production establishments and hence, must undergo regular evaluation and monitoring for ensuring their effectiveness so as to have a successful HACCP system in place.

The EU's Commission's "Hygiene Package" has tackled the application of the PRPs in different food commodities. The Regulation 853/2004 has specific hygiene rules for food of animal origin, which contains a section setting out requirements for fish and fishery products. Similarly, the USFDA has adopted a regulation (21 CFR Part 123) in 1997, stating the compulsory implementation of the HACCP system in all seafood processors as a preventive tool for food safety. The Sanitation Standard Operating Procedures (SSOPs), a prerequisite program, mentioned in Section (123.11) of the regulation is included in the HACCP plan, as per the Section 123.6(b). Hence, the seafood industry in the USA has become the first industry to have the compulsory implementation of this HACCP program. In the USA, the HACCP is required throughout the whole supply chain of fish and fishery products starting from fishing vessels, landing site and shore all the way to the plate of consumers. At each of these stages, hazards have to be identified, and the critical points (CPs) are controlled so as to reduce or eliminate these hazards to the acceptable levels. Guidance is provided to the local seafood industries by the USFDA, and any imports or exports of seafood products must meet the US standards. In terms of the safety of bivalve molluscans sold in the USA, the USFDA has teamed up with the different States to implement the National Shellfish Sanitation Program (NSSP).

In addition to enhancing the safety of food, the HACCP combined with PRPs allow better management of resources, timely responses to food safety issues and can assist the regulatory authorities in delivering efficient inspection and certification services. The full involvement and commitment of various stakeholders in the food supply chain is important for the successful application of HACCP. The PRPs, such as good manufacture practices (GMPs) and GHPs, are necessary for any food production units before the implementation of the HACCP system and have to be applied at different stages of harvesting, handling, processing, storage, and distribution.

20.6.1 ISSUES IN IMPLEMENTATION OF HACCP

The developing countries are required to maintain and enhance their food safety and quality requirements from harvesting to consumption in order to have a wider access to the international seafood market and benefit from the higher revenues. The total adoption of HACCP and PRPs by the major importing countries (US, EU, Japan) has resulted in stringent food safety requirements being imposed on the exporters of developing countries. Failure to meet these requirements results in rejection or detention of imports for various safety and quality reasons from the major seafood importing countries. The implementation of HACCP system as the major requirement by the food regulators and processors within the developing countries has been perceived as a technical barrier to trade. The commitments to the hygiene standards and safety practices during processing and handling are usually below the internationally required standards. The implementation of HACCP in developing nations has major constraints linked to the following factors: Food regulators in developing nations often lack the technical expertise in HACCP implementation and the knowledge of the importance of the prerequisite programs as the basis for implementing the HACCP system. The food control systems in developing countries often have deficiencies, and the systems are generally scattered among different authorities with limited minimal collaboration on food quality and safety issues. The dominant deficiencies mostly exist within legislative frameworks and the non-compliance of such legislative frameworks with international recommendations or standards, unclear responsibility and roles for the various institutions involved, shortcomings within the surveillance and inspection systems, lack of qualified human resources and limitations in the laboratory testing facilities.

A common practice in these regions is the adoption of new laws and regulations without consulting the industry and, when there are no clear time frames for training and implementation, it results in improper enforcement and poor compliance by the industry. One of the major constraints is the absence of communication, trust, and little or no link between the regulators and the food industry that hindered the development of this industry and delays improvements. The successful implementation of the HACCP system in the developed countries is mainly due to the strong communication and collaboration established between the industry and the government regulators.

20.6.1.1 CULTURAL ISSUES

They are often impediments to the advancement of FSMS in the developing countries. This is more related to behavior and perception of the people rather than the implementation of the HACCP system and it is perceived as a troublesome and bureaucratic approach. Within an organization, the common cultural practices are usually reactive, with senior managements being in charge of every activity within the organization and with limited delegation of power and initiatives to other staff. For an effective HACCP system, a proactive approach is a necessity. This perception could only be met through effective education and training and the ability to delegate authority.

20.6.1.2 TECHNICAL ISSUES

They may act as constraints to the development of food safety controls within food businesses. The major concern faced by the food production units during the implementation of food safety standards, particularly HACCP is the lack of administrative, technical, and scientific capacity and, above all, management commitment and motivation in embracing its benefits. Shortcoming in executing the HACCP prerequisite programs in terms of, for example, inadequate personal hygiene, cleaning, and sanitizing and pest control could be due to the poor understanding of these prerequisite programs and their relationship to the HACCP system, or simply having no access to the proper services and tools necessary to perform them adequately. Without the prerequisites programs in place, a risk-based system such as HACCP will fail. Lack of appropriate knowledge, expertise, and personnel training in the hazard analysis and evaluating their risks within the process often leads to complex HACCP systems with too many CCPs and poor documentation. Additional complications include ineffective monitoring procedures, corrective actions, and poor or limited verification procedures of the system.

20.6.1.3 FINANCIAL ISSUES

On an international level, the best way to address food safety practices is through risk-based analysis with a farm-to-table approach. However, in developing countries, this approach is only followed for food products intended for export to lucrative markets with high incentives and with existing stringent regulatory standards. Most developing countries find it difficult to comply with these stringent standards to gain and maintain access to these markets, and the cost of development, as well as maintenance of food safety systems, constitutes a severe constraint to their implementation. Studies on trade flow have shown that when standards of importing countries are much higher than that of the exporting countries, quantities of traded products decline, and the exporters seek alternative markets with less imposed standards. In general, if a food production unit has the willingness to comprehend and commit to the HACCP principles and its PRPs with the rightful resources in place, HACCP shall deliver the promised benefits and rewards with regard to food safety and quality.

KEYWORDS

- **Codex Alimentarius Commission**
- **fish safety**
- **HACCP**
- **international food regulations**
- **marine biotoxins**
- **quality management**

REFERENCES

Ababouch, L., (2005). *Causes of Detentions and Rejections in International Fish Trade*. Food and Agriculture Organization of the United Nations, Rome.

Ahmed, F. E., (1992). Review–assessing and managing risk due to consumption of seafood contaminated with microorganisms, parasites, and natural toxins in the United States. *Int. J. Food Sci. Technol., 27*, 243–260.

CDC, (2014). Foodborne Diseases Active Surveillance Network (FoodNet). *Centre for Disease Control and Prevention, US Department of Health and Human Services*. Retrieved from: http://www.cdc.gov/foodnet/ (accessed on 09 March 2022).

CDC, (2016). *Surveillance for Food-Borne Disease Outbreaks United States 2014*. Annual Report. Atlanta, Georgia: Centre for Disease Control and Prevention, US Department of Health and Human Services.

Clemens, T., Sorensen, K., Rosenkotter, N., Michelsen, K., & Brand, H., (2017). The directorate-general for health and consumers 1999–2014: An assessment of its functional capacities. *Health Policy, 121*, 594–603.

Codex Alimentarius Commission, (2006). *Understanding the Codex Alimentarius* (3rd edn.). Food and Agriculture Organization of the United Nations, Rome.

Codex Alimentarius Commission, (2012). *Code of Practice for Fish and Fishery Products* (2nd edn.). Food and Agriculture Organization of United Nations, Rome.

Codex Alimentarius Commission, (2014a). *International Food Standards*. Food and Agriculture Organization of United Nations, Rome. http://www.codexalimentarius.org/ (accessed on 09 March 2022).

Codex Alimentarius Commission, (2014b). Joint FAO/WHO food standards program. *Codex Committee on Fish and Fishery Products: Discussion Paper on Histamine* (33rd Session).

Directorate-General for Internal Policies, (2013). *Policy Department B Structural and Cohesion Policies: Fisheries and Aquaculture Department.* Compliance of Imports of Fishery and Aquaculture Products with EU Legislation.

European Commission, (2005). Commission Regulation (EC) No. 2073/2005 of 15 November 2005 on microbiological criteria for foodstuffs. *Official Journal of the European Union, L 338/1*, 2005.

European Commission, (2007). Regulation (EC) No. 1441/2007 of 5 December 2007 amending Regulation (EC) No. 2073/2005 on microbiological criteria for foodstuffs. *Official Journal of the European Union L 322/12*, 2007.

European Commission, (2014a). *Health and Consumers.* http://ec.europa.eu/food/fvo/how_en.htm (accessed on 09 March 2022).

FAO, (2006). *Guidance to Governments on the Application of HACCP in Small and/or Less-Developed Food Businesses.* FAO Food and Nutrition Paper No. 86.

FAO, (2014a). *Fisheries and Aquaculture Department.* http://www.fao.org/fishery/en (accessed on 09 March 2022).

FAO, (2014b). *The State of World Fisheries and Aquaculture: Opportunities and Challenges.* Food and Agriculture Organization of the United Nations, Rome.

FAO, (2016). *The State of World Fisheries and Aquaculture.* http://reliefweb.int/report/world/state-world-fisheries-and-aquaculture (accessed on 09 March 2022).

FDA, (2014). *Protecting and Promoting Your Health.* U.S. Food and Drug Administration. http://www.fda.gov/default.htm (accessed on 09 March 2022).

Ferro, E., Wilson, J. S., & Otsuki, T., (2013). *The Effect of Product Standards on Agricultural Exports from Developing Countries.* Policy Research Working Paper 6518. World Bank.

Food Processing Technology, (2014). *The Ten Most Traded Food and Beverage Commodities.* http://www.foodprocessingtechnology.com/features/featurethe-10-most-traded-food-and-beveragecommodities-4181217/ (accessed on 09 March 2022).

Iwamoto, M., Ayers, T., Mahon, B. E., & Swerdlow, D. L., (2010). Epidemiology of seafood associated infections in the United States. *Clin. Microbiol. Rev., 23.* doi: 10.1128/cmr.00059-09.

JAS, (2007). *Overview of the Revised Japanese Agricultural Standards Law.* http://www.maff.go.jp/e/jas/pdf/law06.pdf (accessed on 09 March 2022).

Laurian, U., (2015). Food safety in developing countries: Moving beyond exports. *Glob. Food Secu., 4*, 24–29.

Lehane, L., & Olley, J., (2000). Histamine fish poisoning revisited. *Int. J. Food Microbiol., 58*, 1–37. doi: 10.1016/s0168-1605(00)00296-8.

Lund, E. K., (2013). Health benefits of seafood: Is it just the fatty acids?. *Food Che., 140*, 413–420. doi: http://dx.doi.org/10.1016/j.foodchem.2013.01.034.

Ministry of Health, Labor and Welfare, (2009). *Food Poisoning Statistics in Japan.* Inspection and Safety Division. Department of Food Safety: Pharmaceutical and Food Safety Bureau.

Ministry of Health, Labor and Welfare, (2014). *Food Safety Information in Japan.* https://www.mhlw.go.jp/english/topics/foodsafety/ (accessed on 05 April 2022).

Mortimore, S., (2001). How to make HACCP really work in practice. *Food Control, 12*, 209–215. doi: http://dx.doi.org/10.1016/S0956-7135(01)00017-2.

Novotny, L., Dvorska, L., Lorencova, A., Beran, V., & Pavlik, I., (2004). Fish: A potential source of bacterial pathogens for human beings. *Veterinarni Medicina., 49*, 343–358.

RASFF, (2012). *Annual Report of the Rapid Alert System for Food and Feed.* http://ec.europa.eu/food/food/rapidalert/docs/rasff_annual_report_2012_en.pdf (accessed on 09 March 2022).

RASFF, (2013). *Preliminary Annual Report of Rapid Alert System for Food and Feed.* European Commission Health and Consumers Directorate-General. Veterinary and International Affairs.

Stier, R. F., Posnick, L. M., & Kim, H., (2002). *Constraints to HACCP Implementation in Developing Countries* (p. 36). Food Safety Magazine.

Takahashi, T., (2009). *Laws and Regulations on Food Safety and Food Quality in Japan.*

Weick, C., (2006). Trade impacts of administrative food import regulations: Evaluating the U.S. bioterrorism act. *Paper Prepared for the IATRC Summer Symposium on "Food Regulation and Trade: Institutional Framework, Concepts of Analysis and Empirical Evidence" held in Bonn, Germany.*

WHO & FAO, (2013). *Joint FAO/WHO Expert Meeting on the Public Health Risks of Histamine and Other Biogenic Amines from Fish and Fishery Products: Meeting Report.* World Health Organization. https://apps.who.int/iris/handle/10665/89216 (accessed on 05 April 2022).

Part V

Entrepreneurship Development in the Fish Processing Sector

CHAPTER 21

Entrepreneurship Initiatives in the Fisheries Post-Harvest Sector Through the Agribusiness Incubation Program

GEORGE NINAN

ICAR-Central Institute of Fisheries Technology, Cochin – 682029, Ernakulam, Kerala, India

ABSTRACT

Technology developments in fish processing sector offer scope for innovation, increase in productivity, increase in shelf life, improve food safety and reduce waste during processing operations. There are huge prospects for live fish marketing, chilled, and minimally processed fish products, mince-based convenience products, specialty products, coated products, and shelf-stable products. Fish processing operations generate about 47% of waste in India, comprising of skin, head, viscera, trimmings, liver, frames, bones, and roes. Developing appropriate and scalable technology to recover or isolate the valuable components from rest raw materials for high-value nutraceuticals and health formulations products could be of paramount importance in the context of optimal utilization of the resource. The commercialization of fish products still poses a lot of challenges to the entrepreneur and researcher in terms of optimization of technologies and ultimately developing the technologies into a commercially viable business plan. In 2016 an Agri-Business Incubation Center (ABI) was established in CIFT under the XII plan scheme of National Agriculture Innovation Fund (NAIF) of ICAR. The role of the ABI Center is to facilitate the innovator and the researcher to turn their ideas into commercial ventures with a focus on incubation and business development programs, including entrepreneurship, skill development,

Advances in Fish Processing Technologies: Preservation, Waste Utilization, and Safety Assurance.
Ranendra K. Majumder, PhD & Amjad K. Balange, PhD (Eds.)

and Grassroot innovators activities. This chapter gives an overview of the importance of fish as a functional food, the business incubation initiatives and entrepreneurship avenues in the fish processing sector and a profile of commercially successful technologies from the Fisheries Post-Harvest Sector.

21.1 INTRODUCTION

Fish is a versatile and heterogeneous commodity covering a wide variety of species. It can be prepared and consumed in many different ways and forms for either food or non-food uses. Marked differences also exist in how fish is utilized, processed, and consumed within and between continents, regions, and countries. The bulk of the utilization of fish production is in the form of products for human consumption, and this share is projected to grow from 89% in the base period (2016–2018) to 91% by 2028 (OECD/FAO, 2019). Overall, growth in demand is also expected to be fueled by ongoing changes in dietary trends, which should continue towards a greater variety in food choice along with increased health, nutrition, and diet concerns. Fish, being a concentrated source of protein and of many other essential fatty acids and micronutrients, plays a particular role in this regard by providing a valuable and nutritious contribution to a diversified and healthy diet.

World trade of fish for human consumption is projected to total 45.8 MT live weight by 2028, up 11% on the 2016–2018 base period. In 2010, fish accounted for 16.7% of the global population's intake of animal protein and 6.5% of all protein consumed. Moreover, fish provided more than 2.9 billion people with almost 20% of their average per capita intake of animal protein, and 4.3 billion people with about 15% of such protein (Barik, 2017). Exports of fish for human consumption are expected to concentrate in fewer exporting countries. Live, fresh or chilled is often the most preferred and highly-priced form of fish and represents the largest share of fish for direct human consumption, 45% in 2016, followed by frozen (31%), prepared, and preserved (12%) and cured (dried, salted, in brine, fermented smoked) (12%). Freezing represents the main method of processing fish for human consumption; it accounted for 56% of total processed fish for human consumption and 27% of total fish production in 2016 (FAO, 2018). Developing countries now represent almost 50% of global fish exports, with annual net export revenues exceeding USD 25 billion (FAO, 2012). Throughout the world, post-harvest fish losses are a major concern and occur in most fish distribution chains; an estimated 27% of landed fish is lost or wasted between landing and consumption. When discards prior to landing are included, 35% of global catches are lost or wasted and therefore not utilized. Climate change, weather variability and changes in the frequency and extent of extreme weather events are anticipated to have a major impact on the availability and trade of fish and fish products mainly through habitat destruction, changes in fish migration patterns and natural productivity of fish stocks.

Fish is an important source of protein to meet the food and nutrient security of rural and urban population of costal and non-coastal states of India. The total fish production in India was 129.5 lakh tons during 2017–2018 from marine and aquaculture. Approximately 83% of the total production of fish is available for domestic consumption, 11% is exported and 6% is used for non-food purposes such as reduction to fish meal and oil. However, the monthly per capita consumption of fish is low in India which is 0.27 Kg in rural areas and 0.25 Kg in urban areas (NSSO, 2014).

Among the nations, India ranks sixth in marine capture fisheries and second in inland fish capture and aquaculture production. India ranks sixth in the top 10 exporters of fish and fish products, and the marine products exports generated Rs. 46,589 crores of foreign exchange in 2018–2019 (Anon, 2020). However, less than 10% of the marine exports are traded as value-added products, the rest are mainly frozen items. As of the case with the rest of the World, value addition is the key for the Indian Fisheries sector for diversification and increased unit value realization from the resources. This has to be in line with the Sustainable Development Goals of the United Nations in which one indicator is to measure the value of Sustainable fisheries as a percentage of GDP in small island developing States, least developed countries and all countries (Anon, 2015).

Fish processing sector is constantly being innovated with new technologies for development of convenience products catering to the diversified needs of the market. A large number of values added and diversified products both for export and internal market based on fish, shrimp, lobster, squid, cuttlefish, bivalves, etc., have been identified (Ninan and Ravishankar, 2018). Trends in fish and shellfish consumption demand for unconventional products in the form designer food products, minimally processed products, health supplements and nutraceuticals. The secondary raw material from the fish processing operations is a potential source of high value byproducts. This waste is now termed as rest raw material, which is an important source for two categories of products viz. low value products of mass upgrade and high value products from quality rest raw material. In India, the rest raw materials from fish processing are either disposed of or converted into mass upgrade products viz., animal feed, fish meal and fertilizer. This practice leads to underutilization of raw material and may affect the sustainable utilization of available resources. The disposal of fish processing waste is under strict regulations due to environmental issues, and it adds to the operational cost of the seafood industry. Challenges to full utilization of rest raw material from fish processing are lack of suitable technologies to improve palatability, nutrient utilization and refinement of feed formulation in which protein sources are combined to overcome nutritional liabilities of underutilized ingredients (Sachindra and Mahendrakar, 2015). Complete and optimal utilization of aquatic resources involve technology solutions for isolation and commercial extraction of high value ingredients form the secondary raw materials. Aquatic food processing sector should be able to diversify their product profile by adding process lines for secondary raw materials.

However, for fishpreneurs the main challenge will be to identify the unique selling proposition (USP) in terms of novelty and market demand for aquatic products. With the aim of utilizing the vast research and development facilities and knowledge available with its institutions, ICAR has started a technology management and business incubation drive, designed for the Indian agricultural sector to promote agribusiness. This initiative started with the establishment of Agri-business incubation centers (ABIs) which effectively networked all ICAR research institutions for the management of new and cutting-edge technologies developed by these institutions. The Agri-Business Incubation Center established in ICAR Central Institute of Fisheries Technology cater mainly to fishpreneurs and innovators in the post-harvest sector with focus of technology support, skill development, and mentorship.

21.2 FISH AS A FUNCTIONAL FOOD

Global fish production has increased from 37 MT in 1961 to 200 MT in 2016 with a per capita consumption increased from 9 kg to 20 kg (FAO, 2018). The development of a more diversified fish processing sector with raw and processed fishery products could be witnessed over the years. There had been many studies on the development of by-products from fish and their use in feed, fish silage, fertilizer, biofuels, food packaging, human food or nutraceuticals or pharmaceuticals (Arvanitoyannis and Kassaveti, 2008; Govindharaj et al., 2019; Herpandi et al., 2011; Kim and Mendis, 2006; Mahboob, 2015; Radziemska et al., 2019; Yuvaraj et al., 2019). However, its use in food and nutraceutical field is still challenging because of safety, stability, and bioavailability issues and due to lack of good market potential. In general, foods which are having nutritional and health benefits are designated as functional foods. The American Diabetic Association (ADA) defines functional food as "food for fortified, enriched food that have beneficial effect on health when consumed as a part of diet at effective level on regular basis" (Hasler, 2000). Fish is a source of nutrients and micronutrients viz protein, minerals, vitamins, trace elements fish and fish products can be considered as functional food as they have a positive role to improve health apart from basic nutrition. Fish related functional foods can replace the currently used sports medicines, as these increase the energy and immunity levels and significantly improves the overall health of athletes.

In human body, dietary protein is essential to maintain cell integrity, cell functions and they form an integral part of muscle enzymes, blood, hormones, and membranes. Proteins are made up of amino acids, the right balance of which is needed for better utilization of proteins. Protein-energy malnutrition can result in infant mortality, stunting or wasting due to poor growth. Protein deficiency can affect functions of vital organs like kidney, brain. Amino acids are required for the human body to have growth, better health, reproduction, and prevention of diseases. Peptides from protein are known for bioactivities viz antioxidant, anti-inflammatory, anticancer, and antiulcer. Fish is rich in all the essential amino acids making it equal to FAO reference protein.

Fish is a good source of water-soluble and fat-soluble vitamins, but lack vitamin C. These vitamins play a major role in the function of vital organs in our body, immune, and reproductive systems. Vitamin D is helpful for absorption of calcium for bones; Vitamin E has antioxidant and anti-inflammatory properties and vitamin B complex are helpful for metabolic functions.

Fish contains macro and microelements, which are very vital for human health. Calcium in fish contributes to vital roles in human body viz maintaining required calcium level in blood, bone, and cells and also for the proper secretion of hormones like insulin. Children, pregnant, and lactating women require large amount of calcium and phosphorous, which can be met from enough consumption of fish. The role of phosphorous at the cellular level is immense as in the production of energy through ATP or in activation of enzymes and cell signaling molecules by phosphorylation. Sulfur is a component of amino acids, cystine, methionine, and B vitamin thiamine and biotin. Apart from calcium and phosphorus, fish is rich in sulfur and potassium. Potassium plays a key role in electrolyte balance, maintains membrane potential. Magnesium is an element which is part of more than 300 metabolic reactions in the body (Rude and Shils, 2006), fish is considered as a good source of magnesium. Iron deficiency can alter metabolic processes (Beard and Dawson, 1997), adversely leading to anemia, especially in children, lactating, and pregnant women. Many varieties of fish are good source of iron and copper. For iron absorption in the body, copper plays a significant role, it is also necessary for hematologic and neural functions (Tan et al., 2006). Zinc is another micronutrient essential for our body functions, nearly 100 enzymes depend on zinc for their reactions. Its role varies from cell replication, protein synthesis to wound healing (Malhotra, 1998). Fish is considered as a delivery source for Zn and Mn. Cobalt, chromium, selenium and nickel are other micronutrients in fish, whose beneficial roles in the human body have been established by many.

Omega-3 (n-3) fatty acids such as DHA and EPA are plenty, especially in marine fish, contribute significantly for the functioning of vital organs viz heart, brain, immune, and reproductive systems (Connor, 2000). For ameliorating many diseases, their role has been reported by many: cardiovascular disease, depression, lupus, rheumatoid arthritis (Darlington and Stone, 2001), ulcerative colitis (Alex and Triadafilopoulos, 1992), kidney disorders (Donaldo et al., 1999) diabetes and cancers (Larsson et al., 2004).

Direct consumption of fish for food provides a vital source of protein and a variety of essential fatty acids and micronutrients, such as iron, zinc, and vitamin A. These micronutrients are particularly rich in smaller-sized fish, which are often more readily available to low-income, at-risk populations owing to their low cost and abundant availability. Identified protein rich, oil rich and micronutrient dense fish species could be utilized in community nutrition programs like 'mid-day meal' program in operation in schools in countries like India. Such species include *Anabas testudineus, Amblypharyngodon mola, Puntius sophore,* and *Tenualosa ilisha* which could provide wholesome nutrition to consumers (Mohanty et al., 2019). Fish are an important source of food and nutrients owing to the fact their seasonal availability is often different from crops.

21.3 BUSINESS INCUBATORS

Business incubators provide an attractive framework to entrepreneurs (referred to as incubatee/client) for dealing with the difficulties faced during the start-up stage. Business Incubators help in creating the links that connects the SMEs with the public and private sector, and help in attaining its sustainability and scalability.

Incubators have evolved over the years since the inception of the first incubator in 1957. However, in general, business incubators provide the following support (single-window approach) for a start-up:

- Scientific and technical backstopping;
- Technology transfer from University and Research Institutes;
- Business plan preparation and marketing consultancy;
- Infrastructural facilities like office space, conference rooms, communications;
- R&D lab space;
- Mentoring assistance;
- Access to funding agencies, angel investors, venture capitalists, crowd-sourcing, etc.;
- Training to incubatee-SME management, skill set development;
- Promotional support through exhibitions, conventions, workshops, etc.;
- Operational costs reduction;
- Networking with Industry, Research bodies, Commerce Chambers, Government;
- Assisting in getting clearances with Government regulatory bodies and Licensing bodies;
- IPR management.

21.3.1 STAGES OF BUSINESS INCUBATION

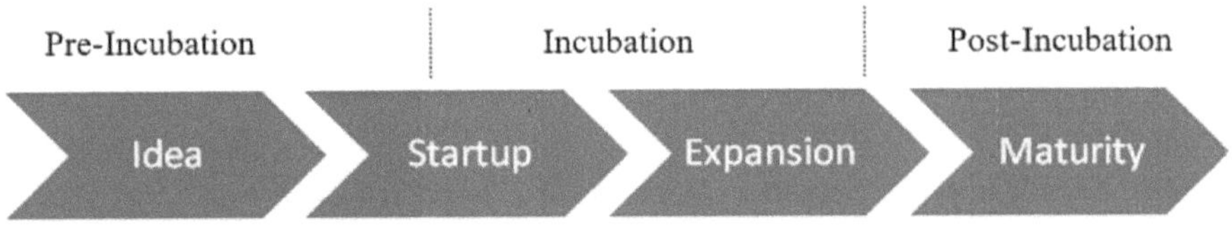

- Pre-incubation stage can be offered for helping individuals who have an innovative idea. These incubators are usually attached to Universities and Research Institutes and have easy access to scientific and technical support. The risk factor will be high at this stage and can be mitigated by the incubator.

- Incubation stage is where the idea transforms into a plan and operationalized. Incubators can help in refining the plan, provide resources and even invest in the company, thereby financially supporting the start-up. This is the stage where incubation happens. The stage will help the incubatee in moving to a more mature stage in the business cycle and usually is for a period of five years.
- Post-incubation stage can be utilized by those agribusiness ventures that are looking for specific support facility.

21.4 BUSINESS INCUBATION INITIATIVES IN FISH PROCESSING SECTOR

Fisheries sector, with its important role played in the socio-economic development of the country, has become a powerful income and employment generator, and stimulates the growth of a number of subsidiary small, medium, and large-scale industries. In order to translate the research results arising from the field of fisheries and other agricultural sectors, ICAR have set up an innovation-based Agri-Business Incubation Center (ABI) at the ICAR-Central Institute of Fisheries Technology (CIFT), Cochin. ABI is managed by Zonal Technology Management Center (ZTMC) and aims at the establishment of food business enterprises through IPR enabled ICAR technologies.

ABI supports operations on business projects as a measure of enhancing the foundation for new technology-based industries and establishing a knowledge-based economy. It focuses on finding new ways of doing business in fisheries and allied agricultural fields by finding doors to unexplored markets. The Center helps prospective entrepreneurs, by providing pro-active and value-added business support in terms of technical consultancy, infrastructure facility, experts' guidance, and training to develop technology-based business ideas and establish sustainable enterprises. It acts as a platform for the speedy commercialization of the ICAR technologies, through an interfacing and networking mechanism between research institutions, industries, and financial institutions. The Incubator at ICAR-CIFT differs from traditional Business Incubators as it is tailored specifically for technology-based industries and is operational at an area with a high concentration of fish production. This industry-specific incubator also allows new firms to tap into local knowledge and business networks that are already in place. ABI offers their services to industries not only in Cochin, but also all over India through virtual incubation. Beyond promoting business growth, the Center is also trying to bring its benefits to all the fisheries communities in India.

21.4.1 PROCESS OF INCUBATION

The Business Incubation Center targets entrepreneurs, from fledgling start-ups in need of basic small-scale processing capacity to sophisticated businesses in need of R&D back up, office infrastructure and pilot/test market processing facility for the development of new products. It possesses good infrastructure facilities suitable for providing direct incubation of nine entrepreneurs in a corporate environment within the premises of ICAR-CIFT, at a time. The purpose of direct incubation is to support emerging companies through their infancy. BIC, apart from being a multi-tenant facility with on-site management that delivers an array of entrepreneurial services to clients operating with the facility, it also serves clients that are not located in the facility through virtual incubation or incubation without walls.

The Center regularly conducts industry interface and technology promotional programs for sensitization of entrepreneurs and to identify interested potential candidates for physical and virtual incubation. The Clients at BIC gets the privilege of meeting Scientists, Business Manager, and Business Associates directly, to discuss and finalize the strategies to be adopted to take the business forward. It is also the peer-to-peer relationships that develop within the incubator, that ensures the delivery of basic services such as how to actually incorporate a business; what are the legal issues; how to take intellectual property protection; how to do basic accounting and cash flow; how to do business presentations, etc. Those kinds of skills are what are transmitted as part of the incubation process.

The residency period for direct incubatees is normally for two years, extendable by another year in special cases, depending on the progress of incubation. As the business venture becomes mature enough, the concessions and the facilities provided to the Incubatee companies will be gradually withdrawn. Each Incubatee of the Unit will have to pay to the Institute a charge for utilization of space, at a rate concessional to the benchmark rate, which is the prevailing market rent realizable. Incubatee mentoring will continue in virtual mode after graduation, on need basis.

21.4.2 AGRI-BUSINESS INCUBATION CENTER (ABI) AT ICAR-CIFT

ICAR-Central Institute of Fisheries Technology (ICAR-CIFT) located at Kochi, Kerala state (India) started the ABI Center as a platform for empowering the fisheries sector by creating new technology-based industries. ABI Center is functioning at a location with high fish production and vital markets, which makes it easily accessible to clients. It operates an important networking mechanism between R&D institutes, private industry, government agencies, academia, and funding agencies. This entrepreneurial support system caters to its clients through strong technical and advisory support, and assists them to orient their resources in the most optimized manner, thereby yielding high productivity and economic value. The Center follows a facile technology dissemination procedure and it enables the entrepreneurs to explore new ways of doing business through a wide spectrum of activities. Pro-active and value-added business services are provided to registered incubatees in the form of technology transfer, contract research, consultancy, contract service, office space, certified state-of-the-art pilot level production facility, on-site guidance and specialized training to establish innovation-based business enterprises.

The Institute has developed a wide range of technologies pertaining to fishing technology, resource/energy optimization, fish processing, value addition, high value by-products, packaging, customized processing equipment, health care products, aquaceuticals, etc. Some of the entrepreneur ready technologies developed by ICAR-CIFT are depicted in Table 21.1. These technologies are further classified according to their Technology Readiness Level (TRL).

ICAR-CIFT allows start-ups as well as established business enterprises across the nation to register as incubatees at ABI Center and get access to new innovations, cutting edge technologies and scientific know-how through direct as well as virtual incubation. The selective, comprehensive service offering and handholding accelerates the growth and sustenance of these incubatees.

21.4.3 SERVICES AND FACILITIES OFFERED BY ICAR-CIFT BUSINESS INCUBATOR

The Center, through its business support services, provides links to supporting industries; upgrade technical/managerial skills; provide scientific/technical know-how; assist in market analysis, brand creation and initial test marketing; protect IP assets; and find potential investors and strategic partners.

Incubation facilities under one roof are:

- Furnished office suites within the premises of ICAR-CIFT, with shared facilities like secretarial assistance, computing, copying, conferencing, video conferencing, broadband internet and communication services;
- Pilot level production lines;
- Culinary facility;
- Access to modern laboratory facilities for product testing and quality control;
- Access to well-equipped physical and digital libraries.

TABLE 21.1 List of Entrepreneur Ready Technologies Developed by ICAR-CIFT

Harvest	Fuel-efficient multipurpose vessel for deep sea fishing
	CIFT turtle excluder device (CIFT-TED)
	Device for juvenile fish excluder cum shrimp sorting
	Fish aggregating devices
	Foldable traps
	Square mesh codend
	CIFT–multi seam trawl
	CIFT semi-pelagic trawl system (CIFT-SPTS)
	Large mesh purse seine
	Short body shrimp trawl
	Cut-away top belly shrimp trawl
	Treated rubberwood canoe
	FRP coated rubberwood canoe
	Treated coconut wood canoe
	CIFT sun boat
Value Addition	Microencapsulated sardine oil
	Seaweed NutriDrink
	Ready-to-serve/ready-to-cook food products
	Fish Kure – extruded product
	Seaweed and fish enriched noodles
	Seaweed enriched cookies
	Value-added products – fish sausage/battered and breaded products/cured and dried fish products/smoked Masmin flakes/fish wafers/pickles
	Laminated Bombay duck
	Diversified products from black clam/mussels

TABLE 21.1 *(Continued)*

Engineering	Solar-electrical hybrid dryer
	Solar-LPG hybrid dryer
	Solar tunnel dryer
	Solar-electrical cabinet dryer
	Solar biomass hybrid dryer
	Infra-red continuous dryer
	Fish descaling machine with variable drum speed
	Tabletop fish descaling machine with fixed drum speed
	Hand-operated fish descaling machine
	Energy-efficient effluent treatment plant
	Modern and hygienic mobile fish vending kiosk
Waste Utilization	Fish ensilage
	Foliar spray
	Collagen peptide
	Collagen chitosan membrane
	Chitin and chitosan from crustacean shell
	Carboxymethyl chitosan
	Fish feed from processing discards
	Surgical sutures from fish gut collagen
	Succinyl chitosan-based hydro-alcohol hand sanitizer
Health Care/ Quality Assurance	Squalene and squalene powder
	Oyster protein hydrolysate (CIFTOPEx)
	Fish protein isolate from Bombay duck
	Glucosamine hydrochloride
	Natural hydroxyapatite
	Protein hydrolysate from tuna red meat
	Deodorant for seafood processing units and fish markets
	Antiseptic ointment for prawn/fish handlers
	Fish calcium capsules
	Test strips for sulfite residues
	Chloritest paper
	CIFTest kit for ammonia and formaldehyde adulteration in fish
	Design and development of modern hygienic fish markets

21.4.3.1 PILOT LEVEL PRODUCTION LINES

A state-of-the-art generic semi-commercial production facility is made available to incubating entrepreneurs for developing value-added products from fish. BIC provides access to these facilities along with support of manpower, and assists the entrepreneurs in the production and testing of new product formulations. For the tenants, the pilot plant is an ideal testing arena to determine the commercial viability of new products. The plant also serves as a process lab, a place to see how processing equipment impacts food products under varying conditions. There are production lines for pre-processing, cooking, retort pouch processing, canning, sausage production, extruded products, chitin, and chitosan, smoking, curing, and drying, breading, battering, and product packaging. By providing access to these resources, the Center greatly reduces one of the major barriers to the commercialization of institute technologies by smaller firms – the high capital cost of intermediate or large-scale process equipment.

21.4.3.2 BUSINESS SERVICES

The business-oriented services offered by ABI include assistance in complying with business regulations and licensing procedures, financing, information services, marketing, and tailor-made services designed for the various tenant enterprises. Incubator clients can also gain a special advantage in terms of tax savings through special regulations for Business Incubators. BIC also offers a wide variety of services, with the help of strong associations throughout the Business Incubation Network.

The ABI, since its inception, had registered 165 incubatees, of which 95 have successfully graduated and established small-scale enterprises, 16 firms have taken out the office facility, and more than 100 clients have utilized the pilot plant facility for product development and trial production for test marketing. More than 850 prospective entrepreneurs have utilized the services of ABI for validation of prototypes and model testing. About 60 product brands developed by ABI are available in the market.

21.4.3.3 PROMOTION OF ICAR TECHNOLOGIES

The ABI, since its establishment at CIFT, has been responsive to the rapid transformation of innovation processes and business needs, and has been continuously trying to enhance the visibility of ICAR technologies through Business/Industry Meets, Exhibitions, Industry Interface Programs, etc. This has helped in strengthening the public private partnerships and to bring together innovators involved in research and development, and entrepreneurs from the field of fisheries onto the same platform. Technology exhibitions are regularly organized, and entrepreneur-ready innovations and technologies developed by the ICAR Institutes specialized in fisheries and aquaculture are exhibited to the Industry. The areas addressed are seed production technologies of fish and shrimp, cost-effective, and nutritious fish feed formulations, diagnostic, and test kits, new, and improved aquaculture methods, harvest, and post-harvest technologies, ready-to-cook/ready-to-serve products from fish, pharmaceutical, and biotechnological products, and techniques for fisheries waste management.

21.5 COMMERCIALLY SUCCESSFUL TECHNOLOGIES FROM THE FISHERIES POST-HARVEST SECTOR: A PROFILE

21.5.1 BATTERED AND BREADED FISH PRODUCTS

Fish and fishery products are one of the most preferred food commodities on account of its richness in nutrients. It has gained much popularity on account of its option for conversion into a wide array of value-added forms that can satisfy the modern consumer demand. The most prominent among the group of value-added products is the battered and breaded products. The technology relates to fish processing for the production of value-added products viz., coated fish products. Fish is processed to mince and further for the production of fish mince based coated products. Fish mince can be used for processing a variety of value-added products. Depending on the type of product, a frozen storage life up to 6 months without any appreciable quality deterioration can be expected for this category of products.

1. **Technology Description:** A coated product is one which is coated with another foodstuff. Coating provides a crispy texture and appealing color/flavor; enhances the nutritional quality; the coatings can be fortified with necessary nutrients; acts as a moisture barrier minimizing moisture losses during frozen storage/microwave reheating and acts as food sealant by preventing natural juices from flowing out. Preparation of battered and breaded products consists of the following steps: portioning/forming, pre-dusting, application of batter, breading, flash frying, freezing, packing, and storage. An array of coated products such as patties, balls, burgers, fish fingers, dehydrated fish minces, cutlets, etc., can be prepared from fish mince.
2. **Benefits:**
   - Value-added fish-based products in ready-to-cook or ready to serve convenient form;
   - Effective utilization of by-catch fishes;
   - Enhances the nutritional quality; the coating can be fortified with necessary nutrients;
   - Value addition – increases the bulk of the substrate thus reducing the cost element of the finished product;
   - The product is protein enriched snack food.
3. **Scalability:** The business margin of value-added product processing can reach 2 to 3-fold from the main input value. The value-added and the business margin of fisheries product processing is very big irrespective of the scale of their operation.
4. **Business and Commercial Potential:** The production and marketing of battered and breaded fish products provides opportunities for new businesses, with associated economic and employment benefits. Main raw material as well as rest raw material such as processing frame meat can be processed into a wide range of these products to increase their economic value. In addition, exported value-added products may generate substantial foreign exchange earnings and further employment benefits.

   There is a major scope for battered and breaded fish products in the domestic as well as export market thereby attracting business ventures on a small scale as well as large scale level.

5. **Major Equipments/Machinery Required:**
   - Meat bone separator;
   - Mincer;
   - Silent cutter;
   - Molding machine;
   - Battering and breading machine;
   - Fryer;
   - Cooker;
   - Blast freezer;
   - Packaging and sealing machine.
6. **Target Market/Customer:**
   i. High end customers and urban population;
   ii. Working couples;
   iii. Executives, travelers, etc.
7. **Investment Required:**
   i. Fixed Capital:
      - Land and building: Rs. 20 lakh.
      - Machinery and Equipment: Rs. 74.20 lakh.
   ii. Operational Expenses:
      - Working capital (per day): Rs. 0.91 lakh.
      - Average cost of production: Rs. 215/kg at 100% capacity.
      - Expected average selling price: Rs. 250/kg.
      - Net profit ratio: 21%.
      - Rate of return: 18.1%.

21.5.2 FISH SAUSAGE

India has very good scope for expanding its value addition basket at both domestic and export market. At present only 10% of fish caught is being exported. The demand for fish products is increasing both at domestic and export fronts due to its proven health benefits. Among the variety of value-added products, sausages have a special place in the market. Sausages prepared from fish has the advantage of health aspects apart from its unique taste. Although fish sausage is highly relished, the availability of sausage on the shelf is limited to chicken, pork or other meat varieties. However, there is a good market for fish sausage if promoted suitably.

1. **Technology Description:** For fish sausage, a widc variety of fishes can be used ranging from lean to fatty variety. Sausages are usually prepared from minced meat mixed with ingredients like fat, binders, fillers, spices, salt, etc., according to the consumer's preference. The advantage of sausage preparation is its adaptability to a variety of ingredients resulting in distinctly different taste which increases the variety of sausage products. Fish mince is mixed with additives (starch, sugar, salt, spices, fat) and stuffed into cylindrical casings, either natural or synthetic casings of desired size and marketed either as fresh, cooked, smoked or dried.
2. **Benefits:**
   - Sausages are ready-to-cook or ready to serve convenient products;
   - Suitable for wide variety of fishes which increases the variety of products;
   - Product can be modified using different ingredients depending on the consumers choice;
   - Provides good quantity of nutrients with calorie value of 250–300 kcal 100 g–1.
3. **Scalability:** The business profit of value added products like sausage processing can be double or higher in comparison to the initial raw material. Utilization of under-utilized species as well as fish processing frame meat can be an added advantage, thus enhancing the business profitability.
4. **Business and Commercial Potential:** Production process for sausage is very well established and can be easily adapted for automation. Sausages can be prepared from a wide variety of fishes with varying amounts of fat, including combination sausages from different fishes. Market for fish sausage is very bright on account of its delicacy, nutritional health benefits as well as convenience. There is a major scope for these value added products in the domestic as

well as export market thereby attracting capital intensive business ventures for the production of an array of sausage based products.

5. **Major Equipments/Machinery Required:**
   - Meat bone separator;
   - Mincer;
   - Silent cutter/bowl chopper;
   - Sausage stuffer;
   - Clipper;
   - Slicer;
   - Cooker;
   - Air blast freezer.
6. **Target Market/Customer:** High end customers and urban population.
7. **Investment Required:** A capacity of 250 kg sausage production per day for a minimum of 250 days per annum will be economical.
   - Fixed capital: Rs. 32 lakh.
   - Working capital (per month): Rs. 3 lakh.

21.5.3 FISH PICKLE

Fish and fishery products have gained significance on account of their nutritional as well as socio-economic relevance it possess. They satisfy the dynamic consumer demand by facilitating the production of an array of nutritious delicacies. Among these one of the most prominent and age-old delicacy is a cured product viz., fish pickle. Fish or prawns have very limited shelf life and they need to be processed immediately to preserve them for a longer period. Fish/prawn pickle, if made properly under hygienic conditions adding requisite quantity of salt, spices, and preservatives, would have a high shelf life. Most of the seafoods like prawn, tuna, seer fish, etc., are ideally suitable for making fish pickles.

1. **Technology Description:** Apart from staple foods, people like spicy adjuncts to make the food palatable and relishing. Pickles are good appetizers and digestive agents. There are several varieties of pickles which are consumed throughout the year. In this context, pickle, a spicy preparation, has gained importance in Indian menu. Preparation of pickles is an old art and a variety of these products are made in Indian homes. The basic principle of pickling is curing of fish by salting, acidifying by addition of vinegar and/or oil and spices. In general, pickling enhances the product shelf life to six months and more.
2. **Benefits:**
   - These are ready-to-serve convenient form of products used as side dishes in Indian menu;
   - Effective utilization of by-catch fishes;
   - A means of value addition – reducing the cost element of the finished product;
   - Shelf-stable nutritious commodity.
3. **Scalability:** There is a huge scope of upscaling for this technology by means of utilizing the less utilized raw material as well as rest raw material to high delicacy cured products with the addition of other ingredients. Based on the scale of operation, the level of profit margin can be improved several folds.
4. **Business and Commercial Potential:** Like vegetable pickles, fish pickle has also gained popularity in the recent past. The demand for these types of ready-to-serve fishery products is increasing day by day among the non-vegetarian population in our country. There is a good demand for fish/prawn pickles in foreign countries too. Driven by a rising demand amongst both the Indian diaspora and native clientele abroad, the country's pickle exports have risen a massive percent.

   There are very few national brands and the bulk of the market is controlled by the local manufacturers. Demand for ready to serve fish products is increasing in the domestic sector as well as having a potential export market. Apart from that, individual households, hostels, restaurants, roadside eateries, canteens, etc., are bulk consumers. Potential investment in this field by considering the aspects like proper product marketing, attractive point of sale, publicity, and handsome commission to retailers can have a wide impact for this technical innovation.

5. **Major Equipments/Machinery Required:**
   - Pulverizer or grinder;
   - Heating and frying arrangements;
   - Bottle washing machine;
   - Pickle filling and sealing machine;
   - Weighing machines;
   - Packaging machinery.
6. **Target Market/Customer:** Urban and rural population.
7. **Investment Required:**
   - Production capacity: 1,000 Kg/month.
   - Fixed capital: Rs. 27.40 lakh.
   - Working capital (per month): Rs. 4.2 lakh.
   - Total cost of investment: Rs. 40 lakh.
   - Average annual cost of production: Rs. 43.63 lakh.
   - Turn over per annum: Rs. 60.75 lakh.
   - Net profit ratio: 28.18%.

21.5.4 FISH WAFERS/CRACKERS

Seafoods play a major role in satisfying the demands of the diverse global customers on account of its availability, accessibility, nutritional as well as quality and safety aspects. These valuable commodities are simultaneously convenient due to its potentiality for high product diversification. Being one of the fastest growing economies and the second-largest consumer market in the world, India offers a strong platform for the processed seafood industry. Dried, ready-to-fry-and-serve wafers/crackers, employing carbohydrate as the main base and incorporating salt and several other ingredients with or without spices are very popular.

Being dried products, they have a shelf life of one to two years. By-catches as well as processing frame meat can be very well employed as raw material for the preparation of these dehydrated commodities.

1. **Technology Description:** Since long there is a high demand for flavored dehydrated products in the Indian menu. Preparation of starch based dried products is an age-old practice in the households and has wide popularity in the domestic market. Incorporation of protein rich ingredients into these products can upgrade their nutritive value as well as enhance the taste as well. The basic principle involves the addition of starch into the fish mince together with spices and other ingredients which is blended, cooked for gelatinization and further cut and dried to a moisture content less than 10%.
2. **Benefits:**
   - Ready-to-fry or ready to serve convenient form.
   - Effective utilization of by-catch fishes as well as processing discards like frame meat.
   - Fortification of carbohydrate-based commodities.
   - Product diversification based on customized demands.
3. **Scalability:** Possibilities of upscaling the production of value added products like seafood based wafers/crackers are enormous as there is a rising demand for such products in the domestic as well as external market. The profitability of this technology can be extended based on the scale of operation and is projected to be high so as to obtain the returns on investment under a short period of time.
4. **Business and Commercial Potential:** The demand for these types of ready-to-fry dehydrated fishery products is on the rise among Indian customers. Individual households, bakeries, hostels, restaurants, roadside eateries, canteens, etc., are bulk consumers in the domestic market.

This technology can be adopted by small-scale entrepreneurs targeting the local market as well as commercial dealers oriented for domestic as well as export market. The investment involved for this technological innovation can be customized based on the targeted production attracting small scale to large scale business ventures.

5. **Major Equipments/Machinery Required:**
   - Pulveriser/grinder;
   - Stuffer/molder;
   - Steamer;
   - Chopper/cutter;
   - Dryer;
   - Fryer;
   - Packaging machinery.
6. **Target Market/Customer:** Urban and rural population.
7. **Investment Required:**
   - Production capacity: 50 Kg/day (crackers).
   - Total fixed capital: Rs. 33 lakh.
   - Total working capital: Rs. 10.55 lakh.
   - Total capital investment: Rs. 64.65 lakh.
   - Average cost of production: Rs. 104.94 lakh.
   - Turn over per annum: Rs. 202.50 lakh.
   - Net profit ratio: 48.17%.
   - Payback period: 0.84 years.

21.5.5 FISH PASTE

Fish, one of the richest and cheapest sources of nutrients, especially protein and omega-3 fatty acids has gained much importance as a health food in human nutrition. They feed the majority of the human population and also are food for many organisms, hence playing a major role in the food chain. Options of value addition in this sector has increased the demand for fish-based commodities among the customers.

Fatty and non-fatty fishes of marine origin and freshwater fishes can be used for fish paste preparation. A variety of fishes with low economic value can also be employed for this value addition. Marine fatty fishes could serve as a rich source of omega-3 fatty acids and impart the characteristic taste of fatty fishes. Incorporation of ingredients help in concealing the identity of the original raw material and improves the consumers acceptability. Fish paste in Ready-to-eat (RTE) form can be stored at ambient temperature with a shelf life of minimum 1 year.

1. **Technology Description:** Fish paste is a value added product prepared from deboned fish by mixing with condiments, salt, and oil. In some cases, starch, sugar, and skim milk powder is also added and chopped in a meat bowl chopper to get fine consistency and taste. Ingredients inclusion could be customized based on the consumer-targeted/needs. Both fatty and non-fatty fishes of marine origin and freshwater fishes of low value are preferable for fish paste production.
2. **Benefits:**
   - Very rich in highly digestible protein with well-balanced essential amino acids;
   - Can be customized to variety of taste;
   - It finds application in processing several 'convenience foods' like salad dressings, spread (bread spread/spread on roti/chapatti), condiments, etc.;
   - Suitable for all age groups including infants, lactating women and aged community.
3. **Scalability:** Value addition in the form of fish paste is estimated to have a high profitability in comparison to its initial raw material. Based on the scale of production, raw material adoption, process, etc., the degree of profitability can be projected.
4. **Business and Commercial Potential:** There is a huge commercial potential for the utilization of rest raw material as well as by-catch to commodities that require minimum investment. On account of the nutritional insecurity prevailing in the present scenario, a bright market is expected for products like fish paste in the Country as well as overseas.

These nutritious by-products simultaneously attract small scale business ventures targeting the domestic market as well as is a good option for large scale production by incorporating an array of species with diversified formulations for domestic as well as export market.

5. **Major Equipments/Machinery Required:**
   - Chiller;
   - Meat bone separator;
   - Meat mincer;
   - Bowl chopper;
   - Paste filling and packing machine for bottle filling;
   - Cooker/sterilizing equipment/retort;
   - Tray sealing machine.
6. **Target Market/Customer:** Urban and rural population.
7. **Investment Required:**
   i. Fixed Capital:
      - Land and building: Rs. 30 lakh.
      - Machinery and equipment: Rs. 60 lakh.
      - Working capital (per day): Rs. 0.8 lakh (for 100 kg paste).
      - Average cost of production: Rs. 800/kg at 100% capacity (for 100 kg paste).
      - Expected average for rice: Rs. 950/kg.

21.5.6 FISH PROTEIN POWDER

Fish could be considered as functional food as it offers health promoting substances other than the major nutrients. On account of this, fish and fishery products are one of the most preferred food commodities. Seafood product diversification by effective processing and value addition of main and rest raw material can augment marketing of these commodities in the food and nutraceutical sector. Simultaneously, it leads to reduced post-harvest losses contributing to global economic growth and nutritional security. Fish powder is a product prepared from prime quality edible fish muscle. It is a protein rich (75–85%) shelf-stable product intended for human consumption. The proteins present in fish powder are highly digestible (90–98%) and can supply all the dietary essential as well as non-essential amino acids. In addition fish powder is also a good source of functional bio-minerals including selenium and zinc.

1. **Technology Description:** Fish powder consists of two variants viz., whole fish protein powder and hydrolyzed fish protein powder including muscle protein hydrolysates and collagen peptide. The hydrolyzed fish protein powder is more suitable for infants, aged population, and for the people having difficulties in protein digestion. The hydrolyzed fish collagen powder is a high value bioactive ingredient having applications in nutraceutical and functional food formulations and essentially recommended for arthritis patients as a supplement for bone regeneration. It can be packed in metalized polyester laminated with polyethylene and HDPE containers, glass bottles, etc., for 6 to 9 months at ambient temperature.
2. **Benefits:**
    - Source of nutrients like amino acids;
    - Possess various bioactive properties viz., Antioxidant, Anti-inflammatory, Anti-hypertensive, Antimicrobial, Immune-modulator, Appetite suppressor, Cholesterol-lowering agents, Anti-obese, etc.;
    - Application as Binder, Dispersing agent, Emulsifier, Ingredient in restructured products (gelling agent), Protein fortification, Thickening agent, etc.
3. **Scalability:** There is a high upscaling possibility for protein powder from seafood sources. The profit margin may vary depending on the process/protocol adopted as well as the scale of operation. However, a high upgradation is projected for the final product in comparison to the initial raw material used.
4. **Business and Commercial Potential:** On account of the rising demand amongst both the modern customers for health-promoting food commodities, a huge potential is expected for products like protein powder in the internal as well as export market.

Small scale business ventures focusing on the local market as well as established ventures targeting the domestic and export market can be potential investors to this technology for exploring its commercial possibilities.

5. **Major Equipment/Machinery Required:**
   - Chiller;
   - Meat bone separator;
   - Meat mincer;
   - Meat washing machine;
   - Reactor;
   - Cooker;
   - Dryer;
   - Pulverizer;
   - Ribbon blender;
   - powder filling and packing machine for bottle filling/pouch packing.
6. **Target Market/Customer:** Urban and rural population.
7. **Investment Required:**
   i. Fixed Capital:
      - Land and building: Rs. 35 lakh.
      - Machinery and equipment: Rs. 70 lakh.
      - Working capital (per day): Rs. 0.75 lakh (for 100 kg paste).
      - Average cost of production: Rs. 850/kg at 100% capacity (for 100 kg powder).
      - Expected average selling price: Rs. 900/kg.

21.5.7 READY-TO-SERVE FISH CURRY IN RETORT POUCHES

The technology relates to a ready-to-serve fish curry in a retortable pouch. The technology provides a method for preparing the ready-to-serve fish curry in retortable pouch with excellent storage stability and quality with a shelf life of more than one year at ambient temperature.

1. **Technology Description:** The preparation of ready-to-serve fish curry includes processing fresh fish, preparing a curry medium and finally mixing the processed fresh fish and curry medium in an appropriate ratio. The processing of fresh fish comprises of cutting fresh fish into streaks of 3–4 cm, washing in potable chilled water and cold blanching of the washed streaks. Curry preparation varies with region. Depending on the requirements, ingredients are processed, and the curry media is prepared in different styles like Mughlai, Punjabi, Kerala, Konkan, Oriya, Bengali, Manipuri, etc. The processed fish curry medium is mixed in an appropriate ratio to obtain the ready-to-serve fish curry and packed in the said retortable pouch. The air inside the retortable pouch is exhausted by steam injection. Subsequently the retortable pouch is heat-sealed and subjected to retort sterilization at 121°C for a specified time.
2. **Benefits:**
   - The technology provides a method for preparing the ready-to-serve fish curry in retortable pouch with excellent storage stability and quality.
   - The ready-to-serve fish curry is thermal processed and do not require any further processing before consumption.
   - The thermal processing conditions have been standardized for this product in order to make it safe for consumers.

- Fishermen will get better returns and consumers will get ready to serve safe products.

3. **Scalability:** The business margin of ready-to-serve fish products can reach 2 to 3-fold from the main input value. The technology can be scaled up for full commercialization.
4. **Business and Commercial Potential:** The production and marketing of ready-to-serve fish products provide opportunities for new businesses, with associated economic and employment benefits. There is a major scope for ready-to-serve fish products in the domestic as well as export market, thereby attracting business ventures on a small scale as well as large scale level.
5. **Major Equipment/Machinery Required:**
   - Preprocessing and processing tables and utensils;
   - Mini Boiler for steam generation;
   - Retort machine;
   - Chiller cum cold store.
6. **Target Market/Customer:**
   - High end customers and urban population;
   - Working couples;
   - Executives, travelers, etc.
7. **Investment Required:**
   i. Fixed Capital:
      - Land and building: Rs. 35.5 lakh.
      - Machinery and equipment: Rs. 50.30 lakh.

   ii. Worki+1p6ng Capital:
      - Raw material and packaging material: Rs. 217.20 lakhs/year.
      - Utilities (water, electricity, steam, etc.): Rs. 14.01 lakhs/year.
      - Other Contingencies: (Maintenance and other expenses): Rs. 4.39 lakh.
      - Cost of production: Rs. 29.91/pouch at 100% capacity.
      - Expected selling price: Rs. 40/pouch.
      - Expected sales: Rs. 324.90 lakhs/year.
      - Net profit: Rs. 55.69 lakhs/year.
      - Net profit ratio = 17.14.
      - Rate of Return = 32.13%.

21.5.8 FOLIAR SPRAY FROM FISH PROCESSING DISCARDS

Foliar feeding is an effective method for correcting soil deficiencies and overcoming the soil's inability to transfer nutrients to the plant. Tests have shown that foliar feeding can be 8 to 10 times more effective than soil feeding and up to 90% of foliar fed nutrients. The application of foliar spray has been advocated in spices like cardamom, black pepper, tea, etc., and encouraging results have been reported. The quick absorption of the nutrients and precise dosage of foliar sprays has resulted in the success of precision farming of costly vegetables and flowering plants. The controlled nutritional supply through praying is an effective method which gives predicted resulted in most of the cases. The optimized supply of required micro and macronutrients results in the maximum productivity of the available space and minimizes the wastage of costly inputs.

1. **Technology Description:** The fish waste collected from fish processing industries and fish markets are homogenized. It is then cooked for half an hour with required quantity of fermentable carbohydrate. It is then cooled and added with starter culture of lactic acid bacteria. The product is then kept under covered condition for a period of 20–3-days and the supernatant is collected and packed in bottles. It is supplemented with potassium and magnesium in the required level.

2. **Benefits:**
   - The entire bulk waste generated in the domestic fish markets will be cleared within the market hours itself.
   - Conversion of fish waste to high value end products is a wiser option for the industry as they can potentially generate additional revenue as well as reduce the cost of disposal of these process discards.
3. **Scalability:** The business margin of foliar spray can reach 6 to 8-fold from the main input value, considering the huge projected demand from production sector in the future.
4. **Business and Commercial Potential:** Recently, the application of natural fertilizers to plants and orchards have gained significance considering the harmful effects of organic fertilizers. Foliar feeding is an effective method for correcting soil deficiencies and overcoming the soil's inability to transfer nutrients to the plant. Foliar feeding can be even 8 to 10 times more effective than soil feeding and up to 90% of a foliar fed nutrients. Fish based foliar spray formulation ensure effective absorption of micro and macro nutrients, support precision farming of crops and minimize the wastage of nutrients through leaching.
5. **Major Equipment/Machinery Required:** The pellet feed production line contains the following subunits:
   - Crusher;
   - Mixer;
   - Ensilation tanks.
6. **Target Market/Customer:**
   - Farming community;
   - Fertilizer manufacturing companies.
7. **Investment Required:**
   i. Fixed Capital: for 10,000 L/month foliar spray production.
      - Excluding land and building (1,500 sq. ft.).
      - Machinery and equipment: Rs. 3 lakh.

   ii. Operational Expenses:
      - Working capital (3 months): Rs. 0.75 lakh.
      - Average cost of Production: Rs. 35/L at 100% capacity.
      - Expected average selling price: Rs. 150/L.
      - Net profit ratio: 32%.
      - Rate of return: 26%.

21.5.9 FISH INCORPORATED EXTRUDED SNACK PRODUCTS

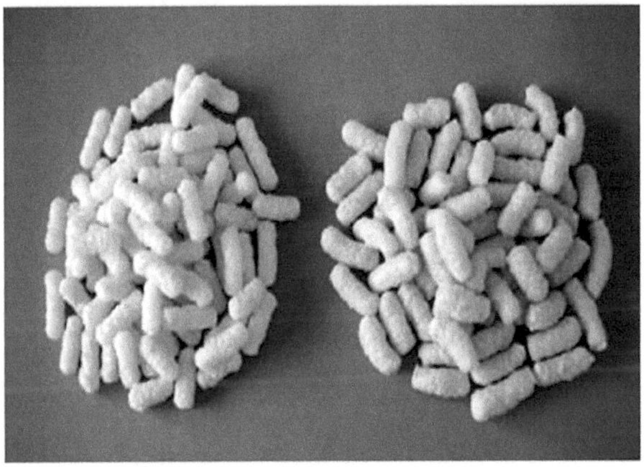

Food extrusion is a size enlargement process where small granular or powdered particles are reinforced into larger-sized particles with different shapes, texture, color, etc. It is used for the manufacture of food products such as ready-to-eat breakfast cereals, snack foods, soft moist pet foods and textured vegetable protein. Extruded snacks are gaining importance nowadays due to their peculiar taste, texture, and convenience. They are rich in calories and contain low levels of protein, which makes it necessary to fortify them with protein-rich diets. One of the possible ways for doing this is to utilize fish and fish proteins. Utilizing fish meat and fish portions and its derivatives like fish protein hydrolysate powder, dry fish powder, etc., will add value to the low-cost and underutilized fish and shellfish, thus promoting their utilization.

1. **Technology Description:** In extrusion cooking, food material is heated either by an external heat source or through heat produced by friction and forced through dies to expand and extrude in desired shapes. Food extrusion using fish provides a great versatility for the development of low-cost, high-nutritive protein-rich and convenient food products. The addition of flavoring is usually carried out on the dry extrudate by spraying or dusting, because of the changes caused by the losses of volatiles during extrusion. Metalized laminated pouches with nitrogen gas are used for the packaging of extruded products and can be stored at room temperature for 4 months.
2. **Benefits:**
   - Versatility: wide variety of products are possible by changing the ingredients, varying the operating conditions and shape of the dies;
   - Low operational costs;
   - High production rate: operate continuously and have high throughputs;
   - Good quality products: involves high temperatures applied for a short time and the limited heat treatment therefore retains many heats sensitive components;
   - No effluents: is a low-moisture process, eliminates water treatment costs and does not create problems of environmental pollution.
3. **Scalability:** The business margin of extruded snacks can reach 2 to 3-fold from the main input value. The business margin of fisheries product processing is very big irrespective of the scale of their operation.
4. **Business and Commercial Potential:** The production and marketing of extruded snack products provides opportunities for new businesses, with associated economic and employment benefits. Fish meat as well as rest raw material such as processing frame meat can be processed into a variety of these products to increase their economic value.

   The combination of various ingredients makes the products in this product category tastier and crispier, and consequently, the segment has been witnessing increasing demand from every section of the society, especially from kids and younger generation.

   There is a major scope for fish enriched extruded snack products in the domestic market thereby attracting business ventures on a small scale as well as large scale level.

5. **Major Equipment/Machinery Required:**
   - Twin-screw extruder;
   - Ribbon blender;
   - Silent cutter;
   - Drier/fryer;
   - Rotary flavor coating pan;
   - IR moisture analyzer;
   - PLC controlled form-fill sealing machine;
   - Stainless steel dressing table, sieves, trays, tubs, knives, etc.;
   - Nitrogen cylinders.
6. **Target Market/Customer:**
   - Teenage and college students.
   - Children.
7. **Investment Required:**
   i. Fixed Capital:
      - Land and building: Rs. 20 lakh.
      - Machinery and equipment: Rs. 33.00 lakh.

   ii. Operational Expenses:
      - Working capital: Rs. 1,235,680.
      - Average cost of production: Rs. 30,350 per day.
      - Expected average selling price: Rs. 56,000 (for 5,600 pouches at Rs. 10/pouch).
      - Net profit ratio: 21%.
      - Rate of return: 18.1%.

21.5.10 LIVESTOCK AND AQUA FEED FROM FISH PROCESSING DISCARDS

Wastes generated by fish processing industries are tremendous compared to other industries. Processing wastes from fishes vary with size and species accounting for around 50–60% of total fish weight. In India the fish farming activity is expanding exponentially, leading to high demand for quality feed. The waste generated from fish processing sector can be used for making good quality fish feeds. This in fact helps the processing industries to control pollution of the environment.

1. **Technology Description:** Bulk waste generate in domestic fish markets and fish processing facilities is collected in wet form. Further, the waste is crushed to fine paste and cooked along with other necessary ingredients to form a dough mass. The dough is further passed through a pelletizer and dried to a moisture content of less than 10%.

   Disposal of process discard is a major constraint in the seafood sector as these discards are easily perishable and often invites public resistance in view of environmental pollution. The present technology introduces a process addressing a value-chain approach for the management of bulk-waste generated in the seafood sector, especially in domestic fish markets, by directly converting to a high-in-demand commodity such as aqua and livestock feed, which at present is at its infant stage.
2. **Benefits:**
   - The entire bulk waste generated in the domestic fish markets will be cleared within the market hours itself.
   - Conversion of fish waste to high-value end products is a wiser option for the industry as they can potentially generate additional revenue as well as reduce the cost of disposal of these process discards.

3. **Scalability:** The business margin of livestock and aquafeeds can reach 3 to 5-fold from the main input value, considering the huge projected demand in future.
4. **Business and Commercial Potential:** Most fish farms in Asia still rely on trash fish and fish meal to meet protein requirements. Hence the long-term sustainability of these farming activities is subjected to the technological inputs on alternative high protein source. As per current statistics, a minimum of 6MT of fishery waste is being generated every year in the country. Fishery wastes contain a significant proportion of protein and other micronutrients, and hence can be considered as an ideal raw material for fish and livestock feeds. In recent years, due to intense farming activities worldwide, there is a major scope for livestock and aqua feeds in the domestic as well as export market thereby attracting business ventures on a small scale as well as large scale level.
5. **Major Equipment/Machinery Required:** The pellet feed production line contains the following subunits:
   - Crusher;
   - Grinder;
   - Mixer;
   - Forming machine;
   - Steaming unit;
   - Pelletizer;
   - Dryer;
   - Packaging and sealing machine.
6. **Target Market/Customer:**
   - Aqua and livestock farmers.
   - Feed manufacturing companies.
7. **Investment Required:**
   i. Fixed Capital: for 500 kg/Day feed production.
      - Excluding land and building (2,500 sq. ft.).
      - Machinery and equipment: Rs. 15 lakh.

   ii. Operational Expenses:
      - Working capital (3 months): Rs. 5 lakh.
      - Average cost of production (35% protein content): Rs. 28/kg at 100% capacity.
      - Expected average selling price: Rs. 75/kg.
      - Net profit ratio: 25%.
      - Rate of return: 19.6%.

21.5.11 SOLAR FISH DRYERS

Fishers catch fish as major aquatic products and are intended mainly for domestic consumption and sale in the local market. However, in case of excess catch and during periods glut, tremendous losses occur because the fishers have neither access to markets in big cities nor to the international market due to poor product quality and absence of good marketing and distribution system. As an alternative, fisherman can convert the catch into value added products viz., dried fishery products, smoked, and salted fish, etc., with enhanced shelf life and market value.

ICAR-CIFT has developed various types of solar dryers with different backup systems suiting different product and client demands. ICAR-CIFT solar fish dryers are designed to supply maximum heat energy for drying from incident solar radiation.

1. **Technology Description:** Solar drying is an alternative which offers numerous advantages over the traditional method, apart from being environmentally friendly and economically viable. In solar drying, a structure, often of very simple construction, is used to enhance the heating effect of the solar radiation. Compared to the sun drying, solar dryers can generate higher air temperatures and consequential lower relative humidity, which are conducive to improved drying rates and hence lower moisture content of the final products. In order to overcome the heat energy supply during off sunshine hours, backup heating source is provided like LPG, Electrical, and biomass.
2. **Benefits:**
   - Ideal for drying of fish, fruits, vegetables, spices, and agro-products;
   - Hygienically prepared premium quality uniformly dried product;
   - Economically viable and eco-friendly – maximum use of solar energy;
   - The drying time is reduced considerably with improved product quality;
   - Lesser operating expenses;
   - Labor requirement is considerably reduced.
3. **Scalability:** The fish dryers are available from as small as 20 Kg to 1-ton capacity. The production size can increase if appropriate investment is made. The dried product business margin is very big irrespective of the scale of their operation.
4. **Business and Commercial Potential:** The production and marketing of dried fish products provide opportunities for new businesses, with associated economic and employment benefits. Various fish species can be commercially dried into a wide range of products to increase their economic value. In addition, there is a potential for export oriented dry fish products which may generate substantial foreign exchange earnings and further employment benefits. Potential

investors in this technology would be Budding entrepreneurs, Fisherfolks, and export market-oriented business people.

5. **Major Equipment/Machinery Required:**
   - Solar dryer with any one backup system – electrical, LPG or biomass;
   - Pre-processing facility with SS cutting table;
   - Dry fish sorting and packing table;
   - Band sealing machine;
   - Fish salting tank;
   - Waste disposal drum;
   - Electronic weighing balance;
   - Storage rack.
6. **Target Market/Customer:** Urban and rural population.
7. **Investment Required:**
   i. Fixed Capital:
      - Machinery and equipment: Rs. 28.05 lakh (for each unit).

   ii. Working Capital:
      - Raw material and packaging material utilities (water, electricity, steam, etc.): Rs. 26.23 lakh/year.
      - Other contingencies: (Maintenance and other expenses): Rs. 26.23 lakh/year.
      - Depreciation of dryer, etc.: Rs. 26.23 lakh/year.
      - Cost of production: Rs. 2,18,583/month.
      - Expected selling price: Rs. 500/kilogram (average rate).
      - Expected sales: Rs. 9,000/day at 100% capacity (approx.).
      - Net profit: Rs. 1,714/day at 100% capacity.
      - B-C ratio: 1.23.

21.6 CONCLUSION

Fish and any value-added product from fish can be an ideal functional food with multiple roles in growth, nutrition, immunity, and disease control. Growth and development of the fishery sector has huge potential to provide employment and entrepreneurship opportunities in many fields such as capture and culture fishery activities, fish processing, value addition, seed production, feed development, transportation, cold chain, packaging. Utilization of rest raw material from fish processing for developing high value nutraceuticals and pharma-grade products is another promising area which calls for new ventures. In this context, the role of business incubators in the domain of fisheries that supports the creation and growth of business through organizational and technical assistance, which at the same time contributes to the reduction of entrepreneurial failure, assumes great importance.

KEYWORDS

- **business incubators**
- **entrepreneurship**
- **fish products**
- **incubation centers**
- **omega-3 (n-3) fatty acids**
- **post-harvest fisheries**

REFERENCES

Alex, A., & Triadafilopoulos, G., (1992). Fish oil fatty acid supplementation in active ulcerative colitis: A double-blind, placebo-controlled, crossover study. *Am. J. Gastroenterol., 87*, 432–437.

Anon, (2015). *Resolution Adopted by the General Assembly of UN on 25 September 2015 A/RES/70/1-Transforming Our World: The 2030 Agenda for Sustainable Development.*

Anon, (2020). *Marine Products Exports from India 2018–19*. www.mpeda.gov.in (accessed on 09 March 2022).

Arvanitoyannis, I. S., & Kassaveti, A., (2008). Fish industry waste: Treatments, environmental impacts, current and potential uses. *Int. J. Food Sci. Technol., 43*(4), 726–745.

Barik, N. K., (2017). Freshwater fish for nutrition security in India: Evidence from FAO data. *Aquacul. Rep., 7*, 1–6.

Beard, J. L., & Dawson, H. D., (1997). Iron. In: O'Dell, B. L., & Sunde, R. A., (eds.). *Handbook of nutritionally Essential Minerals* (pp. 275–334). New York: Marcel Dekker, Inc.

Connor, W. E., (2000). Importance of n-3 fatty acids in health and disease. *Ame. J. Clin. Nutri., 71*(suppl), 71S–175S.

Darlington, L. G., & Stone, T. W., (2001). Antioxidants and fatty acids in the Amelioration of rheumatoid arthritis and related disorders. *British J. Nutri., 85*, 251–269.

Donaldo, J. V., et al., (1999). The long-term outcome of patients with IgA nephropathy treated with fish oil in a controlled trial. *J Ame. Soc. Nephrol., 10*, 1772–1777.

FAO, (2012). *The State of World Fisheries and Aquaculture* (p. 209). Rome.

FAO, (2018). *The State of World Fisheries and Aquaculture 2018 – Meeting the Sustainable Development Goals*. Rome, Licence: CC BY-NC-SA 3.0 IGO.

Govindharaj, M., Roopavath, U. K., & Rath, S. N., (2019). Valorization of discarded Marine Eel fish skin for collagen extraction as a 3D printable blue biomaterial for tissue engineering. *J. Cleaner Produc., 230*, 412–419.

Hasler, C., (2000). The changing face of functional foods. *J. Ame. College Nutri., 19*, 499s–506s.

Herpandi, N. H., Rosma, A., & Wan, N. W. A., (2011). The tuna fishing industry: A new outlook on fish protein hydrolysates. *Compreh. Rev. Food Sci. Food Safety., 10*(4), 195–207.

Kim, S. K., & Mendis, E., (2006). Bioactive compounds from marine processing byproducts: A review. *Food Res. Int., 39*(4), 383–393.

Larsson, S. C., Kumlin, M., Ingelman-Sundberg, M., & Wolk, A., (2004). Dietary long-chain n-3 fatty acids for the prevention of cancer: A review of potential mechanisms. *Ame. J. Clin. Nutri., 79*(6), 935–945.

Mahboob, S., (2015). Isolation and characterization of collagen from fish waste material skin, scales and fins of *Catla catla* and *Cirrhinus mrigala*. *J. Food Sci. Technol., 52*(7), 4296–4305.

Malhotra, V. K., (1998). *Biochemistry for Students* (10th edn.). New Delhi, India: Jaypee Brothers Medical Publishers (P) Ltd.

Mohanty, B. P., Mahanty, A., Ganguly, S., Mitra, T., Karunakaran, D., & Anandan, R., (2019). Nutritional composition of food fishes and their importance in providing food and nutritional security. *Food Chem., 293*, 561–570. doi: 10.1016/j.foodchem.2017.11.039.

Ninan, G., & Ravishankar, C. N., (2018). Entrepreneurship ventures in fish processing. In: Sudheer, K. P., & Indira, V., (eds.), *Entrepreneurship Development in Food Processing* (pp. 269–290). New India Publishing Agency, New Delhi.

NSSO, (2014). *NSS Report No.558: Household Consumption of Various Goods and Services in India, 2011–2012.* Ministry of Statistics and Program Implementation, Government of India.

OECD/FAO (2019). *OECD-FAO Agricultural Outlook 2019–2028* (pp. 196, 197). Chapter 8 fish & seafood. OECD Publishing, Paris/Food and Agriculture Organization of the United Nations, Rome. https://doi.org/10.1787/agr_outlook-2019-en.

Radziemska, M., Vaverková, M. D., Adamcová, D., Brtnický, M., & Mazur, Z., (2019). Valorization of fish waste compost as a fertilizer for agricultural use. *Waste Biomass Valoriz., 10*(9), 2537–2545.

Rude, R. K., & Shils, M. E., (2006). Magnesium. In: Shils, M. E., Shike, M., Ross, A. C., Cabellero, B., & Cousins, R. J., (eds.), *Modern Nutrition in Health and Disease* (10th edn., pp. 193–221). Baltimore: Lippincott Williams & Wilkins.

Sachindra, M. M., & Mahendrakar, N. S., (2015). Fishery byproducts-an overview. In: Sachindra, M. M., & Mahendrakar, N. S., (eds.), *Fish Processing Byproducts: Quality Assessment and Applications* (pp. 1–13). Studium Press LLC, USA.

Tan, J. C., Burns, D. L., & Jones, H. R., (2006). Severe Ataxia, myelopathy and peripheral neuropathy due to acquired copper deficiency in patient with history of gastrectomy. *J. Parent. Ente. Nutri., 30*, 446–450.

Yuvaraj, D., Bharathiraja, B., Rithika, J., Dhanasree, S., Ezhilarasi, V., Lavanya, A., et al., (2019). Production of biofuels from fish wastes: An overview. *Biofuels, 10*(3), 301–307

Index

B

D

E

G

H

I

N

O

S

U